The International Countermeasures Handbook
13th Edition, 1988

Library of Congress Catalog Card No. 75-325108
ISBN 0-918994-19-5

ABOUT THE COVER: *The cover art for this edition was created by Bradley Burns of Watkins-Johnson Co. for the article "Offboard Countermeasures" by Dr. Norman Friedman on page 60.*

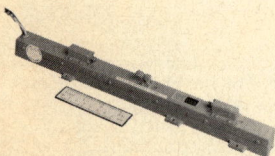

Integrating Electronic Warfare into C²

Under control, on target, or out of control?

By Albert A. Gallotta

If you read the papers any day, everything seems to be out of control: the budget, the deficit, medical care, weaponry, defense contracting, the news media, crime, the stock market. They consume too much, spend all that they have and involve too many people. Yet this is not true of the integration of electronic warfare into command and control. Not nearly enough resources and brainpower are being applied to achieve even nominal integration. Sadly too, a broad service-wide strategy is lacking to bring together the product of our EW resources effectively to help our commanders understand the electronic environment and then to take the proper operational actions. If one understands the scope of electronic warfare, "under control" is the most apt description of EW integration today.

In reality, there is an insatiable need for EW information in modern warfare. It cannot be denied that EW information is a major contributor in helping to exploit the employment signatures of an enemy's military electronic systems. In the first stages of battle, when posturing and positioning of forces is the main activity, EW may be the only timely and available source of information about an enemy. Correlated EW signals permit us to assess the status of an enemy's weaponry and to derive his probable intentions. They also assist us to deploy proper, timely, orderly, and prioritized EW countermeasures. However, these EW contributors lose much of their effectiveness if utilized only in a stand-alone mode. They maximize their value and effectiveness when they are integrated with other friendly weapons systems.

That single step, integrating EW with other weapons systems, changes the name of the older, conventional warfare game. It has become necessary for EW systems to be accounted for and to accommodate to other friendly electronic systems, such as gun laying radars, missile targeting systems, intelligence centers, computers, displays, command centers, communications systems and other sensors and links that serve the commander.

Next Decade. The big changes about to explode upon the EW communities involve participating in the command and control programs that are now being specified for the next decade. They will place important and increasingly more stressful demands upon EW. Indeed these new, large C² programs are acting like huge magnets tugging on all available electronic sensors, including EW, for both positive and negative intercept data.

Late 20th century EW integration infers electronic automaticity, large amounts of data processing, storage and retrieval, correlation of the product of varied EW

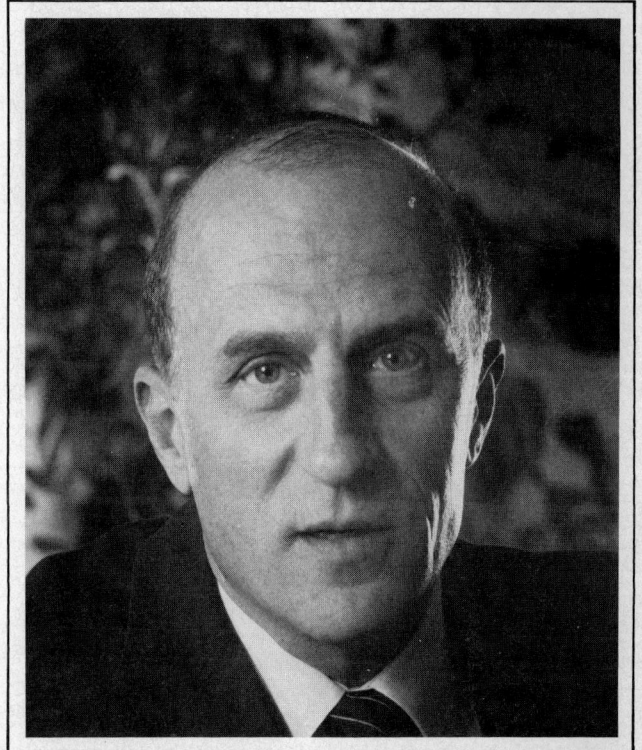

Rear Adm. Albert A. Gallotta, U.S. Navy (retired), has served as president of the Association of Old Crows since 1985. He is the founder and president of AAG Associates, Inc., a consulting firm specializing in electronic warfare and C³I matters. In the course of his 30-year Navy career, he spent 23 years in electronic warfare-related activities. From August 1982 to April 1985, he was vice commander of the Naval Electronic Systems Command (Navelex). Prior to that, he was director, Electronic Warfare Division, OP-944 on the staff of the chief of naval operations. During the period July 1976 to April 1979, Gallotta was manager of the REWSON (reconnaissance, electronic warfare, special operations and naval intelligence) project for Navelex. Before that, he was special assistant to the chief of naval operations for air acquisition and research and development programs, and as EA-6B program coordinator and branch head for tactical air EW systems. He served as the commanding officer of fleet air reconnaissance squadron two and taught navigation and marine engineering in the NROTC program at Princeton University. Gallotta holds a bachelor's degree in international relations from Brown University and a master's degree in management of national resources from George Washington University.

systems and comparison of EW with other sources of information. It also involves decision-aiding for mission execution, hand-off to other weapons systems and feedback on mission progress. Precise and detailed mission scenarios must be meticulously planned and tested in

EW

WATKINS-JOHNSON COMPANY
SYNTHESIZERS, ANTENNAS, SIMULATORS AND TEST SYSTEMS

Featuring many of W-J's field-proven components, Watkins-Johnson's SSE Division offers a complete line of broadband frequency synthesizers, antennas and pedestals for receiving and jamming systems, ECM and radar simulators, and fully computer-controlled automatic test systems for EW receivers and subsystems.

The WJ-47100 Coherent Jammer Simulator (COJAS) is a computer-controlled jamming simulator used to analyze pulse doppler radar vulnerability to known jamming techniques, and to evaluate radar performance. The flexibility of the COJAS to mix and match the various modulation and jamming techniques allows the operator to subject a pulse doppler radar to projected jamming scenarios to determine radar system effectiveness. The modularity of the COJAS allows for easy expansion of the system's capabilities, such as increasing power output, adding new ECM techniques, or operating the system in a transportable configuration.

The WJ-45020 Synthesizer/Sweep Generator is a 10 MHz to 18 GHz source that utilizes advanced synthesis techniques developed by Watkins-Johnson Company. The broadband frequency, +10 dBm output power, excellent phase noise synthesizer is designed in a half-rack configuration. Options are available for frequency extension (up to 60 GHz) and local control is offered by adding a front panel with keypad/display. The WJ-45020 is excellent for transportable RF ATE, EW/ECM simulation, RF system test and alignment, and antenna range frequency source applications.

W-J's LPR 200 Low Profile Spinning DF Antenna is designed for use in direction-finding systems. It is suitable for use in medium-speed (400 knots) aircraft with low underside-to-ground clearance and for naval shipboard use. The unit is 13 inches high and 30 inches in diameter (33 × 76 cm). Its frequency range is 1.0 to 18.0 GHz (optional coverage to 40 GHz can be provided). The antenna may be easily installed or removed from aircraft surfaces, requiring only externally inserted bolts and clearance for the RF and DF connectors. Since the radome and the radiating elements within, radiate about the pedestal, an external radome is not required. All required RF distribution is contained within the assembly. Inclusive RF distribution (preamps, switching, etc.) allows for higher net gains for the output of the RF rotary joint. The antenna's radiation pattern characteristics remain constant about the 360-degree rotation of the antenna. Aerodynamic drag is low. At 10,000 feet and at an airspeed of 250 knots, the drag is 118 lbs. The LPR 200 receives its power and control from the W-J designed EC61 series of antenna control units.

Detailed specifications on W-J's complete line of synthesizers, antennas, simulators and test systems are available from Watkins-Johnson Company, 2525 North First Street, San Jose, California, 95131-1097, (408) 435-1400, ext. 3505, TWX 910-338-0505.

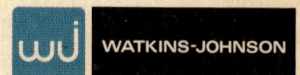

WATKINS-JOHNSON

order to present the best options to commanders who, in the press of battle, will not have time to evaluate every situation meticulously.

This is no small task for either the C^2 systems or the EW systems, because the conventions and rules for the optimal employment of these two disciplines, i.e. their integration, are sorely lacking today.

Briefly, the value, priority, applications, structure, understanding, routing, accesses, classifications, and the correlation of the EW data with command and control systems are essentially unregulated today and are only minimally integrated. Those elements begging for recognition and solution in an architectural sense are:

- Lack of resources (involves sheer numbers of people and devices),
- Quality of equipment (relates to U.S. capabilities versus the threat),
- Automated software (which affects total system responsiveness),
- Correlation methodology (concerns operational employment concepts), and
- Decision aiding (required for mission planning and execution).

The pace, threat numbers and geographical size of the expected battles of the future suggest strongly that an automated command and control battle management system drawing heavily upon sensor data and employing EW countermeasures and deceptions as well as weapons-related equipment, will be an absolute necessity.

It will be the task of future EW designers to accommodate to the demands of command and control in the 21st century. EW data formats; EW information, connections, flow rates and distribution paths; priorities; sequences; decision aiding; common display symbology; correlation functions and preprocessing levels will be new considerations heretofore largely unrelated to EW hardware development. Accounting for the contribution of EW to the whole battle process will be of greater importance as automation begins to match the broader correlation functions being described for command and control systems.

The hitch is that the architecture of such a C^2 system is poorly understood; it is limited by existing EW and C^2 military equipment, and is also handicapped by the present day attitudes of most military and industrial planners. It is absolutely essential that these military-industrial partners achieve a better understanding of this problem. In its purest form, there must be a recognition that EW is a servant of command. There must also be an understanding that the commander can better exercise that command by more optimally controling all his resources with new electronic tools. (This argument tolerates the "uncontrolled" employment of EW in such activities as own platform self-protection when under fire.)

EW and C^3I. While much has been written about the value and importance of EW along a battle timeline, the employment of EW in the Army, Navy, Marine Corps and Air Force mission areas, and the command, control, communications and intelligence systems, not enough of it has been about integration of EW and C^3I.

These mission-related EW/C^3I requirements nonetheless derive from a basic need to fight and win battles. In modern warfare that involves the overwhelming use of electronics by both sides, which in turn necessitates evaluating the actions of the opposition and in part exploiting his electronic systems, calculating a military objective and employing resources to achieve that objective. Today this implies the gathering and distribution of enormous volumes of information, and in using it on the battlefield, there are many opportunities for EW to support command. Indeed these opportunities are almost as numerous as the pieces of EW equipment that exist. Since the EW domain extends from beneath the oceans to outer space, and its use is essential in all phases of battle, the value of EW resources to C^3I is of significant importance.

Unfortunately, these two systems (EW and C^3I) cannot coexist without some mutual interface, nor can they be optimized in an integrated system without significant contributions from one another. Thus it appears that they must be joined at their common cutting edge— operations. The real question is whether they will evolve naturally into an efficient and optimized system without a concious effort to achieve compatibility, commonality and interoperability. Examples of large systems that have achieved the benefits of integration through accommodation (i.e. federated systems) where standardized interfaces abound, are telephone systems, railroads, power grids and communication networks. However, in spite of their successes at integrating, even these systems have grave limitations when they are stressed or overloaded.

The EW and C^3I communities must cope with the differences between peacetime one-on-one (or few-on-few) exercises, and concentrate on the larger conflict with its onslaught of signals and confusion of orders. If we fight again, it will be a "come as you are" war, and what we do now to achieve a warfighting, integrated EW/C^3I system could tip the balance in our favor.

If we continue to keep talking about integration and fail to face up to its complex technical and resource demands we will never get there. What it's going to take is a special task force to size this problem, evaluate practical solutions which minimize the effect of necessary changes on all users and to lay out a financial support plan that requires the contributions of all EW and C^3I sponsors to support and fund those integration efforts that will never get done unless all the users and sponsors participate. ∎

ESM

WATKINS-JOHNSON COMPANY
RECEIVERS

Watkins-Johnson offers the world's largest selection of receiving equipment for surveillance, direction finding and countermeasures. This equipment presently covers the radio frequency spectrum from 0.5 kHz to 40 GHz. Practically all the receivers built by W-J can be used separately or in complex system arrangements.

The receivers outlined below are representative of the technology available from Watkins-Johnson in communications spectrum surveillance.

The WJ-9195C Rapid Acquisition Spectrum Processor (RASP) is a broadband receiver and spectrum display unit, offering exceptional scanning speed and dynamic range. With available frequency extenders installed, it will cover a frequency range of 2 MHz to 1400 MHz. The WJ-9195C will scan user specified segments of the RF spectrum at a rate of 1 GHz per second. For greater resolution, a 5 kHz mode is selectable. The resultant data is displayed on six independently programmable traces capable of being set to show a view of either the entire spectrum or an area of particular interest. The unit is also designed to function as a system controller, with the capability of controlling up to 15 external receivers. The unit's receiver, digital RF processor and display are housed in a single rack-mounted 8.75 in. × 19 in. × 17 in. (22.2 × 48.3 × 43.2 cm) enclosure.

The WJ-9040 Modular Receiving System provides the user with a product line of compatible VLF/HF/VHF/UHF communications receivers and peripheral devices which can be configured to fill a variety of requirements. This equipment exhibits low power consumption, light weight, small size, and other characteristics which contribute to making it highly suitable for a wide variety of applications. Pictured is a dual-receiver configuration featuring the WJ-8626A-4 HF and WJ-8628-4 VHF/UHF receivers, which combine to provide continuous 5 kHz to 1400 MHz frequency coverage using two independent, full-function receivers. Identical control and operating modes facilitate ease of operation and efficient transition for the operator when using both receivers. Up to 300 channels of memory are available for memory (channel) and frequency (F1-F2) scan routines.

The WJ-8610A and WJ-8610A-1 Controllers are the central control units for many of the complex multiple receiving systems being used today. Up to fourteen 21.4-MHz receivers may be controlled locally, via the front panel, or remotely, by interfacing with a system computer. The WJ-8610A-1 incorporates a number of software enhancements, including the capability of driving an external XYZ monitor, scanning of up to 150 preprogrammed frequencies (with 300 lockout frequencies), and automatic handoff of signals to any receiver in a pool.

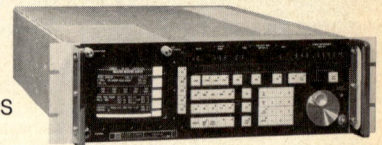

The WJ-8976 Three-Channel Direction-Finding System provides accurate azimuth and elevation bearing information within the 20 to 500 MHz frequency range. Accuracy is not affected by modulation type, thus permitting effective operation on conventional as well as spread spectrum signals. The system consists of a three-channel slave receiver, a WJ-8617B-9 receiver used as a master tuner, and a high-speed digital processor. A variety of hardware options are available which enhance the system's operating capabilities and expand the frequency range to 2 to 1100 MHz.

For further information on W-J's communication receiver capabilities, contact Watkins-Johnson Company, 700 Quince Orchard Road, Gaithersburg, Maryland, 20878-1794, (301) 948-7550, TWX 710-828-0546.

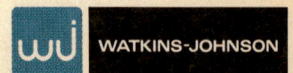

WATKINS-JOHNSON

THE INTERNATIONAL COUNTERMEASURES HANDBOOK
13TH EDITION 1988

Foreword

Integrating Electronic Warfare into C2 3
 Albert A. Gallotta, Jr.

ANALYSIS AND TECHNOLOGY

Electronic Combat: Past and Future 19
 Donald C. Latham
Soviet Signals Intelligence: The Use of Diplomatic Establishments 24
 Dr. Desmond Ball
New Technologies for Electronic Support Measures 47
 Dr. Myron L. Cramer
Extending the Frequency Range of ESM Antennas 54
 Glen R. Gray
Offboard Countermeasures 60
 Dr. Norman Friedman
Artificial Intelligence in Object Recognition 74
 Dr. David L. Milgram and *Thomas W. Miltonberger*
Electronic Warriors Need Training Too 84
 Richard Short
Advanced High-Voltage Power Supply Development 89
 Dale Hollis
Integrated Techniques for Precision Waveform Measurement 97
 Dr. Gregory J. Donaldson
Soviet Ocean Reconnaissance Threat from Space 103
 Nicholas L. Johnson

SYSTEMS OF THE WEST

Electronic Warfare Systems: AN/ Designated Hardware 113
Electronic Warfare Systems: International Hardware 138

ESM

WATKINS-JOHNSON COMPANY
RECEIVERS

For applications in complex tactical and strategic environments, these receivers offer advanced technological solutions from W-J's ESM Division.

The WJ-36000 Electronic Warfare Alert and Collection System (EWACS) combines two complementary receiving techniques to provide detection and recording of exotic (agile) emitters. The system consists of a superheterodyne receiving system and an instantaneous frequency measurement (IFM) receiving system, covering the 0.5 to 18.0 GHz frequency range. This system provides high probability of intercept (POI), while accurately measuring the signal parameters of pulse repetition interval, pulse width, frequency, frequency agility, scan pattern and jitter. The IEEE-488 and R-232C bus structures are available for external computer controlled applications.

The WJ-38000 Portable Elint Receiving System has been designed for rapid deployment in airborne, shipboard, land-mobile or fixed-based operations. This system, housed in a ruggedized case, covers the 0.5 to 18.0 GHz frequency range and operates on 28 VDC or 115/230 VAC. Although compact in size, it can operate either as a manual system or in a fully automatic mode. The system measures the parameters of the intercepted signals, compares these parameters to a user defined library, and identifies known threat emitters. Signal intercepts are displayed in both analog and alphanumeric formats. The IEEE-488 and RS-232C bus structures are available for external computer controlled applications.

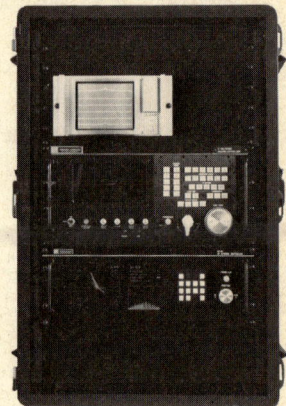

The TN-123 High-Accuracy Microwave Tuner is a fully synthesized multioctave frequency converter that covers the full 0.5 to 18 GHz frequency range with a 500 MHz instantaneous bandwidth and excellent RF performance. It differs from other multioctave tuners by using a unique architecture to provide full synthesizer accuracy during scan and ultralow phase noise (less than 1.0 degree rms) when fixed tuned.

The WJ-8969 Microwave Receiving System is designed for wideband and narrowband applications in the 1 to 18 GHz frequency range. The WJ-8969 consists of an IF demodulator/controller and a tuner unit. These two half-rack units can be installed side-by-side in a 19-inch equipment frame, or the tuning unit can be installed in a remote location. A simplified front-panel keyboard enables the rapid selection of parameters such as RF frequency, IF bandwidth, detection mode, gain-control mode and RF tuning rate, and prominently displays control settings for operator viewing. The system provides eight operator-selectable bandwidths covering 10 kHz to 100 MHz. It also includes AM, FM, CW and pulse detection modes. For computer-controlled applications, an IEEE-488 or RS-232 bus is available.

For information on W-J microwave ESM receivers, contact Watkins-Johnson Company, 2525 North First Street, San Jose, California, 95131-1097, or call (408) 435-1400, ext. 3248, TWX 910-338-0505.

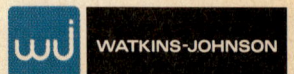
WATKINS-JOHNSON

SOVIET WEAPON SYSTEMS AND ELECTRONICS

Orders of Battle 155
Aircraft 160
   Bomber/Strike Aircraft 160
   Fighters and Attack Aircraft 163
   Helicopters 170
   Reconnaissance/Electronic Warfare Aircraft 176
   Reconnaissance/Airborne Warning Aircraft 176
   Reconnaissance/Anti-Submarine Warfare Aircraft 177
   Cargo/Transport Aircraft 178

Ships 180
   Aircraft Carriers 180
   Guided Missile Cruisers 182
   Cruisers 186
   Command Cruisers 188
   Guided Missile Destroyers 189
   Destroyers 196
   Frigates/Corvettes 198
   Guided Missile Patrol Boats 209
   Small Combatants 212
   Amphibious Ships 214
   Intelligence Ships 219

Submarines 226
   Ballistic Missile Submarines 226
   Guided Missile Submarines 229
   Attack Submarines 232
   Research Submarines 238
   Auxiliary Submarines 238
   Rescue and Salvage Submarines 239
   Training Submarines 239

Ground Combat Vehicles 240
   Tanks 240

Missiles 248
   Intercontinental Ballistic Missiles 248
   Submarine Launched Ballistic Missiles 252
   Land Attack-Theater Missiles 254
   Anti-Ship Missiles 261

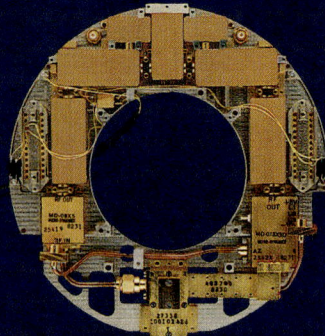

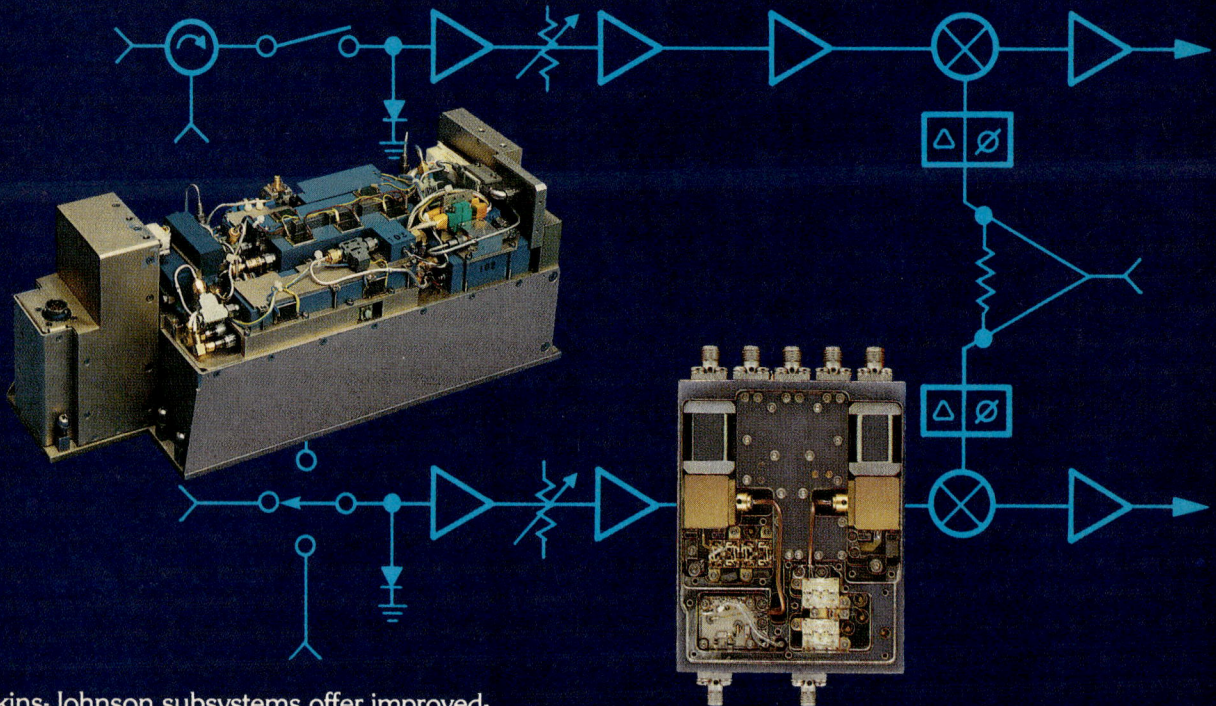

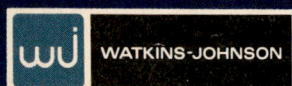

Anti-Ballistic Missiles 266
Anti-Submarine Warfare Missiles 267
Surface-to-Air Missiles 268
Ground Combat Vehicles/Air Defense 275
Air-to-Air Missiles 280
Anti-Tank Missiles 284
Anti-Radiation Missiles 287

Sensors 288
Strategic Radars 288
Tactical Radars 290

$ ELECTRONIC WARFARE BUDGETS

Summary of Projected DOD Spending 327
How To Interpret DOD Program Element Keys 328

OSD Budgets

RDT&E Programs 329

U.S. Army Budgets

RDT&E Programs 330
Procurement Items 332

U.S. Navy Budgets

RDT&E Programs 333
Procurement Items 336

U.S. Air Force Budgets

RDT&E Programs 338
Procurement Items 344

EW COMPANIES AND MARKET

World EW Sales: The Top 50 Companies 349
Profiles of Major U.S. EW Companies 350

INDEXES AND REFERENCES

Master Editorial Index 379
Index of Advertisers 386

MMIC BASE

Avantek MMIC-Based Components and Subassemblies —to 26 GHz—Today . . . for the Next Generation of Electronic Defense Systems.

Avantek's monolithic MIC-based components and subassemblies offer the broader bandwidths, enhanced performance, and added reliability —as well as reduced size and lower power consumption—designers need for the defense electronics systems of tomorrow. Avantek high-performance multi-function components combine the benefits of *all* thin-film circuitry and MMIC construction for optimal size and weight.

Avantek's MMIC-Based Products Include:

MODAMP™ Transistor Packaged Silicon MMICs—.01 to 10 GHz

PlanarPak™ Amplifiers— .01 to 1 GHz

Small-Signal Amplifiers— 1 to 20 GHz

Medium-Power Amplifiers— 6 to 18 GHz

Dielectrically Stabilized Oscillators —4 to 20 GHz

YIG-Oscillators—2 to 18 GHz

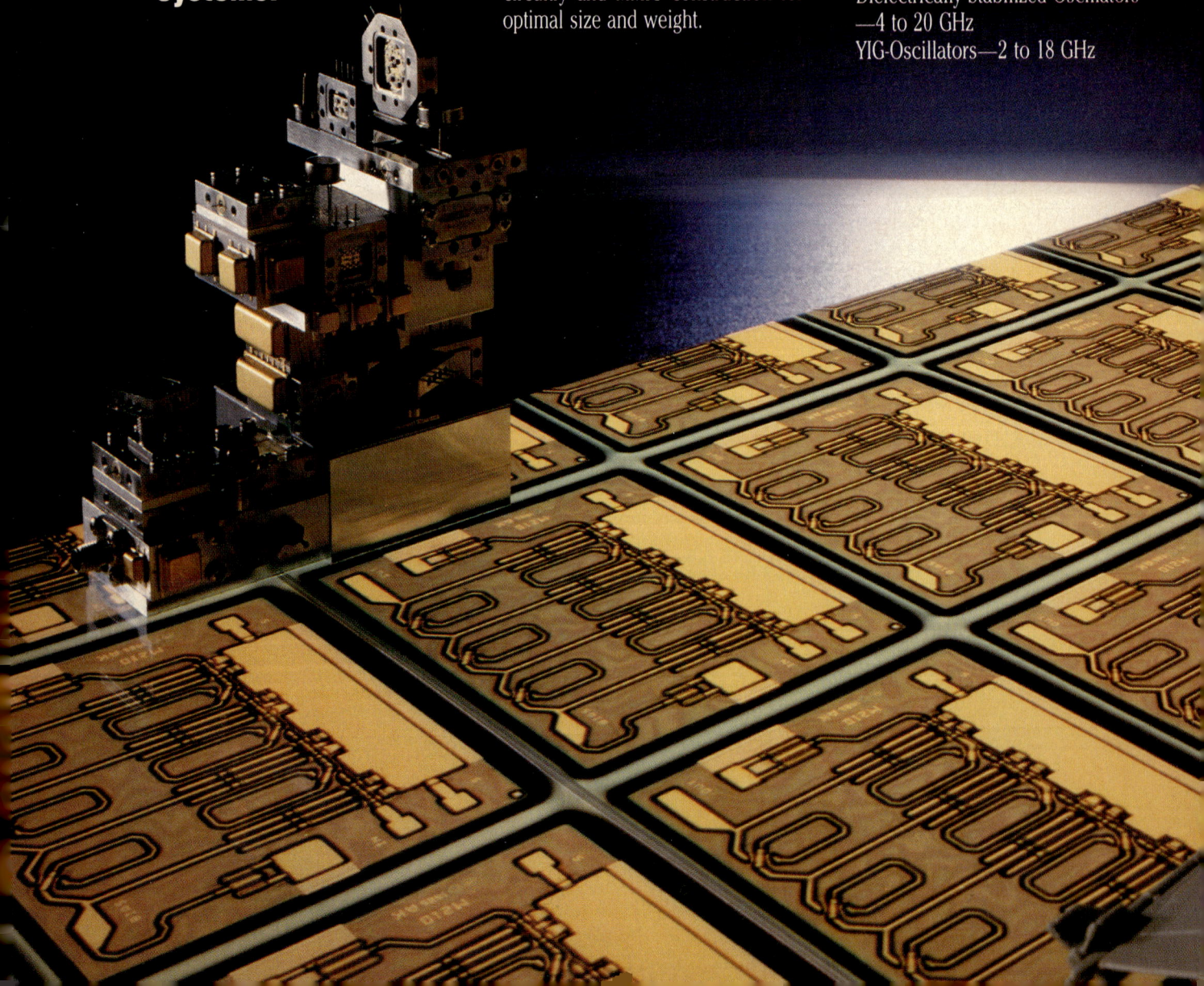

HERE'S A BETTER YOUR MICROWAVE

Components, Subassemblies, Monolithics...

While there may be many approaches to your microwave problem, the best solution can be seen only from the perspective of a broad base of technology and experience. Alpha Industries has long been a recognized leader in manufacturing microwave switches, limiters, attenuators and FET amplifiers. We've also supplied complex hybrid MIC subassemblies to integrate those components. And now we've developed GaAs MMICs to meet today's most demanding performance requirements.

The Right Approach. So when you design your next system, don't lock yourself into a solution that may not measure up. Talk to us first, and let us give you a little perspective. Whether it involves discrete components, multifunction assemblies or monolithics, we'll study your application and help you decide on an approach that suits it best. If your requirement calls for individual *components,* we'll offer you a selection that enjoys a solid record of program proven results. If your needs are more complex, we'll help you develop an *entire hybrid MIC configuration* within a frequency range that's virtually unlimited. And, if the situation warrants, we'll develop a *full MMIC configuration* to meet your size and reliability requirements.

PERSPECTIVE ON SYSTEM PROBLEMS.

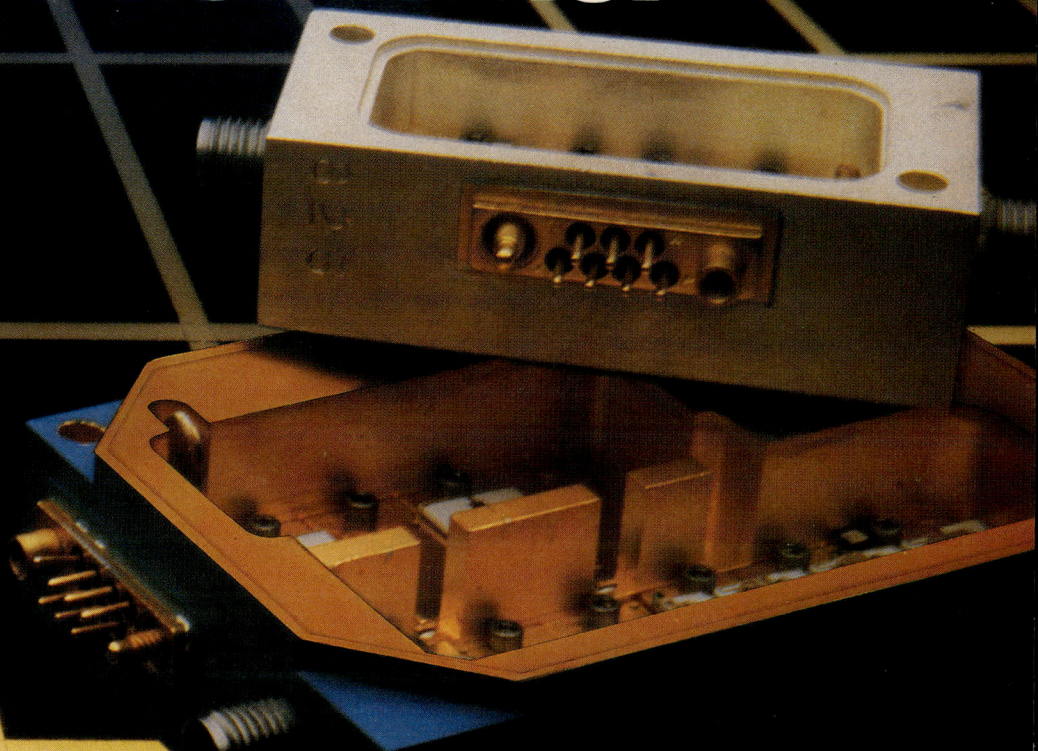

Alpha is developing GaAs technology at .1 micron resolution.

The Right Solution. As a leading manufacturer of microwave components and subsystems, we're able to provide such a wide range of solutions because of our size and experience. For example, we can achieve the *highest levels of component integration* because we have the latest in computer-aided design, engineering and machining equipment. We're able to meet the *tightest operating parameters* because we control the entire manufacturing process from materials to final assembly. And we're familiar with the most *demanding Mil Spec requirements* because we've developed products and subsystems for a large selection of EW programs.

The Right Supplier. The result is an added dimension that can add up to a powerful difference when performance is the only result that counts. So when you're faced with a problem that demands the flexibility of choice, take advantage of the three-way perspective that goes with the Alpha approach. Get in touch with our Advanced Products Division today, and find out why when it comes to system design problems, it pays to look at the situation from more than one point of view.

Alpha Industries, Inc. Advanced Products Division
651 Lowell Street • Methuen, MA 01844 • 617-682-4661 • TWX: 710-321-6434

ai Alpha
The Microwave People

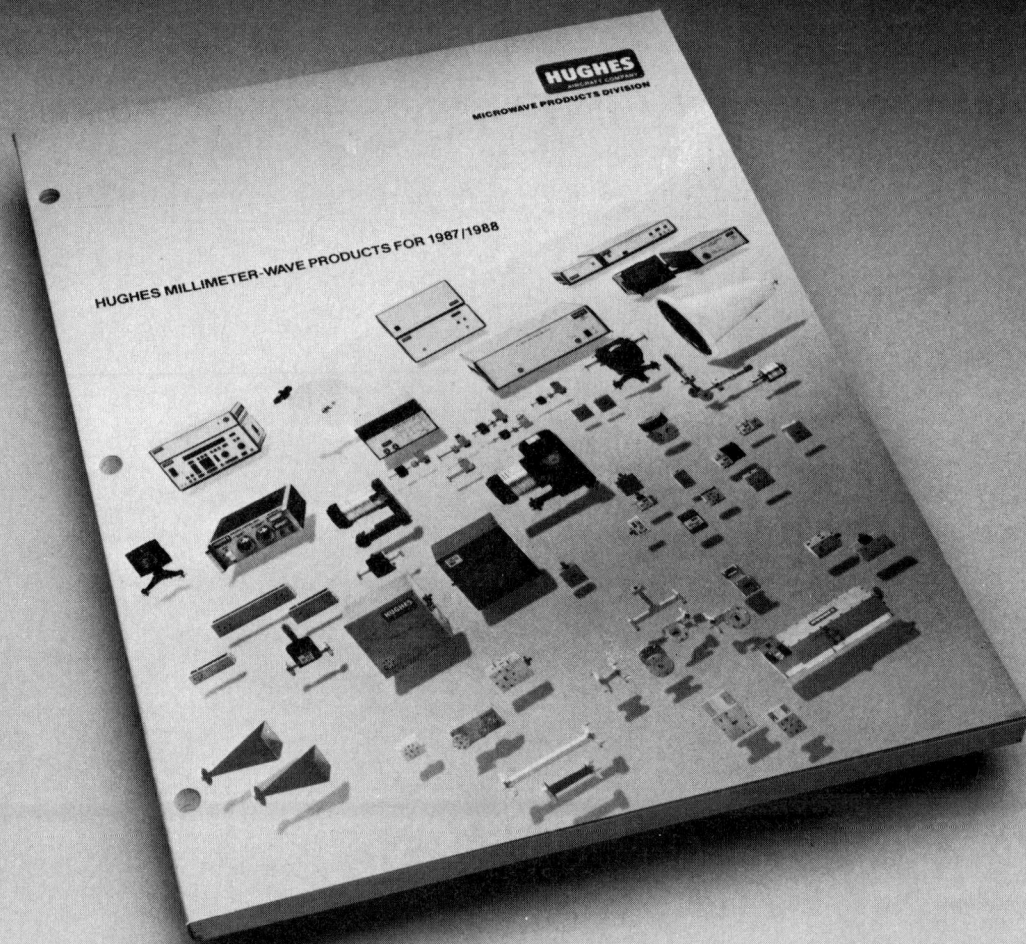

ANALYSIS AND TECHNOLOGY

Electronic Combat: Past and Future 19
  Donald C. Latham
Soviet Signals Intelligence: The Use of Diplomatic Establishments 24
  Dr. Desmond Ball
New Technologies for Electronic Support Measures 47
  Dr. Myron L. Cramer
Extending the Frequency Range of ESM Antennas 54
  Glen R. Gray
Offboard Countermeasures 60
  Dr. Norman Friedman
Artificial Intelligence in Object Recognition 74
  Dr. David L. Milgram and Thomas W. Miltonberger
Electronic Warriors Need Training Too 84
  Richard Short
Advanced High-Voltage Power Supply Development 89
  Dale Hollis
Integrated Techniques for Precision Waveform Measurement 97
  Dr. Gregory J. Donaldson
Soviet Ocean Reconnaissance Threat from Space 103
  Nicholas L. Johnson

Electronic Combat: Past and Future

Remembering the lessons of past electronic conflicts is vital to maintaining and broadening the use of the electromagnetic spectrum.

By Donald C. Latham

The history of electronic combat is one characterized by fast-paced evolution. The term itself is deliberately broad enough to encompass not only electronic warfare, but also rapidly developing assets in support of:

• Command, control and communications counter-measures (C^3CM) strategies, and

• Suppression of enemy air defenses (SEAD) activities.

Ever since Marconi's invention of the "wireless," armed forces around the world have sought ways to use such assets to counter their enemy's effective use of the electromagnetic spectrum. The objective of electronic combat is to gain control of the use of the electromagnetic spectrum by targeting, exploiting, disrupting, degrading, deceiving, damaging or destroying enemy electromagnetic equipment and systems in support of military operations.

The first evidence of the use of jamming in a military conflict occurred in 1903 during the Russo-Japanese War when a Russian radio operator on shore intercepted Japanese signals from small ships used to spot naval bombardment.[1] The radio operator, realizing the importance of these signals, keyed his spark transmitter to jam the signals, resulting in little damage and few casualties from the bombardment.

But, as has oft been said, countries that forget histories of warfare are doomed to repeat their mistakes. Though the first step in radio jamming had been taken, the lesson was either overlooked or soon forgotten. The next opportunity to employ effective electronic countermeasures was missed, resulting in disaster for the commander who failed to recognize its utility. The Russian fleet under Admiral Roxhestvenskiy suffered nearly total defeat in May 1905 during the Battle of Tsushima when the Japanese cruiser *Shinano Maru* was allowed to transmit, uninterrupted, the course, position and speed of the Russian fleet, allowing Admiral Togo to mass his fleet and catch the Russian ships in a deadly crossfire.[2] Although the captain of the auxiliary cruiser *Ural* urged Rozhestvenskiy to use a recently installed radio transmitter to jam the "enemy sighted" message, the commander refused permission and steamed into the Straits of Korea, where his fleet was suddenly met and destroyed or captured by the Japanese.

Evolution to Revolution. The trend of development of electronic measures, followed by rapid responses of countermeasures development, began in earnest at the outset of World War II. The extent of electronic combat

The expanding role of electronic combat is exemplified by the airborne radar jamming system to be carried on UH-60 helicopters, used to disrupt hostile air defense, surveillance and target acquisition radars from standoff range. (Illustration courtesy of Grumman Aerospace Corp.)

exploits during the war and since was a closely guarded secret until recent years. The revelations of a series of electronic developments, which alternately placed both sides on the verge of disaster, offer us an important message. They not only confirm a place for electronic combat in future warfare, but also give insight to its great importance.

The first use of radar in battle occurred at the outset of the war when the German pocket battleship *Graf Spee* used the "Seetakt" fire control radar to sink nine British merchant ships in the South Atlantic in 1939.[3] The device provided remarkable precision at ranges of more than nine miles. This introduction of radar so worried the British that they immediately began research on a comparable naval radar and studied means to possibly neutralize it with appropriate electronic countermeasures.

The importance of the electronic war escalated rapidly in the air war over Great Britain and Germany as each side employed a constant stream of electronic measures, countermeasures, and counter-countermeasures.[4] In what Prime Minister Winston Churchill referred to as the "Wizard War," the British introduced meaconing to combat the Lorenz navigation systems that had enabled the Germans to achieve precision night bombing. Germany came up with their "Knickebein" radio-guidance system, which was soon countered by the British, and the struggle by each side to outmaneuver the other was on.

To combat German air defense acquisition radars, the Allies began to use "window," or chaff. The Germans responded with improved techniques to distinguish aircraft echoes from other reflections and exploited the Doppler effect to determine the velocity of the aircraft targets. They introduced the "Lichtenstein SN2" radar with a range out to 40 miles on their night fighters, allowing the ground control stations to merely direct the fighters approaching from the rear, but the Germans countered with radar warning receivers on their fighters which allowed them to passively home in on their prey.

Similar battles of electronic wits took place in sea warfare as well, directly affecting both surface and sub-surface tactics. The efforts reached huge proportions by the invasion of Normandy, with large-scale jamming used to screen the force of landing craft and to neutralize the German coastal gun control radars. Deception operations were carried out to indicate an invasion at Calais rather than Normandy by employing communications jamming and by using a flotilla of small vessels carrying special metal plates and towing buoys or metalized balloons to create radar echoes similar to large warships.

The first use of "smart" weapons was unveiled in 1943. The United States had conducted a number of experiments with remotely controlled vehicles in the late 1930s that led to development of a crude radio-controlled glide bomb, the GB-1.[5] A number of these 2,000-pound bombs were dropped by B-17s on Cologne, causing heavy damage. Around the same time, the Germans devised a large rocket-powered bomb, the Henschel Hs 293, which was guided by a four-tone radio-control system from aboard their bombers.[6] The Hs 293 proved to be a deadly anti-ship weapon until the Allies produced a jammer capable of leading the bombs off target.

The struggles in electronic measures and countermeasures went on until the end of the war, playing an important role in both sea and air battles, favoring first one side, then the other. What became very clear to both sides was that applications of electronic technology were subject to rapid changes. What also should have been apparent was that waging war in the electronic spectrum had profound influences on the outcome of the war. But for whatever reason, perhaps because of the advent of nuclear capabilities, interests in electronic measures and countermeasures waned in the years following World War II.

In the early stages of the Korean War, this lapse was not of great consequence as North Korean air defenses were fairly rudimentary. However, when China entered the war in November 1950, the lessons were brought home again as early warning radars, located safely over the border, along with gun and searchlight radars, were put into action.[7] The loss of a number of B-29s quickly prompted use of chaff and electronic jamming and evoked landmark changes in tactics. The U.S. Air Force began to apply integrated tactics to suppress air defenses by leading bombing missions with attacks first against the anti-aircraft defenses followed by stand-off jamming during the main strike.

Forgotten Lessons. Following the Korean War, interest in radar detection and jamming once again waned until May 1960, when Francis Powers' U-2 reconnaissance plane was brought down over Russia with a SAM-2 in the first use of a surface-to-air missile. When the United States found itself engaged in raids over North Vietnam, the SAM-2 again appeared as a major threat. The electronic measure/countermeasure challenges characteristic of World War II were on again. Radar warning receivers to detect the "Fansong" radars, combined with evasive tactics against the SAM-2, were reasonably effective until the anti-aircraft gun defenses were strengthened to engage the fighter-bombers as they dove to avoid the missiles.[8] Hasty development programs produced a number of jammers, including the first of the externally attached jamming pods, which were quickly employed for self-protection of our aircraft.

The war in Vietnam also witnessed a resurgence of "smart" weapons with the appearance of laser-guided bombs (LGBs) and Wild Weasel aircraft. A laser beam tracking guidance system on the LGBs could home in on targets illuminated by a laser designator, enabling "surgical" bombing of hard-to-hit targets. The Wild Weasel system combined the AGM-45 Shrike anti-radiation missiles with radar warning receivers which could guide the pilot toward the radar signal to launch the missiles. Aircraft losses to the SAM-2s decreased significantly from one aircraft for every 10 missiles launched in 1965 to one in 70 in 1966.[9]

The mid-1960s saw other events that portended the increasing importance of electronic combat in the future. During the Six-Day War of 1967, the Israeli destroyer *Eilat* was sunk by Styx missiles launched from two Egyptian torpedo boats near Port Said, marking the first time that a warship had been sunk by missiles.[10] The event caused some rethinking about naval warfare tactics and weaponry. Although the Israeli Gabriel missile was more accurate than the Styx, its range was considerably shorter. The Israelis chose to overcome the deficiency by equipping their missile-launching boats with electronic jammers and deceivers and by covering their ships with radar-absorbent materials so that they might close within range of their more effective missiles—a strategy that paid off handsomely when the Yom Kippur War broke out in 1973.[11]

Events involving electronic combat in air warfare in the Middle East arena were equally momentous. The frequent fighting along the Suez Canal provided the setting for numerous changes in air defense and air defense suppression techniques. The Middle East became an electronic proving ground, with Israel benefiting from new electronic warfare equipment on F-4 and A-4 fighters. The equipment was relatively effective against the SAM-2 and SAM-3 missile systems and their associated radars.

But the Israelis were apparently caught off guard both operationally and technologically at the start of the 1973 Yom Kippur War. The Egyptians had used the lull in fighting since 1970 to revitalize their air defense equipment and tactics.[12] The modern SAM-6 missile system was introduced and used in combination with radar-controlled ZSU-23-4 anti-aircraft guns and the shoulder-launched SAM-7 missiles. Together, they presented a formidable and highly mobile air defense barrier. The SAM-6 was using CW signals, which could not be countered by the jammers aboard the Israeli aircraft since the jammers had been designed for use against pulsed radars. The ZSU-23-4 guns were using higher frequencies than any previously used by the Egyptians, and the SAM-7 missiles incorporated a new kind of guidance system employing infrared homing. These technological surprises placed Israeli aircraft in great danger, but the Israelis prevented catastrophic losses by quick reaction with chaff cut to match the new frequencies, infrared flares to counter the SAM-7 missiles, and modifications in tactics to interdict the SAM-6 systems. However, the final toll in Israeli aircraft was high.

Thinking for the Future. The history of electronic combat experiences has taught us many lessons that we should be loath to learn again, both for the sake of our national security and for economic reasons. What should we have learned from all of this? And how can we avoid relearning the same painful lessons at high costs in men and materiel and potentially high risk to the survivability of our nation in some future conflict?

The specific equipment of future electronic combat may be difficult to predict, but the need for emphasis in this area seems clear. Experiences in more recent years in the Falklands, in the Bekka Valley and in Libya continue to bear out the very dynamic and revolutionary nature of electronic combat as well as its impact. If we are to ensure adequate defense of our country, we must be prepared not only to stay abreast of electronic threat developments but also to reliably predict associated technological trends which will be critical to providing suitable countermeasures. It is probably safe to say that any future combat operations—on land, in the air, at sea and some day in space—are going to be dominated by electronic systems, both passive and active, which will directly affect the tactical outcome of these future battles. Therefore, I believe our capability in electronic combat is a critical element of our survival!

As we look to the future, we can also project far more sophistication in the way that electronic combat capabilities will be employed. What could evolve is a high degree of integration of electronic combat functions and supporting functions. Whereas decades ago we thought in terms of effectiveness and survivability of individual platforms operating against singular threats, we now must think of the effectiveness and survivability of multiple aircraft, ships, or even spacecraft in tightly integrated operations against multiple threats. We must consider concepts for applying passive and active techniques across the spectrum, and employing sophisticated countermeasures and counter-countermeasures both on board the platforms and supported by resources external to those platforms.

Technical Challenges. The first technical challenge that we face is how to provide a sufficiently broad, flexible and available technology base to draw from in responding to threat changes. If we have learned any lesson from the history of electronic combat, it is that we must be prepared to respond very quickly to changes in the electromagnetic environment. We can ill afford to endure a development/acquisition cycle that may run for years if an enemy achieves an electronic surprise. Though we must minimize being only threat-reactive and do our best to anticipate threat changes, we must also be prepared to respond quickly to unforeseen changes.

A number of specific technology areas will be critical to us in the near future. One of the most rapidly evolving is the field of lasers and directed energy weapons and countermeasures. Evidence of operational use of laser devices for optical obscuration or to attack the human eye has already appeared on the battlefield. Such uses of lasers could have a devastating effect on military systems reliant on optics and strong psychological, as well as physical, effects on personnel. It is also well known that the Soviets have extensive research ongoing to explore the kinetic destructive capacity of high-energy particle beams.

The development of avionics compatible with low observable aircraft presents a major technical challenge. The integrated electronic warfare system (Inews), the integrated communications, navigation and identification avionics (ICNIA) system and Pave Pillar form the basis for integrated avionics that will give pilots the situational awareness needed to complete their missions while maintaining levels of survivability anticipated for "stealth" type aircraft. Each of the individual technology areas is complex in itself; their integration will be considerably more difficult. In addition, higher reliability of system components is a major developmental objective, and the program intends to provide increased capabilities in less volume—factors that further increase the challenge.

A trend toward emphasis on passive means for acquisition systems and fire control systems also appears evident. Just as the Germans took advantage of radar receivers on their fighters during World War II to home in on the "Monica" radars in the tail of RAF bombers, modern forces are similarly exploring techniques to employ passive techniques to gain the advantage in the electronic environment. In fact, it is expected that significantly greater use will be made of passive intercepts, or electronic support measures (ESM) as they are sometimes called, in future combat operations. This will likely involve equipping platforms with sophisticated signal intercept, recognition, and direction finding capabilities which, coupled with other devices, can result in a passive fire control capability. Stealth technology and anti-radiation homing weapons may have been partially responsible for renewed interests in passive electro-optic, infrared, and signal detection techniques.

The field of remotely piloted or unmanned aerial vehicles (RPV/UAV) is also evolving very rapidly. Perhaps the most popularly known use of RPVs in battle occurred in the Bekka Valley in 1982 when the Israelis effectively employed them to help suppress and destroy Syrian air defenses. RPVs or UAVs offer some intriguing advantages over manned aircraft for application of electronic measures and countermeasures. Efforts are under way in several countries to employ UAVs for reconnaissance, as jamming platforms and as lethal devices to suppress electronic emitters and other targets.

We should also take notice of the multispectral aspects of lessons learned in Vietnam and the Middle East. There seems to be a clear trend to target acquisition and tracking using multiple sensors that apply combinations of radio frequency, electro-optics, infrared, and other techniques. The challenge of devising elective measures to counter them when used in combination is far more difficult than the mere integration of such techniques.

Finally, in tracking current threat developments, we cannot overlook the potential that many of the electronic systems used by a potential enemy will likely have operating modes that are reserved solely for wartime use. An active search must be made for such hidden capabilities, especially in new frequency ranges or modulation techniques that would make our present countermeasures less effective or even obsolete. Technical intelligence gathering for these reserve modes, along with careful analyses and technical estimates or projections, are thus every bit as important as operational intelligence on the outbreak of conflict—perhaps more important.

Economic Challenges. It is obvious that aggressive pursuit of these and all other high-technology areas with potential application in electronic measures and countermeasures would be an expensive endeavor. Resources must be allocated wisely to achieve an effective yet economical balance of capabilities. The Department of Defense has, for the past three years, prepared an electronic combat master plan to accomplish just that. This master plan intends to stimulate and institutionalize cross-service planning and provides a single, concise reference on electronic combat activities. It enables managers to avoid wasteful duplication and guides the services toward the most propitious technology and system development activities in acquiring an optimum complement of electronic systems. It has promoted efforts such as the advanced self-protection jammer and establishment of a joint airborne expendables office to foster economic savings through use of common equipment and technologies.

Our most recent challenge, and possibly one of the biggest economic as well as technical challenges we face is development of common integrated avionics for the next generation tactical aircraft. With escalating development and acquisition costs, we cannot afford several independent, large-scale avionics programs. A unique opportunity exists with three services embarking in the same time frame on separate programs for advanced aircraft. The services have come together to develop a core of technology based on Inews, ICNIA, the Pave Pillar architecture and other programs that will provide a baseline for common integrated electronics and avionics. A joint integrated avionics plan for new aircraft was forwarded to Congress. It outlines joint service efforts for modern digital, integrated electronics, avionics, and embedded communications security for advanced aircraft. The services will take advantage of key technologies such as very high speed circuits (VHSIC), machine intelligence, multi-use sensors and software enhancements to develop modular and interchangeable building blocks for integrated systems.

Reaching For Cost-Effectiveness. Since we cannot afford the risk to our security that would attend a lapse in electronic combat preparedness, we really have a significant cost-effectiveness challenge. But we face many dilemmas when we consider even how to define or really measure cost-effectiveness in electronic combat.

The first dilemma confronting us is that we simply cannot afford all that is necessary to ensure fully adequate preparedness. The task then becomes one of carefully identifying the electronic combat elements that are critical to the preservation and effectiveness of our military assets, and then ensuring that we acquire those capabilities as economically as possible.

When we think of achieving economies, the hue and cry that we first hear is that we must have commonality. As a matter of fact, commonality can indeed help to achieve economy as well as interoperability. But we must be careful not to allow commonality to become a false and misleading goal. To carry an example to extreme, it may be possible to conceive a "common" airplane to fulfill all manner of battlefield roles. But such an airplane would be absurdly complex and costly, and would be obsolete before flown, if it could even be built. The dilemmas we face then are choices as to how we should partition battlefield and electronic functions. We must achieve commonality at the lowest common denominator that makes both economic and operational sense.

One way of approaching cost-effectiveness is to examine a given need in some larger context, or to measure the worth of a component or system as part of a much larger systems context. Synergistic effects can be achieved through an integrated strategy or architecture that attempts to achieve a balanced complement of capabilities that can overcome the limitations or vulnerabilities of individual elements.

The U.S. concept of C³CM is an example of a strategy

that aims for such synergistic effects. It embraces the notion that an enemy's C^3 systems are important targets to attack and that attacks on them should be considered with a balanced approach of physical destruction, disruption, information denial and deception. The concept also includes protection of friendly C^3 assets from similar action by the enemy. As one might guess, much of the concept involves manipulation of the electromagnetic spectrum.

By extending the concept of C^3CM as a balanced and integrated strategy into the broader arena of all electronic measures and countermeasures, we would define a similar electronic combat strategy as one that would strike a balance of electronic capabilities for land, sea, air and space warfare, and one that would apply against all assets, not just C^3. Perhaps our greatest challenge today is to transcend the mindset that basically considers radio and electronic techniques in the traditional view connoted by the term "electronic warfare" and expand our thinking into the broader realm envisioned by the term "electronic combat," which includes EW, SEAD and C^3CM. Capabilities must be planned that provide for flexible responses and that are oriented to strategic goals.

Have we learned our lessons well? We may not know until the next real conflict arises. Even if one were to take a most optimistic view of our current posture in electronic measures and countermeasures, our attention must remain focused on the dynamic nature of electronic combat. Our planning must be founded on well organized and executed gathering and analyses of threat intelligence and on lessons learned from past and future conflicts as well as exercises. Then we must concentrate

management attention on achieving a fully integrated and interoperable complement of electronic capabilities.

References

1. Alfred Price, *History of U.S. Electronic Warfare*, The Association of Old Crows, 1984, pp. 4-5.
2. Mario de Arcangelis, *Electronic Warfare*, Blandford Press, 1985, pp. 11-18.
3. de Arcangelis, pp. 27-29.
4. de Arcangelis, pp. 37-51, 66-81.
5. Paul Dickson, *The Electronic Battlefield*, Indiana University Press, 1976, p. 181.
6. de Arcangelis, pp. 92-94.
7. Alfred Price, *Instruments of Darkness*, Charles Scribner's Sons, pp. 251-253.
8. Price, *Instruments of Darkness*, pp. 259-273.
9. de Arcangelis, p. 168.
10. de Arcangelis, pp. 178-189.
11. de Arcangelis, pp. 196-204.
12. de Arcangelis, pp. 189-194.

Donald C. Latham held the Pentagon's top position in command, control, communications and intelligence for six years until he left office in July 1987 to join Computer Sciences Corp.'s Systems Group in Falls Church, Va. He served first as deputy under secretary of defense (C^3I) in the office of the under secretary of defense for research and engineering beginning in 1981, and then in 1984 became assistant secretary of defense for C^3I. Latham received a bachelor of science degree in electrical engineering from The Citadel in 1955. He also has a master of science degree in electrical engineering and an advanced electrical engineering degree, both from the University of Arizona. Latham served as an Air Force officer assigned to the National Security Agency from 1957 to 1959, and later served in executive positions with the agency in Europe and Washington, D.C. He has also been a manager for Martin Marietta Aerospace Co. and RCA.

Soviet Signals Intelligence: The Use of Diplomatic Establishments

A windfall of information is theirs for the taking.

By Dr. Desmond Ball

The exploitation of communications and other signals by Soviet and Warsaw Pact signals intelligence (Sigint) organizations poses a serious threat to the defense capabilities, political and diplomatic integrity, and commercial and technological activities of the non-communist countries. This threat remains generally under-appreciated. The Soviet Union maintains the most extensive and comprehensive Sigint capability in the world: Some 350,000 Soviet personnel are engaged in Sigint activities, and there are more than 500 Sigint ground stations located within the Soviet Union, Eastern Europe and elsewhere around the world. Other Sigint systems have been deployed on board an extraordinary range of mobile platforms, including submarines, ships, aircraft, satellites, and various trucks and other ground vehicles.

In addition, the Soviet Union has installed Sigint systems in about half of the countries in which it maintains diplomatic establishments (embassies, consulates and/or official residences). Diplomatic missions are currently maintained in about 115 countries, and Sigint operations are presently conducted in about 60 of these countries. Sigint systems are installed in embassies, consulates, trade and other commercial missions, the offices of military attaches, the official residences of diplomatic personnel and official recreational facilities.

It is not possible to ascertain precisely the exact number of countries in which the Soviet Union uses diplomatic establishments for Sigint operations (although 60 must be very close), or the exact nature of such operations: The number of countries in which the Soviet Union maintains diplomatic establishments changes quite frequently, and information provided by defectors and other human intelligence (Humint) sources can sometimes soon lose currency. Examination of the antenna systems installed on the roofs of Soviet diplomatic establishments, the number of antennas, their designs, frequency coverage and orientation, is necessarily inconclusive, and structures are frequently constructed to hide microwave antennas. But enough is known of existing Soviet installations and practices to piece together a compelling argument that they are unexcelled in the world when it comes to espionage and Sigint exploitation. It is my purpose herein to reveal some of the depth and strength of their efforts.

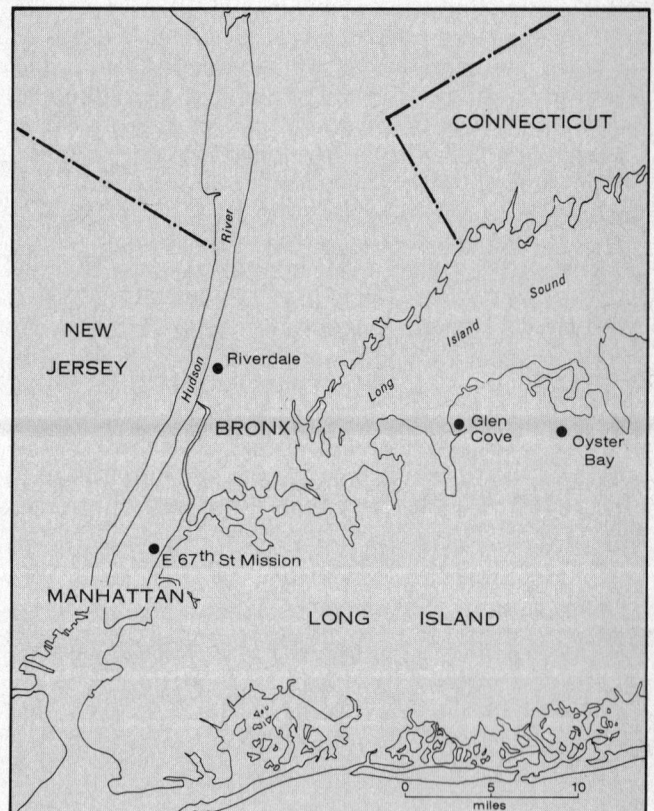

Figure 1. A map of the four Soviet Sigint posts in New York: Riverdale, East 67th Street, Glen Cove, and Oyster Bay.

Increased Use of Diplomatic Posts. Following the October 1973 war in the Middle East, the headquarters of the United States European Command (USeucom) conducted an assessment of "the known communications collection capabilities of the Soviet Union and Warsaw Pact nations" in Europe, the Near East and the United States, and concluded that Soviet diplomatic establishments in 22 countries could have been used to monitor U.S. and NATO communications relating to military and diplomatic activity concerning the war. The assessment concluded that both the Komitet Gosudarstvennoy Bezopasnosti (KGB), or Committee for State Security, and the Glavnoye Razvedyvatelnoye Upravleniya (GRU), or Chief Intelligence Directorate of the Soviet General Staff are known to conduct Sigint collection from a variety of diplomatically immune buildings around the world. Specifically:

• In Europe, known sites are located in the following cities: Vienna, Austria; Copenhagen, Denmark; Potsdam, East Germany; Helsinki, Finland; Paris, France; Rome, Italy; The Hague, Netherlands; Oslo, Norway; Stockholm, Sweden; London, England; Frankfurt, Cologne and Bonn, West Germany and Brussels, Belgium.

• In the Near East, Sigint collection is accomplished from: Cairo, Egypt; Athens, Greece; Tehran, Iran; Baghdad, Iraq; Beirut, Lebanon; Karachi, Pakistan; Damascus, Syria; Aden, Yemen; and Ankara and Istanbul, Turkey.

• Washington, D.C. and New York City are prime collection locations within the continental United States.

In 1975 a review by the Central Intelligence Agency (CIA) of Soviet Sigint organizations and capabilities concluded that Sigint operations were conducted in Soviet diplomatic establishments in "at least 48 foreign countries," while a similar CIA review in 1984 concluded that such operations were being undertaken in more than 50 countries. Over the past 12 years, the number of countries has increased by one per year.

Figure 3. The Soviet residential complex in Riverdale, N.Y.

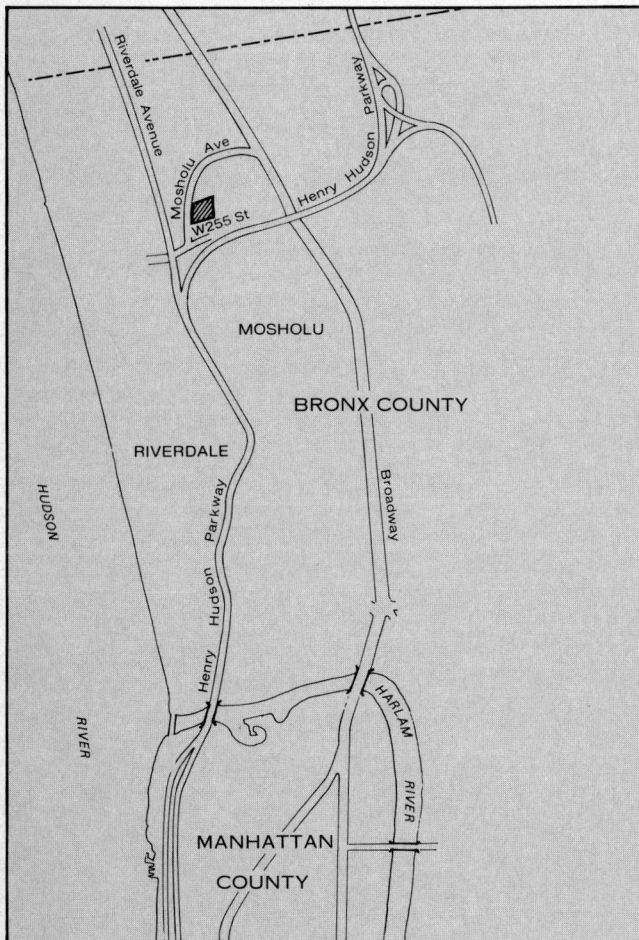

Figure 2. Location of the Riverdale site.

Table 1 lists 60 countries in which the Soviet Union is believed to use diplomatic establishments for Sigint purposes. In many of these countries (such as the United States, the United Kingdom, Japan, France, Belgium, West Germany, Turkey and Pakistan) more than one such establishment is used. The total number of Soviet diplomatic establishments known to maintain extensive Sigint collection activities is more than 80. In addition, other buildings occupied by Soviet diplomatic personnel, such as apartments and commercial offices, are used for more limited Sigint activities as circumstances permit. Given that the average number of personnel engaged in Sigint activities in these establishments is 10 to 12, it can reasonably be estimated that some 1,000 Soviet personnel are currently collecting Sigint from diplomatically immune facilities all over the world.

The KGB and GRU are extremely active at the locations listed in Table 1. Within the KGB, the responsibility for such operations rests with the 16th or Communications Directorate. In the case of the GRU, which in fact is much more involved in these Sigint operations than the KGB, the responsibility rests with Technical Service Groups (TSGs). These are part of the Third Branch (Foreign Radio Intelligence) of the Radio and Radio-Technical Intelligence Directorate, usually referred to within the GRU as

TABLE 1
Soviet Sigint Posts
Located in Diplomatic Establishments

	Country	Site		Country	Site
1	Afghanistan	Kabul	35	Mongolia	Ulaan Baatar
2	Angola	Luanda	36	Mozambique	Maputo
3	Argentina	Buenos Aires	37	Nepal	Katmandu
4	Australia	Canberra	38	Netherlands	The Hague
5	Austria	Vienna	39	New Zealand	Wellington
6	Bangladesh	Dacca	40	Nicaragua	Managua
7	Belgium	Brussels (2)	41	North Korea	Pyonyang
8	Burma	Rangoon	42	Norway	Oslo
9	Canada	Ottawa			Svalbard
		Montreal	43	Pakistan	Islamabad
10	China	Beijing			Karachi
11	Cuba	Havana	44	Peru	Lima
12	Denmark	Copenhagen	45	Philippines	Manila
13	East Germany	East Berlin	46	Portugal	Lisbon
		Potsdam	47	Seychelles	Victoria
14	Egypt	Cairo	48	Singapore	Singapore
15	Ethiopia	Addis Ababa	49	South Yemen	Aden
16	Finland	Helsinki	50	Spain	Madrid
17	France	Paris	51	Sri Lanka	Colombo
		Marseilles	52	Sweden	Stockholm
18	Greece	Athens			Gothenburg
19	Iceland	Reykjavik	53	Switzerland	Berne
20	India	New Delhi			Geneva
21	Indonesia	Jakarta			Zurich
22	Iran	Teheran	54	Syria	Damascus
23	Iraq	Baghdad	55	Thailand	Bangkok
24	Ireland	Dublin	56	Turkey	Ankara
25	Italy	Rome			Istanbul
26	Japan	Tokyo	57	United Kingdom	London (2)
		Osaka	58	USA	Washington, D.C. (3)
		Sapporo			New York (4)
27	Kenya	Nairobi			Pioneer Point, Md.
28	Kuwait	Kuwait			San Francisco
29	Laos	Viangchan			Chicago
30	Lebanon	Beirut	59	Vietnam	Hanoi
31	Libya	Tripoli	60	West Germany	Bonn
32	Madagascar	Antananarivo			Cologne
33	Malaysia	Kuala Lumpur			Frankfurt
34	Mexico	Mexico City			West Berlin

NOTE: The figures in parentheses indicate the number of sites at each location.

the Sixth Directorate. Although there is some competition between the KGB and the GRU in this field, and hence sometimes duplication of Sigint operations within given diplomatic establishments, there is typically also some division of labor between them.

The selection of which Soviet diplomatic establishments should be used for Sigint operations, and the relative responsibilities assigned to KGB and GRU personnel, depend on a variety of factors. The most important of these is the location of the particular establishment in relation to communications and other signals of interest to the respective agencies. The size of the establishment, both physically and in terms of the number of available personnel, as well as the nature of the establishment, are also important. For example, the KGB generally has responsibility for operations conducted from consulates, while the GRU conducts operations from the offices of military attaches (where these are separate from the embassy building). In general, about 40 to 45 percent of the personnel employed in any Soviet diplomatic establishment are KGB officers and perhaps 15 to 20 percent are GRU officers, of which about 15 percent are concerned with Sigint collection. Hence, in a relatively large mission of about 300 personnel, about 20 people are involved in various sorts of Sigint activity. In relatively small missions, Sigint activity is generally conducted only if there are signals of special interest and no other collection system is available within the region.

Types of Soviet Sigint Operations. In relatively large establishments, four types of Sigint operations are commonly conducted, two by the KGB and two by the GRU. These involve both general communications and the particular communications of police and security services in the host countries:

1. GRU Technical Service Groups. According to Viktor Suvorov (pseudonym), who served in the GRU from 1971 until he defected in Vienna in 1977: "Technical

Service (TS) officers are concerned with electronic intelligence from the grounds of official Soviet premises, embassies, consulates, and so on. Basic targets are the telecommunications apparatus of the government, diplomatic wireless communications and military channels of communication. By monitoring radio transmissions, secret and cipher, TSGs not only obtain interesting information but also cover the system of governmental communications, subordination of the different components of the state and the military structure. The military ranks of TS officers are major and lieutenant-colonel."

The TS groups monitor and record clear-text traffic, voice transmissions and encrypted traffic. In those countries where microwave and tropospheric scatter links are in common use, a large number of unsecure voice telephone conversations provide a particularly lucrative source of intelligence. TS officers monitor these systems, identify the interesting and important channels, and then record all conversations carried on them. In those instances where the intercept officers are unable to understand or decipher a particular message locally, the appropriate tapes are sent to the GRU's Sixth Directorate in Moscow for processing.

2. GRU Radio Monitoring Stations. The GRU radio monitoring stations have been described by Suvorov as follows: "In contradistinction to TS officers, these (officials) are concerned with monitoring the radio networks of the police and security services. The technical services and the radio monitoring station are two different groups, independent of each other, both controlled by the GRU resident. The difference between them is that the technical services work in the interests of the GRU Center in Moscow, trying to obtain state secrets, but the monitoring station works only in the interests of the local residency, trying to determine where in the city police activity is at its highest at a given moment and thus where local espionage operations may be mounted, or where they should not be mounted." These stations are generally manned by two or three officers, who are responsible to the First Deputy to the GRU Resident.

3. KGB 16th Directorate Sigint Operations. The KGB also maintains Sigint operations in various Soviet diplomatic establishments for the purpose of monitoring governmental and diplomatic communications. Officers of this directorate are particularly concerned with the interception of western encrypted communications traffic.

4. KGB Operational-Technical Directorate (OTU) Zenith Rooms. Personnel of the Operational-Technical Directorate (OTU) of Department Fourteen of the First Chief Directorate of the KGB are stationed in the KGB's major residencies in the West to serve as resident audio countermeasures technicians and to monitor the radio transmissions of the host country police and security surveillance agencies. These personnel occupy so-called Zenith rooms, and their activities are functionally equivalent to those of the GRU's radio monitoring stations.

It is obvious that some mechanism must exist for the coordination of these four Sigint activities, particularly in the smaller establishments where the physical and per-

Figure 4. A structure atop the Riverdale complex houses microwave and other eavesdropping equipment.

sonnel resources simply would not allow separate operations. It is also likely that in areas where there are several diplomatic establishments engaged in Sigint activities, there is some division of labor with respect to the monitoring of particular signals, with some establishments focusing on high frequency (HF) transmissions, some on terrestrial microwave and tropospheric scatter circuits, some on satellite communication (Satcom) signals, and others on local police and security service communications activity, etc.

In general, very little Sigint processing is undertaken abroad. The Soviet Sigint complex at Lourdes in Cuba, which according to President Reagan is "the largest of its kind in the world," has a processing and analytical capability, and some of the Sigint collected at Soviet diplomatic establishments in the Washington and New York areas is transmitted to Lourdes for some processing and analysis. In virtually all other cases, Sigint collected at Soviet diplomatic establishments, and which is of no immediate local utility, is sent to the relevant KGB and GRU facilities in Moscow. In the case of the GRU, the decrypting service is located on Komsomolkiy Prospekt in Moscow, and a special facility for computer processing of communications intelligence (Comint) is located near Sokolovskiy railroad station, about 25 miles from the centre of Moscow.

Soviet Diplomatic Establishments. The United States is the main target of Soviet Sigint operations, and provides the most diplomatically immune opportunities for such activity. There are currently 10 diplomatic establishments in the U.S. that have been equipped for extensive Sigint collection, and about 150 personnel are engaged in this activity. Table 2 provides a ranking of these ten establishments in terms of the quantity and quality of the Sigint collected.

As shown in Table 2, the Soviet diplomatic presence in New York is the largest in the United States. Until March 1986, when the Reagan Administration demanded that it be reduced by some 100 posts, there were more than 600 Soviet personnel in New York City, including about 525 employees connected with the United Nations. About 250 are assigned to the Secretariat and 275 to its U.N. missions, and some 90 other personnel are employed in trade missions, the Intourist travel bureau, etc. There are various estimates of the number involved in espionage, but the four Sigint posts in New York employ about 65 KGB and GRU personnel, and are ideally located for Sigint collection (see Figure 1).

In May 1985, the Senate Intelligence Committee reported that some 200 others were engaged in espionage. On the other hand, according to a key Soviet defector, Arkady Shevchenko, a total of about 300-350 Soviet personnel living in the New York area are engaged in espionage activities. Shevchenko lived in the Soviet "recreation" complex at Killenworth in Glen Cove from September to December 1958, served as head of the Security Council and Political Affairs Division of the Soviet Mission to the U.N. from mid-1963 to April 1970, and was U.N. Under Secretary General for Political and Security Council Affairs from April 1973 until his defection in April 1978. According to Shevchenko, some 7 to 10 tons of tapes of recorded telephone conversations and data transmissions are collected in New York and sent to Moscow in diplomatic bags for processing each year, in addition to the intercepts which are processed "on the spot" in these posts.

Soviet Sigint Sites in the U.S. Soviet Sigint activities at each of the sites listed in Table 2 are discussed in greater detail:

1. Riverdale, N.Y. The Soviet residential complex in Riverdale, which is located in Shcharansky Square at the corner of West 255th Street and Mosholu Avenue in the Bronx (see Figure 2), and which was occupied by the Soviets in November-December 1974, is over 200 feet high. It is regarded by U.S. intelligence agencies as the most important Soviet Sigint facility in the United States. As the Canadian Broadcast Corp. reported in June 1981: "The newest Soviet residence in New York is a building in the Riverdale section of the Bronx. They got permission to build on one of the highest sites in New York City, situated at an elevation that permits interception of phone messages over the widest possible area." And as *Newsweek* magazine noted in November 1981, the Riverdale building "bristles with electronic gear, and its location on one of the highest points in the (New York) metropolitan area permits eavesdropping on (telephone) calls throughout the Northeast" (see Figure 3).

On Jan. 31, 1985, Senator Daniel Patrick Moynihan (D., N.Y.), a former member of the Senate Intelligence Committee, stated in the Senate that: "We know that the Soviet residential compound in the Riverdale section of the Bronx is a center of electronic eavesdropping,

TABLE 2
Sigint Posts in the United States

	Post	Date Occupied	General Description
1	Soviet residential complex, Scharansky Square, Riverdale, The Bronx, New York City.	1974	17-story apartment building, based on a massive four-story foundation.
2	New Soviet Embassy complex, 2650 Wisconsin Ave., N.W., (Mount Alto), Washington, D.C.	—	12.5 acres. Complex of five buildings totaling approximately 460,000 square feet. Seven-story chancery/administration building. Four-story residential building containing approximately 160 permanent apartment units and 20 visitor units. Total resident population of 350-400 people. Residential building is 90 feet high, with the roof 420 feet above sea level.
3	Soviet Embassy, 1125 16th St., N.W., Washington, D. C.	1933	Four stories.
4	Soviet Consulate, 2790 Green St., Pacific Heights, San Francisco, Calif.	1972	Seven stories, 74 rooms.
5	Soviet Mission to the United Nations, 136 East 67th St., New York City.	1962	Eleven stories; 1,145,000 square feet.
6	Soviet recreation complex, Pioneer Point, Md.	Mid-1970s	Large estate with three story mansion and about a dozen detached dachas and other support buildings.
7	Soviet recreation complex, Killenworth, Dosoris Lane, Glen Cove, Long Island, N.Y.	1946	37-acre estate. Three-story mansion with 39 rooms.
8	Soviet recreation complex, 136 Mill River Rd., Oyster Bay, Long Island, N.Y.	—	Two stories.
9	Polish Consulate, 1530 North Lake Shore, Chicago, Ill.	—	
10	Office of Soviet Military, Air and Naval Attaches, 2552 Belmont Rd., N.W., Washington, D.C.	—	Four stories.

within range of the vast network of financial and other commercial transactions that occur millions of times each day in New York City over the phone."

And in March 1985, Senator Moynihan stated in an interview with NBC television that the Riverdale apartment complex is "just stocked with listening devices." More recent U.S. intelligence assessments have concluded that the Riverdale post commands the clearest and most extensive access to the U.S. telecommunications network of any single site.

There are two large structures on the northern and southern extremities of the roof of the Riverdale building, each of which contains microwave intercept equipment. The northern structure also contains a Satcom antenna system. Mounted on the southern structure are five large antennas, including an HF/VHF dipole, three stacked VHF antennas (oriented south and east), and a UHF microwave dish oriented towards New York City. Figure 4 depicts one of the structures.

2. The New Soviet Embassy, Washington, D.C. The new Soviet Embassy, located at 2650 Wisconsin Ave., on Mount Alto in northwest Washington D.C., has the potential to be the second most important Soviet Sigint post in the United States. It is one of four Soviet posts in the District of Columbia area. The Soviet Union first requested the Mount Alto site for its new Washington Embassy in 1966. In October 1967 the U.S. and Soviet governments agreed in principle to the Mount Alto site, and a formal agreement was signed on May 16, 1969.

The Mount Alto site, which is approximately 330 feet above sea level, is one of the highest points in Washington, D.C. The site comprises 12.5 acres, and has five

Figure 6. Even though the new Russian embassy on Mount Alto is not officially occupied, this has not kept the Soviets from installing Sigint antennas atop the apartment building. From that vantage point, they can access some of the most important communications conducted in the Washington, D.C. area.

Figure 5. The new Soviet embassy on Mount Alto in Washington, D.C., not yet officially occupied by the Soviets, is near the center of an international controversy. Since the discovery of extensive snooping devices in the new U.S. embassy in Moscow, Congress has decided to deny the Soviets official access to the buildings until the American installation is certified to be secure.

buildings with a total square footage of about 460,000, including a two-story reception building, a four-story consulate, a seven-story administration building, and a nine-story residential building (see Figure 5). The roof of the administration building is about 400 feet above sea level, and that of the residential building about 420 feet above sea level. Some Soviet officials have been living on the site since 1980. Construction of the residential building was completed in 1982, and the rest of the complex is now also essentially finished.

The complex has a commanding view over most of the District of Columbia itself, as well as much of Maryland and northern Virginia. As a senior U.S. intelligence official has stated, "From an eavesdropping standpoint, that's one of the most magnificent vantage points in Washington."

The top floors of the administration and residential buildings have a clear line-of-sight to the White House, the State Department, the Pentagon, the Commerce Department, and several other embassies including those of the United Kingdom, West Germany and France. These sites are sufficiently close that some U.S. intelligence officials have raised the possibility that laser listening devices can be directed at their windows from the Mount Alto site to monitor conversations by analyzing the vibrations of the glass.

The Mount Alto site also provides an unobstructed view of several nodal microwave towers that serve as the conduit for most telephone and data-transmission communications from Washington to other cities on the East Coast. The U.S. Navy Telecommunications Command at 4402 Massachusetts Ave. N.W. and the Naval Security Group Command Headquarters at 3801 Nebraska Ave. N.W., both of which have extensive VHF and microwave facilities, are only about a mile north of the post.

The Western Union microwave distribution center for Washington D.C. is at Tenley Tower on Wisconsin Avenue about two miles north of Mount Alto. A digitized

Figure 7. The current Soviet embassy on Northwest 16th Street in Washington, D.C. bristles with antennas.

voice circuit link from the Pentagon to Tenley Tower passes directly over the Mount Alto site. Signals from microwave relay towers at Andrews Air Force Base and Suitland (used by the Naval Operational Intelligence Office and the Naval Intelligence Support Center at 4301 Suitland Rd., and the Naval Intelligence Command and the Naval Intelligence Processing System Support Activity at 4600 Silver Hill Rd.) to Tenley Towers also pass over the site, as do microwave transmissions from the National Security Agency (NSA) Headquarters at Fort George G. Meade, Md., to Tenley Tower. In addition, the site is close to a microwave relay between Arlington, Va., and Gambrills, Md., that serves the primary telephone trunk group for the eastern seaboard.

Even though the embassy is not officially occupied by the Soviets, a wide range of antenna systems have been progressively installed on the roof of the apartment building since the early 1980s. A variety of VHF systems was observed in November 1984, and a photograph taken in September 1985 shows five VHF antennas, as well as a microwave dish directed at the Naval Telecommunica-

tions Command communications tower (see Figure 6). More recently, another four microwave antennas have reportedly been installed in pill boxes on the top floor. It is likely that additional antenna systems will be installed if and when the complex becomes diplomatically operational, especially if a decision is made to close the current embassy building on 16th Street.

In retrospect, it is clear that the relevant U.S. authorities acted with little foresight in agreeing to the Soviet request for the Mount Alto site. Whereas Mount Alto is one of the highest points in the District of Columbia, the United States accepted a site for its new Embassy in Moscow that is about 100 yards west of the present Embassy location, near the banks of the Moscow River, one of the lowest spots in Moscow. Several congressional inquiries have noted the gross inequity of this situation. According to Senator Moynihan, "We just got snookered; it's inexplicable." And according to James E. Nolan, Director of the Department of State Office of Foreign Missions and former chief of counterintelligence in the FBI, "I'm sure if we knew everything then (i.e. 1966-1972) that we do now, we wouldn't have made the same selection. We wouldn't have picked one of the highest sites in the city."

As part of the original agreement, the Soviets cannot officially occupy the Mount Alto chancery until the new American facility in Moscow it ready for occupancy. This proved to be a fortuitous arrangement, because the new U.S. embassy has been discovered to have so many KGB listening devices implanted in it to be unusable. The U.S. allowed the Soviets to construct the new embassy, and the KGB used the opportunity to honeycomb the embassy with these systems. The Congress is now aware that what was wrought during the days of detente should not be consummated, and action has been taken by the House of Representatives to ensure that the Mount Alto chancery is not occupied by the Soviets until the new U.S. embassy is free of these devices. Recent legislation requires that the Reagan administration must certify the new installation is secure before the Soviet chancery can be occupied.

3. Soviet Embassy, 1125 16th St. N.W., Washington, D.C. The current Soviet embassy is a four-story building, purchased in 1933, located at 1125 16th St. N.W., only about 300 yards from the White House. The roof of the building literally bristles with antenna systems. There are 14 identifiable antenna systems with signals intercept capability (see Figures 7 and 8):

• An 18-foot by 15-foot HF log-periodic directional array used for intercontinental transmission and reception, but generally oriented west-northwest toward the Pentagon, State Department and CIA communications facilities near Langley, Va.

• An HF rhombic array for long-distance transmission and reception, oriented toward CIA communications facilities in Virginia.

• An omni-directional, high gain 250-to-500-MHz VHF-UHF antenna capable of monitoring police, Federal Bureau of Investigation (FBI) and some other government transmissions.

• Four standard VHF antennas, capable of monitoring police, FBI and commercial transmissions.

• Two microwave antennas, one directed northwest towards the State Department and/or CIA headquarters.

• A standard S-shaped FM radio antenna.

• Two UHF antennas, one oriented northeast and the other northwest towards the State Department and/or CIA.

• A vertical quarterwave antenna capable of intercepting government limousine communications.

• A large (4-to-5-foot diameter) mesh-type microwave dish, lying flat on its back flush with the surface of the roof, for monitoring satellite communications.

Also on the roof is a wooden structure which reportedly houses sensitive electronic equipment used to monitor the various signals.

4. The Soviet Consulate in San Francisco. The Soviet Union operates a Sigint post on the top floor of its consulate in San Francisco, a seven-story, 74-room building located at 2790 Green St., Pacific Heights, which was purchased by the Soviets in 1972. The consulate has a staff of some 50 diplomats and support personnel. It is estimated that about six to eight KGB officers in the consulate are engaged in Sigint activities.

There are more than 10 antenna systems visible on the roof of the building, including several different HF Marconi wires and whips, and a VHF/UHF system (see Figure 9). According to FBI officers, microwave intercept antennas are located behind various windows on the top floor. There are several other structures on the roof, including a whitewashed structure that most likely also contains microwave intercept and/or satellite communications systems.

The consulate has a commanding view of San Francisco Bay, and is only a few blocks from the U.S. Army base at the Presidio. It commands an unobstructed line-of-sight of The Pacific Telephone Company's microwave tower in Oakland, which provides the microwave link between San Francisco's central microwave station at Bernal

Figure 9. The roof of the Soviet consulate in San Francisco is also covered with antennas. The white structures probably house microwave and satellite communications receiving equipment. From this vantage point, the Soviets can eavesdrop on several military installations in the area, as well as much of the defense-oriented activity going on in Santa Clara County's Silicon Valley.

Heights and the defense and electronics industries in Silicon Valley some 35 miles to the south. It also has a clear view of the microwave relay tower on Grizzly Peak, behind Berkeley. Although the consulate is not in direct line-of-sight of the Bernal Heights microwave tower itself, it can monitor several of the sidelobes emanating from the tower. In addition, it is believed that the Sigint ability within the consulate is supplemented by portable units hidden in vans that can be positioned for full access to the Bernal Heights tower.

There are numerous defense facilities of interest within the San Francisco Bay area in addition to the defense industries in Silicon Valley, and these include: the U.S. Air Force early warning station at Mill Valley, 10 miles northeast of the consulate; the Mare Island Naval Base at Vallejo, where U.S. Navy nuclear submarines are serviced; the Alameda Naval Air Station, homeport for several Pacific Fleet aircraft carriers, other warships and Carrier Group Three; Moffett Field Naval Air Station (a base for P-3 Orion long-range maritime patrol aircraft); the Air Force Satellite Control Facility at Sunnyvale, and the Naval Weapons Station at Concord.

5. The Soviet Mission to the United Nations; East 67th Street in New York. The Soviet Mission to the UN, which was purchased in 1962, is located on East 67th Street between 3rd Avenue and Lexington Avenue in mid-Manhattan. It is 11 stories high, 10 rooms wide, six rooms deep and comprises 115,000 square feet in ground area. Mounted on the roof are two HF Marconi wire antennas and six VHF antennas, each with nine elements. One of these VHF sets is directed at the roof of the RCA building at 30 Rockefeller Plaza, which mounts five antennas used

Figure 8. One reason for the Soviets to move into a new embassy is that they have run out of room for antennas on the roof of the present one.

by the FBI for communications within the New York City area (see Figure 10).

6. Pioneer Point, Maryland. The Soviet "recreational" facility for its Washington diplomats is located at Pioneer Point, at the end of Corsica Neck Road, near Centerville on the Eastern Shore of Maryland, just across the Chesapeake Bay Bridge from Annapolis. The facility is part of what was once a great estate belonging to the late Jacob Raskob, financier and Du Pont associate. The facility consists of a large three-story mansion and more than a dozen detached dachas and other support buildings. The roof of the mansion houses a variety of large antenna structures, including a large HF log-periodic directional array similar to that on the 16th Street Embassy building in Washington, D.C., a large mesh-type microwave dish antenna, and several VHF antennas. Various other VHF and UHF/microwave antennas are mounted on several of the other buildings.

The complex is extremely well placed for Sigint operations (see Figure 11) according to a report based on interviews with Arkady Shevchenko and U.S. Navy sources. In the mid-1970s, Shevchenko says, the Soviets were ecstatic when they were allowed to purchase a beautiful remote estate with several buildings in a strategic setting on Maryland's Eastern Shore.

At first the antennas went up slowly. Later, a Navy source says, more and more antennas grew. There is an extraordinary range of interesting and important signal sources within the purview of the Pioneer Point establishment: The major microwave relay station at Gambrills, Md., is 30 miles directly west across Chesapeake Bay; the NSA Headquarters at Fort Meade is 35 miles directly west; the large Navy communications station at Annapolis is 20 miles west-southwest; and NSA's Propogation and

Research Laboratory at Kent Island is 15 miles to the southwest.

Numerous other military and intelligence facilities are located within a 50-mile radius of the post, including: the U.S. Army Aberdeen, Md., Proving Ground; the U.S. Air Force radio relay site at Silver Hill, Md.; the Naval Operational Intelligence Center and Naval Intelligence Support Center at Suitland, Md.; the headquarters of the U.S. Air Force Systems Command at Andrews Air Force Base, Md.; the Naval Research Laboratory (NRL) Free Space Antenna Range at Pomonkey, Md.; the Naval Research Laboratory (NRL) Radio Site at Waldorf, Md; the U.S. Navy Communications Station at Cheltenham, Md.; the NRL antenna range and U.S. Air Force Satcom ground terminal at Brandywine, Md.; the NRL Research Station at North Beach; the Naval Security Group's Northwest Station at Chesapeake, Md., which controls the Atlantic HF-DF network; the NRL research and development facility at Randle Cliff; and the NRL Test Site at Tilghman Island.

The Naval Air Test Center (NATC) and Electronic Warfare Integrated Systems Test Laboratory (EWISTL) are located just across the Chesapeake Bay at the U.S. Naval Air Station at Patuxent River, Md. Patuxent River is also

Figure 10. VHF antenna mounted atop the Soviet mission to the United Nations. This antenna is probably used to intercept Federal Bureau of Investigation (FBI) communications in New York.

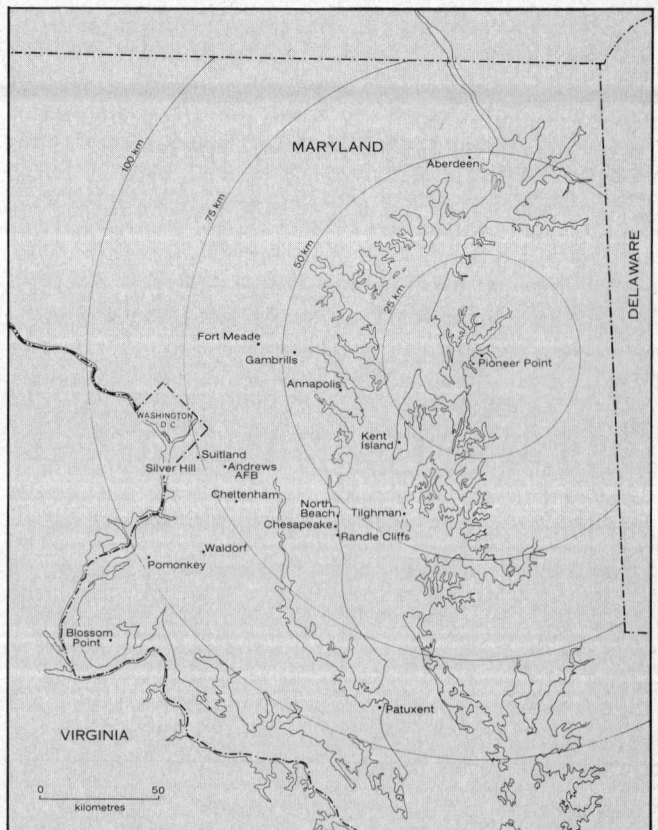

Figure 11. This map shows the ideal location of the Soviet Sigint site at the Pioneer Point recreation area on the eastern shore of Maryland. Many military installations are within communications intercept range of the complex.

the home base of the Atlantic Fleet Tacamo squadron of strategic communications relay EC-130 aircraft. The U.S. Navy operating base at Norfolk, Va., and Langley Air Force Base at Hampton, Va., are further afield, but the communications between these facilities and Washington, D.C., also pass within the purview of the Pioneer Point post.

7. Glen Cove (Killenworth), N.Y. The Soviet establishment at Killenworth in Glen Cove, Long Island, about 26 miles from Manhattan, is a 37-acre estate purchased in April 1946, ostensibly to provide recreational facility for Soviet U.N. personnel. The mansion itself, with three floors and 49 rooms, is very picturesque (Figure 12).

The Sigint capabilities of the Glen Cove post have expanded quite dramatically over the past three decades. According to Shevchenko: "When I first came to the United States in 1958, there were three or four KGB communications technicians and their gear sharing the former servants' quarters in the attic. By 1973, the specialists in intercepting radio signals numbered at least a dozen, and they had taken over the whole floor. Their equipment occupied so much space, in fact, that one of the two large unused greenhouses had been commandeered to store it. These quarters were off limits to other personnel."

By 1978, the number of Sigint personnel had increased. As Shevchenko stated in an interview in June 1981: "All the top floors of the building are full of sophisticated equipment...to intercept all the conversations, telephone conversations on anything which is going on. At least 15 or 17 technicians were working...at...this job."

In addition to the microwave systems reported by Shevchenko, there are also several HF antenna wires and VHF antennas.

The Glen Cove post is ideally located for Sigint purposes, both because of the excellent microwave propagation characteristics of Long Island Sound, commonly known as "microwave alley," and because of the proximity of two microwave nodal points in the East Coast telecommunications network and of several important defense and military-industrial establishments in the region. The FBI, with the cooperation of NSA, has established two monitoring posts adjacent to the Killenworth estate in order to determine the signal receptivity at the post. These have demonstrated that microwave signals generated up to 100 miles away can be clearly received. As one report has noted: "The attraction of the site is that it is in the middle of one of America's busiest transmission zones. Microwaves travel best over water and, with a nuclear submarine base at its eastern end, the dense cluster of the Connecticut and Long Island defense industries on its northern and southern shores, and New York City at its western end, Long Island Sound is a radio spy's paradise (see Figure 13).

In addition to the U.S. Navy's submarine base at Groton, Conn., and the Naval Underwater Systems Center in New London, Conn., there are several major defense companies with operations in the region, including the

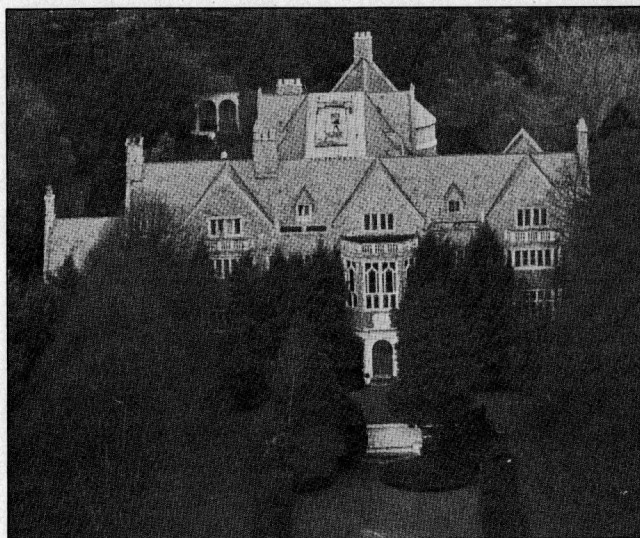

Figure 12. The Soviet recreational facility at Killenworth, N.Y.

Electric Boat Division of General Dynamics Corp. at Groton, which has built all the U.S. Navy's Fleet Ballistic Missile (FBM) submarines; the Manufacturing Division and Commercial Products Division of Pratt and Whitney at East Hartford, Connecticut; the Fairchild Republic Company at Farmingdale, Long Island, which produces the A-10 Thunderbolt close support aircraft; the Government Systems Operations of Eaton Corporation, at Melville, Deer Park and Farmingdale, Long Island, which produce electronic countermeasures (ECM) and command and control systems; Sperry Corporation Defense Products Group at Great Neck, Long Island, which produces surveillance and fire control systems; Grumman Aerospace Corporation at Bethpage, Long Island, which produces the F-14 Tomcat, the E-2C Hawkeye, the A-6 Intruder, the EA-6 Prowler, and modifies F-111 aircraft into the EF-111A Raven. The Glen Cove post is located about 12 miles northwest of Bethpage, and according to Richard Kinsey, formerly deputy chief of the Soviet Desk in the FBI, it has antennas pointed towards both Grumman's main plant and a testing facility that the company has further out on the island.

Two important microwave relay towers are located within close purview of the Glen Cove post. The tower at Stamford, Conn., 12 miles north of Glen Cove, relays microwave signals up and down the New England seaboard for the New York Telephone Company, while American Telephone and Telegraph (AT&T) has a major relay tower at Roslyn Harbor, five miles south of the post. Most of the long-distance calls from the defense companies on Long Island are relayed through one of these towers. In addition to trunk calls, these towers are also used to relay local telephone conversations. For example, about 12 percent of the local circuits in southeastern Connecticut are served by microwave systems. According to Alan M. Parente, former mayor of Glen Cove: "To the best of my knowledge, they (i.e. the Killenworth post) can

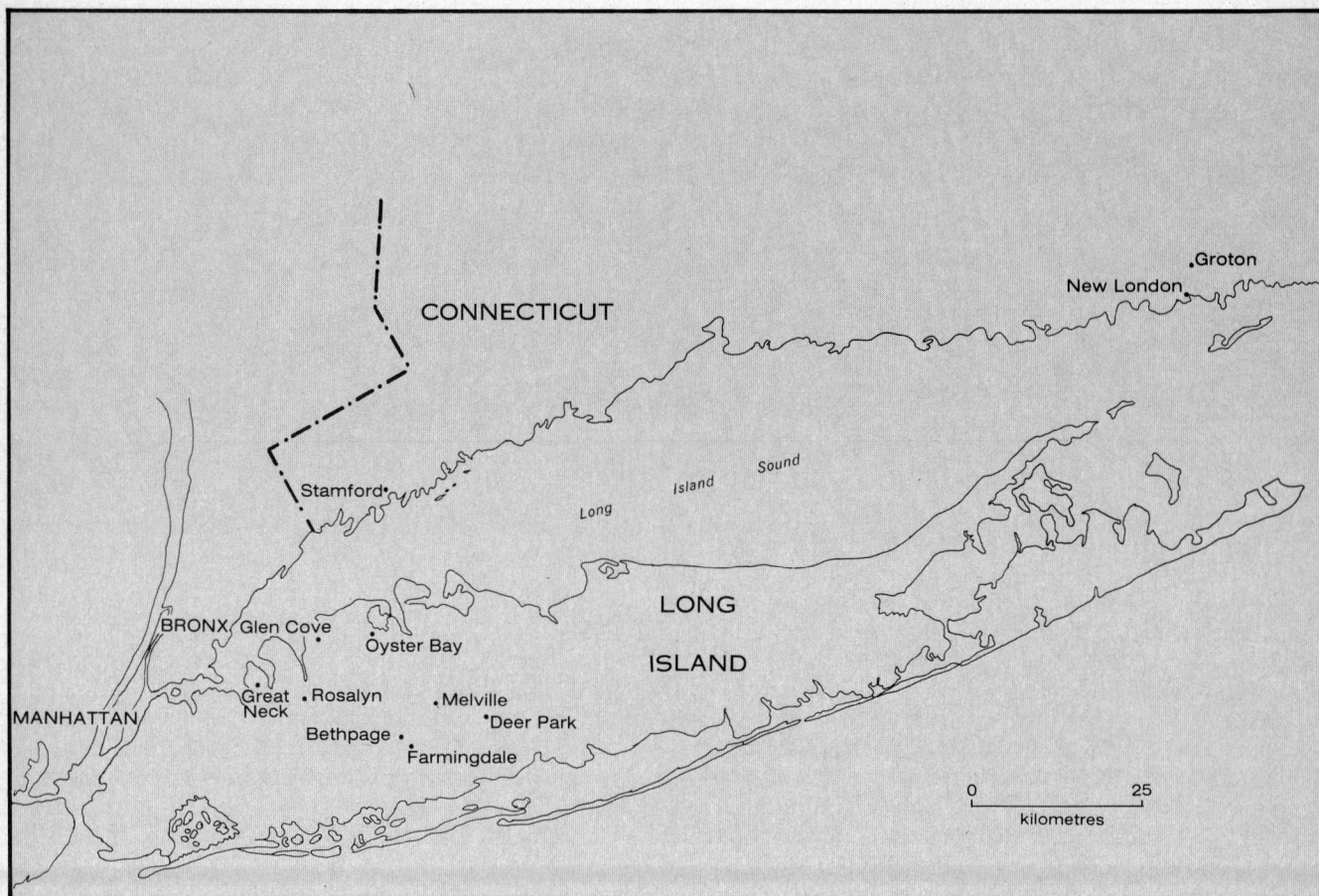

Figure 13. The Glen Cove Sigint site, situated on the southern shore of Long Island Sound near the Bronx, is in an ideal location to intercept communications in one of the United States' busiest transmission zones.

sweep in signals from all the Northeast corridor; New Jersey, New York, Connecticut, Long Island, maybe even Massachusetts."

In addition to its microwave interception capabilities the Killenworth post also uses its HF and VHF systems for Sigint purposes. According to Shevchenko: "Not only does Soviet intelligence use the Glen Cove attic for monitoring open inter-city phone calls relayed by micro-wave along the Connecticut-New York-New Jersey corridor, but it also can pick up Navy radio messages to and from ships, including submarines homeported at the submarine base (at Groton)."

8. Oyster Bay, N.Y. A second Soviet "recreational" facility on Long Island, which is also used for Sigint purposes, is located at 136 Mill River Rd. in Upper Brook-ville just south of the township of Oyster Bay and just five miles east of Glen Cove. It is a large two-story house located in heavily wooded grounds next to the Mill River Country Club. Several VHF antennas are visible from the club, but the woods allow only very limited visibility of the post. Its purview is essentially similar to that of the Killenworth post, and it presumably coordinates its Sigint activities with those at Killenworth.

9. The Polish Consulate, Chicago, Ill. According to

Stansfield Turner (admiral, U.S. Navy, retired), former director of the CIA (1977-81), the Soviet Union also maintains a microwave intercept post in Chicago. This activity is evidently conducted from the Polish consulate, which is located at 1530 North Lake Shore.

10. Office of the Soviet Air and Naval Attaches, Wash-ington, D.C. The Office of the Soviet Military, Air and Naval Attaches is located at 2552 Belmont Rd. N.W., just off Connecticut Avenue. It was first publicly cited as a Sigint collection site by Larry McWilliams, a former FBI counterintelligence officer, in an interview with CBS Television in January 1980. According to other U.S. intelli-gence sources, it is a much less important Sigint collection site than the posts at Mount Alto, 16th Street and Pioneer Point. It is a four-story building that houses a variety of VHF and microwave antenna systems.

In addition to these establishments, there are some half-a-dozen other buildings occupied by Soviet diplo-matic personnel in the Washington area that maintain less sophisticated Sigint capabilities. For example, in June 1986, when Col. Vladimir Izmaylov, the Soviet air attache, was arrested for espionage and expelled from the U.S., it was revealed that his apartment in Wildwood Terrace, Arlington, was being used for microwave reception.

Figure 14. The London Post Office Tower is at the center of microwave communications within the United Kingdom. Both the Soviet embassy and Trade Delegation have line-of-sight access to its signals.

Canadian Listening Posts. In Canada, the Soviets maintain Sigint posts in both their embassy in Ottawa and their consulate in Montreal. There is a variety of sophisticated antennas on the roof of the Ottawa Embassy, which houses both KGB and GRU Sigint personnel, who maintain both general Sigint and local electronic surveillance (i.e. Zenith) operations. The Zenith operation is concerned with monitoring and recording the radio communications of the Canadian Security Service, formerly the Royal Canadian Mounted Police Security Service. At one time the post might also have been able to intercept telephone calls going through Bell Canada's microwave system because its path crossed over the roof of the embassy, but Bell Canada later built a microwave tower that avoids the microwave route around the embassy.

The KGB residency in Ottawa also has portable monitoring equipment of various sorts. During a period in the mid-1960s, for example, a KGB post was established in rooms in a hotel overlooking the East Block Cabinet Room in Parliament House. According to one account: "The KGB monitors were in a perfect perpendicular line for shooting an electronic beam at the cabinet window and have it bounce back to their suite at the hotel. A beam could pick up the vibrations on the East Block

window and thus enable the Russians to understand conversations in the Cabinet chamber."

The consular complex in Montreal consists of four buildings, linked by underground tunnels, on du Musee Avenue just north of Dr. Penfield Avenue, on the southern slope of Mount Royal, overlooking the downtown area of the city, an ideal location for the interception of local microwave telecommunications, including those of the Security Service Montreal office, the U.S. and British consulates, and defense contractors in the area. These buildings are staffed by about 35 Soviet personnel. The main three-story building, which was destroyed by fire on Jan. 14, 1987, housed the Sigint post. Two large vertical antennas were mounted on the roof, and two wooden structures on the roof reportedly contained microwave antennas. The third floor, the windows of which were bricked-up, reportedly contained an extensive suite of electronic surveillance equipment.

Diplomatic Establishments in Europe. As noted in Table 1, there are some 20 countries in Europe in which Soviet diplomatic establishments are used for Sigint operations. In at least half a dozen of these countries, there is more than one Sigint post.

United Kingdom. There are two major Soviet diplomatic establishments in London are used for Sigint operations. Both of these are able to monitor HF and VHF signals, as well as telephone and telex traffic carried in and out of London through the hub of the British microwave network at the London Post Office Tower (Figure 14).

The Soviet embassy is situated in Kensington Palace Gardens, about three miles west of the Post Office Tower. According to one account: "For years Britain's MI5 has been aware of and has warned government departments that from its aerial-roofed embassy in Kensington Palace Gardens the Russians listen to microwave calls being routed through London's Post Office Tower." And according to one report: "It was soon found that the the Russians were taping all the messages and sending the tapes back to Moscow for analysis there. Of course, most of the messages were ordinary domestic calls of no consequence to the KGB, what security men call 'cabbages and kings,' but their next move showed that they were getting enough valuable material to make the operation worthwhile. They imported a computer-analyser into the London embassy via the 'diplomatic bag,' a term that includes large crates as well as leather pouches. This ingenious device recorded only those messages emanating from certain telephone numbers in which the KGB was specially interested."

In addition to the microwave intercept equipment housed within the top floor of the embassy, the roof of the building houses a variety of HF wire antennas and VHF antennas.

The second Soviet Sigint post in London is located in the Soviet Trade Delegation situated high up on the West Hill at 32-33 Highgate West. This site was first obtained by the Soviets in 1946. In the mid-1960s, the original building was torn down and replaced by a modern, four-story building with a variety of complex radio antennas. Some 380 Soviet personnel work in the complex, which reportedly serves as the headquarters of KGB operations in the United Kingdom. It is estimated that about 20 of these are engaged in Sigint activities.

Belgium. The Soviets maintain three Sigint posts in Brussels, two in diplomatic establishments, which have been used for Sigint operations since at least 1962, and one in a Skaldia-Volga automobile plant built in 1967.

The use of the Soviet embassy and Soviet Trade Mission in Brussels for Sigint operations was described in a report by USeucom in December 1973 as follows: "Humint sources have recently confirmed that...Sigint intercept operations have been conducted by both the GRU and KGB from the Soviet embassy and trade mission in Brussels, Belgium, for a minimum of 11 years. These operations include, but are not necessarily limited to, direction-finding, plaintext analysis of voice, teletype, and Morse communications and analysis of the externals of encoded or enciphered messages. Among primary intercept targets are the strategic/tactical air and ground forces of the U.S., NATO and Western European countries. Some examples of exploited information are: combat readiness checks (e.g., emergency action messages), planning/progress reporting of major exercises/operations, real-time status of deploying forces, flight activity by airborne command posts, travel of VIPs, etc.

These same Humint sources confirmed that the GRU participates in real situations by providing the following information: During the Soviet exercise OKEAN 1970, the task of the Soviet TS Group in Brussels was to intercept U.S. military communications traffic pertaining to the exercise. This intercept activity enabled the Soviets to determine what the U.S. and its Western Allies learned about OKEAN, how much they knew, how much they were getting, etc."

One of these Humint sources was Anatoli Tchebotarev, a GRU officer who worked in the trade mission in Brussels from 1967 until he defected on Oct. 3, 1971, and who informed the Belgian government that electronic listening posts had been installed in both the trade mission and the Skaldia-Volga plant to tap and try to decipher coded messages from and to NATO and SHAPE headquarters.

The Skaldia-Volga plant was built in 1967 when NATO headquarters was moved from Paris to Brussels, and is located less than a mile from the NATO complex. The plant has a huge HF aerial on the roof, and its location is ideal for the interception of NATO shortwave radio traffic.

France. The KGB and the GRU maintain offices in numerous establishments in France, at least two of which, the Soviet embassy in Paris and the consulate in Marseilles, are used for Sigint operations. It is estimated that there are about 130 KGB and 50 GRU officers in France,

of which about 30 are engaged in Sigint activities.

The Soviet embassy is located at 40 Boulevard Lannes in the 16th arrondissement, about a little over a mile west of the Eiffel Tower and two-and-one-half miles west of the Quai d'Orsay. The KGB residency occupies the top three floors of the embassy. The antennas on the roof include an enormous vertical HF/VHF log periodic system, a large HF wire aerial, several VHF yagis, and several parabolic microwave antennas, including at least one for monitoring satellite communications.

There are two KGB Sigint posts in the embassy. One is the Zenith room, which is manned by three Technical Operations, or "audio specialists" from Department Fourteen of the First Chief Directorate. The Zenith room is full of VHF and UHF receivers and frequency scanners, and is used to continuously monitor the VHF and UHF frequencies used by the French police and counterintelligence services.

The larger post, known as the Elint room, is manned by personnel from the KGB's 16th Directorate and is concerned with monitoring French government communications. The satellite communications antenna, the microwave antennas and the HF antennas on the roof are connected to this post. The microwave systems provide access to a large portion of the Paris telephone network, including circuits used by the Elysee, Matignon, Quai d'Orsay, Ministry of Defence, etc. The Elint room is also used to intercept computer data networks in Paris.

The Soviet consulate in Marseilles is a private hotel surrounded by an iron fence at 3 Avenue Ambroise Pare in Monticelli Square. It is occupied by some 52 Soviet personnel, and used for both Zenith and general Sigint operations. The Zenith operation is remarkably effective. For example, on one occasion when a Soviet seaman attempted to defect in Marseilles, the KGB's Zenith room was able to monitor the movements of the French police and security officers and to locate and abduct him before the French authorities could themselves find him.

A major activity of the general Sigint operations is monitoring naval movements to and from the French Navy base at Toulon, about 32 miles southwest of Marseilles.

In addition to the large Sigint posts in the Paris embassy and the Marseilles consulate, other Soviet buildings in France. are used for Sigint purposes on a more opportunistic basis. For example, an apartment in Toulon, used by diplomatic personnel from Marseilles, and situated directly opposite the Toulon naval base, was equipped with antennas for monitoring communications to and from the base.

Austria. About 1,000 Soviet diplomats are based in Vienna, of whom some 200 are estimated to be KGB and GRU personnel. According to Viktor Suvorov, who served in the GRU residency in the Soviet embassy in Vienna from 1973 to 1977, there were about 40 GRU officers in the embassy at that time, of which about 10 were engaged in Sigint operations.

The embassy complex, which consists of two large buildings, is located at Reisnerstrasse 45-47, 1030 Vienna. The GRU residency is the 17-room basement under the consular building. Three of these rooms are occupied by GRU TS officers engaged in monitoring Austrian government and military communications, and two by the GRU Radio Monitoring Group concerned with surveillance of Viennese police and counterintelligence service (Staatspolizei). The work of these two groups has been described by Suvorov as follows: "The TS group worked round the clock picking up and deciphering military and government radio messages. The control (or radio monitoring) group was also engaged on radio interception. But it was a quite different kind of work. The TS group was working in the interests of the Information Service of the GRU, picking up scraps of information out of which the command point and the central computer were continually putting together a general picture of what was going on in the world. The radio control group had different functions, though not less responsible ones. They were working only for our residency, keeping watch on what the local police were doing. The group always knew what the Viennese police were up to, how their force had been distributed round the city, and whom their plainclothes agents were following. The radio control could always tell us if, for example, today the police had been following a suspicious-looking Arab at the railway station or that yesterday the whole force had been put onto catching a group of drug peddlers. Very often it was not possible to work out what the police were up to, but even then the radio monitors were always ready to warn us where any particular police activity was going on."

Some 16 antennas are mounted on the roofs of the two buildings. One has a large, 20-foot-high omnidirectional HF antenna; a mast with four horizontal and vertical VHF antennas and an omni-directional VHF antenna; and a HF/VHF rod antenna. The other has two HF wire antennas; a mast with four VHF antennas; two other VHF antennas; and a HF/VHF rod antenna. In addition, other electronic monitoring equipment, presumably including microwave intercept systems, was installed in the attic of the building housing the GRU residency in the mid-1970s

A huge new Soviet diplomatic complex, consisting of some dozen large buildings and surrounded by a seven-and-a-half foot iron fence, is currently being completed in Vienna. Several HF, VHF and UHF antennas have already been installed on the roof of one of the buildings.

Scandinavia. The KGB and GRU currently maintain some eight Sigint posts in Finland, Sweden, Denmark, Norway and Iceland, which, together with numerous Sigint stations in the Baltic states and on the Kola Peninsula (including very large stations at Kaliningrad, Riga, Leningrad and Murmansk) and naval and airborne Sigint operations, provide extremely comprehensive coverage

Figure 15. The Soviet embassy in Tokyo has the ability to eavesdrop on U.S. military installations and space communications.

of Scandinavia, the Baltic Sea and the Norwegian Sea. In Finland and Denmark, Sigint posts are maintained in the Soviet embassies in Helsinki and Copenhagen. Reportedly 60 KGB and 20 GRU personnel are in Sweden, about 15 of which are engaged in Sigint activities in the Stockholm embassy and Gothenburg consulate, which was established in November 1980. In addition, in 1976 the Soviets purchased a 150-acre site in Lulea, in northern Sweden on the northern coast of the Gulf of Bothnia, the ostensible purpose for which was retail sales for Soviet automobiles and tractors, but which was soon equipped with a large radio installation. An associated office in Bromma, a suburb of Stockholm, also has interesting antennas on the roof.

The Soviet embassy in Reykjavik in Iceland is the largest in the country. It is staffed by about 70 Soviet personnel, of whom about 10 are estimated to be engaged in Sigint activities.

There are two Soviet diplomatic establishments in Norway, both of which house Sigint activities. The embassy in Oslo is a relatively large establishment, with some 120 employees. The consulate at Longyearbyen in Svalbard is ostensibly concerned with supporting the 2,100 Soviet citizens working on Spitzbergen, but it is also known to be equipped with radio monitoring facilities. For example, on Aug. 10, 1982, when two Polish scientists working at a research station at Horsnund applied to the Norwegians via radio-telephone for political asylum, the consulate intercepted the transmission and attempted to prevent their defection.

The Middle East. In December 1973, USeucom reported that during the October 1973 Yom Kippur War in the Middle East, the KGB and GRU had conducted Sigint collection from Cairo, Egypt; Athens, Greece; Teheran, Iran; Baghdad, Iraq; Beirut, Lebanon; Karachi, Pakistan; Damascus, Syria; Aden, Yemen; and Ankara and Istanbul in Turkey. Since then, Sigint posts have also been established in the Soviet embassy complexes in Kuwait and Addis Ababa in Ethiopia. In Turkey, there are Soviet Sigint posts in both the embassy in Ankara and the consulate in Istanbul. The Ankara embassy, which is located close to the U.S. embassy, is equipped with a variety of antennas, including a satellite communications terminal, and is in direct line-of-sight of the microwave tower on Elmadag, which relays all inter-city telecommunications into and out of Ankara.

In Iraq, the Soviet embassy in Baghdad, which served as a Soviet Sigint post during the 1973 Yom Kippur War, is located near both the Iraqi Ministry of Defense and the secret service headquarters. During April-May 1978, the Soviets increased the number of antennas on the embassy roof, which Iraqi intelligence determined were for monitoring Iraqi military and intelligence communications within Baghdad, and hence insisted that the new equipment be dismantled.

The Soviet embassy in Beirut is located in a large compound at the intersections of Boulevard Saeb Salam, Rue Moaouiya and Rue Mar Elias in West Beirut. It was used for Sigint collection during the 1973 Yom Kippur War, and in September 1982, when Israeli military forces moved into West Beirut following the assassination of Lebanese President-elect Beshir Gemayel, one of their first targets was the Soviet embassy. On Sept. 16, 1982, Israeli commandos entered the embassy from the roof, dismantled whatever Sigint equipment was transportable, and photographed and then destroyed the rest.

The Soviet embassies in Kuwait and Addis Ababa, Ethiopia, both became major Sigint posts in the early 1980s. By 1984, the embassy in Kuwait had a staff of 135 Soviet personnel, of whom it is estimated about nine to 10 are engaged in Sigint activities. Addis Ababa is well placed for Sigint collection. It not only provides signals surveillance coverage of the Horn of Africa, but is also able to monitor a large proportion of the HF traffic between Western Europe and the Far East.

Africa. The Soviet Union now maintains more Sigint stations in Africa than the United States, Britain, France or any other country. At least nine of these are located in diplomatic establishments. In addition to the embassies in Cairo, Kuwait, Addis Ababa and Aden mentioned above, there are now also Sigint posts in the Soviet embassies in Tripoli, Libya; Luanda, Angola; Nairobi, Kenya; Antananarivo, Madagascar; and Maputo, Mozambique.

Diplomatic relations between Madagascar and the Soviet Union were established on Oct. 3, 1972. By 1978, the Soviet diplomatic and consular staff in Antananarivo had reached about 250 personnel. It is likely that the Sigint activity is particularly concerned with monitoring communications in the Western Indian Ocean.

In Mozambique, the Soviet diplomatic complex in Maputo includes an entire 12-story building on Avenida 31 de Janeiro and two large houses nearby. Together with the embassy in Luanda, Angola, it is well placed to monitor radio communications throughout southern Africa.

In mid-1981, the Soviet Union also attempted to install unauthorized electronic surveillance equipment in Monrovia, Liberia. The attempt was frustrated, and the Soviet embassy was closed in November 1983.

South Asia and the Indian Ocean. There are currently some 10 Sigint posts maintained in Soviet diplomatic establishments in countries in South Asia and the Indian Ocean: Pakistan, Afghanistan, the Seychelles, India, Sri Lanka, Nepal and Bangladesh.

There are two posts in Pakistan: the embassy in Islamabad and the consulate in Karachi. In 1976, it was reported that more than 75 KGB and GRU officers were based in these two establishments, and it was estimated in 1986 that a total of about 20 were engaged in Sigint operations in the two posts.

In Afghanistan, the Soviet embassy is located on Darulaman Road on the outskirts of Kabul, close to the presidential palace, the KGB's main office or "liaison mission" in Kabul, and the headquarters of the Khidamate Aetilaati Danlati (Khad), the Afghani security and intelligence organization. The embassy houses both HF and microwave antennas. (The microwave communications system in Kabul was designed by Soviet technicians in 1979.)

The Soviet embassy in India is located at Shanti Path Ch Puri-21 in the central diplomatic area in New Delhi. In 1976, it was reported that about 175 KGB and GRU officers were employed in diplomatic establishments in India, and in 1986 it was estimated that more than 40 were engaged in Sigint activities. The buildings in the embassy compound house several types of HF antennas as well as two large four-element VHF structures, each about eight-feet high.

The Soviet embassy in the Seychelles has a staff of about 60 personnel. The Sigint post is used to monitor naval activity in the Indian Ocean, U.S. communications to and from Diego Garcia, communications associated with the operations of the U.S. Air Force's Satellite Control Facility (SCF) satellite ground station on Mahe, and internal Seychelles communications.

In Bangladesh, Soviet Sigint activities involving the embassy in Dacca became a matter of public controversy when, on June 20, 1981, about 10 tons of sophisticated electronic equipment, contained in 140 crates, was flown into Dacca International Airport (Kurmitola) on a special Aeroflot cargo flight. The equipment included frequency analyzers, signal generators and electronic filters designed for monitoring radio communications. Although the Bangladesh authorities insisted that this particular consignment be returned to Moscow, it is believed that similar equipment was subsequently delivered piecemeal and installed in the embassy. It was reported in 1976 that about 40 KGB and GRU officers were employed in the embassy, and it was estimated in 1986 that about 10 were engaged in Sigint activities.

In Nepal, also in June 1981, some 84 crates of sophisticated electronic equipment were delivered to the Soviet embassy in Katmandu. This equipment was reportedly intended to be used in the construction of a Sigint post to monitor Chinese nuclear and missile test activities in the Lop Nor region of Tibet as well as communications within Nepal.

East and Southeast Asia. Sigint posts are maintained in Soviet diplomatic establishments in at least 12 countries in East and Southeast Asia: Burma, China, Indonesia, Laos, Japan, Malaysia, Mongolia, North Korea, the Philippines, Singapore, Thailand and Vietnam.

In 1976, the Soviet intelligence presence (KGB and GRU) in Southeast Asia was reported to be about 35 in Burma, 60 in Indonesia, 15 in Malaysia, 25 in Singapore, and more than 25 in Thailand. Relatively small Sigint posts are maintained in the Soviet embassies in each of these countries. The post in Rangoon is more concerned with China and Thailand than with Burma itself. The embassy in Singapore, on Cluny Road, is well placed to monitor HF communications to both Singapore and Kuala Lumpur, as well as naval traffic in the Strait of Malacca. The embassy in Jakarta, which is a 10-story building, has a large HF wire antenna, several HF rod antennas, and VHF antennas, as well as other electronic equipment directed at the Australian embassy in the same block. In the Philippines, a new Soviet embassy compound has recently been constructed in the Forbes Park area of Manila. The main building is three stories. The compound is located next to the Army headquarters at Fort Bonifacio, which also houses the Army's primary radio communications system for communicating with U.S. Air Force and Navy bases in the Philippines. According to one report, the new embassy is in the best possible position to increase electronic surveillance within the Philippines.

The embassies in Ulaan Baatar, Mongolia, and Beijing, China, are considered by the CIA to be major Soviet Sigint posts. Both contain relatively large contingents of Technical Service Group personnel from the Third Branch of the GRU's Radio and Radio-Technical Intelligence (or Sixth) Directorate.

Japan. The Soviet Sigint agencies maintain three posts in Diplomatic establishments in Japan: the embassy in Tokyo; the consulate in Osaka; and the consulate in Sapporo, on the southwest of Hokkaido. More than 60 KGB and GRU officers were reportedly stationed in these three establishments in 1976, and it was estimated in 1986 that about 25 were engaged in Sigint activities.

The consulate in Osaka is located at 2-2 Nishi-Midorigaoka 1-chome in Toyonaka-shi 560, about six miles north of the center of Osaka. It is a large, modern complex, dominated by a seven-story building, which has a single antenna structure with several VHF/UHF antennas and a VHF omnidirectional acquisition antenna. The consulate in Sapporo is located at 826 Nishi 12-chome, Minami 14-jo, Chuo-ku, Sapporochi 060. It is well placed to monitor communications between the

island of Hokkaido and Tokyo, including those to and from the major joint U.S.-Japanese Sigint and radar complex at Wakkanai on the northern tip of Hokkaido.

The Soviet embassy in Tokyo is located on Gaien-Higashi-dori Avenue at 1-1 Azabudai 2-chome in Minatu-ku in central Tokyo. The compound is approximately 110 by 140 metres, bounded by a 10-foot brick wall. There are four buildings within the compound: two two-story buildings, the 11-story embassy, and an 11-story apartment building, which houses most of the Soviet diplomatic personnel stationed in Tokyo.

There are more than two dozen antennas on top of the two 11-story buildings, and several others extruding from balconies on the upper floors of the apartment buildings, including about 18 VHF/UHF Yagi and logperiodic antennas, a VHF omni-directional acquisition antenna, half a dozen HF wire antennas, a 12-foot diameter parabolic satellite communications antenna, and microwave antennas (see Figure 15).

According to a report based on the testimony of Major Stanislav Aleksandrovich Levchenko, a KGB officer who had worked in the Tokyo embassy from 1974 until his defection to the United States in 1979: "The monitoring post on the 11th floor of the embassy building consists of a maze of radio and microwave receivers, tape recorders, sensors, teletypes and electronic equipment. With this array, the electronic surveillance officer eavesdropped on communications from American space satellites and military installations, taped conversations from tapped telephones and recorded those transmitted by microwave. The residency long ago had hooked into the teletype circuits of the Japanese Foreign Ministry and routinely copied all messages sent over them.

The Tokyo embassy also maintains a KGB Zenith Sigint post, which is specifically concerned with monitoring the radio frequencies used by Japanese counterintelligence and police surveillance agencies.

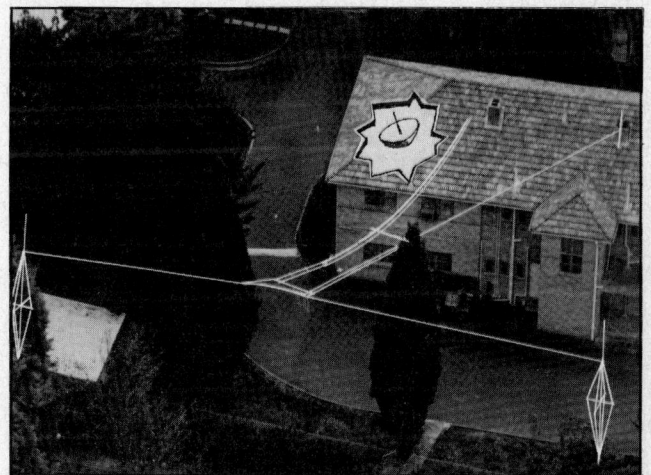

Figure 16. This cutaway concept shows the microwave receive antenna located in the attic of the Soviet embassy in Canberra, Australia.

The compound is located about 450 meters west of Tokyo Tower, the principal communications relay system in Tokyo; it is also close to the Gaimusho (Foreign Ministry); and the Japan Defense Agency.

Australia and New Zealand. The Soviet embassies in Australia and New Zealand are both used for Sigint collection. The embassy in Canberra is located at 78 Canberra Ave. in Griffith, which was purchased in 1943 and which presently houses some 55 Soviet employees, about half a dozen of whom are engaged in Sigint activities.

The Canberra embassy has five receiving antenna systems, including two vertically-polarized omni-directional HF systems (code-named Hay Pole), a large HF wire antenna, an HF/VHF rod antenna, and a microwave receiver housed in the attic (Figure 16). The three HF antennas are evidently used for monitoring diplomatic and other HF communications into Canberra; the HF/VHF rod antenna is used for local surveillance, monitoring police and counterintelligence traffic. The microwave antenna is used to monitor telecommunications to the Telecommunications Tower, which rises 195.2 meters above Black Mountain and is Canberra's most visible landmark (Figure 17). About 70 percent of the telecommunications traffic into Canberra comes via the tower, which also serves as a relay system for most of the telephonic and data traffic between Sydney and Melbourne.

The Soviet embassy in New Zealand is located in Wellington at 57 Messines Rd. in the verdant suburb of Karori, overlooking the inner city area and harbour, and in direct line-of-sight of most of the New Zealand governmental offices. The antennas on the roof include two HF horizontal wire antennas, a two-metre HF/VHF whip antenna, and a mast with three VHF antennas (a simple two-element system, a wire loop, and a VHF log periodic antenna). The building reportedly also contains microwave intercept equipment directed at certain government offices. In 1979, it was estimated that there were 12 to 14 KGB officers employed in the embassy. It is likely that perhaps two or three of these are concerned with Sigint activities.

Latin America. The Soviet Union now has embassies in 15 of the 32 countries in Central and South America, with some 1,000 accredited personnel (as compared to about 700 in 1979). At least five of these embassies house Sigint activities: Buenos Aires, Argentina; Havana, Cuba; Mexico City, Mexico; Managua, Nicaragua; and Lima, Peru. The Soviets in Peru and Nicaragua have special communications linkages with Moscow, which at least in the case of Nicaragua includes a direct link with the GRU headquarters.

The Soviet embassy in Mexico City is a three-story building at Avenida Tacubaya 204, just off the Tacubaya freeway south of the center of the city; it is partially hidden by trees, and has a forest of radio antennas bristling from the roof. It is the largest Soviet embassy in the region, with a staff of about 350 personnel (of whom about one-third are accredited), including at least 150 KGB and GRU officers.

The Soviets have persistently sought to establish consulates elsewhere in Mexico, including Tijuana and Ciudad Juarez. Sigint stations at those sites would be able to provide extensive data on U.S. Navy operations in the San Diego, Calif., area.

Tijuana is about eight miles southeast of the Imperial Beach Radio Station, about 16 miles from the U.S. Naval Air Station at North Island and the Navy's amphibious base on Coronado, and about 26 miles from the Naval Air Station at Miramar.

A post in Ciudad Juarez, just south of the border from

Figure 17. The Black Mountain Telecommunications Tower in Canberra, Australia, is the most visible structure in the capital city. It is the hub of most of the vital communications traffic within the country.

El Paso, Texas, would be ideal for monitoring air defense and Strategic Defense Initiative (SDI) tests at the U.S. Army's Fort Bliss only about five miles away, and the White Sands Missile Range, about 48 miles to the northwest.

The Soviet embassy in Havana is a major GRU Sigint post. It is equipped with VHF antennas for monitoring VHF tropospheric transmissions in the southeast United States, as well as VHF and microwave antennas for monitoring Cuban communications.

Non-Soviet Diplomatic Establishments. Soviet diplomatic establishments are complemented by those of other Warsaw Pact countries, such as Poland, Czechoslovakia, Bulgaria, and particularly East Germany. For example, the Polish embassy in Chicago serves effectively as a Soviet Sigint post, while in Ottawa the antennas (including microwave receivers) on the roof of the Polish embassy are even more remarkable than those on the Soviet embassy.

The East German use of diplomatic establishments for Sigint purposes is particularly noticeable in the African cities of Tripoli, Libya; Antananarivo, Madagascar; Maputo, Mozambique; and Accra, Ghana. According to a report on East German intelligence activities in Ghana in 1983: "These activities have recently been emphasized by the sprouting of an unusually vigorous growth of antennas from the roof of the East German embassy in Accra, rivaling those of the Ghana Broadcasting Corp. and far exceeding normal diplomatic needs."

Collection Operations. The operations conducted at the 100 diplomatic establishments discussed provide an extraordinary volume and range of signals intelligence. Most of these establishments are equipped with HF, VHF and microwave receiving equipment, and some (such as the embassies in Washington, D.C., Ankara and Tokyo) also have Satcom receiving systems. Located in 60 countries, these establishments provide a remarkable geographic coverage. In many cases, it is clear that access to particular communications systems and circuits was a primary determinant of the sites of these establishments. In the United States, for example, most of the 10 identified Sigint posts are located within line-of-sight of important microwave nodal points. According to an analyst at AT&T, "it is most unlikely that these sites were selected for any other reason than microwave interception."

The Sigint collected at these posts pertains to military activities, diplomatic communications, commercial communications, counterintelligence operations and political developments in the host countries.

Military Activity. The KGB and GRU are known to have used Sigint posts in Soviet diplomatic establishments to monitor military signals from the highest national command levels to tactical logistic movements, including communications from the Norad Cheyenne Mountain Complex near Colorado Springs, to the National Military Command Center in the Pentagon; communications among the elements of the National Military Command System in the Washington, D.C., area; Strategic Air Command (SAC) emergency action messages (EAMs) and other combat readiness checks; signals concerning the planning and progress of military operations during the Vietnam and Yom Kippur Wars; flight activity of airborne command posts; and air movements of personnel and supplies to combat and crisis theaters.

The GRU TS Group in the Soviet embassy on 16th Street in Washington has been extremely successful in monitoring official communications, both within the capital and between the capital and other important command authorities and facilities. According to one account, for example, "After the Cuban missile crisis of 1962, Khrushchev complimented the GRU for having provided him with information from telephone intercepts in Washington clarifying the events and discussions in official circles that led to the final resolution of the crisis."

Command frequencies and circuits have been identified by the KGB and GRU Sigint agencies and are continuously monitored. Both SAC and U.S. Navy Current Traffic Message and EAM communications to the strategic nuclear forces are broadcast on some 24 basic frequencies, and there is extensive evidence that the Soviets concentrate on these. In June 1980, for example, following one of the two false alarms of Soviet ballistic missile launches that occurred at the Norad Cheyenne Mountain Complex that month, the E-4A National Emergency Airborne Command Post at Andrews Air Force Base in Maryland was moved to the end of the runway and readied for take-off; the signals relating to this activity were monitored and transmitted to Moscow in real time. One noted authority has argued that, "By listening in on these same conversations during actual hostilities, the Soviets...might learn of American decisions on the launch of nuclear weapons even before our own forces do."

A communications security assessment conducted by USeucom following the Yom Kippur War showed that the GRU and KGB were able to use Sigint posts in diplomatic establishments in the U.S., Europe and the Near East to monitor combat readiness checks from the U.S. Commander-in-Chief, Europe (UScinceur) Airborne Command Post; the real-time status of deploying forces, such as the departure of the fleet ballistic missile submarines Kamehameha (SSBN-642) and Simon Bolivar (SSBN-641) from Rota following the declaration of a heightened defense condition by the United States on Oct. 25, 1973; discussions relating to contingency planning operations and potential task force compositions; traffic concerning airlift departures from the continental United States; the movement of war material from the ports of Nordham and Bremenhaven in West Germany; numerous communications concerning the movement of fuel and other logistic activity in the Mediterranean region; and other related air and naval activities.

It has also been reported by a former CIA officer that the embassy on 16th Street in Washington is constantly used by both the GRU and KGB to monitor Pentagon circuits to its overseas commands, just as it was during the war in Vietnam.

Diplomatic Communications. The fact that Soviet diplomatic establishments are in many capitals located near other embassies, foreign ministries, and other governmental agencies provides extensive access to diplomatic communications. In New York, for example, the Soviet mission on East 67th Street is used to monitor telephone calls from the U.S. mission in the Waldorf Towers some 18 blocks to the south. Senator Daniel Moynihan, who was appointed U.S. Ambassador to the United Nations in June 1975, has testified about a warning that he received from Vice President Nelson Rockefeller. The vice president had chaired a June 1975 conference on CIA activities in the United States, which noted for the first time the extent of Soviet Sigint activities in the United States. Soon after Moynihan's appointment, Rockefeller warned him: "Now, I have something to tell you that you must take with great seriousness. You... have to know that every word you say on the telephone will be listened to by the Soviet mission. In the Waldorf Towers, in the U.S. mission, they will be listening to your phone calls, and you must be extremely careful."

According to one expert, the Soviet embassy in London has been used to monitor Foreign Office communications relayed through the London Post Office Tower: "When the Post Office Tower was being built in London to transmit messages by microwaves, the Foreign Office was warned by the security authorities that the Russians would probably be able to intercept messages, especially as the tower would be in the direct line of sight from the top of the Soviet embassy in Kensington Palace Gardens. It was soon found that the Russians were taping all the messages and sending the tapes back to Moscow for analysis there. This was realized at last by the Foreign Office in the 1970s, when the foreign secretary sent an important secret message via the Post Office Tower to the secretary of state in Washington. In a stupidly short space of time, the Russian ambassador visited the Foreign Office with a complaining document, clearly indicating that the secret message had been intercepted. Only then, when so much had been lost, was it decided to send all such secret messages by undersea cable."

It is also known that Sigint posts in the Soviet embassies in Helsinki and Vienna have intercepted U.S. diplomatic communications concerning the Strategic Arms Limitation Talks (SALT) and other arms control negotiations conducted in those capitals.

Economic and Commercial Intelligence. The KGB has accorded particular attention to the collection of economic and commercial intelligence. Monitoring communications to and from large defense contractors and advanced technology companies is one of the highest priorities. The Soviets also use intercepted material to protect their enormous investments in the West and to manipulate certain commodity markets.

In the United States, it is clear that the Sigint posts in the East 67th Street and Riverdale establishments in New York have a particular interest in communications concerning stock exchange and other financial transactions, while the consulate in San Francisco and the recreational facilities in Oyster Bay, Glen Cove, and Pioneer Point have special interest in defense contractors and high technology companies in Silicon Valley, Long Island and Connecticut, and Maryland, respectively. For instance, in 1978, the Senate Select Committee on Intelligence reported that the Soviet embassy in Washington had probably intercepted a facsimile transmission from the Boeing Company's office in Washington, D.C. to its headquarters in Seattle concerning sensitive aspects of the MX intercontinental ballistic missile program.

According to several former U.S. intelligence officers, the Soviet Union also uses its Sigint posts in diplomatic establishments to collect intelligence for commercial purposes. For example, Raymond Tate, former deputy director of the NSA, has stated, "I firmly believe the Soviet Union has for years manipulated a lot of commercial markets in the world, commodities and other things. That has nothing to do with national security in the military sense. They have a significant cash flow problem, and how do you make money in a cash flow problem? You can turn your intelligence system around and use it to get all sorts of data you can actually use in commercial ventures."

And the Soviets have been fairly adept at doing this. Some particular commodity markets that the Soviets have evidently manipulated with the assistance of Sigint include oil, gold, diamonds, grain and sugar. The most commonly reported case concerns the "Great Grain Robbery" of 1972, when the KGB intercepted microwave telephone communications between the Department of Agriculture in Washington, the Chicago Board of Trade, and other U.S. government agencies, and was able to negotiate a grain purchase on terms not only very favorable to the Soviets ($1.63 a bushel) but which later created grain shortages and higher prices ($2.30 a bushel) for U.S. consumers. As Harry Rositzke, a former CIA officer, has reported, "Perhaps the (KGB's) most lucrative contribution in the economic field was its monitoring of telephone calls into and out of the Department of Agriculture in the early seventies...The KGB coverage of telephonic reports by the grain dealers to the Department of Agriculture clearly helped Moscow time its purchases before the full extent of U.S. grain requirements became apparent in Washington. As a colleague of mine put it, 'The Russians knew more about events in the American grain market than the White House did.'"

According to Bobby Inman (admiral, U.S. Navy, retired) former Director of the NSA and deputy director of the CIA, the KGB also had access to "a lot of economic communications" concerning the sugar market (Commodity Market No. 11 on the New York Stock Exchange) in the 1970s, in order to lower the price of sugar and hence obtain more favorable barter arrangements with Cuba.

Counterintelligence Activities. Almost all of the Soviet diplomatic establishments that maintain Sigint operations include KGB and/or GRU posts concerned with monitoring the communications of the host country security surveillance or counterintelligence agencies. KGB Zenith rooms and GRU Radio Monitoring Stations are used for this activity. In the United States, the Soviet embassy began spectrum scanning of FBI and police radio-telephone channels in the Washington area from the 16th Street embassy before World War II, and during the 1950s and 1960s the Sigint posts in both the embassy and the Glen Cove complex were primarily concerned with supporting Soviet espionage activities in Washington and New York by monitoring local police and FBI surveillance communications. The FBI HF FM radio network in Washington remains accessible to the embassy, and indeed, the channel used to communicate with agents in the field is known colloquially within the bureau as KGB-770!

One example of the effectiveness of this monitoring effort is typical: In 1980, the KGB's Zenith officers in the 16th Street embassy were able to successfully frustrate FBI attempts to identify Ronald Pelton, a former NSA officer who supplied the KGB with information about NSA activities for nearly six years until he was arrested in November 1985. Pelton had telephoned the embassy on Jan. 14, 1980 to make initial contact with the KGB. The telephone conversation was recorded by the FBI, and when Pelton visited the embassy on Jan. 15, the FBI had established a tight cordon around the building, designed to identify the spy and arrest him as he left the embassy. However, a Zenith officer had picked up a burst of radio activity from FBI walkie-talkies and car radios as Pelton entered the embassy and figured that the radio messages had been triggered by his entry. Pelton was then disguised as an embassy worker and departed by a side door surrounded by several other embassy employees. He successfully elluded the FBI watch.

Soviet Quick Response. The role and functions of KGB Zenith organizations in Tokyo, Paris and Ottawa are well known; they are similar to those of the GRU radio monitoring station in Vienna. These counterintelligence radio monitoring activities generally involve only two or three Sigint officers in each post. The purpose of the Zenith room in Tokyo has been described as follows: "In it a technician monitored the radio frequencies used by Japanese counterintelligence and police surveillants. Whenever a (KGB) officer was due to engage in a hazardous meeting with an agent, the technician came on duty and listened. If he heard a flurry or any other abnormality in Japanese communications, he transmitted a signal to a tiny bleeper in the officer's pocket and thereby told him to abort the meeting."

Descriptions of the KGB's Zenith room in Paris are essentially identical.

In Canada, the GRU radio monitoring station in the Ottawa embassy is believed to have successfully frustrated

a major Security Service operation, code-named Gold Dust, in 1966. In the mid-1960s, the watcher service of the Security Service typically maintained surveillance of GRU officers with radio-controlled teams of six cars with two watchers per car, and used only a single frequency every 24 hours. In the case of Operation Gold Dust, the GRU evidently intercepted the watchers' radio communications and terminated a meeting with an agent before the Security Service could complete the operation and arrest the agent in the process of passing material to the GRU officer.

In addition to monitoring police and Security Service communications, the GRU Radio Monitoring Stations and KGB Zenith rooms also monitor, catalogue and analyze all other communications within the relevant urban areas, such as those of fire brigades, ambulance services, construction dispatchers, radio-controlled taxis, etc. This information is used for a variety of purposes, including the selection of frequencies for agent communications which would be difficult for counterintelligence services to distinguish from those of other utility and commercial transmissions.

Private Communications. Included in the communications that are accessible to the Sigint posts in Soviet diplomatic establishments is an enormous volume of personal telephone conversations. The first official public acknowledgement of this activity occurred in June 1975, when the Rockefeller Commission reported that: "The communist countries...appear to have developed electronic collection of intelligence to an extraordinary degree of technology and sophistication for use in the United States and elsewhere throughout the world, and we believe that these countries can monitor and record thousands of private telephone conversations. Americans have a right to be uneasy if not seriously disturbed at the real possibility that their personal and business activities which they discuss freely over the telephone could be recorded and analyzed by agents of foreign powers."

In July 1977, Senator Moynihan, then still a member of the Senate Select Committee on Intelligence, issued a press release in which he stated: "I cannot stress too strongly that modern technology has given to foreign espionage a new dimension which needs to be understood in this country. The targets of Soviet interception of telephone communications now include our business, our banks, our brokerage houses, as frequently as our government agencies. Soviet espionage seeks to penetrate into other aspects of American life, commercial, intellectual, and political, as much as it seeks illegal entry into the councils of governments. This is precisely why the problem is now one of interest to all Americans in their daily lives, not an abstract problem for intelligence operatives in trench coats."

In fact, this Soviet activity represents an invasion of privacy of unprecedented magnitude. It also allows KGB and/or GRU manipulation of personal affairs. Information obtained from monitoring personal telephone conversations is used in blackmail operations, and has sometimes led to the recruitment of western citizens as Soviet espionage operatives. As Senator Moynihan testified in December 1985, "Vice President Rockefeller...warned about blackmail in 1975. He said blackmail is going to be the consequence, and I don't see how you can assume that there is no connection between the number of (Soviet) spies that have appeared in this country in the last couple of weeks and the fact that the Soviets have been listening to telephone conversations for 10 years."

Countermeasures. In most of the countries in which the Soviets use diplomatic establishments for Sigint purposes, the relevant national authorities are undoubtedly aware of the existence of such activities, but they are generally unable or unwilling to comprehend the extraordinary scope, the particular technical capabilities, and the implications of these activities. Most countries have instituted some forms of communications security practices and procedures, but these are frequently rudimentary and generally inadequate when compared to the scope and technical sophistication of the Soviet operations. The United States is the only country that has sought to inform both governmental agencies and the public about the vulnerability of their communications to Soviet interception, and to publicly discuss means of countering the Soviet Sigint threat, although several other countries (such as Japan, France and the United Kingdom) have provided background briefings to journalists in an attempt to generate greater public awareness of the problem. In most other countries, the problem is regarded as being too difficult to publicly address and the relevant national authorities have preferred not to acknowledge its existence.

A fundamental constraint is that, whereas there is a requirement under Article 27(1) of the Vienna Convention on Diplomatic Relations and Article 35(1) of the Vienna Convention on Consular Relations for a mission to obtain the consent of the "receiving" (i.e. host) state for the installation and use of radio transmitters, there is no requirement regarding the installation of receiving equipment. Article 27(1) of the Vienna Convention on Diplomatic Relations states: "The receiving state shall permit and protect free communication on the part of the mission for all official purposes. In communicating with the government and the other missions and consulates of the sending state, wherever situated, the mission may employ all appropriate means, including diplomatic couriers and messages in code or cipher. However, the mission may install and use a wireless transmitter only with the consent of the receiving state."

A second constraint is that numerous Western countries also use diplomatic establishments for Sigint purposes and indeed, both the U.S. and the United Kingdom maintain Sigint stations in their embassies in Moscow. There is an understandable concern that countermeasures undertaken against Soviet diplomatic establishments will provoke reciprocal Soviet moves against their own operations.

U.S. Avoids Confrontation. The unwillingness of the U.S. government to disturb Soviet use of diplomatic establishments for Sigint purposes was highlighted in 1982-83 in connection with the Killenworth post in Glen Cove. In May 1982, following Arkady Shevchenko's revelations about the use of the Killenworth mansion for Sigint collection, the Glen Cove City Council moved to prohibit the use of Glen Cove's recreational facilities by the Killenworth residents. In August 1983 the State Department undertook litigation to force the city to rescind the ban. According to the State Department, the city's action was "incompatible with our obligations under international and United States law." It was the department's responsibility "to maintain control over the conduct of foreign relations without interference from municipal governments." In addition, the department was concerned in order "to protect American diplomats in Moscow from retaliation." As the department noted in November 1983, "The Soviet government has responded to (Glen Cove's) action against its personnel by imposing (in August 1982) comparable restrictions on access to recreational facilities by United States diplomatic and consular personnel in the Soviet Union. This action has had an adverse effect on the morale of our diplomatic and consular staff, who under the best of circumstances must carry out their important work under conditions that most Americans would consider intolerable."

These arguments were somewhat disingenuous, since the U.S. recreational dacha and the "diplomatic beach" at Uspenskoye, on the Moskva River about 15 miles northwest of the center of Moscow, are not comparable to the facilities at Glen Cove. According to a security officer who served at the U.S. embassy in Moscow from July 1981 to December 1982, "the so-called diplomatic beach that was placed off limits to us during August of 1982 is a dismal place, and not often used by the American community." Rather than being disheartened by the Soviet restrictions, the staff of the embassy actually "all cheered" the Glen Cove actions. In April 1984, the Glen Cove City Council succumbed to the State Department and rescinded its ban.

Reducing Vulnerability. There are several measures that can be undertaken to reduce the effectiveness of these Soviet Sigint operations. To begin with, any Soviet proposals for the construction or acquisition of new diplomatic establishments, embassies, consulates, trade missions, official residences, recreational facilities or whatever, should be rigorously assessed according to various Sigint and Comsec criteria. The locations of numerous present establishments could not have been approved if proper assessments had been conducted. The commanding line-of-sight situations of such establishments as Riverdale, Glen Cove, Pioneer Point and Mount Alto in the

United States, Mount Royal in Montreal, Highgate in London, and the embassies in Paris and Tokyo should have been sufficient reason to reject them at the outset. In the case of the new embassy on Mount Alto, the State Department began regular consultations with the FBI, NSA and CIA in 1966-67, and in 1972 the NSA and CIA specifically approved the height of the proposed chancery. Given the extraordinary vantage point offered by the Mount Alto site, it is clear that cognizant U.S. intelligence agencies could not have given sufficient consideration to the proposal.

There is also some scope for reducing the total number of Soviet diplomatic establishments in many countries. For instance, in the Washington, D.C., area alone, the Soviets maintain more than a dozen offices and residential buildings, about half of which are equipped with some Sigint capability. Their occupation of the Mount Alto complex, when it happens, will provide an opportunity to rationalize these establishments. However, under current plans, the Soviets intend to retain most of "their present locations and facilities." In New York, there is no evident requirement for "recreatational" facilities at both Glen Cove and Oyster Bay.

In other cases, there is scope for demanding significant reductions in the numbers of Soviet personnel employed in their diplomatic establishments. The most effective of the GRU and KGB Sigint posts are those in the relatively large diplomatic establishments where 10 to 20 officers are concerned with Sigint activities. Reductions in the total numbers of Soviet personnel in these establishments would have a more than proportionate impact on the effectiveness of these Sigint activities. In the United States, for example, the fact that some 150 Soviet personnel are able to engage in Sigint collection clearly suggests that there is substantial scope for reductions. In many other countries, the size of the Soviet diplomatic presence is greatly in excess of the requirements for genuine diplomatic liaison and consular business.

Limitations In Order. Clearly, the Soviets have blatantly padded their personnel rosters for the purposes of espionage and Sigint collection, and measures should be taken to counter the problem. Some limitations could also be placed on the types of antennas and other telecommunications equipment installed in Soviet establishments. Since October 1985, the U.S. State Department has required Soviet and East Bloc countries to obtain permission to purchase and install telecommunications equipment, and according to testimony by James E. Nolan, director of the Office of Foreign Missions, in December

1985: "The department recently denied a Soviet request to install a parabolic dish antenna at their new embassy site at Mount Alto and at their recreational facility at Pioneer Point, Md.

In practice, of course, the Soviet use of diplomatic establishments for Sigint purposes will remain widespread, and indeed, it is likely that the trend over the past decade of about one additional Sigint post per year will continue. Improved Comsec practices and procedures are therefore imperative. In cases where there is no possibility of relocating the Soviet establishments, either because the properties are actually owned by the Soviet Union or maintained under long-term contractual arrangements, consideration should be given to changing some of the local telecommunication architectures, as the Canadians did during the 1960s.

On the other hand, this is frequently impracticable. It would be absurd to suggest that the Eiffel Tower, the London Post Office Tower, the Tokyo Tower or the Telecom Tower on Black Mountain in Canberra should be relocated. In these cases, land lines should be constructed for particularly sensitive circuits, with fiber optical cables providing the most secure system. Secure telephone systems such as the STU-111 should be used wherever possible. (In July 1986, NSA awarded contracts for the production of 50,000 STU-111 telephone units, about 0.3 percent of the number of telephones currently in use in the United States).

Finally, there must be a much greater public awareness of the vulnerability of telecommunications to Soviet interception. By virtue of their past performance, it is glaringly apparent that the Soviets are adept at using all types of information for their own benefit, to the detriment of the free world. As their espionage activities increase to satisfy their need for high technology, communications involving discussion of commercial and personal matters, will become even more important as a prime target of Soviet Sigint efforts. An informed and aware public is a necessary precondition to the development of effective countermeasures against such Soviet Sigint activities. ∎

Extensive references for this article are available from the publisher on request.

Dr. Desmond Ball is head and senior fellow at the Strategic and Defense Studies Center at the Australian National University, Canberra, Australia. He was previously a lecturer in international relations and military politics in the Department of Government at the University of Sydney, a research fellow at the Center for International Affairs at Harvard University, and a research associate at the International Institute for Strategic Studies in London.

Notes

New Technologies for Electronic Support Measures

Emerging technologies and techniques provide new directions for ESM systems.

By Dr. Myron L. Cramer

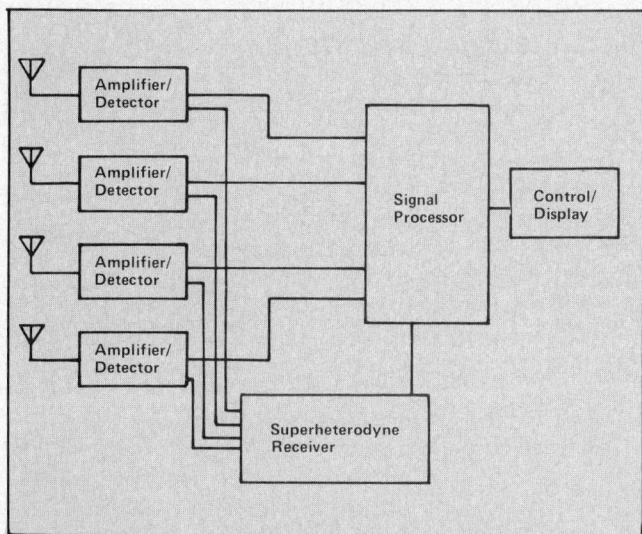

Figure 1. **Radar warning receivers** (RWR) use several antennas to achieve sector coverage, and each one feeds RF into its individual amplifier/detector channel. After an RF-to-video conversion, the pulse trains enter the signal processor, which classifies the emitters and sends the results to the control/display equipment. A tunable superheterodyne receiver provides the capability of detecting continuous wave (CW) radars.

The classic objective of electronic warfare (EW) or electronic combat (EC) systems is to deny an adversary the use of the electromagnetic spectrum while maintaining its use for friendly forces. Electronic support measures (ESM) provide EW/EC systems with their "eyes" into the spectrum, and ESM capabilities are vital to the exercise of control over the electromagnetic environment. In terms of its composition, ESM encompasses a wide variety of tactical and national systems designed to intercept and exploit communications and radar emitters.

New Requirements

The driving force behind the development of new technologies for ESM systems is the need to keep pace with the threat over the near term and to prepare for future battlefield environments. In these environments, electronic systems will be employed in every phase of command, control and communications (C^3) and battle management (BM). Emerging ESM requirements have arisen from such factors as:

• Command, control, communications and battle management (C^3/BM) developments that have proliferated a wide variety of electronic systems.

• Technological advances that have resulted in the increased use of electronic counter-countermeasures (ECCM) and defensive techniques in the design of C^3/BM systems.

• Increased numbers and types of C^3/BM platforms, many having a high degree of mobility.

• Tighter integration constraints that increase the competition between ESM and other electronic systems for platform payload, volume, power, cooling, antenna locations and other requirements.

• The extension of ESM C^3/BM target systems beyond the traditional frequency bands into the electro-optic (EO) and infrared (IR) frequencies.

• Demands by operational users for increased availability, timeliness and user-friendliness.

As a result of these factors, ESM systems are called upon to deal with complex emitter systems over broader frequency ranges and in denser signal environments. At the same time, increasing constraints are being imposed upon the system designer.

Current Designs

A wide range of ESM system designs are currently in use, from simple radar warning receivers to complex signals analysis systems used to collect electronic and communications intelligence. The evolutionary design of current ESM systems reflects responses to the changing electromagnetic environment, incorporation of technological opportunities, design trade-offs and budgetary, scheduling and acquisition constraints. It is anticipated that future ESM systems will develop in a similar fashion.

Radar warning receivers (RWRs) provide their host aircraft with threat warnings. Figure 1 shows the elements of a typical RWR system. Sector or quadrant coverage is provided by separate antennas, each connected to its own amplifier/detector. These amplifier/detectors typically employ wide-frequency-coverage crystal video receivers with filters for the bands of interest. They receive and detect pulsed RF signal energy, then convert it into video outputs that go to the signal processor. The signal

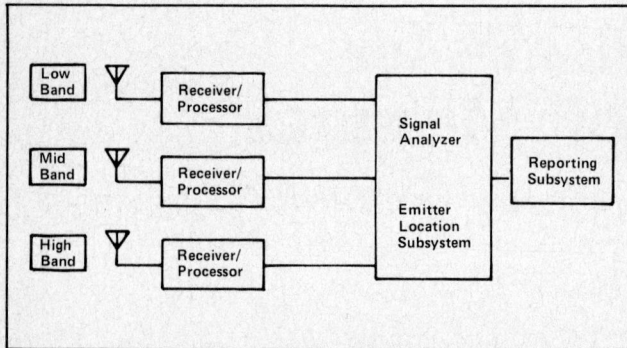

Figure 2. **An electronic intelligence (Elint) system** must scan the spectrum, then locate and identify any hostile emitters operating within it. Because of the wide range of frequencies, most Elint systems deploy separate antennas and receiver/processors against the low, middle and high bands. After signal processing and analysis, the system conveys the results to users via its reporting subsystem.

processor de-interleaves the video pulse trains it receives from the amplifier/detectors, classifies and identifies the emitters present and routes the results to an operator console. Operator interface is accomplished through circular displays presenting angle, range and threat identification; there is also an audio warning. A superheterodyne or other tunable receiver may be added to impart a capability against continuous wave (CW) signals and/or to provide for frequency measurement of selected signals as a pulse discriminant.

Electronic intelligence (Elint) systems, like RWRs, must be capable of receiving and processing the frequency spectrum of threat radars. Unlike RWRs, which serve only a crew warning function, the Elint system must generate and report data on the threat electronic order of battle (EOB). Essential elements of EOB information include emitter location, frequency, times of signal activity, scan mode, polarization, type of propagation (pulse or continuous wave), pulse repetition rate, pulse width, jitter and intra-pulse modulation.

Figure 2 provides a top-level overview of an Elint system. As a consequence of the frequency ranges involved, separate antennas and receiver/processors are used for the low, middle and high bands. For Elint, after received signals are processed and analyzed, the results are reported to the users by the reporting subsystem.

Communications intelligence (Comint) systems include some of the more complex ESM applications. Different systems have unique features based upon their own mission requirements and acquisition history. Figure 3 is a generalized overview of a Comint system that includes elements for signal acquisition, intercept, analysis, emitter location and reporting. The acquisition subsystem acquires signals through spectrum scanning or spectrum compressive approaches. The intercept/analysis subsystem will then classify and identify the emitter. Emitter location subsystems may use direction-of-arrival (DOA) or time-difference-of-arrival (TDOA) design approaches. DOA systems are typically based upon incident emitter signal phase across an ESM system baseline. A TDOA system is

based upon different RF signal times of arrival across a baseline. TDOA systems are sometimes combined with a differential doppler (DD) measurement to resolve ambiguities in the emitter location equations. Reporting subsystems include the elements to prepare, format and send collection results to the user.

Developmental Technologies

A number of ongoing technological developments offer the potential for next-generation ESM systems to address new requirements. The principal developments offering the greatest potential impact include: GaAs field effect transistors (FETs); monolithic microwave integrated circuits (MMIC); very high speed integrated circuits (VHSIC); surface acoustic wave (SAW) devices; array processors; high-speed data buses; and the major disciplines of artificial intelligence and superconductivity. Although each is discussed separately, it should be noted that each application is expected to be highly dependent upon other developments.

Field effect transistors (FETs) have proven their usefulness in RF circuits by providing outstanding low-noise performance. Fundamental physics limits the upper frequency range of the popular silicon FET to approximately 6 GHz, but the upper limit on gallium arsenide (GaAs) FETs is much higher, allowing their use in the higher

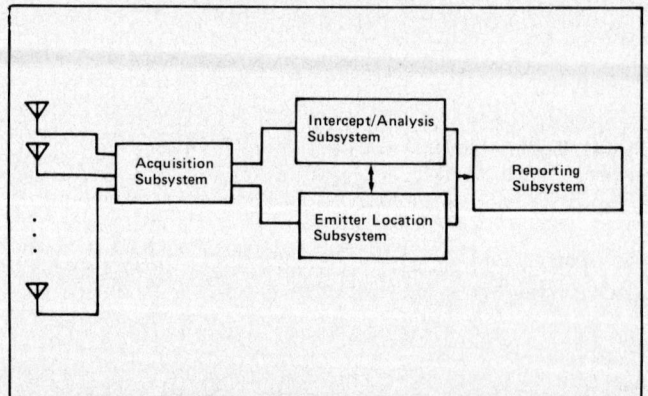

Figure 3. **Communications intelligence (Comint) systems** have features unique to their mission requirements, but, in general, a Comint system will include signal acquisition, intercept/analysis, emitter location and reporting elements.

microwave bands. GaAs also offers additional advantages in lower noise, lower power requirements and smaller size. Since new ESM systems must satisfy significant requirements in these areas, new GaAs FETs are expected to be a factor in future ESM system capabilities.

Monolithic microwave integrated circuits (MMICs) are the latest development in the field of microwave integrated circuits, which have been in use for a relatively long time. These latter circuits receive and process RF in an integrated RF/electronic package. With MMIC, all active and passive elements are formed by standard semiconductor manufacturing techniques on a semi-insulating substrate. The advantages gained through MMIC are

Table 1: Technolgy Applications

APPLICATIONS \ TECHNOLOGIES	GaAs FET	MMIC	VHSIC	SAW	Array Processors	High Speed Data Bus	Artificial Intelligence	Superconducting Devices
Scanning Superheterodyne Receivers	•	•	•					•
Compressive Receivers		•	•	•				•
IFM Receivers		•	•	•				•
Channelized Receivers		•	•	•				•
Acousto–Optic Receivers				•	•			•
Digital Receiver	•	•	•					•
Signal Processors		•	•		•	•		•
Sensor Integration & Fusion			•		•	•	•	•

Table 1. **Technological developments** have a high potential for improving electronic support measures (ESM) equipment. The author, in identifying salient technologies and applications, emphasizes the high degree of interaction and interdependence between developments and applications.

small size, low weight, batch fabrication, improved reproducibility of units and the ability to bring impedance matching elements very close to active devices, further reducing matching and signal losses. Once ongoing manufacturing refinements are completed, we can expect that MMICs will find wider ESM applications.

Very high speed integrated circuits (VHSICs) are important enough for the Department of Defense to dedicate a program to the development of this new class of high-speed circuit. The first chip sets from this program are now becoming available; once incorporated into ESM applications, we can expect to see two orders-of-magnitude improvements in processing throughput, coupled with several orders-of-magnitude reductions in size, weight and power consumption. It is anticipated that VHSIC will allow the incorporation of sophisticated algorithms for emitter signal processing and emitter location.

Surface acoustic wave (SAW) devices are constructed of piezoelectric materials that can couple electronic signals to acoustic waves. SAW devices allow large time-bandwidth products, programmability and large dynamic ranges. Examples of applications include correlators, convolvers and filters. SAW devices are expected to be especially useful for ESM applications against spread-spectrum systems or exotic emitters.

Array processors include a variety of user-programmable, digital computing devices capable of two or more simultaneous processing operations. Array processors find employment in two primary areas: high precision scientific calculations and signal processing. Signal processing operations, such as direction-finding and emitter location, can use parallel, pipeline channels to conduct high speed computations of the type required for high-accuracy emitter location systems. The first such applications are only now being made, but as parallel-processing software

and algorithms are developed, greater use of this technology is anticipated.

High-speed data buses are needed to meet the high speeds inherent in ongoing and future developments. Recently, it became necessary to redefine the term "high-speed" with reference to buses. Rates in excess of a one gigahertz (10^9, or one billion cycles per second) are currently being discussed, and these are orders-of-magnitude greater than the speeds previously thought possible. This increase has resulted from both an expansion in mission requirements and from advances in technology. Increased data rates result from increasing integration of functions, distributed processing and computer communications during time-sensitive computations. Technology advances include fiber-optic cables and components and breakthroughs in data bus architectures and standards.

Artificial Intelligence

Artificial intelligence (AI) is the use of computerized reasoning techniques to handle knowledge. These techniques include problem definition, knowledge base (facts, assumptions, rules), search strategy for the use of knowledge, inference procedures and user-friendly interfaces. In general, AI allows an application to learn information and relational rules and then to apply these adaptively during the course of a program run. AI has existed as a research area since the 1950s, but only now is it becoming mature enough for user applications.

Initial applications of AI to expert systems has been very successful, including battle management decision support systems, signal processing and C[3] network optimization. AI could provide the processing approach enabling future ESM systems to use the other technology advances discussed in this paper. ESM applications of AI could include advanced signal sorting, de-interleaving and emitter recognition and exploitation. A secondary use for AI is in the built-in condition monitoring and fault isolation functions.

Superconductivity

Superconducting devices capitalize on the phenomenon of superconductivity in which materials lose their electronic resistance at low temperatures, a discovery made early in this century. As a result of recent major

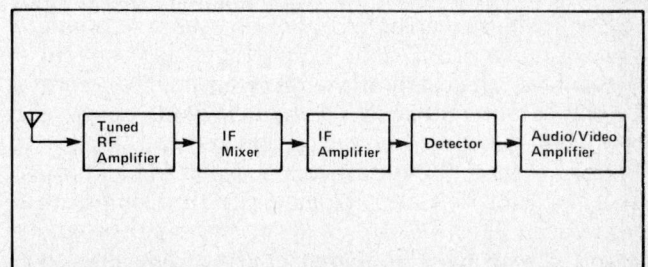

Figure 4. **Scanning receivers based upon the superheterodyne principle** are still the most widely used ESM receivers. Although superheterodyne fundamentals have been known for over seven decades, new technology is expected to improve ESM receiver performance in terms of increased sensitivity, accelerated scan rates, expanded spectral coverage and programmability.

breakthroughs in identifying new superconducting materials, scientists are formulating new classes of super-conductors even as this paper is being written. The break-through is that these materials exhibit superconductivity at temperatures rapidly approaching room temperatures, or at least at temperatures more easily achieved through practical cooling systems.

It is anticipated that these materials will provide the basis for new devices allowing orders-of-magnitude improvement in ESM system sensitivity and dynamic range. New electronic chips could be constructed, free from the constraints imposed by resistance-induced heat damage, and thus could leap ahead of VHSIC technology devel-opments and open up new processing possibilities. The long-range impact of this new technology will totally change the balance of every design trade-off made in ESM systems. The result will open up designs to capabili-ties limited only by the imagination of system designers.

Previously, the only area of superconductivity applica-tions being seriously pursued was the Josephson Junction. While some of this research may have been overcome by events, the applications initiated with conventional superconductors may become even more attractive using the new classes of high-temperature materials, since they overcome the cryogenic cooling requirements and con-straints of these experimental devices. Josephson Junction research has centered about development of signal proc-essors with cycle times of a few nanoseconds. Up until the present, the size of workable devices has been limited by the cooling constraints.

Technology Applications

As noted earlier, the impact of the recent technological developments discussed in this paper will be interactive and interdependent. For ESM systems, Table I shows areas of possible application, considering receivers, signal processors and sensor integration and fusion. Receiver design technology and approaches determine the cap-abilities available for future ESM systems. Within a short while, designers of these systems will have a wide selec-tion of receiver options to consider such as scanning superheterodyne, compressive, instantaneous frequency measurement, channelized, acousto-optic and digital receivers. It is anticipated that these different designs will be used in complementary combinations to optimize their respective capabilities in acquisition and analysis.

Scanning superheterodyne receivers are the principal receiver asset in today's ESM systems. Figure 4 shows the simplified functional design of this receiver approach. Performance of these receivers is expected to improve as a result of new technologies, and this will result in faster scan rates, better sensitivity, increased program-mability, expanded frequency coverage and improved packaging.

Compressive receivers, as depicted in Figure 5, provide an extremely wide and rapid frequency search capability in comparison with other designs. This feature has made compressive receivers popular for spectrum search and

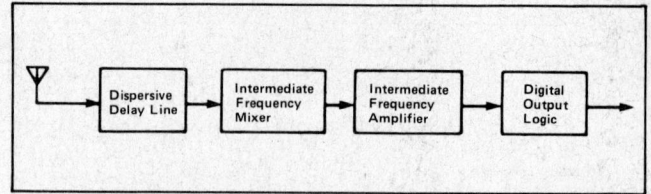

Figure 5. **Compressive receivers** quickly search extremely wide bands and are especially suitable for work against communications equip-ment. New technologies are widening the bandwidth and equipping compressive receivers with higher speed input/output circuits and the ability to search with finer frequency resolution at faster rates.

signal acquisition, especially in the communications bands. As a result of new technologies, improved com-pressive receivers are expected to cover higher radar-emitter frequencies, to incorporate higher speed input/output circuits and to have larger bandwidths, faster search rates and finer frequency resolution.

Instantaneous frequency measurement (IFM) receivers have been developed to provide high-speed frequency determination over a wide dynamic range of signal strengths and frequencies. As shown in Figure 6, this design approach splits the RF power into a set of parallel correlation channels. Frequency-dependent delays are

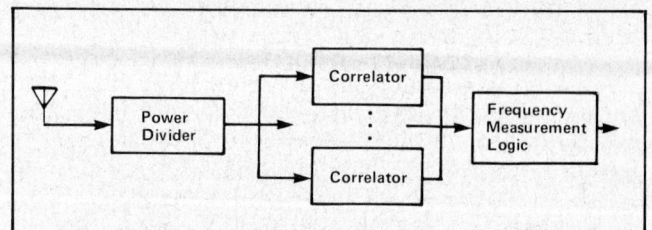

Figure 6. **Instantaneous frequency measurement (IFM) receivers** pro-vide high-speed frequency determination over a wide dynamic range of signal strengths and frequency. Current designs, though offering high sensitivity, are limited in handling simultaneous and overlapping signals. Future designs will offer wider bandwidth performance with increased sensitivity.

used to determine input frequency. Contemporary designs offer high sensitivity, but are limited by their inability to handle simultaneous overlapping signals. Future designs, influenced by technological advance-ments, will improve multiple signal capability, gain wider bandwidth performance and increase their sensitivity.

Channelized receivers have been empoyed in a wide variety of ESM systems. As shown in Figure 7, this design uses multiple parallel channels to filter, amplify and detect incoming emitter signals. This approach allows a wide band of the spectrum to be simultaneously monitored for signal activity with excellent probability of intercept while operating against multiple emitters. Improvements in channelized receivers are anticipated as a result of new filters, SAW devices and MMICs. Improved algo-rithms are expected to increase emitter discrimination capabilities against exotic signals.

LORAL

MIC assemblies for our own systems and for the defense industry.

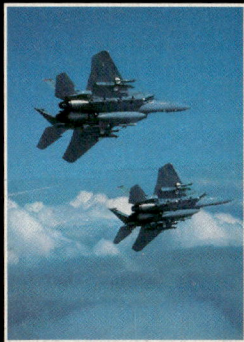

The most advanced, radar warning system in the free world—the ALR-56C.

ALQ-178/RAPPORT system for the F-16 is the first integrated, self-protection system for tactical aircraft from a single supplier.

Expendable flares and chaff distract radar-guided and heat-seeking missiles.

Proprietary cesium lamp generates IR spectrum for deception of heat-seeking missiles.

With Autonomous Synthetic Aperture Radar Guidance (ASARG) "smart missiles" become brilliant.

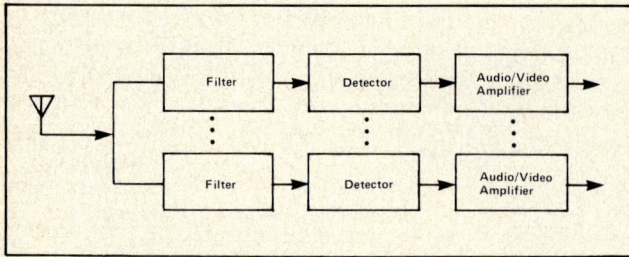

Figure 7. **Channelized receivers** use multiple parallel channels to filter, detect and amplify incoming emitter signals. This provides an excellent probability of intercept against multiple emitters. New filters, SAW devices, MMICs and improved algorithms will enhance their discrimination capabilities against exotic signals.

Acousto-optic receivers are a new approach under development for acquisition and analysis. Acousto-optical receivers offer advantages in throughput and linear dynamic range, and they should be especially effective in working against spread-spectrum emitters in dense environments. While a variety of design approaches is under consideration, Figure 8 shows a concept that implements acoustic, optical and electronic processing with SAW devices, integrated optics and Bragg Cell diffractors. New technologies in SAW devices are of direct

impact on this receiver approach, and performance will be affected by developments in array processors and possibly superconducting devices.

Digital receivers tune and demodulate incoming RF into an intermediate frequency, then they digitize it in the general circuitry depicted in Figure 9. Currently, digital receivers employ digital signal processing techniques using very large scale integrated (VLSI) circuits. Processing includes taking fast Fourier transforms, filtering out noise and inverting. In some proposals, the incoming

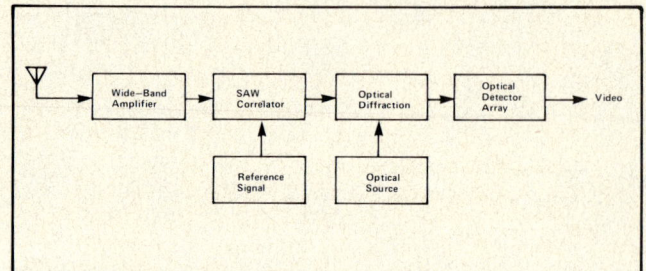

Figure 8. **Acousto-optic receivers** should be effective against spread-spectrum emitters in dense environments. One concept implements acoustic, optical and electronic processing with standing acoustic wave (SAW) devices, integrated optics and Bragg Cell diffractors.

Superconductivity

Superconductivity is the name applied to a class of remarkable phenomena that occur as a result of a phase-transition in selected substances below a critical temperature. This phase transition affects the behavior of the conduction electrons that carry the electric currents, allowing these currents to flow with no measureable resistance. Curiously, the metals that are normally considered to be the best conductors do not become superconductors at low temperatures (e.g., silver, copper and gold). Figure 10 contrasts the behavior of silver and tin. The graphs show that the resistivity of silver decreases with temperature down to approximately $10°$ K; from that point on, the resistivity remains practically constant. The resistivity of tin also drops with temperature until it reaches the critical temperature of $3.7°$ K, where it suddenly disappears in a plunge to zero. One result of zero resistivity is that an electrical current established in a superconductor will flow indefinitely as long as the material remains below its critical temperature.

Meissner Effect. Resistivity, the measure of a material's behavior in electric fields, is not the only way to detect the superconducting phase transition. Another observable phenomenon is a superconductor's response to an externally applied magnetic field. This phenomenon is called the Meissner effect, and it consists of the superconductor's exclusion of magnetic fields. As a result of the Meissner effect, a magnet brought over a superconductor will be repelled, and this provides a visual demonstration of the material's exclusion of the magnetic field.

History. The Dutch Physicist Heike Kamerlingh Onnes discovered superconductivity in 1911. For most of the period since then, known superconductors could operate only at temperatures a few degrees above absolute zero ($473°$ F). Consequently, liquid helium served as the coolant to achieve the required low temperatures. The difficulty in achieving and maintaining low temperatures with liquid helium has precluded many applications.

Between 1930 and 1973, research uncovered alloys such as niobium tin and niobium titanium that exhibited superconductivity at temperatures up to $23°$ K. Until the last few years, though, progress in superconductors has been slow.

The first of the recent breakthroughs came in 1985, when Karl Muller and Johannes Bednorz discovered that a class of metallic oxides, known also as ceramics, exhibited superconductivity at temperatures up to $35°$ K. In February 1987, Paul Chu of the University of Houston upped this to $98°$ K. This discovery was especially significant because it brought superconductivity to within the cooling range of liquid nitrogen.

The next four months produced dramatic developments that brought the superconductivity phenomenon within the cooling ranges of ordinary refrigeration. In May 1987, Paul Chu reported initial indications that his research group had created yet another superconducting material with a critical temperature as high as $225°$ K (or $-54°$ F). The following month, June 1987, scientists at Energy Conversion Devices Inc. reported observations of the Meissner Effect at $260°$ K (or $9°$ F). The fast pace of these breakthroughs has prompted speculation that a room-temperature superconductor will be found by 1988.

RF will be directly digitized and processed with programmed operations to detect, identify and characterize emitters. High resolution and the flexibility possible with arithmetic operations offer exciting possibilities. Constraints are those currently imposed by contemporary VLSI microchips and algorithms. Device technologies in GaAs FETs, MMICs, VHSIC and possibly superconductors will drive the performance envelopes of these receivers.

Signal processors for ESM systems should benefit from almost all of the technology advances discussed in this

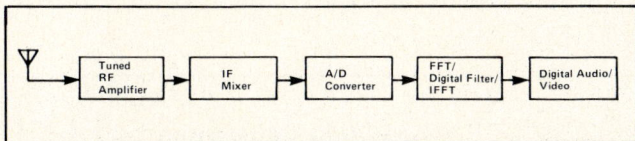

Figure 9. **Digital receivers** demodulate RF into an intermediate frequency signal, then digitize it. Thereafter, very large scale integrated (VLSI) circuits take over the processing. Current VLSI technology and algorithms constrain performance, but the use of GaAs FETs, MMICs, VHSIC and possibly superconductors should enhance it.

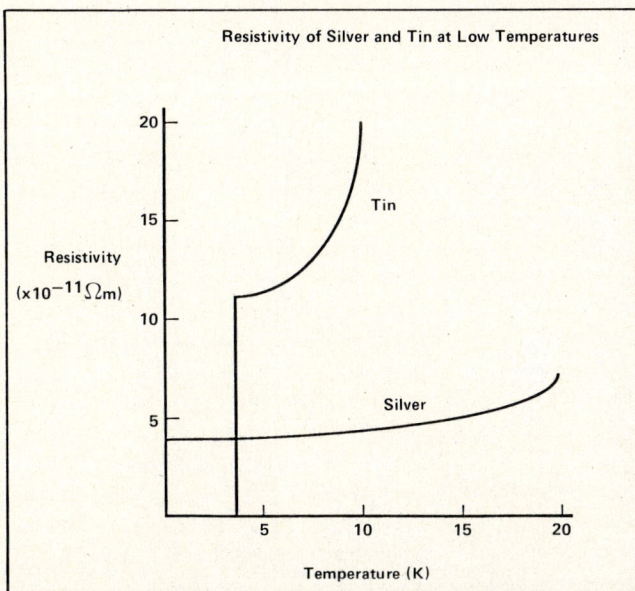

Figure 10. **Superconductivity** occurs when low temperatures cause the electrical resistance of a substance to cease. Performance varies with the type of substance used. Comparing silver to tin, silver's resistance decreases with temperature until around 10° K, then remains constant. Tin's resistance declines until around 3.7° K; it then drops to zero. In the past, most materials required immersion in liquid helium to become superconductive, but ceramics recently superconducted at the temperature of liquid nitrogen. The search for a room-temperature superconductor is now attracting scientific attention.

paper. Digital signal processing will have faster, more capable integrated circuits, parallel processing techniques and improved algorithms. Many of these algorithms will use adaptive techniques to optimize the processing steps, and some will have a greatly improved capability of rapidly classifying signals by incorporating AI's expert techniques.

Sensor integration and fusion at the platform level is one of the most promising applications for advanced technology in ESM systems. The platform level is where the variety of installed sensors (ESM, IR, radar, laser) can be integrated to enhance warning and weapon system performance. Correlation of data from several sensors can increase the probability of correct identification and signal analysis. Given the extension of ESM system requirements into the IR and EO regions and the increased use of optical processing techniques, the potential exists to increase ESM and imagery system integration to the potential benefit of both. In the long term, there may even be a merging of RF and imaging sensors and processors with fusion taking place up front.

ESM Trends

New technologies and techniques are expected to provide new capabilities for future ESM systems. These new capabilities will offer a variety of options to improve receivers, signal processors, sensor integration and fusion. Among the more important trends are the expansion of ESM system performance envelopes in frequency range, signal acquisition, speed, sensitivity and dynamic range. In hardware, there has been increased incorporation of embedded processing into the ESM receivers and the addition of modern displays and controls including digital and graphic readouts. Improved packaging has reduced volume and weight, lowered power and cooling requirements and resulted in the increased use of standard, built-in interfaces for receiver control and reporting. Finally, the inclusion of integral condition monitoring and fault detection for ESM system functions as well as input/outputs has improved reliability. ∎

Myron L. Cramer manages space electronic warfare at Booz, Allen & Hamilton, Inc. The holder of a Ph.D. in physics from the University of Wisconsin, he has over 12 years experience as the manager of programs concerned with electronic warfare, C^3 countermeasures, communications, intelligence, electro-optics, simulation, modeling and operational requirements analysis. Prior to joining Booz, Allen & Hamilton, Dr. Cramer managed the electronic warfare group at ARINC Research Corp. In preparing this paper, he surveyed the ESM industry and wishes to express his appreciation to the respondents who contributed their thoughts and ideas. He also wishes to acknowledge the contributions made by Mac McLaurin, Stephen R. Pratt and Megann L. Hester.

Extending the Frequency Ranges of ESM Antennas

Some antenna problems in the higher frequency bands can be solved by an aperture type ESM system.

By Glen R. Gray

Many different types of antenna systems are employed in performing the electronic support measures (ESM) mission. The primary function of these antenna systems is to provide signal intercept and the direction of emitted signals. Some antenna systems are suitable—and more easily implemented—for VHF/UHF communications intelligence (Comint) intercept functions while others are more suited to microwave- and millimeter-wave-frequency electronic intelligence (Elint) functions.

As Comint signal frequencies are extended into the microwave frequency bands and above, and Elint frequencies are extended into higher millimeter wave frequencies, problems arise that render present techniques less desirable. The primary problem is that the high gain and narrowbeam antennas that solve the problems imposed by higher frequencies do not have the instantaneous spatial coverage necessary for present monopulse techniques. The aperture type ESM system, although not monopulse, provides the solutions to some of these problems.

ESM Antenna System Summary. The various types of ESM antenna systems used for signal intercept, with particular emphasis on emitter direction finding (DF) are as follows:
- Spinning aperture antenna dish and feed
- Amplitude comparison monopulse antennas (Skewed-beam antennas used for amplitude monopulse)
- Phase comparison monopulse antennas
- Switched-beam antenna arrays
- Mode-forming circular arrays.

These types of antenna systems represent a complete range of monopulse systems, non-monopulse (spinning reflector) systems, amplitude comparison antenna arrays, antenna arrays for phase comparison, arrays with simultaneous amplitude beams that are selected to give emitter location in various sectors, and circular arrays that use the phase relationships of mode-forming networks to determine direction of arrival. Most of these techniques can be used across the low frequency (VHF/UHF) and high frequency (microwave) bands. Some implementations of a technique will use large dipoles in the lower frequency bands and spirals or horns in the higher frequencies. Which type of system is selected is often determined by

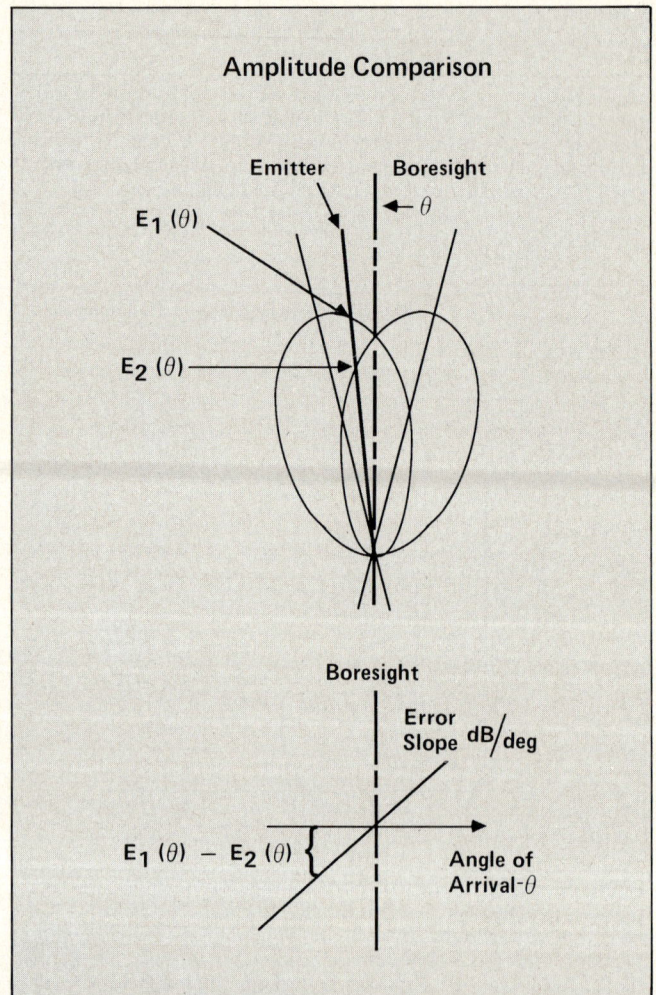

Figure 1. In this example of the amplitude comparison monopulse technique, the direction of the emitter is determined by a comparison of the amplitude received at the two apertures.

the type of platform on which the system is mounted as well as system gain and DF accuracy requirements.

Spinning DF Antennas. Spinning direction-finding antenna systems for microwave ESM systems have been in use for more than 40 years. Relatively small reflector antenna systems are combined with single or multi-axis

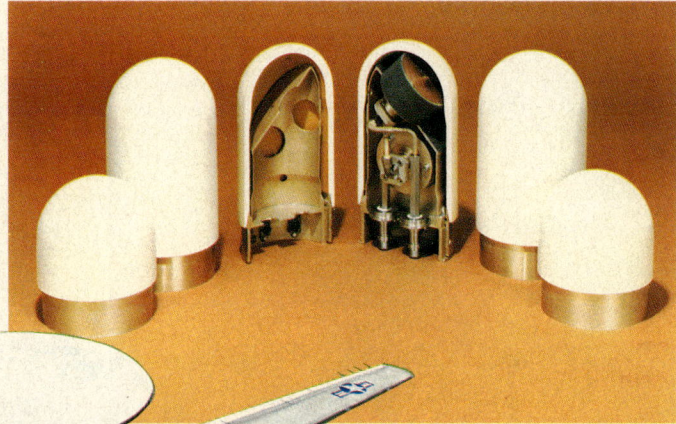

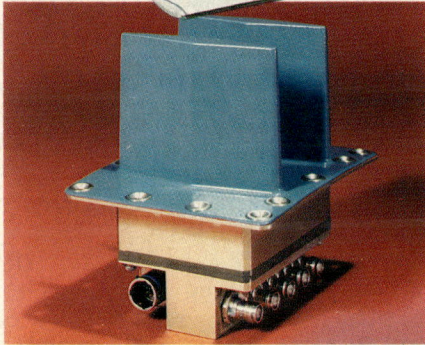

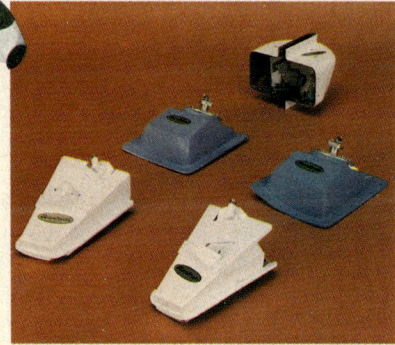

Phase Comparison

Boresight

Emitter

θ

S.SINθ

S

$$\theta = \frac{2\pi}{\lambda} \, SIN \, \theta$$

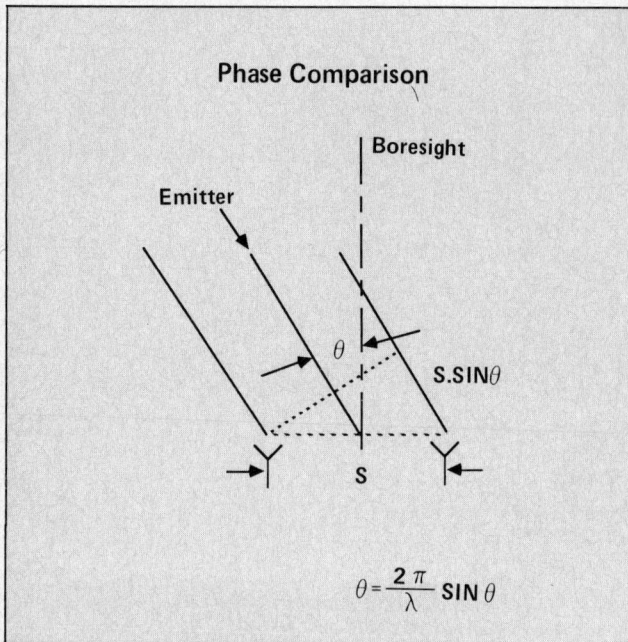

Figure 2. In a phase comparison using monopulse (interferometer) antennas, to determine the angle of arrival, the phase difference and the phase center separation of the two antennas is all that is required.

pedestals to provide a scanned beam. Developments in pedestal components (drive motors, position sensors, rotary joints and slip rings) and the increased mechanical integrity of these components make them popular candidates for many applications. A primary advantage of this approach is the relatively high antenna gain that is achieved, which in turn enhances system sensitivity.

The technique of DF using a spinning narrow-beam-width, aperture-type antenna is conceptually simple. As the antenna is rotated, an emitted signal is received by the antenna and passed to a receiver. The video output of the receiver is fed to a cathode ray tube (CRT). The maximum signal level displayed on the CRT is in the direction from which the signal is coming. As the rotating DF antenna scans through the emitted signal, this signal, if strong enough, will be received through the DF antenna's sidelobes as well as the main beam. Therefore what is displayed on the CRT is a picture of the DF antenna's radiation pattern at a particular fixed-frequency of the antenna.

A significant advantage of this type of system is its low cost and relative simplicity compared to some of the monopulse antenna systems described in subsequent sections. Other less desirable performance attributes of monopulse systems are: (1) limited field of view, (2) lower antenna gain, and (3) poor sidelobe performance. These limitations cause non-monopulse types of systems to be traded off against monopulse antenna systems for higher frequency (millimeter wave) applications.

The direction finding technique just described is manual—that is, the operator places a cursor in the azimuth direction corresponding to maximum received amplitude. But this technique can be automated. The

use of sidelobe-blanking techniques is necessary for this automation. One such technique uses a lower-gain omni-directional antenna to compare received strength with the DF antenna to verify that the received signal is located in the main beam of the DF antenna rather than its sidelobes or backlobes. In order for this comparison technique to be useful, the omni-directional antennas must have relatively uniform radiation patterns over frequency and polarization, and the directional antenna needs to have low-sidelobes. Achieving this performance over a broad frequency range can be a costly and difficult task. Other automatic DF processing techniques that compare received signal video over several scans of the DF antenna are also useful to validate emitter DF. These video comparison techniques are cheaper to implement, but rely on multiple scans of the DF antenna, which may not occur for short burst transmissions.

A shortcoming of spinning DF systems is their inability to look in more than one direction at any instant, even though the antenna may be rotating rapidly—200 rpm is typical, while 400 rpm has been used in some instances. This problem is compounded if the emitter is also spinning, or if the receiver is also scanning in frequency to create a multiple scan-on-scan condition.

Monopulse systems, which simultaneously look in all azimuth directions, are necessary to instantaneously locate emitters. Different types of systems have been developed which are described in the following sections. Most of these systems, however, have much lower antenna gain than the spinning DF approach. This is expected due to their wide-open field of view requirements. Additionally, lower antenna gains—and thus lower system sensitivities—require the platform to be closer to the emitter.

Monopulse. Monopulse antennas provide an instantaneous emitter direction on a single pulse. The angle of arrival is determined by a comparison of amplitude, phase, or a combination of both from two or more antenna apertures. These apertures are positioned symmetrically about the system's boresight axis. When the emitting source is on the boresight axis, the received signal at both apertures is of equal phase and amplitude. When the emitting source is off boresight, the direction of the emitting source may be determined by comparing the phase or amplitude at each receiving aperture. To obtain 360-degree spatial coverage, four to six array elements are usually symmetrically located in a cylindrical array to provide skewed beams. To obtain instantaneous bearing measurements, a receiving channel is required for each element of the array. To achieve polarization diversity, spirals or horns with polarizing grids are popular candidates as array elements.

Amplitude Comparison Monopulse Systems. For the amplitude comparison case, each pair of antennas in the array has a common boresight axis. Figure 1 is a diagram of the technique. The direction of the emitter is determined by a comparison of the amplitude received at the

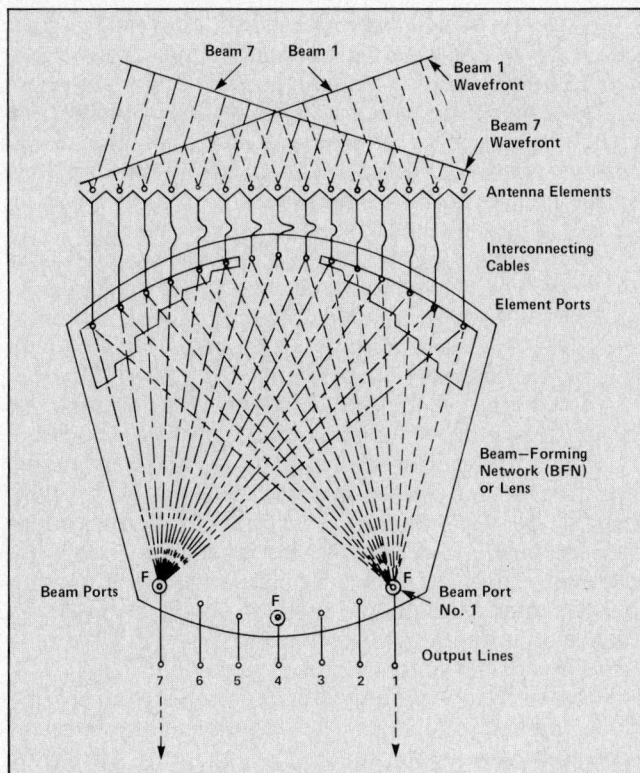

Figure 3. The switched-beam antenna array is a simpler method of obtaining multiple beams, compared to the Butler matrix. It uses lens techniques to achieve a beamforming network without the multiplicity of components. The Rotman lens consists of a homogenous transmission region between the lens ports which feed the array elements through proper delay lines and feed ports on the focal arc of the lens.

two apertures. If the amplitudes are identical, the emitter is located on the boresight axis. If the emitter is off the boresight axis, the direction is determined from the system's error slope, which is proportional to the received amplitudes. In the amplitude system, this slope or amplitude gradient is determined by the angular separation of the antenna elements and the antenna element's beamwidth; the slope is nominally linear between the three dB beamwidths of the antenna pair. This linear error slope represents the idealized case.

In the actual case, the slope is an envelope developed to include errors associated with the overall system. These errors include amplitude imbalance of the antennas and receiving system, polarization errors, changes from the nominal beamwidth of the antenna itself and boresight shift inherent in the individual elements. These errors are extremely critical since a typical error slope for broadband spirals is on the order of 0.3 dB/deg. The most significant errors are in the receiving channel balance and spiral element polarization for broadband systems. Typical measurement errors associated with broadband (multi-octave) amplitude comparison monopulse systems are 5 to 10 degrees (one sigma). Narrowband systems (one octave or less) utilizing scalar horns with greater aperture control have achieved errors on the order of 2 degrees, one sigma.

Phase Comparison Monopulse (Interferometer) Antennas. Phase comparison monopulse is based on antenna separation to obtain an error signal proportional to the direction of arrival. Unlike the amplitude monopulse system, the individual axes of the antenna elements are normal to the plane of the array. In its most simple form, the phase monopulse system senses path length differences between the two elements and the emitter. To determine the angle of arrival, this phase difference and the phase center separation of the two antennas is all that is required as shown in Figure 2. For unambiguous measurement over a 180-degree field of view, the baseline spacing between antenna elements must be less than one-half of a wavelength. With this spacing, measurement accuracies on the order of 5 to 7 degrees may be attained.

As the baseline spacing increases, the phase difference received at the two elements increases, thereby reducing the angular error to phase error ratio. By increasing this baseline separation of elements to 2.5 wavelengths, accuracies on the order of one degree may be attained. However, ambiguities result and must be resolved to determine the correct angle of arrival. These ambiguities may be eliminated with multiple baseline systems. Non-harmonic multiple element interferometers have been built covering bandwidths of 9:1, without ambiguities over 180-degree sectors.

Switched Beam Antenna Arrays. Multiple-beam antennas are also used in covering large sectors in space. This type of antenna is associated with high aperture gain and narrow beamwidths for precise direction finding. These phased arrays consist of a large number of antenna elements coherently fed from a single source through beam forming networks to provide complete spatial coverage from the array's boresight to angles as great as 60 degrees from boresight. To achieve this flexibility in beam positioning implies control of both phase and amplitude at each element of the array. These arrays have used parallel feed networks with phase shifters, and more recently beamformers, such as the Butler matrix or Rotman lens, to achieve simultaneous beams that may be sequentially sampled or continuously monitored depending on the complexity of the receiving system. Each beam uses the full aperture of the array, since the beams formed are orthogonal in space.

The Butler matrix is a parallel array feed, feeding a linear array of antenna elements, which consists of 90-degree hybrids and fixed-phase shifters. This arrangement allows the simultaneous forming of high-gain beams from the array's aperture. The matrix forms orthogonal beams that cross over approximately 4 dB down from the beam peak. The aperture distribution is nominally uniform with -13 dB sidelobes. However, techniques summing adjacent beam terminals have been used to suppress the inherent high sides of the matrix-formed beams.

In this array the number of elements must be equal to a power of two and the number of beams formed by the matrix is equal to the number of array elements. The

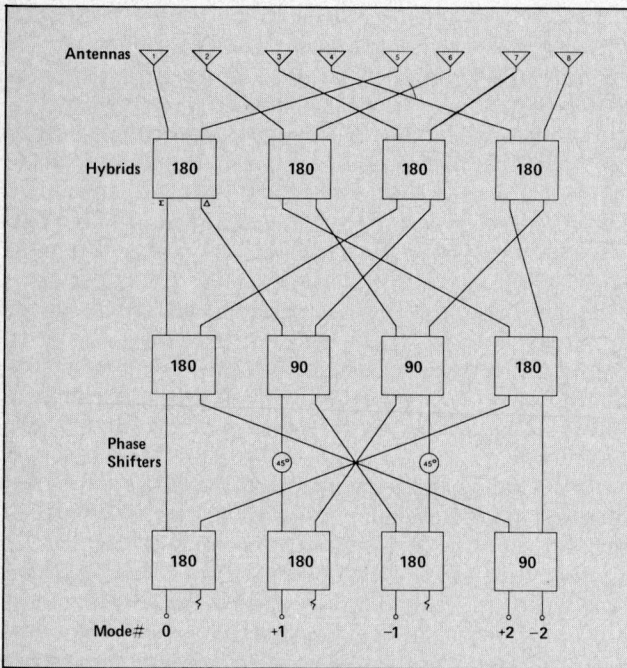

Figure 4. For circular array implementation, the key mode-forming structure consists usually of a Butler type matrix of constant phase-shifting elements specifically selected to produce the required mode values. This mode-forming network is designed for use with eight antenna elements

Mode-Forming Circular Arrays. Mode-forming circular arrays are an instantaneous monopulse ESM antenna system technique which uses straightforward constant-phase mode-forming networks and phase discrimination to significantly increase DF accuracy. In this case the mode-forming network provides linear phase progressions versus angle of arrival values (modes) to accurately (within 1 or 2 degrees) determine the line of bearing to an emitter. Comparing the phase difference of any modes differing by one (for example +1 and +2) yields a one-for-one linear correspondence between phase and bearing angle which is non-ambiguous, but not very accurate due to the "shallow" phase slope.

Comparing the +2 mode with the –2 mode provides four cycles of phase for one cycle of bearing. This result is ambiguous (four bearing angles for one phase value), but it is also four times as accurate. Using both results (which are simultaneously available from the mode-forming network) produces a more accurate and unambiguous bearing determination. Depending on the number of antenna elements used (4, 8, 16, or 32), even higher-order modes are produced by the mode-forming network to yield potentially even higher accuracy.

The key mode-forming structure consists usually of a Butler type matrix of constant phase-shifting elements specifically selected to produce the required mode values.

major advantage of this technique is the multiple lossless beams, excluding the insertion loss of the beam-forming network. The system is relatively narrowband due to limitations of the components and antenna array effects. A second limitation is that the beam pointing direction is a function of frequency similar to arrays using progressive phase steps in forming beams. Furthermore, the complexity of the system grows rapidly with the number of elements used in the array. For example, an array of 64 elements requires 192 directional couplers and 160 fixed phase shifters.

A simpler method of obtaining multiple beams, compared to the Butler matrix, uses lens techniques to achieve a beamforming network without the multiplicity of components. This technique is shown in Figure 3. The Rotman lens consists of a homogenous transmission region between the lens ports which feed the array elements through proper delay lines and feed ports on the focal arc of the lens. The design of the lens employs optical properties and differential path lengths to form beams where required in space. In the case of the lens, therefore, the beams do not scan with frequency. Furthermore, the number of beams formed are not limited by the number of aperture elements as was the case for the Butler matrix. The design of the lens is relatively straightforward; however, as with many microwave devices, the practical development requires experience and time. Once developed, the lens is a candidate for mass production stripline techniques at relatively low costs, particularly when compared to the multiple component beamformers such as the Butler matrix.

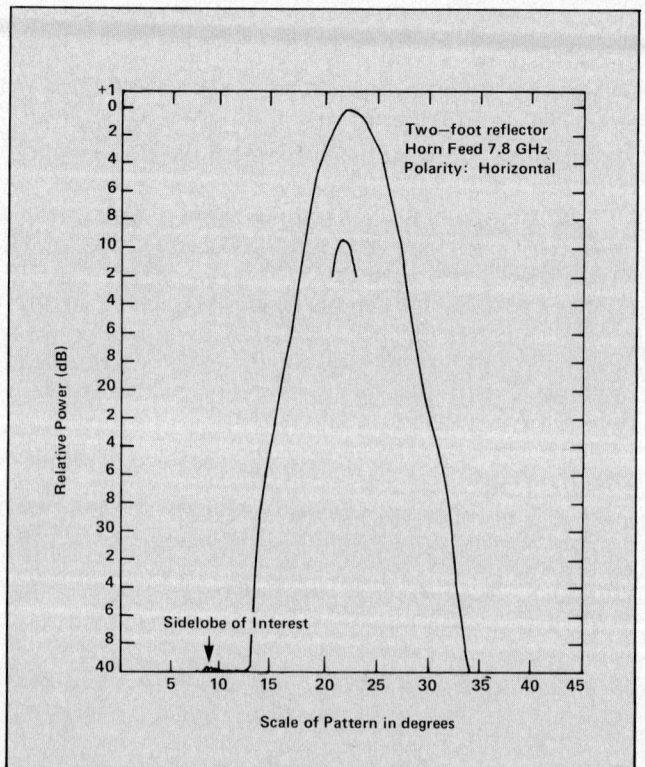

Figure 5. This low sidelobe antenna pattern is the radiation pattern of a three-foot offset-fed reflector, which was designed to cover the communication band of 7.1 to 8.5 GHz. The combination of offset feeding the reflector and using a scalar horn for the feed allowed sidelobe levels of 39.5 dB to be achieved.

An example is shown in Figure 4. This mode-forming network is designed for use with eight antenna elements and yields the following output modes:

MODE NUMBER	PHASE SHIFT OVER 360-DEGREE BEARING	
0	0 degrees	(for all bearing angles)
+1	+360 degrees	(relative to 0-mode)
−1	−360 degrees	(relative to 0-mode)
+2	+720 degrees	(relative to 0-mode)
−2	−720 degrees	(relative to 0-mode)

Not all of these modes need to be used, but the use of the various plus and minus modes can further reduce bearing determination inaccuracies, particularly if multiple look-up tables are used and "best-fit" algorithms utilized.

Higher Frequency Considerations. The particularly important factor that needs to be considered as communication signals are extended into the microwave frequency bands (microwave Comint) is the increasing need to receive PCM-type signals. It is very difficult to lock on to these signals in a high-multipath environment unless more directive antennas with extremely low sidelobes are used. Fortunately, due to the higher frequencies involved, these types of antennas have become practical.

When it comes to ESM antenna systems required to intercept millimeter wave signals (millimeter wave Elint), the salient consideration is the sensitivity of the receiving system, including antennas. Sensitivity is a concern because of the inherently low power levels of the emitters in these bands, and the greatly increasing space loss and atmospheric attenuation of signals at these frequencies. The following two sections discuss the microwave Comint and millimeter wave Elint considerations.

Microwave Comint. Antennas suitable for microwave Comint, particularly of PCM-type signals, need to have more antenna gain—and therefore a more narrow beamwidth—in order to restrict the field of view of the antenna. This restricted field of view will reduce the number of reflected signals received by the antenna system due to multipath reflections from the emitted signal. However, interfering multipath signals can also enter the antenna through its sidelobe structure, and therefore it is extremely important that this antenna also have extremely low sidelobes.

Figure 5 shows a radiation pattern of a three-foot offset-fed reflector, which was designed to cover the communication band of 7.1 to 8.5 GHz. As can be seen from the radiation pattern, the combination of offset feeding the reflector and using a scalar horn for the feed allowed sidelobe levels of 39.5 dB to be achieved.

The scalar horn is a particularly useful reflector feed when low-sidelobe performance is required. The scalar horn allows an extremely low reflector edge illumination because the scalar horn has virtually no sidelobes of its own. If a standard horn was tried, to illuminate a reflector with extremely low edge illumination values, the sidelobes of the horn itself would illuminate the edge of the reflector at some frequency in the band. The resulting phase reversal would destroy the low-sidelobe response. Two additional benefits of using the scalar horn are that its feed pattern remains more constant than a standard horn, and it has a rather flattened beam peak to produce a more uniformly illuminated aperture. This illumination is more efficient and thus the antenna achieves more gain.

Although this discussion is primarily related to antennas, another very important consideration for interception of PCM signals is that a receiver be employed that has extremely low phase-noise characteristics since this allows the receiver to lock up on complex pulse-coded signals. It has been found that even some non-PCM signals of complex characteristics can be detected by receivers with extremely low phase noise.

Millimeter Wave Elint. When it comes to ESM antenna systems required to have increased sensitivity in order to intercept millimeter wave signals, two important techniques are:
- Use high gain aperture antennas for increased antenna gain.
- Incorporate a frequency downconverter at the antenna to minimize path loss through the system.

As discussed previously, aperture antennas at millimeter wave frequencies allow one to maximize antenna gain with a reasonably sized aperture. Also, aperture antennas are easy to shape in the elevation plane to yield constant elevation patterns over the frequency band, or a cosecant-squared elevation beam shape which is particularly useful for airborne emitters. Another advantage of the cosecant-squared elevation beamshape response is that the cosecant-squared response is a good compromise between the narrow beam necessary for increased sensitivity and the broad beam, which may be necessary for spatial coverage or effective radiated power (ERP) considerations of a jamming antenna. ∎

Glen R. Gray is director of antenna systems marketing for Condor Systems, Inc., San Jose, Calif. He has a broad background in the design and development of microwave antennas, direction finding and feed systems. He received his bachelor of science degree in electrical engineering from the University of California at Berkeley, and his master of science degree in electrical engineering from the University of Santa Clara.

Offboard Countermeasures

A wide variety of devices offer protection to combat platforms.

By Dr. Norman Friedman

Electronic countermeasures fall into two broad categories: systems aboard platforms, such as jammers or deception repeaters, and offboard systems, which the platform may eject or launch, or which may operate quasi-independently. Thus the spectrum of offboard systems ranges from devices intended to save their platforms from a homing weapon to devices intended to avoid attack altogether by providing convincing alternative targets. In each case the idea is the same: it is better to build an expendable decoy or countermeasure than to risk loss of a valuable combat asset. The offboard concept, then, shades from what might be considered conventional tactical countermeasures into tactical and strategic deception.

The simplest offboard system, chaff, has been standard since World War II. However, until recently the more elaborate (typically active) countermeasures have generally been limited to onboard use, because their complexity entailed considerable size. The two major exceptions have been decoys (such as the Air Force Quail, GAM-72) carried by strategic bombers and by missiles (in which case they are generally called penaids, or penetration aids); and submarine-launched simulators, such as the current Navy Moss (mobile submarine simulator). It seems likely that, as electronics become more compact, sophisticated offboard countermeasures will become more common, simply because good ones are probably a more reliable means of ensuring the safety of the platforms they protect. Onboard countermeasures can be more sophisticated, but if they fail, they fail catastrophically: their emissions may themselves lead a sophisticated attacker to the platform they are intended to protect. In many cases, too, a sufficiently large offboard decoy can respond to a variety of sophisticated signals beyond the knowledge of its designers. By way of contrast, sophisticated onboard countermeasures rely for their efficacy on detailed knowledge of enemy sensors.

For example, a deception repeater on an airplane will attempt to confuse a tracking radar by returning mimic pulses with the wrong intervals. It may be defeated by a sufficiently complex frequency-hopping scheme. By way of contrast, if the platform under attack can launch a simple corner reflector, the tracking radar may be seduced to follow the reflector, whose effectiveness does not much depend upon the details of the enemy radar signal, as long as the radar is within the frequency range for which the corner reflector is effective. It is much easier to guess a frequency/wavelength range than to be certain of radar operating patterns. For example, a radar may be designed with a switch to shift it from the peacetime patterns (which electronic surveillance can

Self-propelled decoys launched from submarines can seduce away homing weapons, and noisemakers and bubble screens created by the vessel can confuse the sensor systems and weapons of sub hunters.

detect) to special wartime patterns. It cannot, however, be designed to use the same antenna for radically different wavelengths and frequencies.

Realism Paramount. Thus, for a large enough decoy, the issue will often be whether its dynamic behavior (e.g., speed) matches that of a realistic target. Matters are not altogether this simple, because any large decoy imposes considerable costs (e.g., in weapons capacity) on the platform carrying it. Therefore the goal will be the smallest possible offboard device, and it will have to make up in clever electronics what it gives up in size or effective reflecting area. The electronics (such as blip enhancer for radar) in turn will require some knowledge of the enemy sensor. Even so, it should be markedly simpler than an onboard countermeasure, which often requires detailed knowledge.

Even a successful onboard deception device, moreover, is limited in the extent to which it can alter the apparent position of the platform it is protecting. Moreover, it is

Tanks and other battlefield vehicles are the targets of autonomous sensors and "smart" submunitions. They are likely candidates for the protection afforded by offboard expendables such as chaff, flares and aerosols.

because of the way in which missile economics developed. When warheads were so expensive that the decoy payload might have paid off, rockets themselves were too expensive to waste this way. When rockets became cheaper, so did warheads, and realistic decoys were not really so much cheaper as to be worthwhile. These arguments have been revived in the current debate over the Strategic Defense Initiative (SDI), or "Star Wars," as it has come to be known.

Surveillance Seduction. One might think of surveillance seduction as a sort of strategic countermeasure, in that it might mislead an enemy into misunderstanding major planned operations. The famous World War II deception before Normandy used what amounted to a massive offboard countermeasure, the simulated First U.S. Army Group (FUSAG). Because the Germans believed that FUSAG was being withheld to make the main assault in France, they held back reserves during the early, vulnerable, phase of the Normandy landing. Similar autonomous decoying was used in many other World War II amphibious operations. The Navy created special "beach-jumper units" specifically to simulate landings on nearby beaches so as to draw off enemy defensive action.

One might imagine a future case: a simulated carrier battle group. To the extent that the simulated group could overshadow a carefully operated real group, the Soviets might be induced to attack it, and so to sacrifice large numbers of valuable missile bombers (which would be subject to ambush). Similarly, during the 1950s, the great fear of U.S. air defense planners was that the Soviets would launch a large force of ground-based decoys that would draw off limited air defense assets.[1]

Soviets are Vulnerable. Soviet tactical practice is particularly vulnerable to decoying. The Soviets try to centralize tactical decision-making, leaving little or no initiative to the tactical units that actually fight. Everything is planned from above, and in many cases stereotyped tactics are employed because they make such planning easier. The Soviet air defense system is typical of this type of hierarchical organization: There is very little communication between the lower-level fighting units, and also very little real-time feedback from the fighting units to the central commander. One might visualize Soviet style as ground-controlled intercept tactics applied to radically different situations, such as anti-submarine warfare or anti-carrier attacks or even land combat. Such tactics have the advantage that they permit the central commander to force units of different arms to cooperate, even though their communications and tactics may not be entirely compatible.[2]

However, a great deal depends on the judgment of the central commander. He depends heavily on sensors reporting directly to him, sensors not really available to his subordinates. If his perception of the battle situation comes to diverge from the reality, then his tactics may become counterproductive, even fatally so. Conversely, if he begins to doubt the veracity of the sensors on which

difficult for an onboard device to multiply the apparent number of targets, whereas each offboard device simulates another target or enlarges the size of the "real" target. Offboard devices become more efficient as the platforms launching them become more stealthy, since it is easier for small decoys to generate sufficiently strong signatures to overshadow the signatures of the real target. That is already evident in the case of submarines.

Perhaps the greatest level of sophistication in offboard countermeasures is to be found in the long and linked histories of ballistic missile decoys and of ballistic missile defense. Many readers will be familiar with the basic ideas of such countermeasures, because they figured so prominently in the debate which killed off the U.S. anti-ballistic missile (ABM) system. The object of such countermeasures was to raise the cost per warhead killed to an uneconomical level, by forcing the ABM system to engage many decoys per real warhead. On the other hand, the designers of the ABM system could try to discriminate between simple, cheap decoys and real warheads. The better their discrimination, the more expensive the decoys (in terms of weight and volume as well as dollars) sufficient to deceive them, and the less explosive power which could be packed in a given limited throw-weight and volume. One might also imagine a kind of ultimate offboard decoy, a rocket carrying nothing but smaller decoys. It seems never to have been seriously considered

he depends, then he may be unable to make tactical decisions; he may be paralyzed. In that case, the careful training of subordinates, which causes them to eschew individual judgment (for fear of upsetting centrally planned coordinated attacks), could tie up all operations.

For example, Soviet naval operations far out to sea are controlled from land sites. The central commander relies on sensors such as ESM satellites and land-based HF/DF stations, using active radars (aboard statellites and aircraft) to fill in details before ordering actual attacks. Ships, bombers and submarines are ordered into position on the basis of this centrally-collected information, and the attack platforms are not important collectors of data. In distinct contrast to Western practice, the "man on the spot" is by no means considered best equipped to make tactical decisions. In some cases, he may be firing at a target, well over the horizon, which he cannot even detect with his own organic sensors.[3]

Incidentally, large-scale air and missile defense systems (in the West as well as in the East) generally require central decision-making, because they have to deploy limited resources against threats that may appear from many quite different directions. That is why the SDI debate turns so often on issues of deception and decoying.

Saturation Tactics. At the next level down, the offboard deception device confuses a local targeting sensor, such as an air defense search radar. In that case the object is to saturate the defensive system, which can handle only so many targets simultaneously. If the decoys arrive first, and suffice to saturate the system, the real attackers can expect to operate relatively freely. On the other hand, if the radar can watch the decoys being launched, it may be able to track them and so to distinguish them from the aircraft. Much therefore depends on the sophistication of the decoys. For example, the Israelis (and, more recently, the U.S. Navy) use a radar-decoy glider, Samson, launched by tactical aircraft.

One counter-countermeasure, then, is for the system designers to increase their engagement capacity. For example, Typhon, the predecessor of the current U.S. Navy Aegis fleet air defense system, was designed specifically to overcome a predicted Soviet air-launched decoy threat. Even so, the defensive system must actually engage each decoy. It has only so many missiles immediately available, and sufficient numbers of decoys can exhaust it. The ultimate counter-countermeasure, then, must be sufficient sophistication to distinguish the decoys from the real weapons.

Blanket Shields. One might contrast such individual decoying devices from blankets designed to deny all detailed information about an incoming force. Chaff is the classic air defense case: an incoming bomber force can be shielded by a chaff corridor. A tank force might

Aircraft can be protected by offboard countermeasures that are forward fired, ejected or trailed (and recovered). The purpose of all such devices is to seduce the enemy into unsuccessfully engaging the wrong target with scarce weapons.

be covered by smoke impenetrable to missile seekers. Such blankets are limited by their geometry: it is often possible to overcome them by moving a sensor. For example, a chaff corridor sufficient to defeat a shipborne radar probably would not be effective against an airborne or spaceborne radar. The greater the variety of sensor positions, the less useful the blanket. In such cases offboard decoys are still viable, as long as they can deceive the range of available sensor types. For example, a small expendable decoy might be able to deceive a wide range of radars, but it might have little impact on an infrared sensor.

Finally, there are last-ditch deception devices that confuse homing devices. The classic example would be an infrared flare, launched by an airplane under missile attack. The flare is intended to seduce away the incoming missile. Seduction must last long enough to exhaust the energy of the weapon. It does not protect the airplane against the defense system as a whole, because it neutralizes only the weapon immediately at hand.

These generalities aside, tactical situations determine which specific capabilities are required. The three main current categories of combat platforms to be protected in battle are aircraft, ships, and submarines. It appears likely that in future these assets will be joined by tanks and other vehicles, because future autonomous sensors will be able to guide missiles (or "brilliant submunitions") into them.[4]

Aircraft Offboard Concepts. As indicated previously, offboard decoys for large strategic bombers are not a new idea. What is new is electronics compact enough for a decoy incorporating it to be carried by a tactical airplane, which must carry all of its ordnance (and countermeasures) externally. Since the object of the exercise is to saturate defenses, the decoy must be extremely small: otherwise it enforces too severe a limit on the amount of munitions the airplane can carry.[5]

The simplest airborne decoy is a corner reflector in a streamlined dart or glider; the corner reflector would sufficiently enhance the radar return from the decoy to deceive a surveillance radar. However, the efficacy of the reflector depends on its size. The next step, then, is an active repeater or blip-enhancer. Very compact electronics is quite helpful; the U.S. Navy, for example, has recently called for industry proposals for an offboard decoy small enough to drop out of a flare or chaff launcher.

Decoy tactics are defined by propulsion. The ideal offboard countermeasure simulates an airplane long enough to seduce an air defense system into firing at it. If it is small enough, the anti-aircraft missile may not be able to engage successfully, and a single decoy may absorb several scarce weapons. Sustained simulation requires continual maneuvers at airplane-like speeds, and that in turn requires both an organic powerplant and some form on onboard guidance. That in turn makes for a larger decoy, which cannot be carried in great numbers. Quail was self-propelled, but both Samson and

the new compact decoy appear to be gliders, relying on launch speed and momentum for realistic motion.[6]

Note that to some extent the need for decoy endurance is defined by the range of surveillance radars. Quail was designed to help a B-52 penetrate Soviet air defenses at high altitudes, where radars could observe the bomber for several hundred miles. They could then watch the bomber launch its decoy. Moreover, long-range air defense weapons could appear at any point over a very long distance, and it was therefore necessary for the decoy to be in the air for an extended period. Quail was credited with a range of about 250 miles and an endurance of about 30 minutes, during which time it roughly simulated the performance of a B-52. It was discarded in the 1970s, perhaps because it could no longer be expected to mimic a B-52 in the face of improved Soviet radars.

Armed Decoys. The armed decoy is an interesting variant of this theme. If weapons are cheap enough, then it is worthwhile to arm the decoy itself. In that case, even if the air defense system can distinguish it from the bomber, it cannot afford not to engage the decoy. The bomber still carries most of the tonnage, as well as the intelligence to decide which mobile or semi-mobile targets to hit. The U.S. Air Force first proposed a supersonic cruise armed decoy in the mid-1960s, and it sometimes appears that such stand-off missiles as the current SRAM (short-range attack missile) are thought of as semi-armed decoys (they have also been described as defense-suppression weapons).

There is also another important category, the decoy designed to force an enemy to reveal the location of his defenses and so to make them vulnerable to attack. A modern tactical air attack will include defense-suppression aircraft carrying anti-radiation missiles (ARMs). The usual countermeasure is to wait for the ARMs to be launched, then turn off the radars long enough to make them miss. A carefully coordinated air defense system may alternately turn different radars on and off, and so continue to function while frustrating attacks on any one radar. ARMs can be designed to incorporate memory devices to give them a fair chance of overcoming such tactics, but they are inherently limited.

There is an alternative: a cruising ARM that waits for the radars to be turned back on, then attacks. It is a kind of armed decoy, in that the defensive radar may be turned on to track the ARM. The Israelis reportedly used a simple version of this system in the Bekaa Valley of Lebanon: they simulated their aircraft by using small remotely piloted vehicles (RPVs). The presence of the RPVs in turn forced the Syrians to turn on their radars. Once the radars had been located, they could be destroyed by ARMs or even by command-guided rockets. In 1987 the United States revealed its own cruise ARM, the Northrop Tacit Rainbow, which apparently combines the decoy and attack roles.

A great deal clearly depends on the frequency of the radar being countered. Land-based air defenses currently rely very largely on centrimetric radars, and it is not difficult to package a centrimetric antenna (for a deception repeater) in a very small decoy. However, reports of the new stealth technology have encouraged re-examination of much longer-wave radars. Decoys, whether passive or active, would have to be much larger to interact effectively with such radars. The result would not necessarily be fatal. For example, one can imagine a small inflatable decoy that would expand to the appropriate dimensions.

Depleting Air Defenses. From the airplane's point of view, stealth would seem to entail a wide variety of offboard concepts. First, autonomous decoys with more easily detectable signatures could mask an attack by stealth bombers, drawing off relatively scarce defensive resources. Some modern equivalent of Quail should be effective because the signature of the stealth bomber itself is so small.[7] An offboard countermeasures vehicle equipped with a blip-enhancer would seem particularly well equipped to protect the stealth bomber itself by drawing defensive fire. By way of contrast, one might argue that virtually any active countermeasure aboard the bomber itself would carry a risk of revealing the bomber's presence to a sophisticated frequency-hopping radar or to a network of such radars. The decoy might also act as the transmitter of a bistatic ground-mapping radar. This would have two advantages. First, it would increase the relative signature of the decoy. Second, it would provide the bomber with a useful active radar at almost no cost in stealthiness. The only real cost would be the provision of a receiving antenna, which necessarily would be highly reflective.

One might then imagine a stealthy bomber releasing a cruise countermeasure, something like a cruise missile without a warhead, which would be expected to cruise with it in loose formation. The decoy could be controlled from the bomber by means of a very short range (extremely high frequency) radio or even by a fiber optic cable.

Threats to Ships. Surface warships employ (or have employed) a wide variety of offboard countermeasures, primarily against radar (as in bombers and missiles) and against sonar (as in torpedoes). Some missiles home on ship infrared emissions, and many warships now carry rocket flares intended to counter such weapons.

Large-scale fleet decoys lie at one end of the spectrum. In the mid-1950s the Soviets began to deploy air-launched anti-ship missiles, designed specifically to attack aircraft carriers. They reportedly believed that it would take several hits to disable a large armored warship, and accordingly concentrated their bombers into substantial formations. From the U.S. point of view, it was important to deny the Soviets information as to which ship in the formation was the carrier. To the extent that they could not be sure, they would have to press their attack relatively close to the fleet, and so become quite vulnerable to fleet air defense fighters. Although new missile systems might perhaps deal with the Soviet anti-ship missiles, it was clearly preferable to dispose of the bombers.

The offboard countermeasure was the chosen mechanism. fleet escorts were fitted with "blip enhancers." which were expected to make them appear (to Soviet bomber radars) to be carriers, at least at long range. From the point of view of the carrier, the destroyers themselves, then were offboard countermeasures. They might also be considered missile absorbers, in rough analogy to the air-launched decoy described above. To the extent that this role was understood, it was not entirely popular within the destroyer force, and reportedly some ships' companies tended to disable their blip enhancers. The mission was not altogether suicidal, in that the destroyers themselves had point-defense onboard countermeasures such as deception repeaters.[8]

In peacetime, the Soviets trail U.S. battle groups specifically to overcome this tactic. In theory, their "tattletale" should be able to signal back to base which ship in the formation is the carrier; it then turns and fires its own weapons, to add to the general confusion. The U.S. point of view would be that war is unlikely to break out instantly, and that any worthwhile U.S. commander should be able to evade his tattletale during a period of growing tension. Indeed, he might use simpler offboard countermeasures specifically for that purpose. Failing that, he can operate under the KATAH principle: Kill All Tattletales At H-Hour.

Size Draws Attention. The basic idea that a small ship can attract the attention intended for something larger remains valid. In 1971 the Indian Navy bombarded the Pakistani port of Rawalpindi with Styx anti-ship missiles. One Pakistani frigate captain reportedly saved his ship by passing behind a line of merchant ships, causing the incoming missile to shift to those larger and hence more attractive targets.[9] In 1982, during the Falklands war with Argentina, the British formation included pickets and, interposed between the pickets and the valuable carriers, fleet oilers. There were some suggestions at the time that the oilers were placed there to act as decoys, to absorb weapons fired at the carriers. There has also been some speculation that HMS Sheffield acted as an inadvertent decoy; the corner reflector formed by her uptakes and her upper deck produced a very bright radar echo, which the Argentine attacker may have mistaken for the much larger carrier.

More recently, it has been reported that tankers in the Persian Gulf have towed corner reflectors on rafts, in the hope that centroid-seeking missiles would pass between reflector and ship, or that radar-guided missiles coming from astern would home on the rafts. Presumably a more sophisticated towed decoy, incorporating a blip-enhancer, would return so large an echo that missiles approaching from any direction would tend to home on the raft.

Realism Expensive. In a larger sense, it is not difficult to imagine ships fitted out so that they (electronically) resemble carriers, with characteristic radar emissions and perhaps with sufficient blip enchancers and corner reflec-

tors to appear, at least to a non-imaging radar, actually to be large ships. For example, numerous large merchant ships are currently laid up, the market in merchant ships having collapsed so badly that the U.S. government has been able to buy up modern ships for its reserve fleet. It would not be difficult to fit such ships out as carrier decoys. The idea is not really new: in both world wars the Royal Navy built fleets of dummy capital ships specifically for strategic deception.

Such decoy fleets are necessarily relatively expensive. The fundamental requirements are that the decoy have sufficient radar reflectivity and that it be able to mimic fleet maneuvers, i.e., to maintain speed in fairly rough weather (in extremely bad weather sea clutter will reduce search radar effectiveness to the point that decoying may be unnecessary). In the early 1960s the advent of hydrofoils seemed to promise that a small inexpensive ship could indeed maintain her speed in rough weather. The U.S. Navy considered building a class of counter-measures hydrofoils, in effect offboard countermeasures or decoys to protect carrier battle groups (which were then called carrier task groups).

Seakeeping is the key issue here. A much simpler buoy with a radar reflector atop it can simulate a ship, particularly when all it must convince is a simple-minded missile seeker. However, the buoy is unlikely to move like a ship. Its useful life would, therefore, be limited. However, one might well imagine a fleet dispensing such buoys during or just before an air attack, in hopes of confusing incoming bombers. Buoys could, for example, be dropped by air so that they formed a large screen around the major units of the force, perhaps attracting incoming missiles. The Australians have developed a hovering rocket that comprises an electronic decoy package. It moves across the water at a low altitude, and at ship-like speeds, emitting signals which are modulated to simulate the rocking motion of a ship. This expendable decoy system is under evaluation, and may provide a viable defense against anti-ship missiles.

Chaff lies at the low end of the anti-radar spectrum. A ship under direct missile attack can fire clouds of chaff at long range to blanket itself. The attacking missile will generally fly through the cloud and try to reacquire the ship, but if the cloud is large enough and the ship fast and nimble enough, by that time it may be outside the area the missile will search. This is a non-decoy tactic. Closer in, the missile will have locked onto the ship. If she then fires chaff, her object will be either to seduce

One of the most effective shipborne defenses against cruise missiles with active radar terminal guidance is the creation of a false target with chaff.

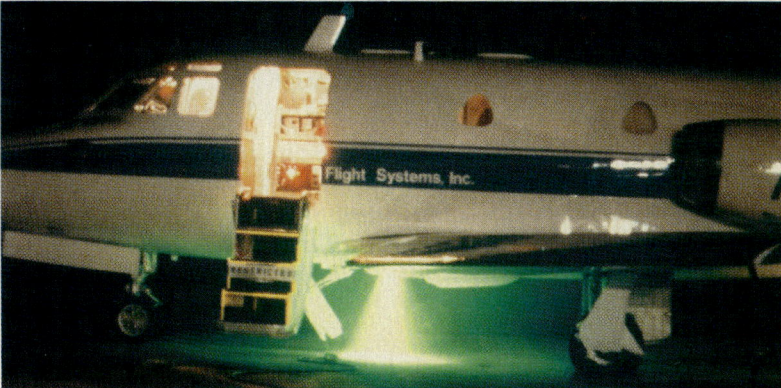

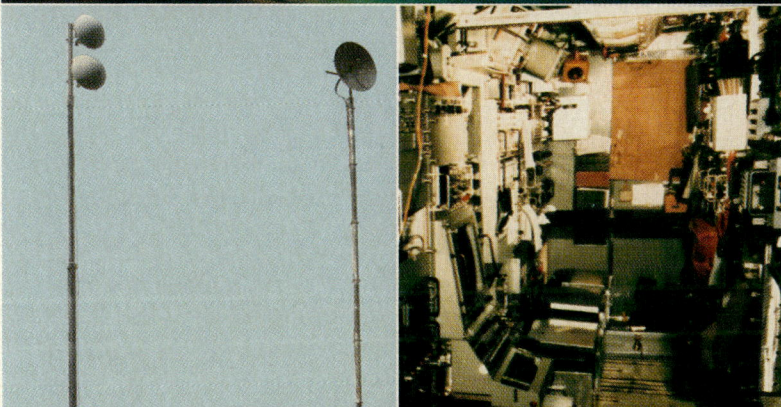

Total Systems From A Single Source.

At the Electronic Defense organization of GTE Government Systems, our mission is to provide total systems for EW, intelligence, electro-optics, and secure communications applications.

A leader in electronic defense for more than 30 years, GTE produces strategic and tactical reconnaissance systems, EW and ESM systems, electro-optical systems, and secure communications systems designed to meet a wide range of defense needs.

We provide comprehensive systems life-cycle support, including threat analysis, concepts, research and development, operational test and evaluation, production, staging, and field support.

That's total systems from GTE Government Systems. One good reason why we've delivered better airborne, shipboard, and ground systems since 1953.

Creative solutions from GTE Government Systems, 100 Ferguson Drive, P.O. Box 7188, Mountain View, CA 94039.

Consider the Source

GTE **Government Systems**

Noisemakers towed by ships can simulate the acoustic signature of the vessel and distract homing torpedoes. This is a critical facet of naval warfare today, since homing weapons are becoming more sophisticated. They may even possess re-attack logic that will reject the decoy after the initial pass.

the missile or to confuse it by enlarging the apparent target. For example, many missiles are centroid seekers. There are numerous recorded examples of such weapons flying between two equally attractive targets. The chief danger is that the chaff decoy generally cannot trigger the incoming missile. Most modern weapons have some type of reattack logic built into them, and they will seek a new target after having been cheated.

Ship Defenses Against Torpedoes. Surface ships often also have countermeasures designed to distract homing torpedoes. Most anti-ship torpedoes are passive homers, seeking the noise of propellers or machinery. Since World War II, ships have towed noisemakers intended to attract such weapons. If the torpedo homing acquisition range is substantially shorter than the distance from ship to decoy (i.e., offboard countermeasure), then the decoy will probably seduce the torpedo. The latter may well have reattack logic, and much will depend upon whether it has sufficient endurance left to find the ship itself.

This consideration makes it attractive to place the countermeasure far from the ship, and also in the direction from which the torpedo is likely to appear. In the late 1940s the U.S. Navy placed acoustic decoy rocket launchers aboard several of its anti-submarine ships. The hope was that the decoy would trap an incoming homing torpedo, or at the least that the submarine would be

unable to fire effectively from beyond surface ship sonar range. The system was discarded because it required warning of the approach of the torpedo, which often could not be obtained, or at least could not be obtained reliably enough. The launcher itself was unsafe, training automatically and creating a dangerous and unpredictable back-blast. However, the idea remains valid; the French Navy is now buying a rocket torpedo decoy, which can create not only noise but a cloud of bubbles which the torpedo may mistake for something solid.

Torpedo decoying has been considerably complicated by the appearance of an alternative form of homing, wake-following. Wake-following schemes actually date back to World War II, but they have become operational only during about the last five years. The torpedo detects the edge of the ship's wake. In the simplest version, it crosses the wake, detects the other edge, and then automatically turns each time it crosses an edge. If it is fired in approximately the right direction, it describes a ladder up the wake and ultimately hits the ship. A more sophisticated wake-homer will detect both edges of the wake, position itself approximately between them and run up the wake to hit the ship directly.

Original illustrations for this feature were created by Mr. Bradley Burns, who works in the art department of Watkins-Johnson Co., San Jose, Calif.

In either case, the torpedo is seeking a signature much more fundamental, hence much more difficult to mimic, than machinery noise. To the extent that it is detecting small bubbles in the wake, the new French rocket torpedo countermeasure might point the way to an appropriate decoy. However, a simple static rocket would produce only a fixed cloud, and the torpedo would seek a more realistic wake when it passed out of the cloud. Perhaps a large self-propelled bubble-maker would be effective—an offboard (and largely autonomous) countermeasure.[10]

Autonomous Decoys. It is also possible to envisage an autonomous surface ship acoustic decoy, which would appear (to a submarine) to be another surface ship. It could be self-propelled, and it would rely on what would amount to a tape recording of typical ship sounds. However silent, a submarine necessarily risks exposure when it fires a torpedo or missile. Its commander will try to limit such exposure. Knowing that a realistic decoy is present, he may be reluctant to risk an attack. In one American exercise some years ago, the known presence of just such a decoy tended to inhibit submarine commanders from firing simulated missiles. In one case, a commander tracked the real carrier but decided that he was actually tracking the decoy; he backed off. He might perhaps have resolved the ambiguity by using his periscope, but that would have entailed an additional risk.

Note that this situation, in which tactical decisions must be made on relatively flimsy sensor evidence, seems somewhat analogous to that of a Soviet central commander confronted with deceptive tactics. Like the submarine commander, he too risks the loss of forces if he attacks the wrong targets.

Submarine Protection. Submarines are a particularly interesting case. Alone of all major current platforms, they have become markedly more stealthy over time. The more stealthy the submarine, the easier it is to simulate, and hence the smaller (and, probably, cheaper) the simulator. Submarines generally use two kinds of offboard countermeasures. One is a full self-propelled simulator, intended to confuse another submarine's detection and tracking sonar. If such deception fails (or is not even attempted), one or more torpedoes will be fired at the submarine. It can respond with short-range non-self-propelled decoys to seduce away or confuse the weapons, and possibly also with some type of blanket (e.g., a bubble screen) to cover its high-speed (hence noisy) retreat.[11]

Note that a submarine (at least in the U.S. Navy) generally relies on passive sonar for detection and tracking, attempting to hear without being heard. An anti-submarine torpedo, on the other hand, almost always homes actively, Thus the echo-enhancer which might well seduce away a torpedo will have no effect on longer-range detection, and a submarine might conceivably use wire-guidance to overcome such countermeasures. Not, too, that it is often suggested that the Soviets, working

under the constraints of much less efficient electronics, would prefer to use active sonar, risking detection. In that case a self-propelled noisemaker would have little effect. it would seem to follow that a successful decoy has to combine echo-enhancement with noisemakers mimicking the sounds of a submarine, a difficult assignment even within the envelope of a full-size torpedo.

All of these schemes are played out against a major constraint: internal volume is extremely limited. As a result, no submarine in the world carries more than 30 torpedoes, and most carry many fewer. Just as in the case of the attack airplane, decoys may, therefore, occupy space which otherwise can go for weapons. There is an additional complication: the submarine has only a limited number of torpedo tubes, and reloading can be relatively slow. To the extent that it takes a torpedo tube to launch a sophisticated decoy, then, that tube is not generally available for a weapon.

The United States currently uses a small-diameter self-propelled decoy, but launches it from a torpedo tube. Thus it does not take up much torpedo space, but it can cut the number of available tubes, especially if the submarine must keep "snap-shot" weapons like the new Sea Lance constantly available in other tubes. The U.S. Navy also uses torpedo-sized submarine simulators as training aids, and at least one former submarine commander suggested that they might represent the minimum required of a really effective self-propelled decoy. One might speculate that the miniaturization represented by the Moss was a consequence of Western electronic sophistication, and that the Soviets might feel compelled to use something considerably larger, with consequences for the total number of weapons they can carry.

Sophistication vs. Brilliant Weapons. In recent years enormous attention has been paid to "brilliant" weapons capable of overcoming conventional (usually onboard) countermeasures and even of selecting appropriate aim points on a complex target such as a ship or airplane. The potential of offboard countermeasures, decoys, and blankets such as smoke and chaff suggests that future reality is likely to be substantially more complex, that even the smartest of weapons may be seduced into attacking the wrong target. The same electronic sophistication which promises "brilliance" may also promise realistic offboard countermeasures which greatly devalue that "brilliance."

This possibility tends to be neglected because, unlike onboard countermeasures, offboard devices derive their effectiveness from a combination of technical sophistication and tactics. Tactical ingenuity may account for most of the value of the decoy, yet it is extremely difficult to fit into the usual models of electronic warfare effectiveness. It may follow that forcing us to confront the importance of tactical ingenuity in warfare may be the greatest contribution of the rise of offboard countermeasures. ∎

Footnotes:
1. Reportedly this threat virtually killed off the North American Air Defense Command. Ironically, it was entirely notional; The

Soviets never had the requisite decoy systems, nor does it appear that they ever contemplated such tactics. Proponents of ballistic missile defense have sometimes compared the decoy threats arrayed against them with the notional threat of the '50s.

2. The Soviets are well aware of the dangers of lack of individual initiative, and frequently write about them. However, their centralized tactics seems to be inherent in their entire society, which combines centralized decision-making with severe penalties for individual error. At least in civilian life, it is much easier for the average individual to go "by the book;" as long as he does so, failure carries few consequences. To the extent that Soviet tactical practice mirrors so disparate a feature of the country as Soviet economic practice, it is difficult to see how anything can change. In this context, Mikhail Gorbachev's calls for individual initiative in the economy might be seen as the (probably futile) equivalent to the many calls for initiative in the Soviet Army, which one can find in the Soviet military press. This issue is important because it defines the future value of many possible offboard countermeasures.

3. These tactics open the real possibility of red-on-red attacks, i.e., catastrophic IFF failures. In the past, the Soviets have often been able to assume that they could enjoy inherent IFF, e.g. using submarines against surface ships. In such engagements, mistakes would hardly be dangerous, since no Soviet surface ships would be present. However, the Soviets are now confronted with situations in which IFF is hardly inherent. Examples include submarine-vs.-submarine in the "bastions" and surface-vs.-surface ships in areas far from the Soviet Union. Alternatively, one might say that the Soviets try to avoid red-on-red engagements by deciding in advance just where their units will be. They should, then, be vulnerble to the "friction" of warfare, quite aside from the issue of decoying.

4. This possibility would seem to limit claims that brilliant munitions will permanently change warfare, by guiding relatively small, inexpensive warheads into expensive armored vehicles. If the brilliant submunitions turn out, unsurprisingly, not to be all that inexpensive, and offboard decoys appear, the economics of such systems may prove far less attractive. Much would depend on timing: before the appearance of the decoys, the submunitions will perform extraordinarily well. If they languish for some years before being used in combat, they will probably encounter deception and decoying in service. One might imagine a future stealthy tank equipped with rocket-fired decoys.

5. Note that all countermeasures to air defenses carry a cost in explosives delivered on target. Because total weapons weight and volume are limited, "smart" stand-off munitions cannot carry the same explosive payload as "iron" bombs. Stealth demands that weapons be carried internally, and internal space is much more limited than that which is available on external pylons.

6. Chaff might be considered the first, most primitive, offboard aircraft passive countermeasure. It could not simulate aircraft motion at all; the usual counter-countermeasure was a moving target indicator that filtered out the slow-moving chaff cloud.

7. There is an important caveat. If it turns out that the stealth airplane has a very un-stealthy signature at longer wavelengths, then the decoy may be ineffective, since, because it is physically quite small, it may have a relatively small signature at such wavelengths.

8. The idea was not altogether novel. Before World War II, the Germans deliberately designed their cruisers and battleships to resemble each other. As a result, when confronted with a German cruiser accompanying a German battleship in the Denmark Strait in May, 1941, the British battlecruiser Hood mistakenly engaged the cruiser—leaving the battleship, the Bismarck, free to sink her.

9. Later it was suggested that Pakistan planned to place corner reflectors aboard all fishing vessels, partly in the expectation that the proliferation of what amounted to decoys would frustrate future Indian missile attacks. The reflectors were described as a means of improving the radar visibility of the boats for easier seach and rescue—which was of course a valid purpose.

10. The problem is clearly a serious one; it accounts for the publicity accorded the recent Soviet 65-centimeter wake-homing torpedo. Previous Soviet torpedoes had been virtually excluded from the Western unclassified military press. In the absence of an effect decoy, the only likely countermeasure would be the physical destruction of the torpedo. Past efforts in this direction, which date back to World War II, have often been disappointing because it is so difficult to detect, classify, and localize an incoming torpedo.

11. A submarine might combine offboard countermeasures with a shot in the direction of the attacker (assuming that it can detect the incoming torpedo in time). In that case, the attacker would be busy evading the counter-shot, and a combination of decoys and a bubble screen might make it difficult for the attacker to re-acquire its target.

Norman Friedman is a noted author and defense consultant. He is a frequent contributor to The U.S. Naval Institute *Proceedings*, and frequently accomplishes studies for the secretary of the Navy, the chief of naval operations, other Navy staffs and defense contractors. He is the author of *U.S. Battleships: An Illustrated Design History*, and other books about warships and weapon systems. He is an expert on Soviet military systems, tactics and strategy.

AIRBORNE COUNTERMEASURES

THE *ONLY* SMART OFFBOARD SYSTEMS WITH OPERATIONAL HERITAGE

Tracor Aerospace

Tracor Aerospace, Inc.
6500 Tracor Lane
Austin, Texas 78725-2070
Telephone 512: 929 2418
Telex: 77 6410
TWX: 910 874 1372

FIRST IN COUNTERMEASURES TECHNOLOGY

Artificial Intelligence for Object Recognition

*A model-based AI approach to recognizing military objects offers
a number of advantages over conventional techniques.*

By Dr. David L. Milgram and Thomas W. Miltonberger

Recognizing objects of military value plays a vital role in automatic target recognition (ATR), situation assessment and intelligence. Because of the volumes of data involved, and the need for a timely response and autonomous operation, the object recognition problem needs to be automated.

A model-based artificial intelligence (AI) approach to object recognition has been under development over the last 10 years by AI vision researchers and has recently been making its way toward military applications. This approach maintains a consistent interpretation of the data in terms of the underlying physics and models of the target, sensor and environment. This approach promises many advantages over conventional statistical techniques, including:

- Increased robustness
- Increased computational efficiency
- Elimination of training sets.

Assumptions in the problem domain often allow for a closed world—that is, a limited number of known objects. Sometimes modifications to the nominal model and/or the possibility of an unknown classification are also required.

Conventional statistical template approaches—that is, pattern recognition—use simplifying assumptions to reduce object recognition to a statistical decision based on a single multi-variant distribution. However, these assumptions fail to hold in real-world situations, thus invalidating the classification process.

The model-based AI technique maintains a consistent interpretation of the data in terms of the underlying physics and models. The validity of any assumptions is guaranteed by reasoning about the explicit uncertainties and errors introduced into the system.

This article not only presents the AI approach to object classification, but alleviates some confusion that has resulted from use of the term *model-based*. In recent years conventional statistical decision techniques, often called *model-based* (which is technically correct since the distribution, is in fact, a model), commonly have been recast as model-based AI techniques, when in fact they are not. We hope to differentiate these two approaches by illustrating the content and use of the models as well as the reasoning processes that are used in both approaches.

Comparison of Techniques. Almost all early attempts at object recognition from imagery and signature data

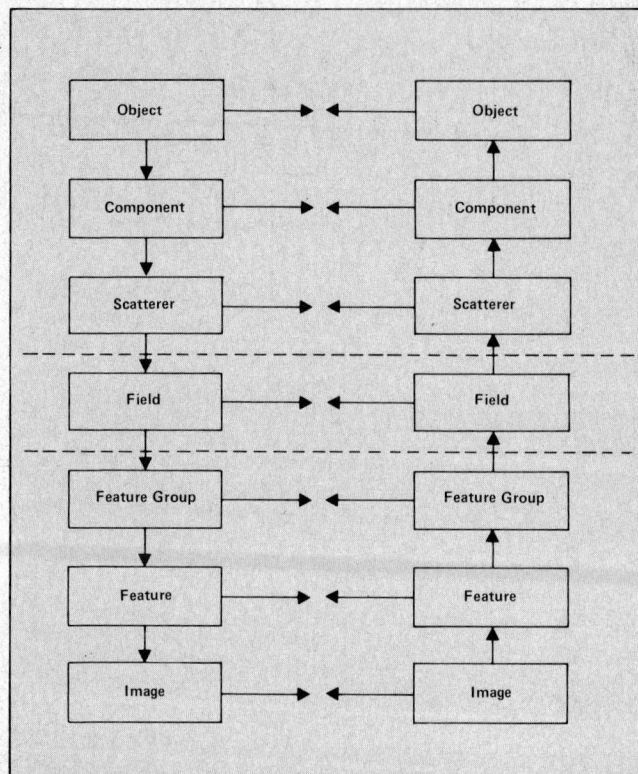

Figure 1. The logical architecture of an AI model-based vision system uses top-down, bottom-up and matching processing.

used some form of statistical template matching technique. This technique has its roots in classical detection and estimation theory as an extension of matched filtering. The technique, as applied to object recognition, consists of storing a set of known templates of each possible target for many different orientations. The data to be classified are compared against each stored template and a measure of similarity is computed. Classification is based on these distance measures. Typically, the stored templates are derived from controlled measurement programs, or possibly from standard prediction codes and models of the target. Both of these cases represent a difficult data acquisition problem: for training sets, it is difficult to get adequate coverage of a foreign target; for prediction models, it is difficult to adequately model the targets, sensor and related physics.

The variant to this approach most common today is to use a feature vector derived from the data. Instead of

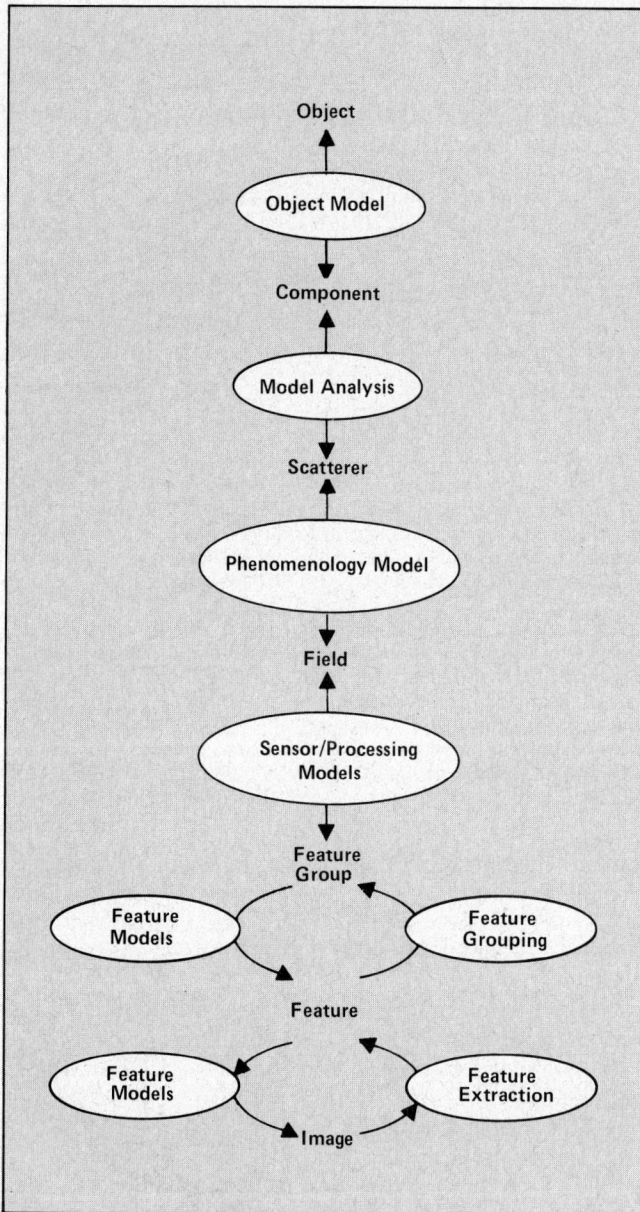

Figure 2. A set of well-defined operations manipulate the various hypothesis structures in object recognition. The operations illustrated represent the single-level transformations.

comparing raw data, a set of predefined statistical measures are calculated from the data. The stored templates also consist of feature vectors as opposed to complete data realizations. The prime advantages of using feature vectors are the reduction in required storage and the speed of the classification process. But these and other similar approaches suffer many of the same pitfalls as the template approach.

The statistical approach to object recognition fails in many real-world applications. This failure is due primarily to inconsistencies between underlying assumptions and real situations. As mentioned above, the template

approach is really an extension to the standard matched filter. The feature vector approaches are really just an approximation of a method of equivalent statistics. These approaches are optimal only when certain limiting assumptions are true—for example, when:

• The mean of the expected signal/image for each possible hypothesis is known.
• The corrupting noise is additive.
• The corrupting noise is white gaussian.

Clearly, these assumptions will almost always be violated. To begin with, the noise distribution in the real world is almost never additive white gaussian noise. Nonetheless, the effects of non-white, non-gaussian noise can be compensated for by filtering and non-linear quantization, assuming that the true statistics are known or measurable.

The biggest pitfall comes from the first assumption. In general, the object can be in any orientation relative to the sensor; therefore, a method that considers only a finite set of discrete orientations can present problems. This objection is usually countered by a sampling and continuity argument that the signal/image does not vary appreciably between the stored samples. This situation sets up a tradeoff between the number of stored templates—which can quickly become overwhelming—and the performance of the system. However, for sensors such as radar, that are highly aspect-dependent, it is questionable whether this argument holds at all.

Another objection is the variability of the actual targets. For example, military objects, such as a tank of a particular type, tend to vary significantly from one to another as time, modifications and wear and tear take their toll. This unmodeled variability makes it impossible to accurately predict the expected signal. Thus, the computed distance between the real data and the stored representation is biased and forms a poor error metric for real-world events. This presents a major source of error,

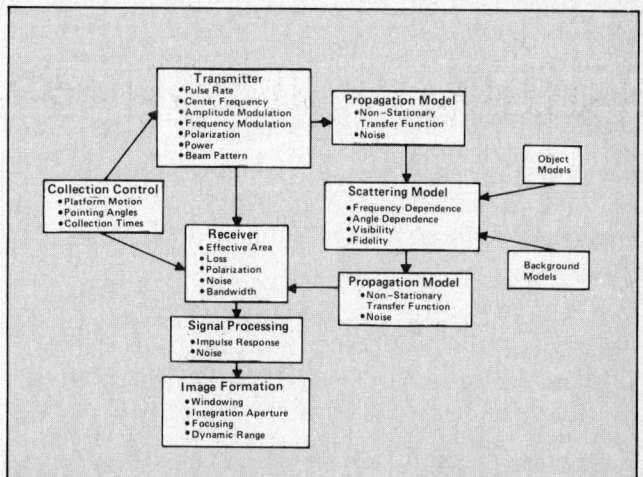

Figure 3. Many operations employ explicit models of the target, environment, phenomenology, sensor and data processing. The various models of the system are related as shown above.

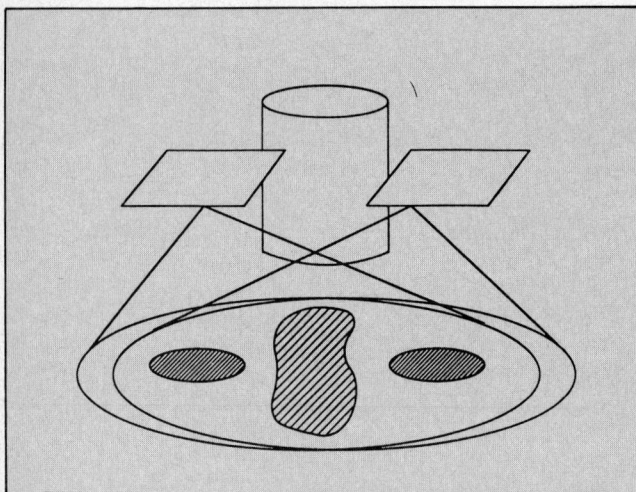

Figure 4. As an example of the use of backconstraints, consider the satellite hypothesis and data depicted here. The hypothesis has uncertainty associated with the translational position of the target in the image. This uncertainty is illustrated by the two ellipses, which indicate the possible locations of the two panels. Assume that the model-based prediction process has generated a prediction of a bright region for each of the panels. The data includes two regions that satisfy the predicted constraints on size, shape, and intensity of the regions.

Figure 5. Three geometrically consistent hypotheses can be generated by matching predicted features with the extracted features. Note that in two of the cases, one panel is left unmatched. This unmatched model component lowers the likelihood of those hypotheses. Also, the uncertainties of translational position have been reduced for each of the hypotheses because the matching process produced a back-constraint between the location of the regions in the image and the translational position parameters of the hypotheses. Because of this refinement, there are extracted features that cannot be explained by two of the hypotheses, even when the uncertainties in the hypotheses are accounted for. This would further decrease the likelihood of those hypotheses. The correct hypothesis has increased confidence due to the consistent matches between predicted and extracted features.

especially as the number of possible target models increases and the signal space distances become smaller.

Of course, criticism serves little purpose unless a viable alternative is presented. Here we are presenting the model-based AI approach to object recognition. This approach has been studied in AI vision research[1, 2, 3] recently and is now moving toward military applications[4, 5, 6].

In model-based AI object recognition, models of the sensor, environment, and 3-D objects are used for prediction and matching to the data. Symbolic, semantically-meaningful features replace the exact representations of the data or statistical features used in conventional approaches. Another key difference from conventional approaches is in the way models are represented and used. This approach recognizes the inherent errors introduced in the modeling process and models these errors as uncertainties in the models. Furthermore, the AI approach combines and propagates the uncertainties of the models, orientation and prediction methods throughout the classification process. This allows the system to use only highly reliable information. In doing so, the classification process becomes more robust to sources of error.

Application Areas. AI model-based approaches have been used in research vision systems for some years now. The first widely known system has been ACRONYM[1] which was developed at Stanford. Newer systems have since been developed by other research center[2, 3]. However, these systems were designed for vision research and are neither robust nor efficient enough to make a direct transition into military applications. Another drawback has been the primary concentration of academic researchers

on fairly high resolution visible imagery, while the military is often interested in lower-resolution, non-visible imagery (IR, synthetic aperture radar (SAR), and inverse synthetic aperture radar (ISAR)) and signals (range-only radar, IR signatures, etc.).

Military applications that are currently being worked with AI model-based approaches include:
• Anomaly detection of satellites from an ISAR image sequence
• Ship classification from ISAR imagery
• Terrain vehicle classification from SAR imagery.
While all of these applications use some form of radar imagery, other sensor modalities and data types should lend themselves to the same basic approach.

The anomaly detection problem lends itself quite well to the model-based techniques described in this paper. The goal of the system[5] is to alleviate the workload of the analyst by reducing the amount of data that must be reviewed. A sequence of ISAR images is supplied as input. In this scenario, information is available as to the object type and motion. It is the system's job to screen the data to determine whether the a priori expectation is correct. If it is, nothing needs to be done. If an anomaly is detected, the system will alert the analyst so that a more detailed analysis can be performed. The system reasons about matches between data features and individual object components—for example, the body or panels—and will provide the analyst with information concerning the components that were not adequately confirmed.

Because the system is doing only anomaly detection and not object classification, the hypothesis generation step (which will be described in the next section) is not required. When an entire sequence of images is available, the prediction process is free to select which images to use for each component. Thus, the system is opportunistic in that it selects the best data for each step of the conformation process.

The ship classification system is currently envisioned as functioning with a human. Here, the roles of the human and automated system are somewhat reversed from those in the anomaly detection paradigm. In this scenario, the user selects a good quality image from a continuous sequence and sends it off to the classification system. Context and other sensor information generally limit the number of possible ship types to a manageable 30 or 50 models. The system then classifies the ship in the image from the possible ship models. It is interesting to compare the method of modeling the targets that were used in the satellite and ship systems. The satellites are fairly small and relatively uncomplicated. They can be modeled down to the individual component level. Ships, on the other hand, are much larger and enormously more complicated. They cannot be modeled accurately to any fine degree of detail. The ships are modeled at a coarser level of detail and finely detailed structure is represented by aggregate properties attached to the coarse geometric representation.

Similar model-based object recognition work has been undertaken to classify terrain vehicles in SAR imagery. One such effort[6] has also included the exploitation of information on military doctrine and terrain. The terrain vehicle problem is different from the other two applications in that clutter and terrain play a much larger role. Terrain vehicles also tend to accumulate the effects of abuse and, as a result, are impossible to model exactly. Again, aggregate approaches to modeling seem in order.

Model-Based AI Object Recognition. The model-based AI approach to object recognition attempts to find causal relationships between the 3-D object models and the observed data. It does this using models of the target, environment, sensor and physics. The logical architecture of an AI model-based vision system is shown in Figure 1. This system uses three kinds of processes to infer the type, position, and orientation of an object in an image:

• Top-Down Processing: Shown on the left side of the diagram. This class of processes maps from hypothesis at one level of abstraction to predictions about properties of those hypotheses at a lower level.

• Bottom-Up Processing: Shown on the right side of the diagram. This class of processes maps from observations at one level of abstraction to hypotheses about properties of those observations at a higher level.

• Matching: Processing activity moves between the bottom-up and top-down columns in the diagram by matching prediction hypotheses with observation hypotheses.

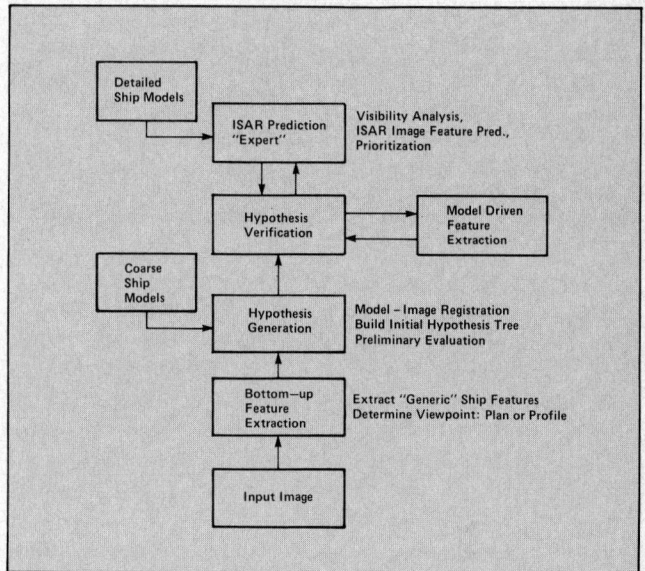

Figure 6. At the heart of the ship image classification system architecture are two large processes: hypothesis generation and hypothesis verification.

As shown, the actual data—either signals or imagery, or in the case of SAR and ISAR, imagery derived from raw radar signals—occupies the lowest level and object models occupy the highest level. One can actually extend this hierarchy to include global situation assessment and force structures above the object level[6], but in this paper we are concerned with recognition only up to the object level.

Notice that processing also occurs at many intermediate levels. While Figure 1 has been labeled for SAR/ISAR data, the same general structure and levels apply for other sensor modalities. Note that there are three general domains for the hypotheses: the data domain, the (EM) field domain, and the target or world domain. These are called levels of abstraction.

The system recognizes an object by generating, verifying and reasoning about hypotheses at each of these levels. This is accomplished with a set of well defined operations that manipulate the various hypothesis structures. Some of these operations are illustrated in Figure 2. These operations represent the single level of transformations. Transformations that skip levels are also possible, as will be shown in the next section. Note that many of these operations use explicit models of the target, environment, phenomenology, sensor, and data processing.

The first step in the exploitation process is hypothesis generation. This usually begins with extracting significant features from the data. Significance is determined by the uniqueness of the feature, how well the feature can be matched to some element of the target models, and the amount of information that the feature represents. These features are used to generate hypotheses by matching them against the target models. A hypothesis includes the related model and constraints on the orientation of the model. The system dynamically generates the hypotheses it

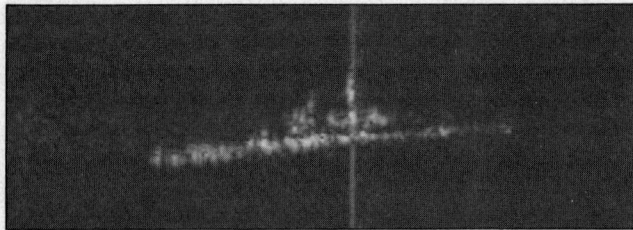

A: *Belknap*-class ISAR "profile" image.

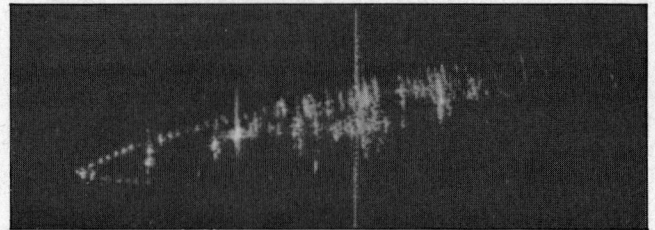

B: *Bainbridge*-class "plan" ISAR image.

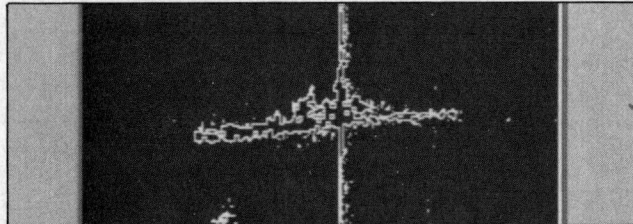

C: Extracted radar antenna streak.

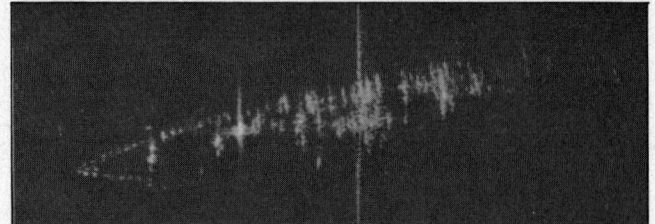

D: Extracted bow.

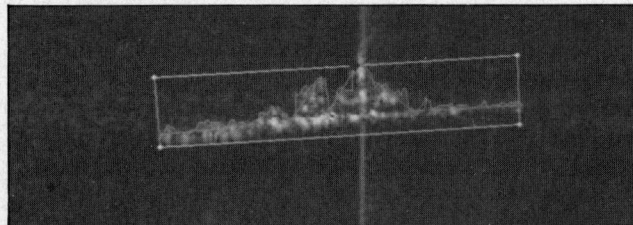

E: Extracted superstructure peaks and endpoints.

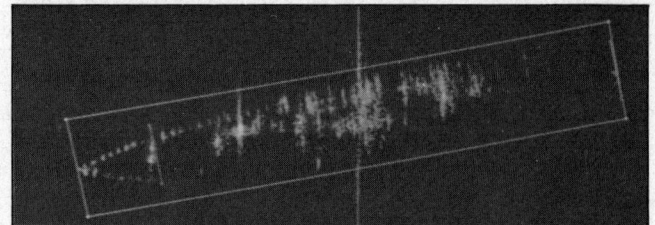

F: Extracted stern locations (2).

Figure 7. Image features extracted from sensor information.

will consider based upon features in the data. Context and prior beliefs can aid in the hypothesis generation process by constraining the possible hypothesis considered.

The next step, called hypothesis verification, uses the object, sensor and environment to predict features that should be present in the data if a given hypothesis is true. A prediction is a symbolic, composite description of the data, not an exact prediction of the data values. This prediction process includes explicit consideration of the uncertainties for the given hypothesis and models.

Prediction uses both analytic and heuristic methods. If analytic methods are used, heuristics are usually required in order to check the validity and ramifications of underlying assumptions and conditions. In general, either form of prediction can be implemented as a rule-based expert system. Essentially, the prediction function is given a hypothesis at one level—for example, a target component and range of orientations—and is tasked with predicting hypotheses, such as bright spots, at a lower level.

There are two kinds of uncertainly that are part of the prediction process:

1. Uncertainty parameters of the predicted feature caused by uncertainty in the initial hypothesis and from modeled error sources.

2. The likelihood of the prediction itself, which is caused by an underlying uncertainty in the prediction process—for example, the phenomenology model.

Both of these uncertainties are computed with each prediction. Because possible errors and uncertainties are

explicitly considered, and because aggregated instead of exact predictions are generated, this method of prediction is much more robust than conventional prediction techniques.

Once a prediction has been generated, it can be used to guide further processing. By using a predicted data feature to guide feature extraction, the feature extraction algorithms can be automatically fine-tuned. This limits the amount of data that needs to be processed to a small region around the location of the predicted feature.

Given a predicted and extracted feature pair, a match can be generated. A match serves to relate evidence in the data to a hypothesis. Typically, there may be several extracted features that satisfy the predicted constraints. Each of these alternative matches represents a refinement of the initial hypothesis. To resolve the ambiguity among alternative hypotheses, the matching process also generates *backconstraints* for each match. A backconstraint is a relationship between the parameters of an extracted feature and the parameters of the hypothesis and associated models. These are used to eliminate uncertainties in the new hypothesis and to check the consistency of each hypothesis.

In addition to the geometric backconstraints illustrated in Figures 4 and 5, other types of backconstraints can be generated in the interpretation process. For example, a symbolic backconstraint can be generated that relates the amplitude of an IR signature to the temperature and emissivity of a surface. Additional backconstraints of this

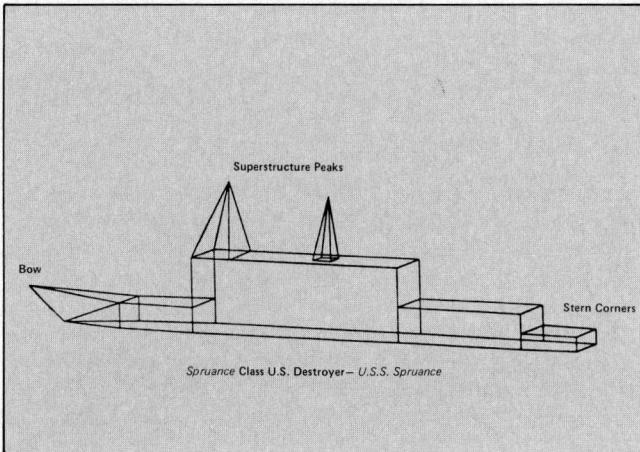

Figure 8. Coarse ship models are represented as a set of 3-D polyhedrons and a set of 3-D registration points.

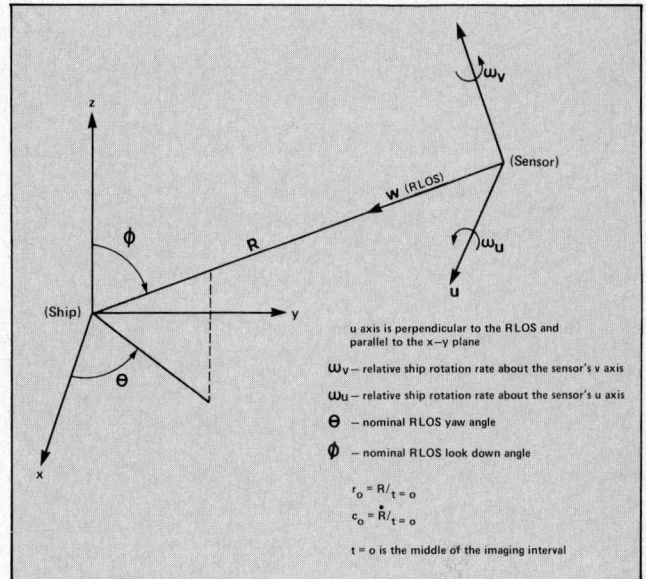

u axis is perpendicular to the RLOS and parallel to the x–y plane

ω_v — relative ship rotation rate about the sensor's v axis

ω_u — relative ship rotation rate about the sensor's u axis

Θ — nominal RLOS yaw angle

ϕ — nominal RLOS look down angle

$r_o = R/_{t=o}$

$c_o = \dot{R}/_{t=o}$

t = o is the middle of the imaging interval

Figure 9. The position and motion of the ship are estimated in the sensor and ship coordinate system defined here.

nature could then be used to solve for the temperature distribution and surface characteristics of the individual components of a target. Constraints over time can be used to relate parameters of the target over the duration of a single collection and even between collections.

This technique of using the most dominant feature first to generate possible hypotheses and then to predict finer features to refine and discriminate the hypothesis is called coarse-to-fine processing. It is very important for limiting the search space while still extracting very fine detail. Compare this approach to the pattern recognition approach which requires a trade-off between the number of complete hypotheses tested and the fidelity of the predictions.

An Example System. This portion of the article is a description of a prototype ISAR ship image classification system developed at ADS. This system automatically classifies an ISAR ship image in terms of a database of three-dimensional ship models. The ship models are based on readily available sources: Jane's[7] and Weyers

Flottenbuch[8]. The system was tested with ISAR images collected at the NOSC radar in San Diego.

Figure 6 is a diagram that shows the overall architecture of the ship classification system. The processing portion of the ship classification system can be viewed as two large processes: hypothesis generation and hypothesis verification. Hypothesis generation matches the output of the bottom-up feature extraction module with a set of three-dimensional point ship models and generates an ordered set of tentative model-to-image matches called hypotheses. Hypothesis verification gathers additional evidence for each hypothesis by using the ISAR image feature prediction module to predict additional image features. Then it uses the model-driven feature extraction module to find instances of the predicted features in the image. The hypothesis verification module reorders the set of hypotheses provided by the hypothesis generation module according to new evidence extracted from the image.

The hypothesis generation portion of the systems uses coarse 3-D ship models and the verification portion of the system uses relatively detailed shape and surface models. Hypothesis generation will create many hypotheses—typically about 10 or 15 for each ship model—and quickly evaluate each one in terms of the coarse model. Hypothesis verification will process a small subset of the best hypotheses in terms of the detailed models and evaluate them relatively slowly.

The following sections trace the ship classification system in the order implied by the system architecture diagram, Figure 6:

Narrowing the Search Space. A realistic ship classification system will need a data base of about 750 military ship models and at least twice as many commercial ship

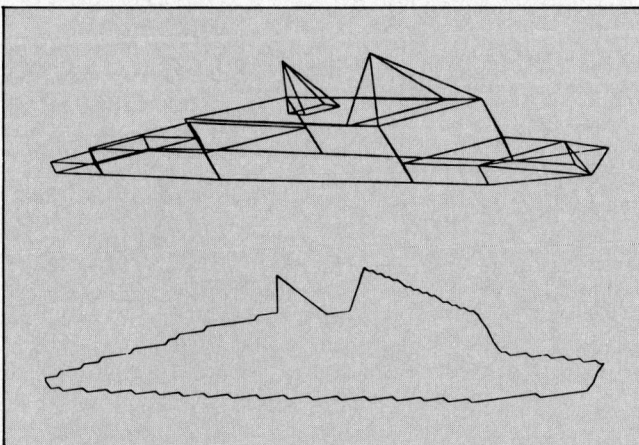

Figure 10. The hypothesis evaluation is the sum of the distances from each extracted feature to the projection of a compatible registration point or to the extremal boundary of the ship projection.

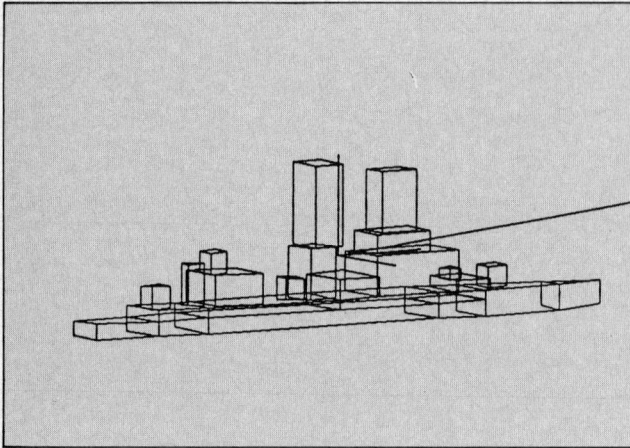

Figure 11. Prediction of image features is done using more detailed models of the ships than were used in hypothesis generation. Instead of modeling the fine detail in a purely geometric representation, the aggregate backscatter properties of individual model components are modeled.

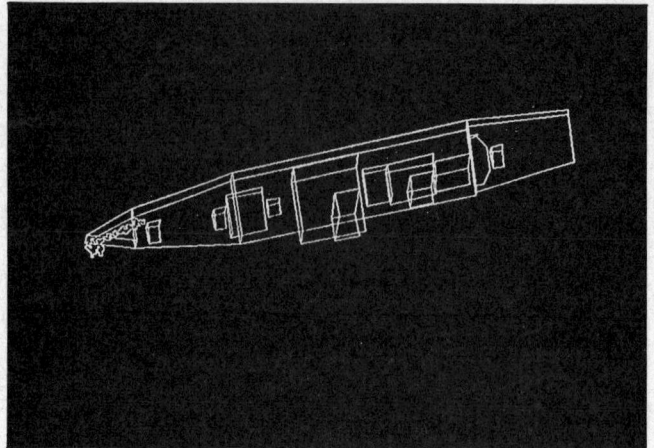

Figure 12. Here a model-based predicted ship feature is matched to a corresponding image extracted from sensors.

models. This number is narrowed down by considering the geographic location where the image was collected and reports from other intelligence sources.

If the ship is being tracked, the system may be able to rule out various ship classes based on estimates of the ship's velocity and maneuverability. An estimate of the ship's heading at the time the image was collected can be used to distinguish the bow from the stern, thus reducing the number of match hypotheses that must be considered. A lower bound on the ship's length can be measured directly along the range axis of the ISAR image. Models that represent shorter ships can be discarded.

Feature Extraction. The bottom-up component of the classification system builds a symbolic representation of the ISAR image in terms of a simple generic ship model. This implicit generic ship model includes only large, easily recognized ship features that are common to all of the ships in the model database and can be recognized for large ranges of sensor viewpoint and ship motion parameters.

The following table lists the generic ship model components and the corresponding image features that are extracted to build the symbolic representation of the image. The feature extraction algorithms are applied to the set of connected image regions that contain pixels whose value is greater than a threshold. The threshold is computed by sampling portions of the image that are likely to correspond to the open ocean (usually the extreme perimeter of the image) and the ship itself (usually the center portion of the image).

- **Endpoints**

If the entire ship is visible, then the endpoints of the ship usually correspond to the extrema of the high intensity portion of the image. If the ship is being imaged nearly stern-on, then the tip of the bow may be obscured. In most cases a nearly bow-on view will not obscure all of the stern.

- **Bow and Stern**

In a plan view, it is often possible to distinguish the bow endpoint from the stern endpoint. If the ship was imaged nearly stern-on, then both corners of the stern may be visible and the extreme tip of the bow may be difficult to detect. If the ship was imaged nearly bow on, then the taper of the ship near the bow makes it distinct and relatively easy to extract.

- **Superstructure Peaks**

In a predominantly profile view, the superstructure peaks appear as extrema in the cross-range dimension of the ISAR image. It is often difficult to distinguish the points we have labeled superstructure peaks from other extrema in a complex superstructure. The classification algorithm is designed so that peaks that match the profile, but do not correspond to the ship model locations we have labeled as superstructures, will not affect the coarse classification evaluation.

- **Rotating Radar Antenna**

A rotating radar antenna will often appear as a distinct, high-intensity streak. Streaks are detected by finding peaks in the projection of the connected components onto the range (horizontal) axis of the image. The boundaries between the streak and the ship are located with a horizontal edge detector.

Hypothesis Formation. The hypothesis formation component forms hypothetical matches between the symbolic representation of the image and a set of coarse ship models. Each hypothetical match or "hypothesis" specifies the parameters of the projection of a 3-D ship model to 2-D ISAR image coordinates. Each hypothesis is evaluated in terms of how well all of the extracted image features match the projection of the coarse ship model.

Coarse ship models are represented as a set of 3-D polyhedrons and a set of 3-D registration points. The polyhedrons are a rough approximation to the shape of the surface of the ship and the points specify locations on the surface that will be used for model-to-image registration. The registration points correspond to the components of the symbolic image representation already described.

ISAR image formation geometry can be viewed as a transformation that maps ISAR returns reflected by visible points on the surface of the ship to (range, cross-range) image locations. This geometrical transformation can be described by a set of equations that depend on six registration parameters.

The position and motion of the ship are estimated in the sensor coordinate system defined in Figure 9. The mapping from a three-dimensional model point, (x,y,z), to a two-dimensional ISAR image location (r,c) is given by the following registration equations:

(1) $\quad c = (-s_\theta y - c_\theta x)w_v + (-c_\phi s_\theta x + c_\phi c_\theta y + s_\phi z)w_u + c_0$

(2) $\quad r = -s_\phi s_\theta x + s_\phi c_\theta y - c_\phi z + r_0$

The ISAR sensor measures range and range rate—that is, motion with respect to the radar line of sight (RLOS). The registration parameters represent the six degrees of freedom: w_u and w_v are the relative rotation rates of the ship with respect to the sensor coordinates axes u and v; s_ϕ, c_ϕ and s_θ, c_θ represent the sines and cosines of the direction of the RLOS relative to the ship. The terms r_0 and c_0 are the nominal (middle of the imaging interval) range and range rate of the ship expressed in image coordinates.

We can solve for one set of registration parameters for each set of three associated model registration points and image features. The hypothesis formation algorithm creates a new hypothesis for each reasonable match between a set of three image features and model registration points. Each hypothesis contains an association table and the corresponding registration parameters. These registration parameters represent the geometric backconstraints that relate a model and the image according to the transformation shown in equation 1 and 2.

Each hypothesis is evaluated in terms of a measurement that captures how well all the extracted features fit the projection of the ship model specified by registration parameters. This evaluation is the sum of the distances from each extracted feature to the projection of a compatible registration point or to the extremal boundary of the ship projection.

Hypothesis Verification. Hypothesis verification attempts to separate the correct hypothesis from the others by gathering additional evidence for or against the hypotheses. The system gathers additional evidence for a hypothesis by predicting ISAR image features that are consistent with that hypothesis and then extracting those features from the image. Predicted image features are consistent with a hypothesis if they satisfy the backconstraints that are known for that hypothesis.

Prediction is done using more detailed models of the ships than were used in hypothesis generation. An example geometric representation is shown in Figure 11. Note that the actual geometry is still quite coarse. Instead of modeling the fine detail in a purely geometric representation, the aggregate backscatter properties of individual model components are modeled. Each facet on the model is tagged with properties that reflect the underlying backscatter of the surface. For example, a facet on the hull will be labeled as "smooth," "dense" and "gently curved" while a facet representing a scaffolded superstructure would be labeled as "complex" and "sparse." Individual junctions (outside seams formed by two facets) and dihedrals (inside seams formed by two facets), trihedrals and volumes can also be labeled with backscatter properties. Note that the actual geometric model is used primarily to provide position, orientation and crude visibility information. This differs from conventional techniques that require all target information to be represented geometrically. By modeling at an aggregate level, sensitivity to the exact geometry is lessened.

The system generates predictions using a rule-based sensor model. The prediction process generates a symbolic feature that describes a small portion of the image. The feature extraction module is then tasked with finding a feature that matches the prediction. A matched feature will increase the likelihood of the hypothesis while a feature that fails to match will decrease the likelihood of the hypothesis. An example of a predicted feature matched to a corresponding extracted feature is shown in Figure 12. If there are multiple extracted features that can match the predicted feature, the hypothesis can be split, one for each match. The resulting child hypotheses are refined using new estimates of the registration parameters dependent on the matched feature.

Processing continues as more and more evidence is extracted from the data. Incorrect hypotheses are pruned away as their predictions continue to go unmatched and the correct hypothesis rises to the top. This iterative approach is opportunistic because only the minimum processing required for classification is performed. ∎

Footnotes

1. R.A. Brooks, "Model-Based Computer Vision," UMI Research Press, Ann Arbor, Mich., 1981.
2. D.G. Lowe, "Three-Dimensional Object Recognition from Single Two-Dimensional Images," Robotics Research Technical Report No. 202, New York University, Courant Institute of Mathematical Sciences, New York, February 1986.
3. A.R. Hanson and E.M. Riseman, "Computer Vision Systems: A Computer System for Interpreting Scenes," funded by the National Science Foundation under grant DCR75-16098, Academic Press, New York, 1978.
4. R.J. Drazovich, F.X. Lanzinger, H. Mesiwala, J. Chan, and T.O. Binford, "Radar Target Classification System Design," AI&DS TR-1011-01, September 1981.
5. T.W. Miltonberger, A.S. Cromarty, D.T. Lawton, and H.E. Muller, "Preliminary Results on a Model-Based Image Understanding System for Detection Space Object Anomalies from Inverse Synthetic Aperture Radar (ISAR) Images," Proceedings of The Fifth Annual International Phoenix Conference on Computers and Communications '86, Scottsdale, Ariz., March 26-28, 1986, pp. 506-516.

6. T.S. Levitt, R.L. Kirby, and H.E. Muller, "A Model-Based System for Force Structure Analysis," *Proceedings of SPIE*, The International Society for Optical Engineering, Applications of Artificial Intelligence II, Vol. 548, Arlington, Va., April 9-11, 1985, pp. 169-175.
7. Capt. J. Moore, *Jane's Fighting Ships 1979-80*, ISBN 0 531 03913 7, Jane's Yearbooks, Franklin Watts, Inc., New York, N.Y.
8. G. Albrecht, *Weyers Flotten Taschenbuch Warships of the World: 56.Jahrgang 1982/83*, Bernard & Graefe, Verlag, Munich, West Germany.

Dr. David L. Milgram is program manager for image understanding with Advanced Decision Systems (ADS), Mountain View, Calif. He has a bachelor or arts degree in mathematics from Harvard College, a master of science degree in mathematics from the Courant Institute of Mathematical Sciences, New York University, and a Ph.D. in computer science from the University of Maryland. Thomas W. Miltonberger is a research engineer with ADS. He received his bachelor of science and master of science degrees in electrical engineering from the University of Illinois.
The authors thank the Naval Ocean Systems Center for sponsorship of the ship classification work, and Lincoln Laboratory, a research center operated by Massachusetts Institute of Technology, for supporting the satellite anomaly detection work, sponsored under a DARPA contract. Advanced Decision Systems is grateful to Dr. David Drake of the Naval Research Laboratory for his assistance and advice in formulating the model-based approach to ship classification. The views and conclusions contained in this document are those of the authors and should not be interpreted as necessarily representing the official policies of the U.S. Government.

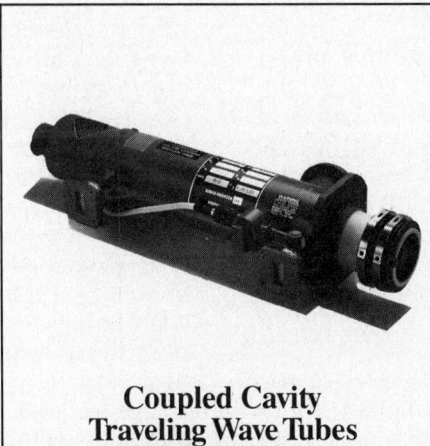

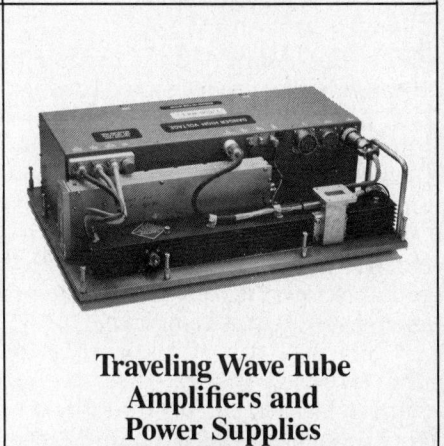

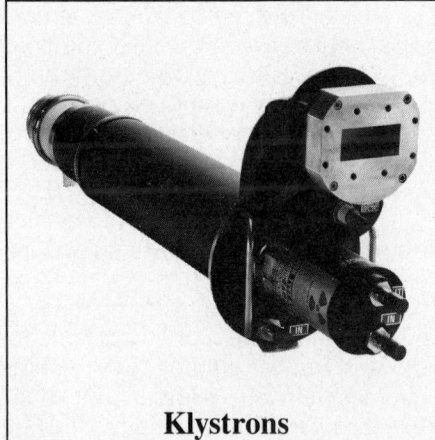

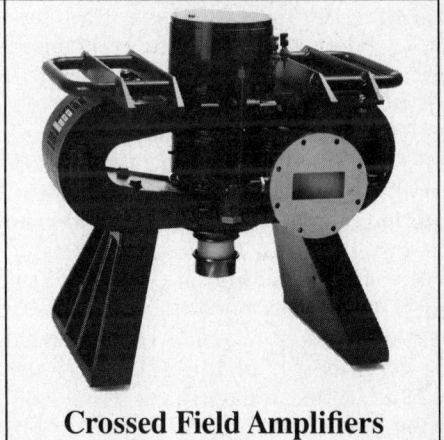

Electronic Warriors Need Training Too

The resources expended on EW training don't match those devoted to acquiring the hardware—but they should.

By Richard Short

The briefings lasted about four hours. They would be given daily during our transit from Pearl Harbor to Subic Bay and represented the Navy's final effort to prepare its aircrews for the missile threat in North Vietnam. The ship and air wing had finished their work-up and pre-deployment inspections. During the final hours before departing Pearl, the Cincpacflt EW QRC group had assembled the latest missile data, which included the discovery of large boxes mounted on selected Fan Song radar pedestals. They were the subject of many discussions, but it was up to the intelligence guys to determine their purpose.

Cmdr. Bob Ashford of the Cincpacflt EW QRC group was carrying that intelligence data when he boarded the carrier just before we sailed, and he would formally and informally brief the air wing during the transit. This was the extent of the Navy's Tacair EW training program. Of course, we were just moving into the tactical EW arena, so it is understandable that effective training systems were not yet in place. However, for the majority of tactical aircrews, realistic EW training requirements still exist and still are not being met.

The Evolution of EW. Those were the days of the Shoehorn Project. We realized that if we were going to reduce aircraft attrition over the beaches in Vietnam, we had to outfit them with radar warning receivers, missile launch detectors, self-protect jammers and chaff dispensers. The question was where to put them in an aircraft already stuffed with subsystems. The answer was the EW compartment on the outside of the upper fuselage just aft of the cockpit on the A-4, into which the equipment was shoehorned. This was a great idea, but it was impossible to manufacture EW suites fast enough for installation in all air wing aircraft. In fact, a depot had to be established at Cubi Point to manage the EW equipment. The departing air wing would offload its EW hardware so it could be picked up by the arriving air wing. Much of the equipment was actually cross-decked. It became apparent immediately that Conus air wings could not receive operational training without operational equipment.

Two of the primary functions of the CARGRU EW officer were to assure that the EW equipment operated as advertised and that the air wing had adequate EW training. Neither of these responsibilities was accomplished easily. One major equipment problem was simply determining if the equipment was actually working during a mission. Pre-flight inspections indicated that the black boxes were operating. They had operated on the bench and appeared to be operating in the aircraft, but there was no way to determine the status of the antennas or transmission lines. Prior to pre-flight, the output of the deception electronic countermeasures (DECM) on the bench or at the transmitter after being installed was within specification, but only milliwatts dripped out of the antennas. Where was the loss?

Then one day a magnificent machine arrived on the carrier evaluation: the Alford Sweeper Cart. The Alford Sweeper Cart was good news and bad news. The bad news was that it was impossible to know, with any degree of confidence, whether or not the DECM system would be emitting energy when over the beach. The good news was that, if it was not emitting energy, we could determine why by using the Cart.

The A-4 had seven different types of transmission lines—some coax, some semi-rigid and some rigid—between the antennas and the DECM/TX. Naturally there was a connector between each coax type. Between missions the lines were swept using the Cart, and the connectors were tweaked/replaced until line loss was at an acceptable level. The aircraft was launched and recovered. The lines were swept again and found to be clogged with losses. When did the mismatches occur? At launch? At recovery? Messages flew between the fleet and chief of naval operations. Action was taken. The transmission lines were redesigned and an A-4 was placed on a shake table to assure that the fix was effective. Finally, the aircrews could launch with a certain degree of confidence that deceptive energy would flow from the DECM antennas instead of just dripping. These types of mechanical fixes were multiplied many times over as we improved the systems on our aircraft.

Unfortunately, the same cannot be said for training. There seemed to be no limit to the money available for those EW QRC programs, but we did not have the time or money to adequately train our pilots to use them. The final training exercise prior to departing Conus for the Gulf of Tonkin was to fly over "ECHO Range" at the Naval Weapons Center, China Lake, Calif. Those few aircraft that had warning receivers were sometimes fortunate enough to see the lamps flash and the strobes jump around on their displays and to hear the sobering warbles. But this could hardly be called training. Apparently there

Type of Training/ Elements of Training	Reading and Memorizing	Training Film	Sound on Slide	Desk Top Trainer	EWPTT*	OFT/WST**	Electronic Warfare Ranges
Cost	Inexpensive	Inexpensive	Inexpensive	Moderate Expense	Moderate Expense	Expensive	Very Expensive
Supporting Operational Personnel	None	Possibly One	Possibly One	None	None	Numerous	Extensive
Trainees per Period	Many	Many	Many	One	One	One	Variable
Space Required	Anywhere if Unclassified	Closed Quiet Room	Closed Quiet Room	Quiet Room	Any Room	Special Spaces	Hundreds of Square Miles
Motivational Quality	Very Poor	Poor	Poor	Moderate	Excellent	Excellent	Excellent
Visual Cues	No	Yes	Yes	Yes	Yes	Yes	Yes
Audio Cues	No	Yes	Yes	Yes	Yes	Yes	Yes
Interaction	No	No	Perhaps Limited	Some	Yes	Yes	Yes
Realism	No	No	No	Moderate	Excellent	Excellent	Excellent
Teaches Transferrable Concepts	No	No	No	No	Yes	Yes	Yes
Training Quality	Poorest	Poor	Moderate	Moderate	Excellent	Excellent	Excellent
Availabilty for repetitous Training	Best	Good	Good	Good	Good	Poor	Moderate
Specialized Training	Good	Good	Good	Good	Good	Poor	Good
Available	Yes	?	?	No	No	No	Limited

* Electronic Warfare Part Task Trainer for Electronic Warfare
** Operational Flight Trainer/Weapons System Trainer

Figure 1. Matrix comparing types of training with elements of training.

was not enough time to be concerned with training. We hardly had time to build the system, let alone a simulator or training device to teach pilots how to use them.

Where's the Training? Overall, Tacair EW training is still in its infancy today. Our priorities must be wrong. For example, the F/A-18 was designed, built and delivered to the fleet with no EW equipment, which is one of the reasons given for not having an EW training capability on the operational flight trainer (OFT). However, we did develop a Part Task Trainer (PTT) called the HOTAS (hands on throttle and stick) to teach pilots stick and throttle switchology. But who needed an EW training device? There simply was no money allocated for EW training. Furthermore, there was no provision for a training device to teach pilots how to optimize the F/A-18's HARM missiles. There just wasn't enough money.

The danger of this mentality is that we could spend the last dollar for one more black box while the majority of our Tacair pilots do not even know how to operate the EW equipment they already have. This ineffectiveness was illustrated often when a section of aircraft approached their bombing target on the training range at NAS Fallon. If one of them began wallowing in the air, it was common to say, "He must be looking for his chaff switches." Consider the tremendous force multiplier that could be realized by buying one less platform or a few less HARM missiles or airborne self protection jammers (ASPJs) and instead allocating that money for effective EW training. It isn't that difficult to understand. We simply have our priorities in the wrong place.

The Prowler community is the exception to inadequate EW training. They are specialists. They have the schoolhouse, part task trainers (PTTs), and even an RF threat simulator to fly against. They must be experts: entire strikes depend on their expertise. An attack/fighter pilot is also expected to put bombs on target and safely return to his ship, and yet he does not receive the same intense EW training.

EW Training Levels. Several levels of training are available, including the R&M method (reading and memorizing), observed fire trainers (OFTs), weapons system trainers (WSTs), and flying on a range. Each level has definite advantages when compared at the element level. If low-cost training is the primary element, the R&M method is the logical choice. Provide an Optevfor Tactics Guide or Natops and tell the student to memorize it. Of course, if survivability is the primary goal, then use an OFT, WST or range.

Since the advent of the personal computer and the proliferation of software, there have been many advocates of desktop trainers, and rightly so. It is a definite asset in the classroom and for personalized individual training. However, it is inadequate in many areas. They can be used to teach basic switchology and decision making processes for normal flight, EW and emergency procedures, but they generally do not provide the integration and synergism of systems and actual flight simulation. That must be accomplished in a flight simulator. Pilots may be able to learn EW switchology and procedures on a PC, but this must be followed with a training device

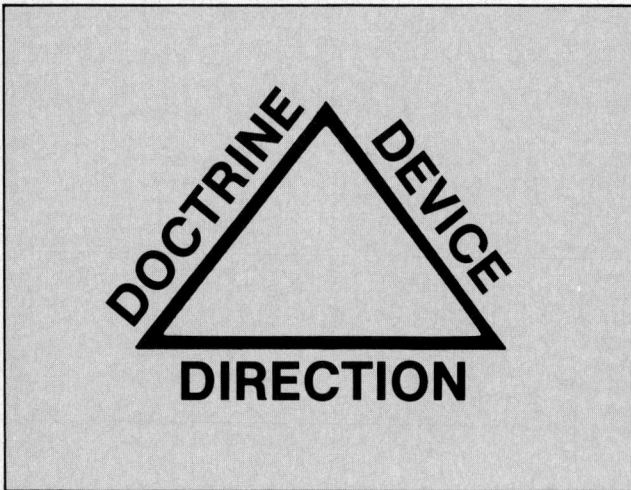

The Three D's of Operational Training

that can integrate initial maneuvers with countermeasures reactions to simulated threats, and there must be a provision for repetitive training so that EW responses become second nature.

All attack aircraft OFTs and WSTs have a limited EW training capability. However, at least five major problem areas preclude adequate EW training:

• Time is at a premium and most of it will be used for flight and instrument training and emergency procedures training.

• When an aircrewman drops chaff or flares, it only provides indication of occurrence at the instrumentor's station, and the instructor must make a subjective decision on whether or not it was effective.

• There are no track-break computer models being used to determine if the pilot made the correct maneuver and/or dropped the correct expendable countermeasures within the critical time frame or had other countermeasures turned on or in the correct operational mode.

• There is no capability to drop new expendable countermeasures currently in production or in the R&D pipeline.

• EW training is limited to where the OFTs and/or WSTs can be located.

There have been plans to add complete EW training suites to the A-7E, A-6E and F/A-18 OFTs or WSTs. These additions will require major changes to software and hardware and will be extremely expensive. Yet, they still will not provide adequate EW training.

Compounding the training problem is the fact that OFTs and WSTs do not go to sea. When the air wings deploy, they leave these training assets at home. This limitation does not significantly affect an aircrewman's flight proficiency as there appears to be adequate flight time available while deployed. In fact, after workup in Hawaiian or Caribbean waters the pilots are quite proficient at launches and traps. However, they get little or no effective EW training.

An EWPTT has numerous advantages over a desk top

trainer, an OFT/WST and even an EW range. Of primary importance for survivability is the development of "transferable concepts," which are not learned well with a desk top trainer. Also, the EWPTT can be used for specialized repetitious training, whereas the OFT/WST normally will be scheduled for longer periods of time and may also allow for only a single EW engagement during the entire mission. Finally, the EWPTT will go aboard ship; the OFTs/WSTs used currently in the attack community do not.

There are those who base their EW training hopes on ranges such as the aircrew electronic warfare training range (AEWTR) at NAS Fallon or Green Flag exercises at Nellis Air Force Base. Although some effective tactical EW training is being accomplished at these ranges, they still have many deficiencies, including the lack of signal density and realism. Beyond that, there is a more fundamental problem—how often does the pilot fly on the range, and what percentage of that time will he spend solving an EW problem?

Perhaps even more significant than the inadequacies of the ranges and OFTs/WSTs for EW training is the fact that they cannot be carried aboard the carriers or moved to remote advanced bases. This inadequacy has always existed, and unless someone picks the up gauntlet it will continue to exist.

A recent study performed by the Air Force Human Resources Laboratory at Williams Air Force Base, Ariz., indicates that as little as four months without practice reduces an aircrew's EW response capability approximately to where it was before beginning EW training. Thus it does little or no good to train a pilot in EW if you are going to send him to an advanced base with no further training. After four to six months he will be back to square one.

Retired Rear Adm. Julian Lake brought the issue into focus when he recently stated, "Even with vast improvements in simulators and programs such as Red Flag, EW training is far from adequate. Operators do not have the facilities to achieve realistic readiness. Even for those fortunate enough to use Red Flag, or similar facilities, it is at best an annual event when the need is for almost daily training or at least as often as other training. The subtleties of EW are often lost without constant use."[1]

Thus it appears there is a requirement for an EWPTT that can be placed on the carriers. This does not mean we should neglect an EW training package in our OFTs and WSTs or install an inadequate system due to lack of funding. There needs to be a continuum of EW training that includes classroom training using desk top devices for tutorials and perhaps signal recognition, EWPTTs, OFTs/WSTs and EW ranges. Basic intelligence information, signal recognition and tactics should be taught in the classroom. The student should practice on the EWPTT and then integrate and test his knowledge on an OFT/WST and an EW range. But this approach costs money and requires an increase in the EW training equipment budget. It requires commitment.

Addressing the Problem. In 1981 the lack of adequate EW training for deployed Tacair crews was addressed by a Navy attack squadron at NAS Lemoore in a letter that was staffed up to Commander, Naval Air Force, U.S. Pacific Fleet (Comnavairpac). Comnavairpac endorsed the letter and forwarded it to the Naval Training Center (NTEC) for action. The letter stated that A-7E pilots based at Lemoore were capable of establishing an adequate level of proficiency in ECM during turnaround cycles if they used all the training courses and devices available. However, at that time there was no means for deployed pilots to remain proficient in interpreting and reacting to visual and aural cues provided by the A-7E ECM suite— a problem that still exists. The letter concluded that a portable simulator was needed that could be taken on the carriers for EW training, and requested that NTEC procure one.

The Naval Training Equipment Center did a commendable job of addressing the problem. They confirmed that the EW ranges and training devices did not provide a complete file of portable EW signatures that could be encountered during deployment, and that those training facilities were not responsive to simulating the changing electronic signatures of the threat weapons systems. The problem was accentuated by the lack of suitable portable EW training devices. It was stated that the use of a portable EWPTT would alleviate this eroding EW proficiency.

The conclusion reached was that there were no EWPTTs in the government or civilian inventory that met the training needs of the attack/fighter aircraft community. There is no indication that any subsequent action was taken to fulfill this critical training need. No one picked up the gauntlet. In fact, because there were no Tacair EW training equipment requirements coming out of Navair, there was no group addressing Tacair EW training equipment at NTSC for a number of years.

However, the lack of adequate EW training devices cannot be blamed on Opnav, Navair or even NTSC. The operational commands and squadrons also must bear the responsibilty, because they have not demanded adequate EW training equipment. Why? It certainly was not a concern for the cost. Rather, it was because EW had once again been relegated to the bottom of the requirements ladder, and with it the need for EW training devices. Until our operational commanders recognize the value of EW and the requirements for adequate EW training, this inadequacy will continue to exist.

In general, operational training can be viewed as an equilateral triangle, with the base being *direction*, and the two sides *doctrine* and *device*. All three elements are necessary, but most important is the direction given by the commander. Doctrine can become useless if not practiced, and it probably will not be practiced without direction. Even with poor doctrine and inadequate devices, a concerned skipper can still provide effective direction that will result in some positive training. Unless the EW officer is a charger and has the skipper behind him, there will not be much squadron EW training.

EW Training Requirements. Tacair EW training can be realistic and profitable with relatively low-cost part task trainers if our operational commanders at at all levels will take the time to generate operational requirements. Basic trainer requirements include:

• Visual and aural cues, mechanical controls and switches. They must provide the realism necessary to ensure that what the student sees and hears teaches him "transferable concepts," or concepts that can be transferred directly to the cockpit in a real threat engagement.

• The student must be in an environment where he can concentrate, where he can be task-loaded and absorbed in the training problem, and where he will indeed be living the training engagement.

• The training device should provide real-time feedback and repetitive exercises. The student must know when he succeed and when he failed and why. If he failed, a similar scenario should be presented where he can try again. Countermeasures should become second nature to aircrews. They should be able to glance at the warning receiver display, correlate it with their audio and automatically determine what maneuver is called for, what expendables should be released, and what should be the status of their DECM.

• The student must be motivated to improve his responses. This motivation can come from three different areas: training device design, self motivation, and command motivation (direction). However, the foundational motivation is command direction. Our commanders must demand devices that will motivate the students, and they must provide clear direction. An operational requirement must be written clearly specifying the device needed. The command that generates a training device requirements needs to be ready to support it through the development phase. If not, the end result may be a desk top trainer that will not do the job and will not be used.

• The student must be rewarded when he is successful.

There are currently three operational expendable countermeasures payloads for aircraft self-protection that are available to fleet squadrons: chaff, flares and POET. Within the next two to three years, expendable countermeasures payloads will also include Generic, AAED, and others which are on the drawing boards. These devices are well worth the millions of dollars being spent on their development because, when properly used, they will significantly increase the survivability of our airmen. However, we must train our airmen to use them. There is no training device available that integrates the overall EW suite with aircraft maneuvers.

Current training at the wing level is primarily classroom oriented, with the majority of our aircrewmen flying on an EW range once a year or less. The value of this flight training, when compared to cost, is highly questionable. The aircrew makes one or two passes through the range, observes RWR displays and hears the associated audio; however, there is currently (nor will there be in the near future) no opportunity to measure maneuver or expendable countermeasure effectiveness.

In summary, neither the wing nor the squadrons currently provide adequate electronic warfare training due to limited funds and time. The EW training that is provided generally applies only to signal recognition, flares and chaff. There are no adequate training devices available for overall EW training.

A final example from Rear Admiral Lake may help focus on the EW training problem: "EW readiness is a constant battle. The Royal Navy paid a price in 1982 for decisions that started years before in budget delays and program stretch-outs. The U.S. Navy ran the same risk for years, and fortunately for them there were no wars. Today some forces do not procure enough EW systems to fill the aircraft that use them, to the detriment of training, maintenance and a future crisis. Just as it is with testing, EW training is more difficult to accomplish than the other forms of warfare. Again the problem of no bang, no smoke, no swoosh leaves the trainee with a low sense of accomplishment. EW is nearly impossible to conduct in peacetime due to security, interference with civil communications, TV or air traffic control. Often it is simply that a realistic electromagnetic environment cannot be duplicated. Thus because of its very nature EW depends on simulation for realistic training."[1]

It should be apparent that the least expensive way to multiply our forces is to provide effective Tacair EW training. Yet the first programs to get cut during a budget crunch are training, EW simulators and/or EW training packages. Purchasing one less weapons platform and investing that money in an effective Tacair EW simulator would multiply the effectiveness of the remaining platforms far beyond the operational value of buying that one extra aircraft. ∎

Richard Short is an employee of AAI Corp. and has been working in the area of training and simulation since 1982. During a Navy career of more than 20 years he was a flight officer in both attack and reconnaissance squadrons. In private industry he has done EW engineering and analysis since 1973. His article is based primarily on his naval experience and reflects his own views, not necessarily those of AAI Corp. or the U.S. Navy.

Reference
1. Rear Adm. Julian Lake (USN, Ret.), excerpts from a paper given at the second NATO Electronic Warfare Symposium, 1987, "EW Lessons Learned Since 1965."

Advanced High-Voltage Power Supply Development

Using a building block approach to design promises to improve the reliability and maintainability of increasingly complex ECM and radar architectures.

By Dale Hollis

Electronic countermeasures (ECM) systems have become increasingly dependent on traveling wave tube amplifiers (TWTAs) to provide the broad bandwidth and high power emission levels needed to counter today's and tomorrow's threats.

TWTAs typically consist of a high-voltage power supply, some form of modulation, logic and control, and a TWT that has been customized to provide pulse or continuous wave (CW) amplification. Table 1 defines the key specifications for typical power supplies for CW and pulse TWTAs. These power supplies are usually switching systems, incorporating a variety of regulation techniques. They are constructed utilizing solid state devices, as they most often must be very compact and light in weight.

Fluids, Gases, Solids. There are three principal methods of constructing high-voltage power supplies to ensure their survivability when operating at high altitude and in high moisture and/or cold environments. One method employed is to fabricate a container that may be sealed so that it can contain a dielectric fluid such as fluorinert. The power supply components are held immersed in the fluid to ensure that no arcing occurs and to enable heat to be transferred to the walls and base plate of the containment vessel, where it can be transferred to the cooling system of the chassis housing the power supply.

Some method must be provided to compensate for the changes in the volume occupied by the fluid as the temperature varies. The most common method of accomplishing this task is to use bellows, which are immersed in the dielectric fluid along with the electrical components. Their volume varies to accommodate volumetric changes in the liquid, keeping the high-voltage circuit elements insulated from one another at all times.

The use of a fluid as a dielectric is wonderfully forgiving. It permeates everything uniformly and, if it has been properly prepared and applied, it ensures that failures due to arcing or thermal problems are unlikely to occur.

The problems associated with the use of fluid are that it is very difficult to ensure that the vessels won't leak as they are subjected to vibration, shock and thermal cycling. Additionally, the fluid must be carefully degassed to eliminate the possibility of bubble formations in the fluid

Replaceable submodules of a high-voltage section.

TABLE I		
TWTA POWER SUPPLIES		
Electrical		
Specifications	**CW**	**Pulse**
Cathode load reg.	Max 1 percent	Max .5 percent
Droop during pulse	<50V	<50V
Cathode ripple cont.	Max .5 percent	Max .5 percent
Cathode	–10 KV	–11 KV
Collector 1	+5 KV WTC	+6 KV WTC
Collector 2	+2.5 KV WTC	+3 KV WTC
Helix	GND	GND
Heater	6.3 V	6.3 V
Anode	+200 V TYP	+200 V TYP
Efficiency	>85%	>85%
Environmental		
Specifications	**CW**	**Pulse**
Alt	70 K	70 K
Temp	–54–+85°C base pl	–54–+85°C base pl
Vibe	8G's RMS	8G's RMS
Shock	50 G	50 G
EMI	Mil. Stnd. 461	Mil. Stnd. 461
Temp cycling	20 cyc. fail free	20 cyc. fail free

chamber. If the bubbles migrate to the proper place an arc will occur.

Building blocks assembled into a high-voltage section.

TABLE II SILICONE POTTING MATERIALS			
Characteristics	GE RTV 615[1]	GE RTV 11[2]	E.C. 4952[3]
Color	Clear	White	Red
Dielectric strength	500 V/Mil	500 V/Mil	550 V/Mil
Thermal expansion	$27 \times 10^{-5}/°C$	$25 \times 10^{-5}/°C$	$16.2 \times 10^{-5}/°C$
Thermal conductivity (BTU)(IN)/(HR)(FT)2	1.33	2.05	7.5
Service temperature range	-60 to 204°C	-54 to 204°C	-65 to 260°C
Viscosity	3,500 CFS	12,000 CPS	30,000 CPS
Specific Gravity	1.02	1.18	2.29

A second method employed in the manufacture of high-voltage power supplies is to place the various electrical components such that, when operating at one atmosphere of pressure, no arcing will occur. The problem then becomes one of insuring that the high-voltage portions of the power supply are always at one atmosphere or more of pressure. This is usually done by constructing a container that may be pressurized to approximately 15+ PSIA with either dry air or a dry inert gas, such as nitrogen or sulphur hexachlorophene.

This method of protecting the high-voltage components from arcing at altitude can be quite effective. There are, of course, problems associated with this method. The valves used to pressurize the container with gas or air are prone to failure with repeated usage. Secondly, the sealing method most commonly used is an O-ring gasket. When the vessel is properly machined, such that the mating surfaces of the pressure container are very well matched, the containers will hold pressure for long periods of time. Unfortunately, as the seals are broken and re-established during repairs, the surfaces become distorted and the pressure vessels become leaky. It is also difficult to maintain the seals when the container is

subjected to repeated shocks, which mechanically distort the container. Although the gas provides the necessary dielectric strength, it is a very poor thermal conductor. Heat extraction must be made via the base plate and walls of the container. This necessitates the extensive use of high-voltage spacers and insulators, such as beryllium oxides and ceramics, between the heat-producing components and the thermal conducting surfaces. This presents new problems with the handling of hazardous materials and, of course, added costs.

The third most commonly employed method of construction is to encapsulate the high-voltage portions of the power supplies in solid silicone potting materials. With this method, molds are constructed of adequate volume to contain the high-voltage assemblies and enough potting material to protect the components from arcing at operating altitudes. The electronic assemblies are then suspended in the mold and the mold is filled with silicone potting material in a liquid state. A vacuum

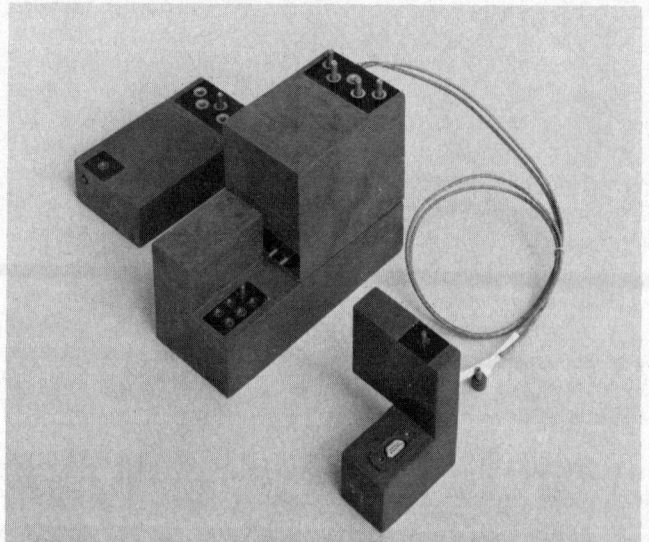

Connectors are installed in each submodule and placed so that they mate with adjacent submodules, allowing for ease of assembly and disassembly.

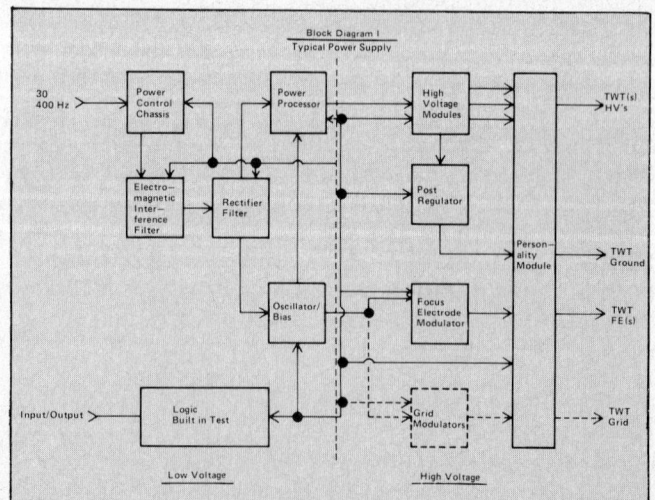

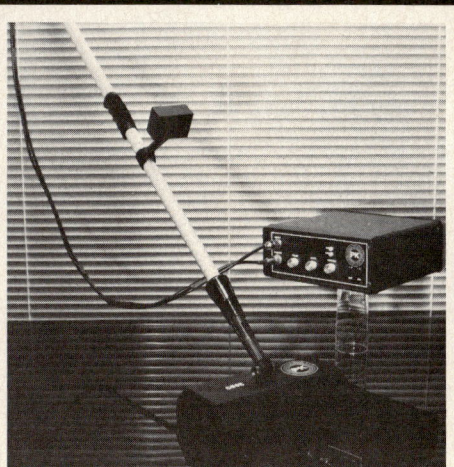

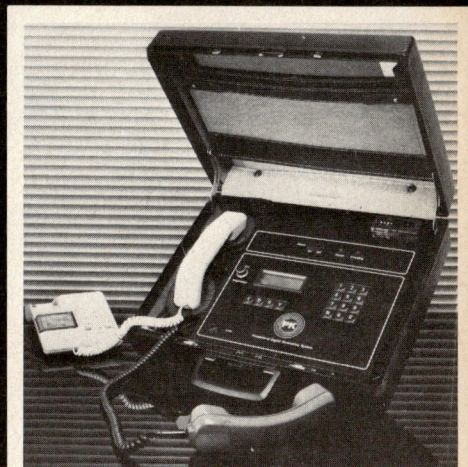

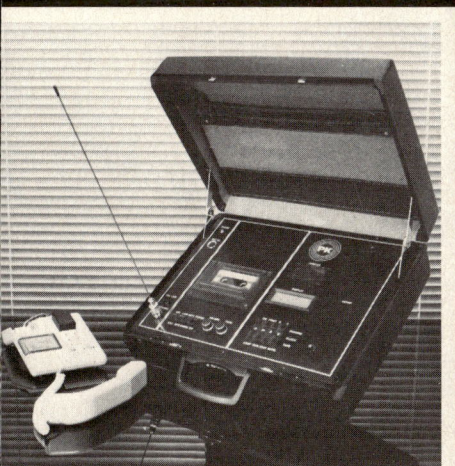

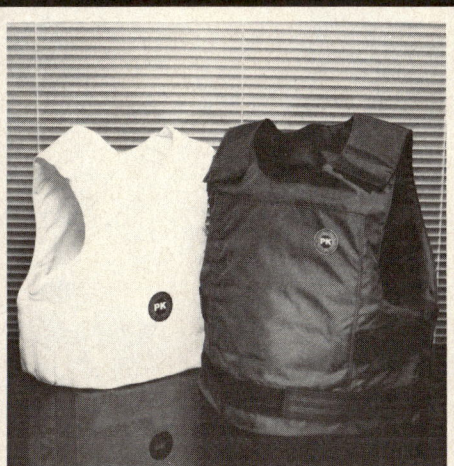

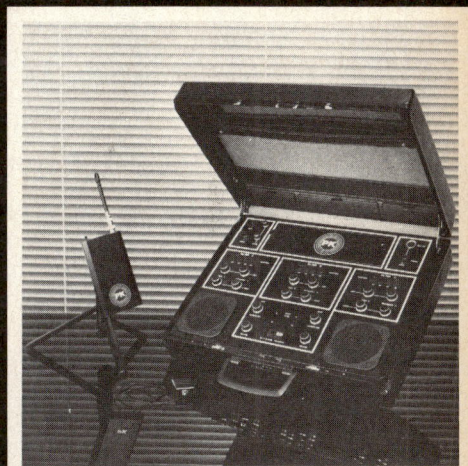

SUB-MODULES PRIOR TO ASSEMBLY

BUILDING BLOCK
HIGH VOLTAGE ASSEMBLY

STEP I

CAPACITOR SUB-MODULE ASSEMBLED
TO TRANSFORMER SUB-MODULE

STEP IV

COLLECTOR SUB-MODULE
INSTALLATION

STEP II

SENSE MODULE ASSEMBLED TO BASE

STEP III

DIODE SUB-MODULE BEING ASSEMBLED

This sequence illustrates how the submodules are assembled into a module, in this case the high-voltage section of an airborne ECM power supply. Normally this module would be considered an expendable. Its conventional counterpart is shown on the next page. One can readily see the difficulty of repairing this module.

is applied to remove air bubbles introduced during mixing. Once the potting material is degassed, it is cured in a temperature-controlled environment. This method is very process-intensive, but has some significant advantages over those previously discussed.

Silicone Potting Problems. Once encapsulated, the components are permanently protected from arcing. There is no fluid to leak or outgas; there is no gas to escape. The silicone materials are relatively forgiving of shock and vibration. No specialized containers are required to house the material, so that the high-voltage assemblies may be easily handled and stored.

There are, of course, some problems with the silicone materials most commonly used for high-voltage potting. The remainder of this article will explore these problems and what we believe to be viable solutions to them. Teledyne Counter Measures Equipment (TCME) exclusively employs silicone potting as the dielectric in the power supplies we currently manufacture. We have found this method to be the most flexible and economical to use when properly applied.

Early in 1984, the company set about the task of defining a long-term research and development program to create our next generation of high-voltage power supplies. This advanced power supply development program is ambitious, in that its long-range goal is the development of a family of high-voltage power supplies that eliminate many of the failure modes associated with generating high voltages in harsh EW environments. The impetus for this program was the increasing performance requirements of the ECM systems where our equipment is employed.

All high-voltage power supplies used in ECM systems share common problems. With the demand for higher power levels and increasing system complexities, it has become apparent that there is a need for increased

The conventional counterpart of the modular unit (previous page).

power handling capability in the power supplies. This has been brought about because there has been little or no increase in the volume allocated to the power supplies despite increasing the performance requirements.

The solid potting materials that are commonly used as

dielectrics in the high-voltage portions of the power supplies are poor thermal conductors. Consequently, increasing power densities present the need to use higher density potting formulas.

Typical potted power supplies in use today have power densities in the range of three to five watts per cubic inch. Present trends are pushing this requirement to six to eight watts per cubic inch and beyond. Along with the requirement to manage increasing power densities comes the need to develop new construction techniques, to allow for the fact that silicone potting materials expand rapidly with increasing temperature. This property of the potting materials requires that special care be taken to ensure that as the potting material expands, the components to which it is adhering are not damaged. With increasing power densities this property of the commonly used materials becomes more difficult to control.

Potting materials are chosen primarily for their dielectric qualities. Generally, the higher the dielectric strength (volts/mil) of the material the better. High dielectric strength gives the designer more flexibility in the mechanical layout of the high-voltage assemblies. Current silicone materials have dielectric strengths in the range of 500 to 550 volts/mil. Typical potting processes seldom achieve the full dielectric potential of these materials.

The key objective in preparing the assemblies to be potted is the removal of any contaminant that might prevent the potting material from adhering to the component and to etch the surfaces of the components so that the potting may adhere firmly to them. Should adhesion be incomplete, creep paths will be created that will result in the assemblies arcing. Further, the potting materials must be carefully degassed after mixing and pouring to ensure that no voids are created.

Another area of concern on the part of the end users is that the portions of the power supplies that are most prone to failure, the high-voltage modules and the modulators, are seldom successfully repaired.

A problem with using solid potting materials is that once an assembly is encapsulated, the probability of successfully replacing a failed component is significantly reduced. The currently accepted method for removing a component from an encapsulated module is to mechanically "dig out" the component. In the process of removing the potting material overlaying the failed part, it is highly probable that other components will be damaged. In our experience, the success rate associated with repairing encapsulated, multiple layer, high-voltage modules is very small. After all, the processes that ensure good adhesion of the potting material to the components make it very difficult to remove the material. Particularly prone to damage are components assembled using silicones and silicone insulated wires. As a result, post potting failures of these modules significantly affect the cost of manufacturing and repairing the power supplies.

Commonality. One commonly expressed desire on the part of the system support people is for commonality

Block Diagram II
High Voltage Section

Collector Supply

High Voltage Out

Magnetics

Diode Module

Filter Block

High Voltage Out

Regulator and Protector

Sense Module

the reader to gain a point of reference with which to relate to the remainder of this discussion.

Block Diagram I shows a typical high-voltage power supply. The low-voltage section provides the bias oscillator necessary to drive the power processor. The EMI filter assures that the noise components of the power supply are isolated from the power source. Control logic interfaces and protection circuitry are also located in the low-voltage portion.

The high-voltage section houses the modules that convert the processed primary power to whatever high voltages are required. The post regulator assures fine control of ripple and controls the switching droop of pulse systems to acceptable levels. The modulators provide the control element pulses required by the TWT under power.

Many improvements have been made to the low-voltage portions of the power supply and control systems. These have been primarily in the form of improved magnetics, more responsive protection circuitry, improved efficiencies and better thermal management. The real breakthroughs, however, have been in the high-voltage portions of the power supplies.

The submodular building block technology shown in the accompanying photographs offers unique solutions to many of the problems discussed previously concerning solid potted modules.

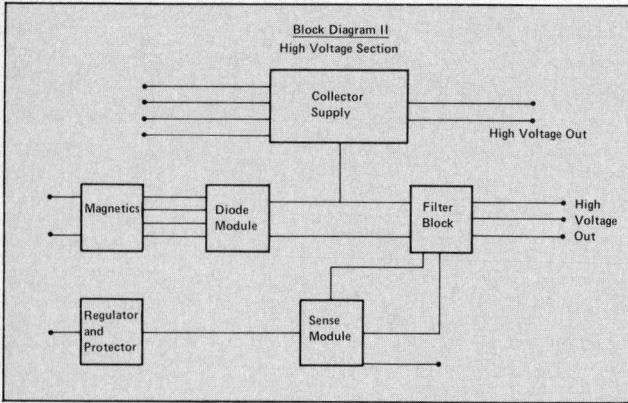

between power supplies. Since no two ECM power supply designs are alike, logistical support of these critical system elements becomes a complex task. Commonality between power supplies has to some degree been achieved at the component level. The application of military standards to the specifications for the systems force manufacturers to use many of the same family of parts. The commonality between supplies, however, ceases when these various electronic parts are made into assemblies and integrated into power supplies. In other words, there are not currently any standards defined for power supply subassemblies that would result in modular portions of one power supply being interchangeable with another, even when both power supplies were designed to operate in the same system with the same TWT.

With the advent of octave-plus bandwidth transmitters, the active portions of ECM systems are seeing ever-increasing life spans. This means that the power supplies will require long-term support and retrofits to accommodate changing mission requirements. Accomplishing the task of physically reconfiguring an existing power supply means replacing whole portions of the supply. This means that lead times are long, as extensive redesign efforts are required. Where extensive reconfiguration efforts are needed, it may be more cost effective to replace the power supply rather than extensively re-engineer the existing one. It is desirable to have a much more flexible design.

The level at which repairs can be effected is a topic of much concern with the end users of ECM systems. Since a power supply can deliver lethal amounts of power and since little in a current power supply can be repaired without depotting and patch potting the modules, it is usually replaced with a spare and returned to a depot for repair. This practice necessitates the stocking of spares that would otherwise not be required. A design that allows for lower levels of maintenance means that the ECM equipment is available during more of the mission time.

Before discussing some solutions to the problems outlined so far, it is necessary to briefly discuss the topology of a typical power supply with modulator. This will help

Thermal Management. One obvious solution to the problem of increasing power densities is to improve the thermal management of the power supplies. The power loss of the power supply is dissipated in the form of heat. In the development of high-density, high-voltage power supplies, the designer usually resorts to high frequency power processing. This is done to keep the magnetics and filter elements small. Additionally, FET technology has been extensively employed for simplicity in the design of the driver circuits required to control the switches. There are, of course, practical limits to increasing switching frequencies. Magnetic core materials must be chosen for a number of characteristics, which compromise the final switching frequency chosen. Ideally, the volume available will allow for a wide selection of alternate materials and methods. This allows the designer to make component selections in favor of efficiency. Unfortunately, the ECM systems in which the power supplies must operate are often carried on board aircraft where the volumes available are severely restricted.

While some combination of design alternatives will yield a switching power supply with a relatively high efficiency, it is still necessary to mechanically design the supply such that acceptable junction temperatures are maintained, that is, less than 125°C.

While there are many design alternatives available to remove the heat from the low-voltage portions of the supplies, heat extraction becomes much more difficult in the high-voltage portions of the system. The problem occurs when high-voltage isolation must be maintained, while at the same time the heat generating components must be put in close mechanical contact with low thermal

Building blocks can be backfitted into existing power supplies.

resistivity heat sinking paths. Within the high-voltage sections of the supply, this task is accomplished by mounting heat producing components to thermal transfer materials with good high-voltage insulating properties. However, since there is typically not adequate surface area upon which to mount these components to provide good heat extraction, some of the heat must be extracted via the potting material itself.

The potting materials most commonly used today are poor thermal conductors. In order to improve the ability of the silicone materials to conduct heat, they must be loaded with solids, usually aluminum oxides. Potting materials with relatively good thermal properties are available. One example is EC 4952 (See Table II).

TCME chose to use this as the new potting material of choice in the building block power supply designs. A problem that had to be overcome was that the loaded material had a very high viscosity. This meant that conventional potting molds could not be used. The use of full-sided molds required developement of a material handling technique that allowed for evacuation of the mold and the material during injection. TCME has developed this technique; an example of the flexibility of this method is pictured.

The full-sided molds used to encapsulate the high-voltage portions of the supplies mean it is possible to very accurately control the dimensions of the individual blocks. This is very necessary because modules are broken down into readily replaceable submodules. For this to be feasible, the submodules have to be accurate physical replicas of one another.

One other necessary feature of the high-density material is that it is dimensionally much more stable over temperature than the lighter loaded materials. This is key to the fact that the submodules are physically interchangeable from module to module.

A new method of post potting inspection is necessary to ensure that the molded pieces are clear of voids and that there is adequate potting material between the components and between the components and the surfaces of the block. Previous inspection methods required visual inspection and high pot testing to ensure that the potting process was successful.

The loaded materials are opaque, so visual inspection is impossible. X-ray inspection is therefore the preferred method for analyzing component placement within the potted submodule. High pot testing is done as before.

The dielectric properties of the loaded silicone materials are slightly better than their unloaded counterparts; in this particular case 50 volts/mil or 10 percent. This gives the system added flexibility in that component placement is a little less critical than when using the other materials. This has importance not only in the general sense that it gives added arc protection, but it also adds a degree of design flexibility to the submodules.

The Right Connections. The submodules become rather like "macrocomponents." Each submodule has a particular function that it performs in the high-voltage module or modulator. Relative functions of the submodules of a high-voltage module are shown in Block Diagram II. Ready installation and removal from the parent module is a desirable attribute of such a component. This means that connectors must be installed in each block and placed so that they mate exactly with those other submodules adjacent to them. The volume required by these connectors results in a proportionate reduction in the total amount of potting material available in each module. The added dielectric strength of the loaded materials more than adequately compensates for the potting losses.

The connectors needed to interface the submodules to one another are very special. Conventional high-voltage connector designs are too large for this application. As designs of the submodules vary from one power supply to another, owing to outline requirements, the interconnect scheme is very flexible. Each miniature connector is able to function independent of its counterparts in the submodule. Each connector is readily placed in a plate that provides the mounting surface for the connector and the necessary pin alignment between the blocks.

The conventional pin and socket assembly is not employed in the building block power supply design. Transportation and handling of a submodule with pins protruding will almost certainly result in the connectors being damaged. Instead, each submodule employs the socket portion of the connectors. Specially insulated feedthrough pins are then used to complete the connection. As seals are installed in all of the sockets, very secure fluid and humidity protection is afforded to the actual electrical interface.

The submodular building block design is making it possible to fulfill another frequently occurring requirement of power supply users, the ability to repair the power supplies. As discussed previously, it is very difficult to successfully repair a potted assembly, but it is very easy to replace a failed submodule. The submodular approach to designing and building power supplies accepts the fact that no repairs are to be attempted within the submodules themselves. Only the modules will be repaired by replacing the failed submodules.

Isolating the failed submodule is accomplished by one of two means. Where passive failure indicators have been employed as built-in test (BIT) in each submodule, the technician need only visually inspect the module, once the cover has been removed, to locate the failed submodule and replace it.

Where BIT is not installed, the module is disassembled and each submodule is plugged into a test fixture to identify the failed block. With the exception of submodules containing magnetics, this type of testing may be accomplished using low voltages, with a high probability of successfully locating the failed block(s). Magnetics typically fail as a result of corona breakdowns, which cannot be detected except at high voltages.

The ability to isolate faults without lethal high voltage being present makes it possible to do high-voltage module repairs at levels below and at the depot. The building block design also makes it possible to share submodules between power supplies. As as example, given a common frequency it is possible to use the same filter block between supplies. Rectifier blocks designed for high-power duty may be employed in more than one design. Adoption of a submodular design concept brings us one step closer to interchangeability of modules. This has obvious implications for the logistical support of the ECM systems and the consequent mission capability of the weapons carrier.

The submodules also make the mechanical reconfiguration of a power supply much easier. The individual blocks may be rearranged and interconnected by a high-voltage ribbon cable assembly. Although no volume savings are realized, it is possible to alter the profile of a power supply. Conventional modules require major mechanical redesigns to accomplish the same reconfiguration.

The building block technology may be readily backfitted into existing power supplies. The form fit factor is one-for-one with conventional designs. This is made possible by minimizing the interconnects between submodules. Careful circuit analysis is required to identify logical groupings of circuit functions. Consideration is given to function, number of elements, risk of failure and number of connectors required. Location of the individual submodules is dependent on their need for close proximity to the cold plate.

Continuation and reliability engineering, which are necessary to all product improvement programs, are enhanced through the use of building block designs. Failures are easily isolated to a particular submodule. Unlike their counterparts, it is possible to electrically isolate functional areas of a particular module in order to study failure modes and to test product design improvements.

It is necessary to make a comment at this point. The building block approach to designing high-voltage power supplies has inherent advantages over conventional designs, as previously discussed. However, it is not a substitute for good electrical design and engineering practices. In other words, it will not forgive a poor design. It will only make it maintainable.

Today's ECM and radar architectures are increasing in complexity. As more and more sophisticated electronics are added to weapons platforms, it becomes obvious that modularity of the designs will be key to reliable and maintainable systems. The building block approach to high-voltage power supply designs brings full modularity a step closer. ∎

Dale Hollis is vice president for logistics programs at Teledyne Counter Measures Equipment Co., Santa Clara, Calif. Before joining Teledyne, Mr. Hollis managed a national service organization for Anderson Jacobson Inc. He has an extensive background in military systems gained with Westinghouse M.L.H. on the Polaris and Poseidon programs and as a member of the U.S. Army 32nd AADCOM. He received his bachelor of science degree in management from San Jose State University.

Integrated Techniques for Precision Waveform Measurement

Laboratory instrumentation technology is improving techniques for measuring microwave signal parameters in critical signal acquisition and analysis applications.

By Dr. Gregory J. Donaldson

Advances in laboratory instrumentation technology are in turn leading to improved techniques for measuring microwave signal parameters in strategic reconnaissance, countermeasures equipment evaluation, training missions and other critical applications. By intelligently integrating appropriate subsystems, signal acquisition and analysis that traditionally has taken months to accomplish can be done in days, or even in real time.

The primary hardware ingredients are a control computer to set up the instruments and a digital storage medium that allows the raw data to be immediately available to analysis programs. De-emphasizing custom electronics in favor of commercial equipment exploits the advantages of relatively inexpensive hardware and allows modular, low-cost upgrades as instrumentation evolves. At the same time, however, there are performance areas where custom design, such as low-noise front ends, broadband tuner/demods or surveillance/DF receivers, is a necessary augmentation.

Measurement Applications. Strategic reconnaissance typically has used non-engineering personnel at the receiver controls to record signal activity on analog tape for later analysis. By the time the decision is made to analyze the tape, the knowledge of what was recorded and the signal path configuration of the receiving system may have evaporated. In contrast, an engineer working with a digital storage and analysis capability could record much the same data, annotate the recording conditions, and perform the analysis immediately. The major hardware challenge is developing a receiver with sufficient sensitivity to obtain a useful signal to noise condition for a distant signal. In addition, interesting classes of emitters exhibit frequency agility and broad signal bandwidths as well as exotic modulations. This dual requirement calls for a sensitive system with broadband capability.

Electronic combat equipment evaluation presents a similar receiving challenge: to illustrate the interactions between a victim radar and a particular electronic countermeasure. Often one of the two emitters is at a much higher power level at the receiver. Although high sensitivity is still needed to process the weaker signal, sufficient dynamic range must be provided to accommodate both. Equipment evaluation involves detailed questions about the quantitative values of modulations and signal levels. This demands an accurate knowledge of the receiving

Integrated mobile equipment is used to support precision measurement requirements. A multi-band high-gain directional antenna system can be slaved to target tracking inputs and is equipped with a large-magnification television camera for optical tracking. The taller three-dipole array is an intercept and DF system for HF through UHF. The spinning antenna at the far end has its radome removed to display the antenna for a surveillance and direction-finding receiving system. Other mast-mounted antennas include omnidirectional antennas and radiated self-test signal sources.

system transfer curve and its response to frequency and amplitude modulations.

In ECM training and mission evaluation, emphasis shifts to how well the air or ground crew reacted to the electronic threat, who won, and what the score was. Since many ECM and ECCM equipments are hands-off boxes, there is little information to be had beyond, "Did the lights come on?" The capability to quantify the signals and their interactions is the basis for an accurate decision, and there is an ascending scale of difficulty. One ECM vs. one victim is the least difficult, since two signals can be processed simultaneously or else gated to display the features of one signal or the other. More difficult is the question of many emitters or techniques against multiple threats, but this data can be acquired sequentially if the timescales are long enough. Very difficult is the possibility of many ECM-equipped platforms against multiple ground threats, since the receiver must attempt to record an extremely volatile environment where critical para-

Installation of the systems into the equipment van is shown during the final phases of integration. The HP9000 control code is user-modifiable to allow customizing in the field. In this case, HP BASIC was selected in order to provide the best mix of structure, rapid flexibility, and instrument control.

meters can change over microsecond timescales. Finally, and difficult to the point of absurdity, is the requirement to quantify many ECM platforms in "free play" against both air and ground threats. Low signal levels, platform mobility, and high degrees of freedom in tactics all out-distance a general-purpose receiver.

In all of these applications, frequency management and emitter control often are overlooked and underfunded. This activity must take into account the verification of expected emitter characteristics, the susceptibility of all receivers to all known emitters, the possibility of un-known emitters, and the expected signal levels implied by the emitter locations. Obviously a receiving system is required to support this activity by identifying signals, determining their location if unknown, and reporting interfering emitters in real time. In the context of electronic warfare evaluation, there is an exceedingly high cost to ignorance: missions are run with nothing gained, misleading interpretations are made of corrupted data, and equipment is deployed based on invalid or insufficient data.

A New Modus Operandi. Computing power and instrumentation have come to the point where a different modus operandi is feasible. For example, consider an approach where one engineer is assigned to a project such as equipment evaluation or Elint collection. He is trained to understand the important features of the signals and to determine what data must be collected and analyzed. This depth of understanding allows the engineer to coordinate a collection plan with the information

user, considering the technical capabilities of the receiving system vs. the signal behavior and mission objectives. When the mission is run, this engineer, perhaps with an equally skilled associate, gathers the relevant data and ascertains in real time that the data is adequate. This data is stored digitally on a large-volume disk.

Immediately following the mission, the control computer becomes an analysis station, using standard computer tools such as histograms, scattergrams, FFTs and other transforms to convert raw data into useable information. Proper annotation of graphs with titles and classification markings is straightforward for the computer to generate, when guided by the engineering intelligence. The product is a timely, quantitative report in which each result has supporting evidence presented and which has an identifiable author who is responsible for the validity of the results.

The computer assists with two important appendices: a data log showing the contents of every record taken, and an automatic equipment performance verification showing that every signal path was tested and performing within tolerance at or near the time when the data was recorded. The problem with this scenario is that the required assets are unavailable. Data is taken and muddled through a diffuse process of analysis or is taken without equipment verification, losing its credibility and rendering it worthless.

Given this scenario in which talent and technical continuity are the key elements of the information quality, it is possible for one engineer to plan, collect, analyze and report on the selected events. The entire process from planning (mission minus five days) to report (mission plus five days) consumes 10 days. Compare that to current reporting timescales.

There are corollaries. EW technology is not standing still, and emitter diversity is increasing rapidly. This means that the receiving system of a few years ago is now unable to capture newer modulations. Keeping up requires a continuous thrust to upgrade collection and recording technology. The only way to achieve a low-cost upgrade is to rely as heavily as possible on non-custom instrumentation that can be replaced on a unit by unit basis to take advantage of performance developments.

In addition, users continue to ask for more processing and more analysis of previously unconsidered or unavailable aspects of the data. In response, there must be a continuous process to upgrade the analysis capability as emitters, modulations and users' queries diversify. This means that the software for system control, as well as for data acquisition and analysis, must be suitable for on-site maintenance and development, which in turn means high-level user-modifiable code.

Measurement Requirements. The most obvious requirement is to cover the frequencies of interest with high sensitivity and high processing bandwidth. For radar intercept receivers, that usually translates to a receiving

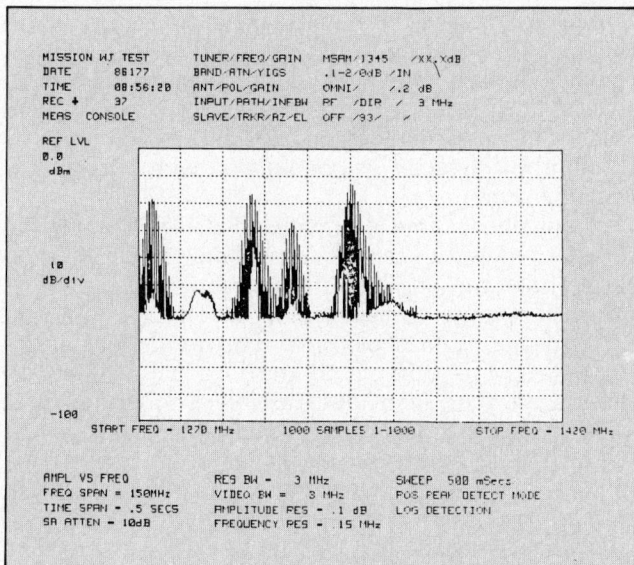

Amplitude vs. Frequency Plot. Several radars as well as CW signals in the San Jose, Calif., area are illustrated using the MAX HOLD feature of an HP8566 spectrum analyzer. Annotation of the RF path configuration as well as the recording device parameters are stored and posted on the plot. The time corresponding to this record is read from a satellite receiver.

system covering frequencies from 100 MHz to 18 or 40 GHz, with sensitivities in a 1-MHz bandwidth of roughly –105 dBm (–135 dBm including the antenna gain). Two processing bandwidths are needed: narrow (2 to 20 MHz) for best sensitivity and selectivity, and wide (500 MHz or greater) for coverage of frequency-agile emitters. A more subtle point is that a collection system can use directional antennas to achieve remarkably low system noise floors, although the antenna must be pointed at the signal.

One attractive possibility to direct the high-gain system is to combine a general-purpose high-probability-of-intercept surveillance receiver with automatic direction-finding capability. This serves as the environmental monitor while precision measurements are made using other equipment. Sensitive receiving systems with automatic signal processing and automatic DOA determination are available and may be controlled remotely by the same computer that controls the precision measurement and recording system.

The need to make measurements on simultaneous signals, not necessarily at the same RF, requires several independently tunable channels and appropriate demodulators, reporting instrumentation such as a multichannel digital oscilloscope, and a storage medium such as a large-capacity hard disk. One area that calls for special attention is the design of the RF front end to support the multiple-receiver processing. The flexibility of a distribution system, possibly with tracking preselector filters, is an instance where custom design is inevitable. The payoff, however, is enormous in that front ends can be made which have noise figures of 6 to 10 dB that will support dynamic ranges of order 70 dB. A low-noise

front end transforms a relatively insensitive spectrum analyzer into a highly sensitive receiver.

Signal measurements can be categorized into data types, including:
- Amplitude vs. frequency
- Amplitude vs. time
- Frequency vs. time
- Predetected amplitude vs. time
- Special purpose demodulator output vs. time.

Amplitude vs. Frequency. Spectral occupancy, or amplitude vs. frequency, can be provided by a spectrum analyzer in its sweeping mode, from which emitter frequency and power spectral distribution is observed. In the case of ECM, the bandwidth characteristics of noise are illustrated. Taking this data in rapid succession shows the time evolution of broadband frequency excursions. Also, by using a display feature such as MAX HOLD on a Hewlett Packard 8566 spectrum analyzer, one can allow

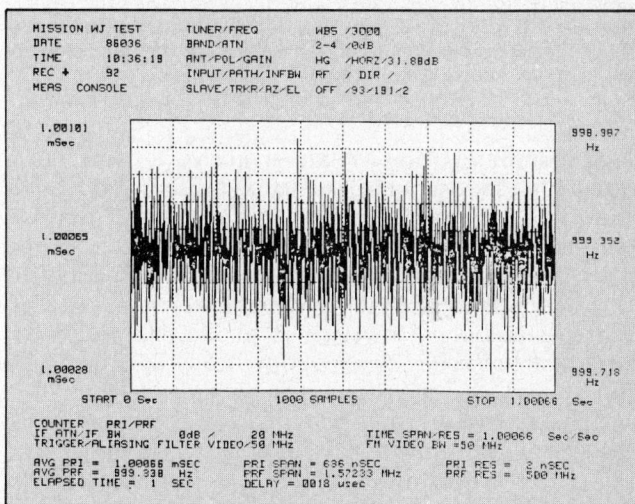

PRI Sequence vs. Sample Number. Data for 1,000 inter-pulse intervals vs. interval numbers generated by a pulse generator shows that a clean signal produces an extremely accurate estimate of emitter PRI. In the case of an emitter using stagger or jitter, a straightforward calculation can illustrate the intervals in use, their sequence and their variance.

the sweeping receiver to display the results of many sweeps, thus mapping out the total frequency envelope of successive intercepts.

An alternative way to present related data is to use the output from a scanning intercept receiver and present the results on a "pan display" covering the frequency range under surveillance. This data can be updated rapidly, thus giving a dynamic picture of the active signals. Under certain conditions, the same data can be presented in a three-dimensional "waterfall" display, which can show a consecutive time sequence of the pan display. This receiver can be the same ESM equipment that is providing emitter angle of arrival information to the precision measurement system.

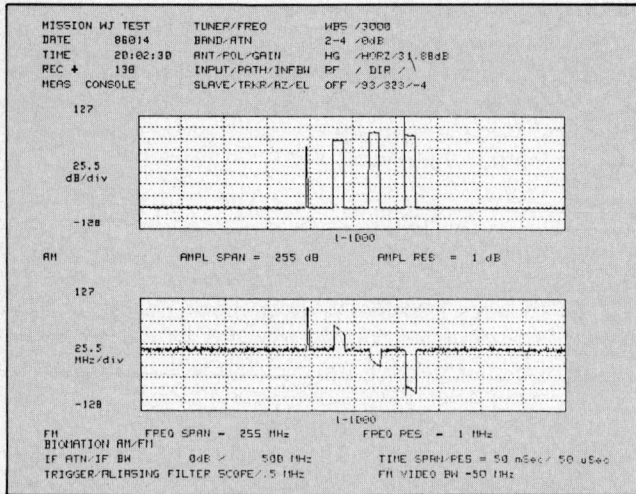

Dual AM/FM vs. Time. The plot shows AM/FM detached data in a 500-MHz bandwidth through a tuner and 500-MHz frequency discriminator. This frequency agility was produced using a sweeping RF signal generator together with a pulse modulator. The sweep excursion was 500 MHz, at a slow sweep repetition rate. The pulser was used to simulate a frequency-hopping emitter. Both pulse-pulse agility and on-pulse chirp are easily recognizable. To measure chirp excursions or slope, an expanded version of similar data is preferred.

Amplitude vs. Time. When a receiver is tuned to a signal (or a spectrum analyzer in SPAN ZERO mode), it is possible to demodulate the signal to recover the AM waveform. Normally a spectrum analyzer is considered a narrowband device; for example, the maximum displayed bandwidth of an HP8566 is 3 MHz. However, the HP8566 makes available an IF having 25 to 50 MHz bandwidth when the tuned frequency is above 2.5 GHz. By processing a sample of this IF in an amplitude detector it is possible to gain a wideband function without disturbing the narrowband functionality. Returning the other split signal to the spectrum analyzer at 0 dB gain preserves normal operation. When the demodulated signal is recorded by a digital oscilloscope, the signal-to-noise ratio, rise time and other on-pulse parameters are determined by using the calibrated response (counts out vs. input power) to find the appropriate levels on the pulse.

Using the spectrum analyzer in the fixed tune mode allows measurements of antenna scan parameters: scan type and rate, beamwidths, sidelobe levels and a variety of longer term measurements such as modulation depth and rate. To quantify ECM techniques frequently demands that many seconds of modulation be recorded to illustrate the interaction between the victim radar and the ECM signal. This is done by reading out the spectrum analyzer display continuously, possibly several times per second. In the ECM vs. radar scenario, a useful technique to isolate one or the other signal is to apply a gate to the wideband IF path so that the displayed signal is that which is passed through the gate. Two signal paths, one gating the other, is not hard to arrange and provides either a spectral or time domain display of only the selected signal.

Measurements in the time domain traditionally have been regarded as more difficult than those in the frequency domain. However, multichannel digital oscilloscopes have made it possible to record not only pulse shapes but also simultaneous AM and FM on pulse by using a log detector in conjunction with a frequency discriminator.

Amplitude vs. time information certainly illustrates the emitter pulsewidth as well as amplitude modulation on pulse, but it can also illustrate pulse group characteristics for such purposes as signal identification.

By measuring threshold break time over a large sample of pulses, it is possible to measure pulse repetition interval, stagger, jitter, and possible wobbulation provided that the time of each threshold break is stored. Using the demodulated amplitude as an oscilloscope input is a convenient way to generate a clean gate which can be applied to a time-interval counter. Recording sequential intervals for 1,000 pulses allows a detailed exposition of quite complicated staggers.

Frequency vs. Time. The use of frequency discriminators to obtain frequency information in the time domain applies to two widely divergent bandwidth regimes: narrowband and wideband. Narrowband (tens of kHz) quantifies fine-grained frequency modulation on pulse such as Doppler shift, low-level chirps, or ECM velocity walkoff, and wideband (500 MHz or greater) covers wide frequency excursions such as hopping, broadband chirp or noise-like ECM. As in the case of amplitude, fine frequency measurements may allow emitter identification to be based on the spectral coefficients of the frequency modulation on pulse. Also, there is the possibility of quantifying ECM velocity deception programs.

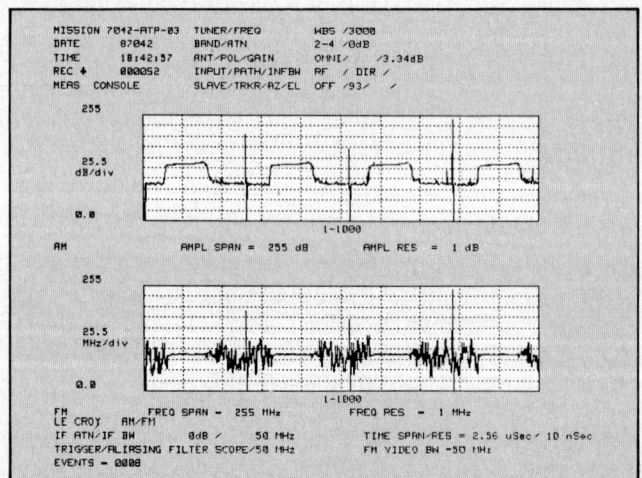

Multi-Event Dual AM/FM vs. Time. Dual channel AM/FM data is shown corresponding to a signal having a one-meter/second PRI and a one-microsecond pulsewidth recorded on a custom multi-event segmented memory digitization and storage unit. Interpulse time is suppressed. The vertical lines on the plot are actually TOA bits for each threshold break (10 nanosecond resolution). Signal input is from built-in test equipment.

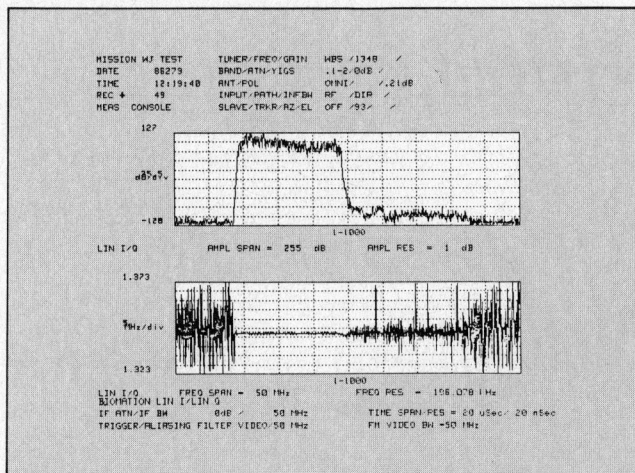

Linear I/Q vs. Time. A reconstructed AM/FM plot of a local radar using linear I/Q data. This data has an information bandwidth of roughly 100 MHz, even though the video bandwidth and digitizing bandwidth is 50 MHz. Reconstructed amplitude is formed by adding the I and Q channels in quadrature, and the reconstructed phase is calculated using the inverse tangent of the Q/I ratio. Frequency is determined by differentiating the instantaneous phase. Since this is a linear detection process, the dynamic range is limited to slightly more than 20 dB, which is adequate for modulated-carrier signals, but minimal for pulsed signals. For improved dynamic range performance, one may use Limited I/Q processing, which looks quite similar, except that amplitude information is largely destroyed by the limiting process. Phase information is preserved, so that this becomes the preferred technique for illustrating phase and frequency modulations on pulsed signals.

One configuration that makes this demodulation straightforward is the routing of the 21.4-MHz IF from a spectrum analyzer (such as an HP8566) to a communications receiver (such as a WJ8617B) which has a narrow-band AM and FM demodulator. This makes demodulated bandwidths from 10 kHz to 4 MHz available to a digitizer which can generate the familiar dual-channel plots of AM and FM vs. time.

On the other hand, frequency-hopping emitters and broadband chirps are increasingly common and can be studied using wideband receivers and frequency discriminators. As before, one can use the "high IF" of the spectrum analyzer to provide a 25-to-50-MHz bandwidth output. This bandwidth is not adequate for modern hoppers, and a better solution is to use a stable broadband tuner followed by a discriminator having a 500-MHz bandwidth. This allows the frequency of every successive pulse to be recorded. Additionally, a broad filter often allows an entire ECM "spot" to be quantified as a band of noise, possibly swept. Using a continuous wave trigger, it is possible to record the time evolution of various sorts of swept noise.

In order to record successive pulses, it is necessary to employ a multiple event digitizer which can segment the digital memory in such a manner that each successive event is stored but the uninteresting inter-pulse time is not stored. Such digitizers also allow control of the "pre-trigger" ratio—the fraction of time preceding the threshold break that caused the trigger, relative to the length of the recorded interval—in order to view the leading edge of the pulse. The duration of the interval stored when the system triggers can be set by reading the time-base from an analog oscilloscope, where the signal has been manually optimized. Obviously, prior knowledge can be used in predictable situations.

Control over the sample rate and sample size allows either few samplings per pulse and a long sequence of pulses to fill the local memory, or a large number of samples per pulse but fewer pulses to fill. Experience has shown that a reasonable range is from 16 samples per pulse for 2,000 pulses to 1,024 samples per pulse for 32 pulses. Digitizer rate is evolving extremely quickly, with reasonably-priced units available that digitize at 100 megasamples per second. There are other limitations, but illustrating a pulse train having 30-nanosecond pulses at several million pulses per second is readily achievable.

Analysis of I/Q Data. Pre-detected data provides several unique features of recording that are not possible to recover from other demodulators. For example, the direct recording of the waveform preserves the phase information and therefore allows a determination of the phase evolution from one pulse to the next. Certain classes of emitters are interrupted CW, in which a single-frequency sine wave is gated by a modulator. In this case, the emission is said to be coherent, in that each successive pulse carries a specific phase depending on the frequency of the master oscillator and the elapsed time between pulses.

One of the important techniques for recording pre-detected waveforms is the use of "in-phase/quadrature-phase" (I/Q) demodulation. This technique produces two copies of the signal, converted to baseband frequencies which are 90 degrees out of phase with each other. The signals are then bandpass filtered and digitized. The use of both signals allows recording of positive and negative frequency components and extends the information bandwidth beyond the Nyquist limitation of digitizing a single channel.

The utility of using I/Q demodulation for Elint is that the technique can cover broad bandwidths and yet is accurate for narrowband applications as well. For example, a system having 50 MHz of video bandwidth, when digitized by a 100 Msample/second digitizer, would be Nyquist-limited to 50 MHz. The I/Q process maps the RF spectrum into frequency components from –50 MHz to +50 MHz, and the entire information bandwidth of 100 MHz is preserved.

I/Q demodulation is done either in a linear mode having perhaps 25-dB dynamic range, or in a limited mode in which the amplitude information is destroyed but the phase information is preserved. Thus the strengths and weaknesses of this technique are complementary to those of log AM detection. Possibly the most interesting application is the demodulation of simultaneous phase and amplitude modulated signals.

Recovery of signal amplitude information is done by adding the square of the I and Q components in quadrature. Recovery of the instantaneous phase is done by taking the arctangent of the Q/I ratio. Each digitization produces a record of amplitude and phase, from which successive values of phase can be differentiated to calculate the instantaneous frequency. The power of the approach is that amplitude, phase and frequency are all recovered, with double the available Nyquist bandwidth.

Any Elint system has custom requirements such as spread spectrum demodulation or PSK. Typically, the outputs can be made into a baseband waveform that is conveniently recorded in the same time-domain digitizer as outlined above.

Data Products. Much emphasis has been placed on the inexpensive, near-real-time capability to gather and present raw data, such as spectrum analyzer data (both scanning and fixed tuned), AM/FM to time domain digitizer, AM(1)/AM(2) to time domain digitizer, I/Q to time domain digitizer, and PRI interval counter/storage. These are used primarily to verify that the appropriate data has been recorded, but detailed analysis is usually not possible during the frenetic collection of a typical mission.

The primary distinction of off-line data analysis is that the time pressure to acquire raw data no longer exists. Therefore, the analyst can move back and forth throughout the collected data, sorting and selecting those records that are pertinent. One of the first steps is to remove known instrumentation effects by calibration corrections. The next step is to apply the analysis tools that turn the data into information: markers, windows, histograms, scattergrams, and various forms of digital filtering and transforms. Also important is the ability to select events meeting certain criteria, in order to perform statistical analysis on an ensemble of data. Much of this is done to support individual requirements, and tends therefore to be custom-coded to support the information requirements of the end user.

The remaining necessity is to take data to establish that the receiving system was in fact operational, and that the conversion from RF energy level to digitizer counts is understood to the required level of precision. There is no other way to achieve this than to supply an RF signal source at a calibrated power level into the upstream end of the receiver. The response can be evaluated at constant frequency, since the input power level is varied in order to describe the system noise floor and compression dynamic range. The dynamic range of the system depends not only on the properties of the front end, but also on how the video outputs are presented to the digitizer. It is possible to misalign a system, such that the digitizer saturates well before the front end and the usable dynamic range is far less than is supported by the RF section. It is also possible to mismatch a receiver such that the actual dynamic range is high but the output response is mapped into only a portion of the digitizer counting range. This degrades the available amplitude resolution.

To characterize the system response, an expression for the gain is needed to relate the power out to the input power. The problem is that the gain is frequency-dependent and will be different for each signal path leading to the digitizer. In order to determine gain, using the same signal source at constant power while stepping in frequency generates a "cross-band gain curve" from which a fit determines parameters to approximate the cross-band gain (both in and out of band). The equation for output counts then can be inverted to find the input power.

The use of a frequency discriminator also requires that information be gathered on the response across the passband. The voltage output of the discriminator is transformed into frequency by inverting an expression relating the digitized response to the discriminator slope plus possible correction terms for non-linearity or offset. Data to illustrate the response may be acquired by sweeping a signal source across the discriminator, marking the nominal band edges, and recording the output of both AM and FM detectors. This has the advantage of mapping the AM passband response as well as characterizing the FM. Differences are to be expected both as a function of RF and input power level, thus a complete characterization will step both these parameters as well. The process generates a great deal of data. However, all can be done under computer control, using machine-calculable criteria for such things as linearity and offset to reduce the output avalanche.

The application of precision waveform measurements to support strategic reconnaissance, countermeasures equipment evaluation, training missions scorekeeping and range emitter management share certain technical requirements. Among these requirements is the ability to quantify modulations and signal levels. Much of this activity can be achieved using standard laboratory instrumentation: spectrum analyzers and digital oscilloscopes. Modern emitters require two important additions—low-noise front ends followed by precision broadband demodulation, and a sensitive automatic direction-finding receiver, which can steer the directional antennas to signals of interest.

Bringing the ensemble under the control of a single computer has the advantage of making the data available digitally and subject to immediate manipulation. A trained engineer working with software analysis tools can convert the raw data to information suitable for the end user in a matter of hours or days, depending on the complexity. Such an approach makes the most of inexpensive instrumentation rather than relying on custom electronics. The result is that highly sophisticated measurements can be acquired, analyzed and reported quickly and effectively. ∎

Dr. Gregory J. Donaldson works with detailed emitter waveform measurement, combat identification and monopulse direction finding systems at Watkins-Johnson Company, San Jose, Calif. He received his bachelor of science degree in physics from the University of Washington, and master of science degree and Ph.D. in physics from Stanford University.

Soviet Ocean Reconnaissance Threat from Space

The Soviet Union's space-based ocean surveillance network is being expanded and integrated into Soviet military strategy.

By Nicholas L. Johnson

After 20 years of experimenting with more than 50 spacecraft, the Soviet Union's space-based ocean surveillance systems appear poised to enter a new era of tactical and strategic operations. A flurry of unusual activity during 1986 and 1987 in the network, normally characterized by its highly regimented and predictable behavior, suggests an imminent transition to an advanced Block II system. With its heavy reliance on naval power around the globe and in the presence of a growing Soviet naval threat, the United States must be prepared to meet this new challenge or face a further adverse tilting in the balance of power.

The objectives of this ocean reconnaissance network are to detect, identify and track U.S. and Allied naval forces and to relay this information in real time directly to Soviet naval and air elements for targeting purposes. In peacetime and periods of world tension, this information enables Soviet military leaders to monitor the movements of Western naval forces and to warn of unusual or threatening formations. In wartime, ocean reconnaissance data will help direct Soviet weapons platforms or the munitions themselves against enemy vessels. The devastating effects of even single missile hits on the *HMS Sheffield* and the *USS Stark* vividly illustrate the need for modern warships to remain unseen and elusive.

Although known in the West since the 1970s, the Soviet Union's space-based ocean reconnaissance system was not officially acknowledged by the U.S.S.R. until 1984. Naturally, the Soviet Union views the system as purely defensive in nature and justifies it by the perceived threat of U.S. aircraft carrier task forces to the Motherland and to other regions of Soviet interest or protection. The Soviets have pointedly stated that they "do not believe the monitoring of these aircraft carriers and maintaining capability to oppose them in the event of war to be a destabilizing function."

Rorsats Make Their Debut. The first and best known Soviet ocean surveillance satellite debuted in the final days of 1967 and is known as the radar ocean reconnaissance satellite (Rorsat). As the name implies, Rorsats work on the simple principle of emitting a strong radar beam and analyzing the reflected energy. Reportedly, Rorsats can identify destroyer-size vessels 50 percent of the time,

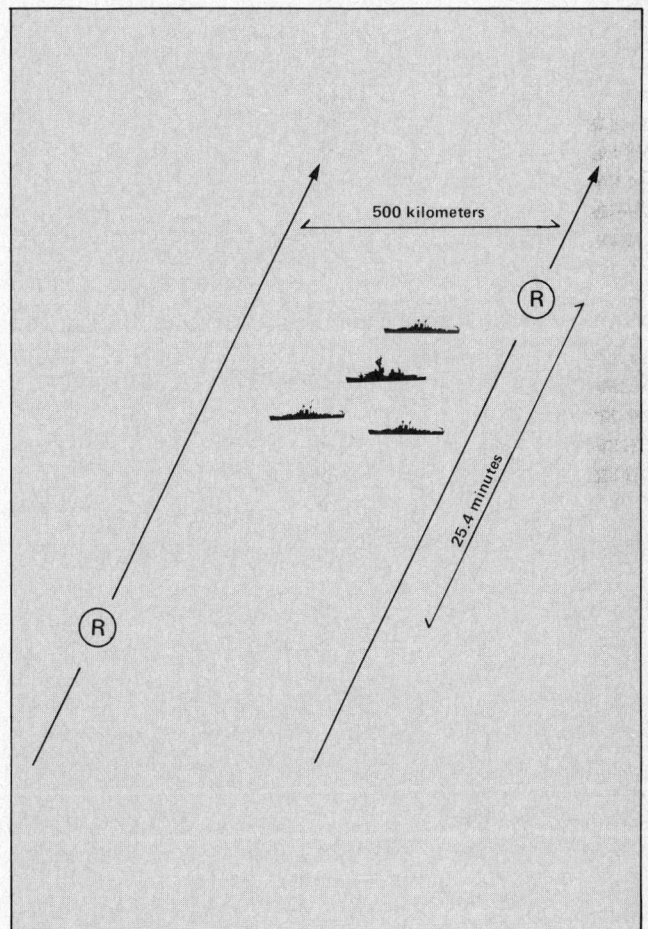

Figure 1. The groundtracks of a typical pair of Rorsats will be displaced 500 kilometers for naval targets near 45 degrees north latitude. Some Eorsat pairs have exhibited very similar patterns.

cruiser-size ships 75 percent of the time and aircraft carriers as much as 90 percent of the time. Once they detect and identify a naval group, Rorsats can estimate the bearing and speed of the targets. This latter information is susceptible to the deliberate change of course just prior to the highly predictable overflight of the Rorsat.

To increase the probability of detection, Rorsats circle the Earth at very low altitudes—250 to 265 kilometers—

and at inclinations of 65 degrees. By restricting their flights between 65 degrees south and 65 degrees north latitude, Rorsats increase their observation time over the major oceans of the world where U.S. and Allied forces are deployed. Since the probability of detection is also a function of the power of the radar beam, the Soviet designers looked at methods of maximizing the electrical power available to the radar. Batteries alone were out of the question due to the radar's high power drain, and solar panels large enough to quickly recharge the batteries presented atmospheric drag difficulties due to the satellite's low operating altitudes.

The solution was innovative as well as technically challenging. Even though the United States and U.S.S.R. were still experiencing considerable launch and payload reliability failures, Soviet spacecraft engineers committed Rorsats to a nuclear power supply. Nuclear power supplies offer compactness and high power densities compared to other electrical generation techniques, but they require additional safety measures throughout the launch and life of the satellite. An additional attribute normally associated with nuclear-powered devices is longevity. Curiously, however, Rorsats have consistently exhibited very short operational lifetimes with an average of eight to 12 weeks and a maximum of 20 weeks. Consequently, Rorsats are flown only sparingly for test and evaluation periods and during times of increased Allied naval activity or world crisis. Whether the cause for the short lifetimes lies in the power supply or elsewhere in the payload cannot be determined from available unclassified sources.

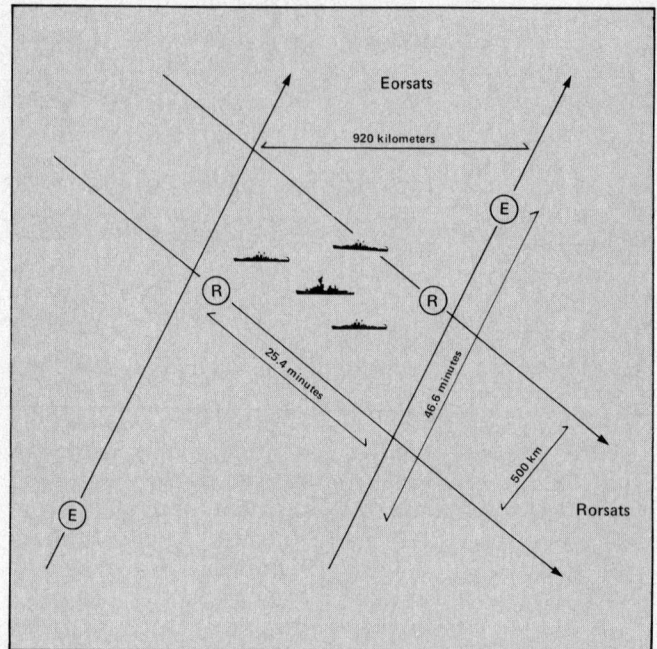

Figure 3. Within one hour, a naval target at mid-latitudes might be observed by two Rorsats and two Eorsats. The 47-minute Eorsat spacings are preferred when working with a pair of Rorsats, and the orbital planes are selected to permit the intersection of ascending Eorsats and descending Rorsats at mid-latitudes in the Northern Hemisphere.

A Rorsat's radioactive power supply could reenter the Earth's atmosphere within weeks of mission termination or system malfunction. To prevent such an occurrence, a separate engine is attached to the power supply which boosts the hazardous compartment to a much higher, circular orbit of about 900 kilometers, where it will remain for several hundred years. This boosting operation failed only twice in 30 flights between 1967 and 1986. The first failure with Kosmos 954 in 1977 eventually resulted in an uncontrolled reentry over Canada which left a small region of tundra contaminated with radioactive debris. A redesign of the spacecraft ensued, and flights resumed in 1980. When the second boost failure occurred in 1982 on Kosmos 1402, the radioactive fuel still returned to Earth but the deadly material was dissipated high in the atmosphere where it posed no significant threat to life.

Enter the Eorsat. The Soviet Union developed a complementary satellite system that relies on more conventional techniques and power supplies to accomplish its ocean surveillance mission. Dubbed Elint ocean reconnaissance satellites (Eorsats), these spacecraft perform classical electronic intelligence (Elint) missions concentrated over the same ocean areas as the Rorsats by passively listening to radio and radar emanations from the ships themselves. The growing U.S. emphasis on naval systems to defend against cruise missiles and aircraft-delivered ordnance means that powerful search radars are operated routinely. This effectively provides a beacon to any Eorsat flying nearby.

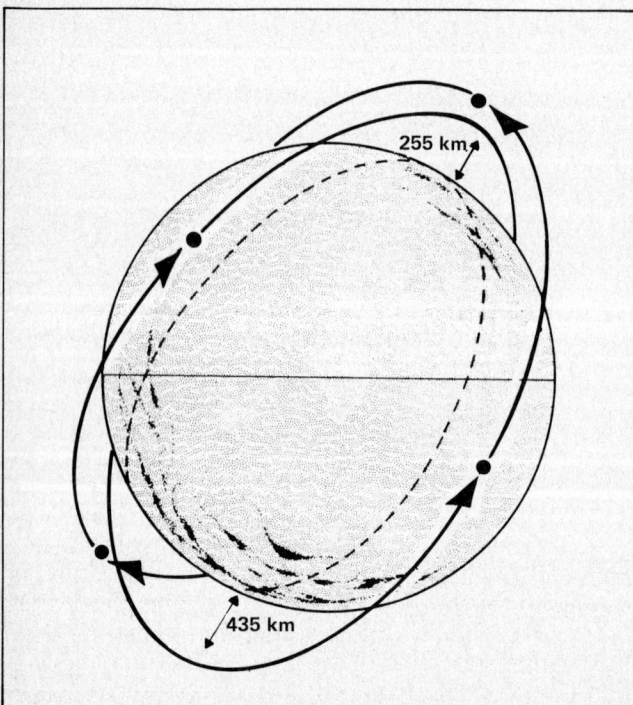

Figure 2. A pair of Rorsats (255 kilometers) and a pair of Eorsats (435 kilometers) appear to be a minimum constellation for reliable ocean surveillance.

Eorsats did not appear until 1974 but have rapidly become the major element in the Soviet space-based ocean reconnaissance network. During the past four years at least one Eorsat has been operational at all times. In contrast, a total of six Rorsats accumulated less than 14 months on station time during the same period.

A single overflight by a Rorsat or Eorsat may not be sufficient to ensure detection or provide an accurate location. Therefore, the Soviets often construct constellations of Rorsats and Eorsats. The first Rorsat pair was formed in 1974 several months before the maiden flight of the Eorsat. Unlike other Soviet satellite constellations which employ widely separated orbital planes, the Rorsat pair (Kosmos 651 and Kosmos 654) drew immediate attention because they resided in the same orbital plane but crossed the equator more than 25 minutes apart. Due to the Earth's rotation (one degree every four minutes), the groundtrack of the second Rorsat was displaced more than six degrees to the west compared to that of the leading Rorsat. This equates to 500 kilometers in the mid-Atlantic latitudes (45 degrees north). From these different perspectives the location and bearing of a task force can be defined more accurately. Rorsats also have experimented with 38-minute separations, resulting in 750-kilometer groundtrack displacements. The possible spacings are not arbitrary but rather are dictated by orbital mechanics which allow the two satellites to maintain the same set of rigid equator crossings in a pattern which repeats every seven days.

Like the Rorsats, Eorsats began flying in pairs after the initial solo trials. The first Eorsat pair (Kosmos 1094 and Kosmos 1096) operated in 1979 when the Rorsats were grounded following the Kosmos 954 accident. Eorsats also are flown in virtually identical orbital planes with intra-plane spacings of 23.3 and 46.6 minutes. Although different from the Rorsat pairings, Eorsat constellations adhere to similar relationships that allow a set of equator crossings repeating every four days. The wider displacement of Eorsat groundtracks in the 47-minute constellations is partially offset by the their higher, 435-kilometer altitudes and the resultant greater field of view.

Since Rorsats and Eorsats share the same tactical missions, the notion of joint operations is natural. In fact, since the second Eorsat mission, Rorsats and Eorsats have always worked together when in orbit simultaneously. A single Eorsat (Kosmos 777) and a single Rorsat (Kosmos 785) first tested the concept in 1976, and by the next year an Eorsat (Kosmos 838) flew in formation with a pair of Rorsats (Kosmos 860 and Kosmos 861). The latter network was the preferred constellation until 1981 when a single Rorsat worked with a pair of Eorsats for two weeks. Finally, in 1982 a full constellation of two Rorsats and two Eorsats was formed. This appears to be the minimum desired configuration. An even larger network of two Rorsats and three Eorsats was tested in 1985, followed by one Rorsat and four Rorsats in 1986.

The U.S. Department of Defense considers the Soviet space-based ocean reconnaissance network to be one

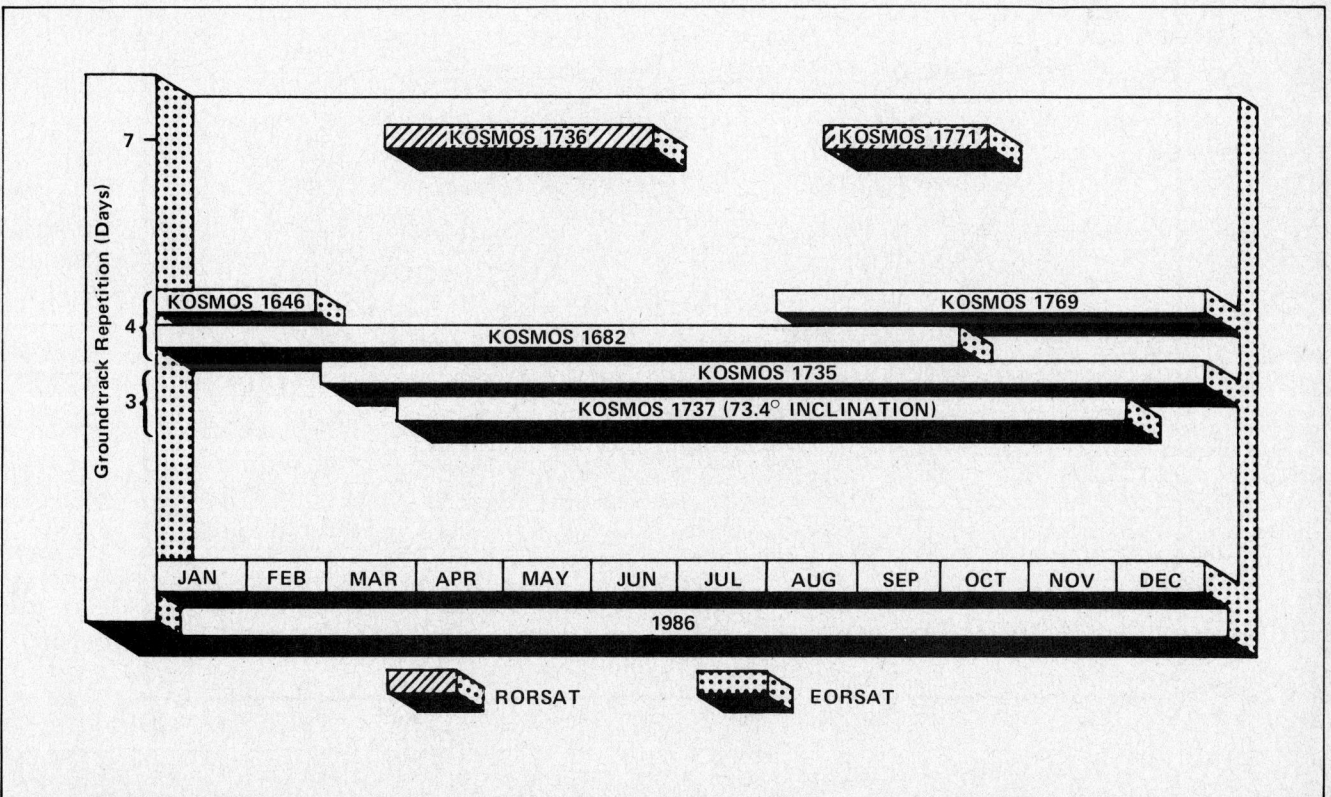

Figure 4. The year 1986 witnessed an unprecedented level of unusual activity in the Soviet ocean surveillance satellite network.

of the highest value tactical space assets available to the Soviet Union. The U.S. anti-satellite (Asat) program is in large measure rationalized by the threat of Soviet Rorsats and Eorsats. In testimony before Congress in April 1987, Gen. John Piotrowski, commander-in-chief of U.S. Space Command and Norad, said:

"For example, if the United States were ever called to enforce the free flow of oil to the Western world through the Straits of Hormuz or to deploy military force to fulfill our obligations to our NATO allies, Soviet radar ocean reconnaissance satellites could provide time-critical tracking and targeting of U.S. troop reinforcements in ports and of U.S. convoys and battle groups during their movement across the oceans . . . Undeterred by a U.S. capability to put these hostile systems at risk, these Soviet capabilities have the potential to seriously jeopardize our ability to project and sustain U.S. forces and to fight once engaged."

In addition to physical destruction of Rorsats and Eorsats by a U.S. Asat weapon, other countermeasures are available. Rear Adm. Richard Macke, commander-in-chief of Navy Space Command, has stated that, while captain of the aircraft carrier *Dwight D. Eisenhower,* he hid his ship from the Soviets for four days using radio silence techniques. When applied judiciously, this pro-

cedure can be effective against Eorsats. However, radio silence does not deter Rorsat detections, and extended periods of radio and radar silence are unwise when within range of Backfire and Blackjack bombers. Electronic jamming of a Rorsat radar may prevent precise position fixes but is a tip-off of general location. Furthermore, data from a pair of Rorsats subjected to jamming could be used to localize the position of the jamming source. Finally, as the number of operating Rorsats and Eorsats increases, the observation opportunities will likewise multiply. Hence, temporary changes in speed and bearing by a task force when a Rorsat or an Eorsat comes within range will quickly be seen through after several plottings of actual position.

Quick Response. The maturation of the Soviet space-based ocean surveillance program since 1980 is evidenced not only by increasing numbers of operational spacecraft but also by the quick reaction to changing conditions, apparent deployment during times of crisis, and new tactics which may be aimed at extending operational envelopes and enhancing survivability. One of the impressive aspects of the program is the Soviet Union's ability to promptly replace failing satellites. For example, an Eorsat (Kosmos 1260) was launched less that 24 hours

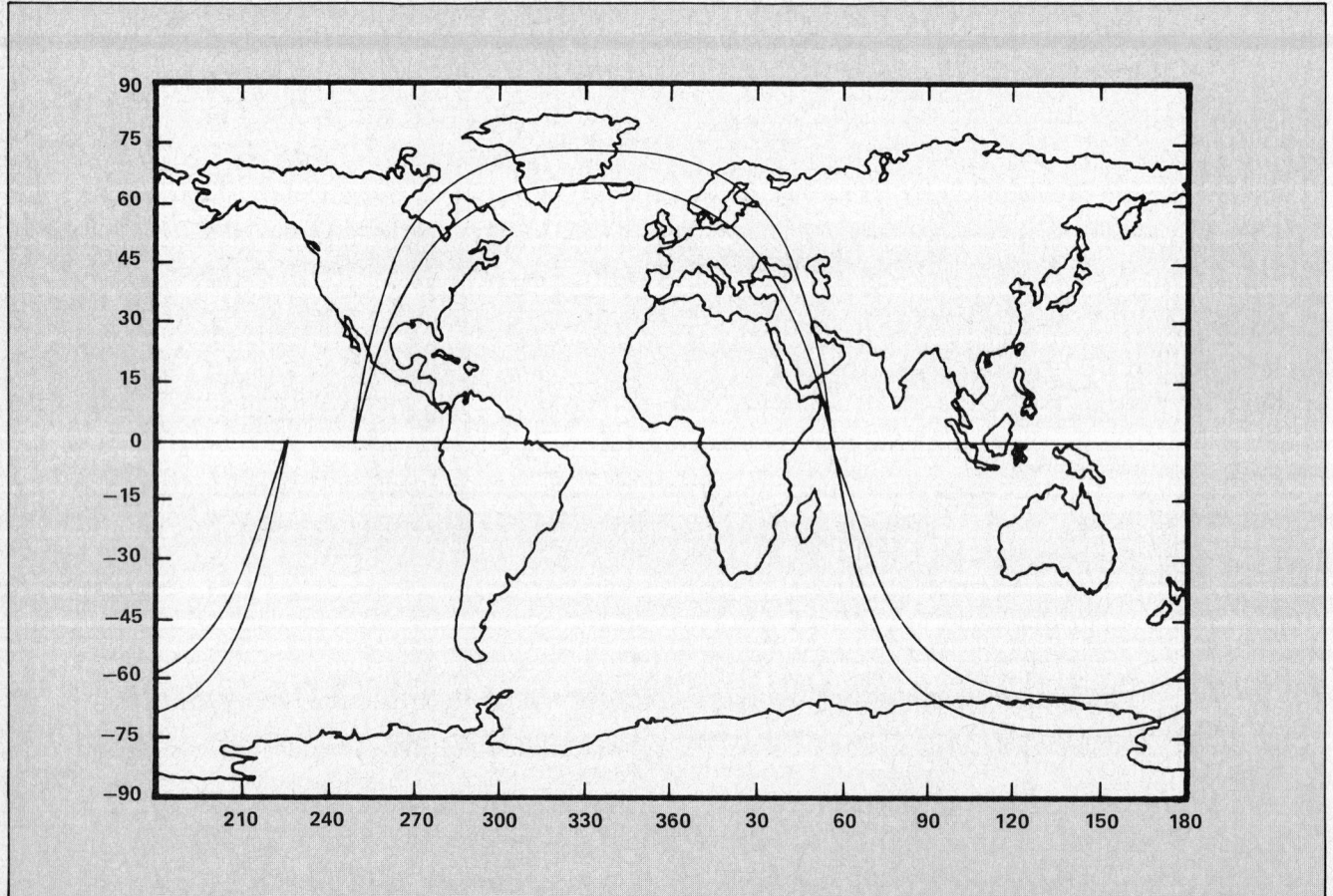

Figure 5. The precise groundtrack coordination of the Kosmos 1735 and Kosmos 1737 Eorsats in 1986 left little doubt as to their objectives.

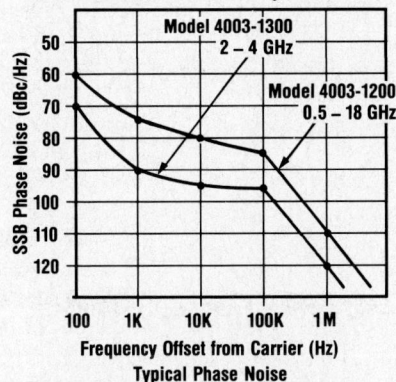

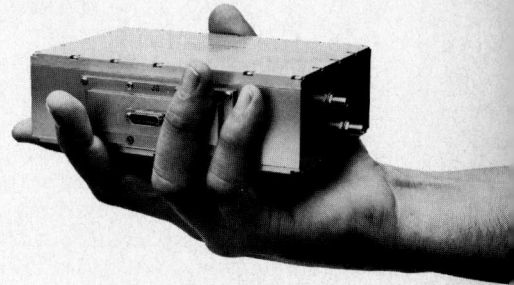

after a sister satellite, Kosmos 1167, terminated its primary mission in early 1981. Kosmos 1260 in turn was replaced within a day by Kosmos 1306. In 1982 the Rorsat Kosmos 1372 was replaced within 19 days of its premature end in order to preserve a Rorsat pairing. When the other member of the original Rorsat pair, Kosmos 1365, finally ceased operations, it was replaced in only five days. The U.S. landing in Grenada in October 1983 was followed within four days by the launching of the Eorsat Kosmos 1507, which immediately went to work with the Eorsat Kosmos 1461, the only Soviet ocean reconnaissance satellite operating at the time of the crisis.

Increased levels of Soviet satellite activity also are noted regularly during large-scale NATO exercises. A new Eorsat and a new Rorsat were launched in August 1981, just in time to join a resident Eorsat to observe a NATO exercise billed as the largest naval war games since World War II. The formation of a pair of Rorsats and a pair of Eorsats in the summer of 1982 immediately preceded the initiation of the Autumn Forge NATO war games which included naval maneuvers by 160 ships. The launch of a Rorsat to join a pair of Eorsats late in 1984 was at least circumstantially related to U.S. landing exercises that began the next day on the western coast of Africa. The second time a pair of Rorsats and a pair of Eorsats were operational occurred in the summer of 1985 just prior to the NATO Ocean Safari exercise, involving nearly 200 warships from 10 countries in a trek from New England to Europe. This unprecedented naval maneuver presented the Soviet Union with a rare opportunity to monitor NATO strategy and test its own ocean surveillance satellite network. The Soviet ocean surveillance network was beefed-up again in the summer of 1986, as the largest peacetime naval armada formed in the Sea of Japan for exercises off the east coast of the Soviet Union. For the first time, two pairs of Eorsats and a single Rorsat combined to make a formidable spy network.

One of the best opportunities to test the Soviet ocean surveillance network during a crisis came in the spring of 1982 as England prepared to respond to the Argentine invasion of the Falkland Islands. Due to the malfunction of an Eorsat within a week of launch in February 1982 and the subsequent failure of an older Eorsat, the Soviet Union was without any ocean surveillance satellites for the first time in two years. Thus, when the British fleet headed for the South Atlantic in April, Soviet naval and naval air forces were called in to keep track of the convoys. Not until the British fleet had arrived near the Falklands were the Soviets able to launch an Eorsat (Kosmos 1355). Two weeks later a Rorsat (Kosmos 1365) was orbited, followed by a second Rorsat (Kosmos 1372) after another two-week interval.

How the Soviets used the Eorsat and Rorsat observations during the conflict is open to conjecture. In general, the position of the British fleet was well known to the world. The most intriguing—yet completely unsubstantiated—hypothesis is that Kosmos 1355 was somehow involved in the Exocet attack on the HMS Sheffield. The Soviet Eorsat passed directly over the Falklands at 1220

GMT and continued on a bearing that brought it within radio range of Soviet territory 35 minutes later. Within a few hours an Argentine Super Etendard fired the deadly missle that mortally wounded the Sheffield. The Sheffield had been standing picket duty for the British fleet and should have been easily detected by Kosmos 1355. The information could then have been relayed to Argentine forces either by elements of the Soviet fleet in the region or by a command center in the U.S.S.R. In actuality, the Argentines probably needed no help to locate the Sheffield. However, the above scenario illustrates the potential power a space-based ocean surveillance system could wield.

A New Era. Recent activities in the Soviet satellite network suggest the Soviets are trying to expand their capabilities and may be moving into a second generation. From 1967 to 1985 Rorsats and Eorsats had scrupulously followed very particular orbits. In fact, station-keeping maneuvers were always required every few days to prevent orbital perturbations from degrading mission effectiveness. Then in 1986 three new Eorsat profiles were demonstrated and a new paring technique was tested. In 1987 this experimentation continued with a new ocean surveillance satellite of unknown type circling the Earth in a much higher than normal altitude and with the switch of a resident Eorsat from one surveillance pattern to another.

The new Eorsat test program apparently started in February 1986 when Kosmos 1682 moved from its nearly circular orbit into an eccentric orbit with the same orbital period. This permitted the satellite to retain its strict timing requirements but experiment with observations at different altitudes. A little more than a week later, a new Eorsat, Kosmos 1735, was launched into a slightly lower than normal orbit which allowed a surveillance pattern of three days instead of four days. This may foreshadow an eventual constellation of three Eorsats working in three-day patterns.

A month later the second Eorsat of the year (Kosmos 1737) was launched in the heretofore unused inclination of 73.4 degrees. This mission may have considerable significance for future Soviet ocean surveillance operations. Until Kosmos 1737, every Rorsat and Eorsat had been launched from Tyuratam into an inclination of 65 degrees. The new inclination permits deeper coverage into the northern oceans where U.S. and Soviet maneuvers have increased in recent years. Perhaps more importantly, the inclination of 73.4 degrees is remarkably close to the 73.6 degree inclination used by geodetic satellites launched from Plesetsk. The Plesetsk geodetic satellites use an up-rated version (SL-14) of the Rorsat/Eorsat launcher (SL-11) at Tyuratam. The implication is that Eorsats may in the future originate from Plesetsk as well as Tyuratam to provide the Soviets with greater launch flexibility. Currently, Eorsats and Rorsats may have to share launch facilities at Tyuratam with the Soviet Asat and all use the SL-11 launch vehicle. In time of tactical war, the ability to replenish Eorsats from Plesetsk would be valuable.

Another interesting facet of Kosmos 1737 was its operations with Kosmos 1735. Although they flew at different inclinations, their orbits were designed to allow a unique coordination of groundtracks. Both satellites crossed the equator at the same points but reached different latitudes. This surveillance pattern offers new possibilities for tracking ships in the strategic North Atlantic. In December 1986, Kosmos 1737 again exhibited unusual behavior when it was abruptly brought back to Earth. No other Eorsat had ever been deliberately deorbited, and the Soviet Union has offered no explanation.

In February 1987, the mystery surrounding innovations in the ocean surveillance program deepened. Kosmos 1818 was initially placed in a parking orbit used only by Eorsats. However, instead of moving to a nearly circular orbit at about 435 kilometers the satellite climbed to a circular orbit near 800 kilometers. Since that time, Kosmos 1818 has made periodic orbital adjustments identical to those of Eorsats and Rorsats and has followed a seven-day surveillance pattern like that of Rorsats. The type of surveillance technique employed by Kosmos 1818— passive Elint or active radar—has not yet been verified nor has the rationale for the new higher orbit. One hypothesis is that the higher altitude makes the satellite less vulnerable to the U.S. air-launched Asat.

On April 8, 1987, the Soviets placed a new Eorsat, Kosmos 1834, in a three-day surveillance pattern and teamed with Kosmos 1735 to form the first pair of this type. After less than two weeks of joint operations, the older satellite moved to a slightly higher altitude into a four-day surveillance pattern where it joined Kosmos 1769. The previously undemonstrated ability to shift from one surveillance pattern to another could be valuable when evaluating countermeasure options in a time of war. Specifically, evasive maneuvers are usually considered to be potentially effective responses to Asat attacks. If the target satellite can design an evasive maneuver to permit continued operations, mission effectiveness need not be sacrificed.

Today, the Soviet space-based ocean surveillance network is unequaled anywhere in the world. Activities in recent years clearly suggest this high value tactical program is being expanded and integrated into Soviet military strategy. With an increasingly capable Soviet Navy appearing in all the oceans of the globe, the United States would be wise to learn from the lessons now being taught by the U.S.S.R. and be prepared to meet this growing challenge. ■

Nicholas L. Johnson, an internationally recognized authority on the Soviet space program and space debris, is advisory scientist at Teledyne Brown Engineering, Colorado Springs. From 1983 to 1986 he directed all space defense, space surveillance and space debris analytical activities. He also played an extensive role in supporting Norad and NASA space surveillance requirements, and was responsible for a NASA-sponsored technical report that lists all satellite breakups. From 1979 to 1983, Johnson conducted space defense analyses for the U.S. Air Force Space Division, and played a principal role in creating space defense command and control system sizing scenarios for the Aerospace Defense Command. He is the author of TBE's internationally acclaimed annual report on Soviet space activities, *The Soviet Year in Space*, and serves as a guest lecturer at the U.S. Air Force Academy and the Naval Postgraduate School. Many of his articles have been published in scientific and military journals, and he is the author of the authoritive two volume set (1979-80) of the history of the Soviet deep-space and manned space flight programs for the American Astronautical Society. He is also a contributor to *Jane's Spaceflight Directory*. Johnson has served in both the U.S. Air Force and the U.S. Navy. He is a fellow of the British Interplanetary Society and a member of the American Institute of Aeronautics and Astronautics.

Technologies from AEG help the Air Force to ensure national defense.

The intensive cooperation between the Air Force and AEG covers the electronic equipment of airborne weapons systems

Tornado nose radar

which affords superiority in the event. Special proof of our efficiency: the unique precision in the production of the Tornado nose radar.

For decades AEG has designed and produced airborne radars and ECCM systems for aircraft. For their materialisation an extensive line of key components, equipments and subsystems is available.

Self-evident at AEG: the effective research and upgrading of equipments and systems as well as the reliable technical and logistic support services.

The electronic equipment of airborne weapons systems is only one example of the efficiency and the innovative power of AEG. Other areas of concentration in the field of defense technology are reconnaissance and surveillance, command, control and communication, electronic warfare, firepower and fire control, underwater warfare, airborne, naval and ground equipment,

training, systems management, support, service and quality assurance.

AEG Aktiengesellschaft
Theodor-Stern-Kai 1
D-6000 Frankfurt 70

AEG

A1.WAK 3566 E

SYSTEMS
OF THE WEST

Electronic Warfare Systems: AN/ Designated Hardware 113
Electronic Warfare Systems: International Hardware 138

Electronic Warfare Systems

AN/ DESIGNATED HARDWARE

AN/AAQ-4 Infrared Countermeasures System

First introduced as an internal configuration on U.S. Air Force EB-66 aircraft for deployment in Southeast Asia. Subsequently the system was reconfigured for helicopter application and updated with multithreat capabilities for redeployment on U.S. Air Force H-53s. In the helicopter configuration, the dual transmitter provides protection on bands I and II without engine suppressors. The system electronically modulates a visual cesium infrared source to produce a highly effective jamming signal. An updated version, AAQ-4(B), is under development.
MANUFACTURER:
Northrop Corp.
Defense Systems Division
Rolling Meadows, Illinois USA

AN/AAQ-8 Infrared Countermeasures Pod

Multithreat infrared countermeasures system capable of operating in supersonic environment. This pod is a second-generation system updated to meet new and continuing threats. Mounted in an aerodynamically faired pod, this system has been extensively deployed on Air Force F-4, A-7 and C-130 aircraft. Pod can be configured with a ram air turbine allowing protection independent of aircraft prime power and cooling resources.
MANUFACTURER:
Northrop Corp.
Defense Systems Division
Rolling Meadows, Illinois USA

AN/AAR-34 IR Receiving Set

Cryogenic, tail-mounted unit provides passive detection for the F-111 family of aircraft, self-protection against threats from heat-seeking missiles. The detector is insensitive to clouds, sun and ground clutter due to closed-cycle cryogenics converter. It contains multiple-threat detection and discrimination modes; has augmented resolution and data rate in tracking mode and is coupled with automatic pilot warning and countermeasures command. Consists of four units: a scanner and a cryogenic converter normally mounted on the tail unit, a processor and a cockpit controller. Displays include CRT, threat warning indicator and external countermeasures command module. It is insensitive to moisture, operates at temperatures from –57 to +132°C and weighs 63 pounds.
MANUFACTURER:
Cincinnati Electronics Corp.
Cincinnati, Ohio USA

AN/AAR-44 Infrared Warning Receiver

Warns of ground-launched missile attacks for aircraft and helos. It searches the lower hemisphere for threats while tracking and verifying missiles. The system uses a scanning lens sensor system. Provides automatic pilot warning and automatically controls countermeasures. Primary threats envisioned are the Soviet SA-7/9 and other missiles similar to Redeye. System weight is 45 pounds.
MANUFACTURER:
Cincinnati Electronics Corp.
Cincinnati, Ohio USA

AN/ALE-29A/29B Countermeasures Dispensers

Mounted within the aircraft and consists of two dispensers with sequencing switches, a chaff programmer and a cockpit control unit. Each dispenser carries 30 RR-129/RR-144 chaff units or 30 MK46/47 flares. It is used with A-4, A-6, A-7, RA-5C, F-4 and F-14 aircraft. AN/ALE-29B, an improved version of 29A, allows payload flexibility of selectively choosing up to three types of expendable payloads (chaff, flares, jammers) in any increment up to 10. It also can eject multiple flares simultaneously for a tailored response best suited to a particular situation. It is predecessor of the AN/ALE-39 system.
MANUFACTURERS:
Goodyear Aerospace Corp.
Akron, Ohio USA

Tracor Aerospace, Inc.
Austin Austin, Texas USA

Lundy Electronics & Systems Inc.
Pompano Beach, Florida USA

AN/ALE-36 Countermeasures Dispensing Pod

Modified AN/ALE-38/41 adapted to carry RR-136 and RR-137 chaff units. The QRC/TBC-600 version is a modified AN/ALE-38 pod carrying 600 RR-170 chaff units for saturation/corridor chaff applications. Payloads are carried in 10 dispenser blocks of 60 payloads each and include an integral programmer. Each dispenser can be loaded with chaff units or flares. Fully loaded, it weighs 680 pounds.
MANUFACTURER:
Tracor Aerospace, Inc.
San Ramon, California USA

AN/ALE-38/41 Pod-Mounted Bulk Chaff Dispenser

High-capacity operation for path finder or corridor seeding roles. The Navy ALE-41 is based on the Air Force ALE-38 using the shell of the old ALE-2 dispenser. Units provide continuous dipole dispersal and instantaneous bloom for laying chaff corridors. It can also be used for aircraft self-protection. The bulk chaff dispenser rolls have been designed to support both systems. Precut dipoles are sandwiched between two wraps of Mylar film, providing instantaneous bloom and continuous dipole dispersal. Six 50-pound rolls are carried in each pod. Loaded weight is 500 pounds. Following type rolls are currently in use: RR-155/AL, RR-145/AL, RR-163/AL, RR-17A/AL, RR-167B/AL, RR-171/AL and RR-172/AL. Airborne platforms cleared for carriage of ALE-38/41 include: F-105F, F-4, A-3, EA-4, A-6, EA-6, A-7 and AQM-34H. The system can be automatically turned on by the radar warning receiver.
MANUFACTURER:
Tracor Aerospace, Inc.
San Ramon, California USA

AN/ALE-39 Countermeasures Dispenser

Internally mounted system can dispense mixed loads of chaff, flares and primed oscillator expendable transponder jammers. A total of 60 can be loaded in any combination in multiples of 10. It is able to dispense the three payload types independently or concurrently. The system includes an automatic programmer enabling the pilot or EW operator to select the number of cartridges and drop intervals. It is standard equipment in U.S. Navy A-4, A-6, A-7, F-14, and F-18 aircraft and CH-46, CH-53, AH-1 and UH-1 helicopters. Used by several foreign countries. For retrofit applications, the electronic controls for reprogramming during flight can be used with dispensers of any configuration that are designed to eject payloads using electrically initiated gas generation squibs. Electronic controls can also be integrated with radar warning receivers for true automatic operation.

MANUFACTURER:
American Electronic Laboratories Inc.
Lansdale, Pennsylvania USA

AN/ALE-40 CM Dispenser System

Militarized version of the TBC-120 countermeasures dispenser. It exists in internal, semi-internal, external and pylon mounting. The system consists of four dispensers, a chaff/flare programmer and a cockpit control unit. One dispenser is mounted on each side of the inboard armament pylons. Each dispenser accommodates 30 RR-170 chaff cartridges for a total of 60 cartridges per pylon and 120 per aircraft. The outboard dispensers can carry 15 MJU-7/B IR flares. System drag is comparable to a Sidewinder missile and launcher. Loaded weight is less than 132 pounds. Various configurations have been developed for adaptation to other aircraft types. Installation configurations vary from internal, flush mounted, skin mounted or scab-on, to semi-internal mounting. The AN/ALE-40(V) 4, 5, 6 is installed on F-16 aircraft; the AN/ALE-40(V) 7, 8, 9 semi-internally mounted systems installed on F-5E/F aircraft. The AN/ALE-40(N) is the adaptation of AN/ALE-40 for skin mounting on the NF-5 (Royal Netherlands AF) aircraft. Other variants of the AN/ALE-40 are operationally deployed on Hunter and Mirage aircraft. The advanced generation AN/ALE-40(V)X is an exact mechanical replacement for AN/ALE-40(V) systems and uses solid-state technology to incorporate a threat adaptive programmer. The microprocessor-controlled programmer accepts information from the aircraft radar warning receiver and tail warning system if available, air data computer and throttle quadrant power setting, and enables the optimum deployment routine for available expendables. The AN/ALE-40(V)X is capable of the following operating modes: (1) Off/standby. (2) Manual—in this mode the pilot initiates a preselected dispensing program. (3) Automatic—this is a fully automatic mode and causes automatic selection of the appropriate expendables as well as automatically selecting the optimum dispensing program for the threat engagement at hand. (4) Semiautomatic—this requires pilot initiation of the automatically selected optimum dispensing routine.

MANUFACTURER:
Tracor Aerospace, Inc.
Austin, Texas USA

AN/ALE-43(V) Chaff CM Dispenser Set

New-generation bulk chaff dispenser that cuts chaff dipole lengths in flight. Counters radar frequencies from A through L bands for chaff corridor seeding, area saturation and self-protection applications. Unique design allows long resonate dipoles in the A and B bands to be generated, as commanded, when the chaff cutter is cutting dipoles on one of three cutter rollers selected in flight. This yields a specific combination of dipole lengths from single frequencies to multiple broadband frequencies as required by the user. The chaff supply is made up of metalized glass roving packages (RR-179), each weighing 40 pounds. AN/ALE-43 comes in three versions: (V)1 pod used on the EA-4F, EA-6A and TA-7C aircraft; (V)2 internal installation in the ERA-3B; and (V)3 internal installation in the NKC-135. Dispenser output rate is up to 8.1-by-10.6 dipole inches per second for 660 seconds over an aircraft speed range to 550 knots. No ram air is used, thus allowing ALE-43(V) use on helicopters. AN/ALE-43(V) provides for a continuous or pulsed chaff cutting dispensing mode. The pulse mode provides one to nine seconds cutting on time in one-second steps. The (V)1 pod version mounts on standard 14-inch and 30-inch store

suspension stations. Uses 38 VDC for control and 115V, three-phase, 400 Hz for cutting power. Pod (empty) weight 305 pounds. Chaff weight 320 pounds.

MANUFACTURER:
Lundy Electronics & Systems Inc.
Lundy Technical Center Division
Pompano Beach, Florida USA

AN/ALE-44 Dispenser Pod

Used with drones and tactical aircraft, it is a lightweight system that dispenses chaff (RR-129) cartridges and/or IR flares (MK-46) at subsonic and supersonic speeds. A typical system consists of a control unit and two pods. Each pod has a capacity of 32 rounds. Dispensing modes are: units/burst (1 or 2); bursts/program (1, 2, 4 or 8); and dispensing rate (4, 2, 1 or ½ bursts/second). It can be mounted on a wing tip, stores pylon or fuselage, and uses 28 VDC power.

MANUFACTURER:
Lundy Electronics & Systems Inc.
Lundy Technical Center Division
Pompano Beach, Florida USA

AN/ALE-45 CM Dispenser System

Solid-state, microprocessor-controlled countermeasures dispenser. It is automatically responsive to threat notification from the radar warning receiver or pilot. An optimally computed dispensing program provides aircraft protection without expending unnecessary stores. Dispensing programs are easily changed by replacing a plug-in, shop-replaceable, threat matrix memory. It has comprehensive, built-in test compatible with F-15 maintenance concepts.

MANUFACTURER:
Tracor Aerospace, Inc.
Austin, Texas USA

AN/ALE-47 Countermeasures Dispenser

Threat-adaptive expendable ECM system; upgrade of the AN/ALE-40 installed on F-4, F-5, F-16, A-7 and A-10 aircraft. The LRUs are electrically and mechanically interchangeable with the existing ECM dispenser. Modification to the aircraft wiring that connects the warning sensors (RWR, TW and MLD) and aircraft sensors (air data computer and afterburner switch) is required to provide the threat adaptive and automatic capabilities.

MANUFACTURERS:
Tracor Aerospace
Austin, Texas USA

Loral Defense Systems-Akron
A Division of Loral Corp.
Akron, Ohio USA

AN/ALQ-99 and ALQ-99E Tactical Jamming System

Used in the U.S. Navy EA-6B and USAF EF-111A electronic warfare aircraft. The system generates high power jamming signals to jam early warning/GCI and acquisition radars in support of strike aircraft. The system consists of a receiver processor subsystem for threat detection, suitable displays and controls to display threats and jamming responses and permit operator control of system operating modes, exciters to generate the jamming techniques and transmitters with associated antennas. The original configuration for the EA-6B was developed in the 1960s, and the aircraft was configured to carry five jamming pods with two transmitters per pod. The Air Force version of the system, developed in the 1970s, was based on

THEY SEEK.
THEY FIND.
THEY TEST,
TALK AND
JAM.
IN
MILLIMETER
WAVES.

the Navy configuration except for a single operator position in the EF-111A, as opposed to two operators in the EA-6B, and internal mounting of 10 transmitters as opposed to pods. The Navy system has been updated several times over the years, and the service is improving the jamming techniques and the receiving subsystem. The ALQ-99E is undergoing its first major upgrade.

MANUFACTURER:
Eaton Corp.
Deer Park, New York USA

AN/ALQ-119(V) ECM Pods

Family of jamming pods on USAF F-4, F-15, F-16, F-111, A-7 and A-10 aircraft. The system evolved out of the early QRC-335 program in the Vietnam era. Its main purpose was to counter Soviet SAMs such as SA-2/3, and eventually replaced the ALQ-101(V) pods first used by USAF in Southeast Asia. Exported to Israel, West Germany, Japan, Turkey and Egypt. Through various engineering change programs, the ALQ-119(V) evolved into an early prototype dual-mode jamming pod and was the test bed for many component and subsystem developments. Program to upgrade the pod reliability, maintainability and threat jamming capabilities completed. ALQ-119(V) being replaced by ALQ-131. Update program to increase jamming effectiveness, implement power management, integrate RWR operation with pod, and improve maintenance and reliability.

MANUFACTURER:
Westinghouse Electric Corp. Defense Group
Baltimore, Maryland USA

AN/ALQ-122 B-52 ECM Set

Multiple false target generator system used in the B-52. ALQ-122 automatically searches for, acquires and tracks threat signals, and generates a narrowband, low-duty-cycle ECM program to deny range and azimuth. Low-level signals are linearly amplified by AN/ALT-16 transmitters aboard the B-52. Primary victim radars are ground controlled intercept, height finders, early-warning and acquisition radars.

MANUFACTURER:
Motorola Government Electronics Group
Scottsdale, Arizona USA

AN/ALQ-123 Infrared Countermeasures

Provides aircraft protection from IR-homing missiles. Pod-mounted, powered by a ram-air turbine. Pulsed cesium lamp causes error signal in IR seeker. Operates in two modes: pilot activated or autonomous. Pod weight: 378 pounds. Designed for A-6, A-7. Potential application to RF-4B, F-14, F-15, F-16 and F-18 aircraft.

MANUFACTURER:
Loral Electro-Optical Systems
A division of Loral Corp.
Pasadena, California USA

AN/ALQ-125 Tactical Electronic Reconnaissance Sensor (Terec)

Establishes and maintains hostile electronic order of battle (EOB) in a tactical situation. The system provides rapid threat recognition and location and dissemination of this information to tactical commanders. Terec features two options: (1) cockpit display, which provides on-board EW operators with the ability to pass along data to other users, and (2) data link transmission of information to selected ground sites. Internal or pod mounted. Applicable to RF-4, F-15, F-16, EF-111, C-130, MRCA

P-3C or RPVs. The Terec airborne sensor can be mounted on manned or unmanned platforms to provide long stand-off ranges. Airborne tape recording of data and a complete ground processing element allow detailed analysis of emitter operating characteristics and location. Terec can locate surface-to-air missile sites in near-real time.

MANUFACTURER:
Litton Systems, Inc.
Amecom Division
College Park, Maryland USA

AN/ALQ-126A

More than 1,100 ALQ-126/ALQ-126A produced and all converted to the ALQ-126A configuration. It is in service on Navy and Marine Corps A-4, F-4, A-6, EA-6B, A-7, RF-8 and F-14. Foreign military sale to the Netherlands air force. It is no longer in production. See ALQ-126B.

MANUFACTURER:
Sanders Associates Inc.
Nashua, New Hampshire USA

AN/ALQ-126B DECM Systems

Replaces AN/ALQ-100 and AN/ALQ-126A series of internally mounted jammers for Navy and Marine Corps fighters and attack aircraft. Fully compatible with aircraft having provisions for AN/ALQ-126A. Initial production was awarded in August 1982; first production unit was delivered May 1984. Systems have been delivered and continuous production is planned to reach a total of more than 1,100 sets of this version by the end of 1988. Foreign military sales to the Australian (F/A-18) and the Spanish (EF-18) air forces. Direct commercial sale to Canada. Multimode, power-managed, reprogrammable defensive electronic countermeasures (DECM) system for the protection of Navy/Marine Corps A-4, A-6, A-7, F-4, F-14, F-18 aircraft. Resulted from ALQ-126A product improvement program and is thus compatible with current aircraft installation provisions. The use of distributed microprocessors, solid-state amplifiers, microwave integrated circuits, a digital IFM receiver, advances in digital circuitry and LSI provide the required performance against current and projected radar threats in a highly compact design. Operates either autonomously or when fully integrated with the AN/ALQ-162 CW DECM system, the high-speed anti-radiation missile (HARM), and other on-board avionics. The AN/ALQ-126B has replaced the ALQ-100 and ALQ-126 family. Frequency coverage expanded to include I/J band and power output increased to over one kilowatt per band at a four percent to five percent duty cycle. Techniques available include: mainlobe blanking, inverse conical scanning, range-gate pull off, swept square wave. While antenna coverage differs with the aircraft, fore and aft coverage is typically 60 degrees beamwidth that are tilted downward approximately 15 degrees. System receiver is essentially wideband video with set-on receiver available for the transponder variant. 100-nanosecond response characteristic with about four-microsecond memory duration. Transponder mode is essentially zero delay. Unit is 2.3 cubic feet; 190 pounds; 3 kVA power, off 400-Hz/three-phase supply. The receiver/transmitter element provides the transmit stage outputs and accepts vehicle blanking inputs to achieve ECM with other on-board avionics. Tag outputs and blanking inputs are provided individually for each band. In production and in service.

MANUFACTURER:
Sanders Associates Inc.
Nashua, New Hampshire USA

AN/ALQ-128 Warning Receiver

As a part of the F-15 tactical EW system (TEWS), this multi-mode threat warning receiver is mated with the ALQ-135 internal countermeasures set, ALR-56 RWR and ALE-45 chaff dispenser. Warning coverage goes beyond the H/I/J-band range of the ALR-56. Improvement development to increase the range of threats is under way.

MANUFACTURER:
Magnavox Electronic Systems Co.
Fort Wayne, Indiana USA

AN/ALQ-131 ECM Pod

Designed as the standard ECM set for all Air Force tactical aircraft, replacing the AN/ALQ-119. Used in practically all AF attack and fighter aircraft except the F-15, which employs Northrop's ALQ-135 and Loral's ALR-56. Modular and is planned to be available in up to five-band jammer bands when fully configured with eight jamming modules to meet different mission requirements. The low-band jammers are supplied by Motorola, with Loral supplying the receiver-processor. It weighs 831 pounds. Two improved versions of the system are currently in development: Seek Ram (ALQ-131 Block 3) and Have Charcoal. The ALQ-131 is a candidate for early application of very high-speed integrated circuits. Westinghouse has produced almost 700 ALQ-131 pods and expects total sales to reach 1,500 units. The ALQ-131 will eventually be replaced by the ALQ-165 ASPJ system in F-16s and on a number of Navy fighters and attack aircraft. In addition to U.S. sales, Egypt has ordered 40, Pakistan 21 and the Netherlands 60 of the Block 1 version.

MANUFACTURER:
Westinghouse Electronic Warfare Division
Baltimore, Maryland USA

AN/ALQ-131(V) ECM Pod

Dual-mode jammer (noise/deception), Air Force's standard jamming pod asset. The system is constructed using modules inserted into a canister that provides its own cooling, structural support and environmental protection. An I beam with integral flourocarbon cooling forms the center of a canister and travels its entire length, creating equipment bay locations on three sides. There are 16 different structural configurations to match mission requirements. Basic pod is a three-band 1½-canister terminal threat arrangement, 111 inches long. Smaller and larger options are available for a variety of mission and threat scenarios. Functional core of the ALQ-131 is the control and interface (C/I) module containing a programmable digital computer. The C/I module contains a digital waveform generator that can supply simultaneous waveforms for jamming modulations. Modularity and computer control are the basis of the ALQ-131's mission flexibility. When selected frequency coverage at minimum weight and/or drag is required on a high-performance aircraft, the pod offers frequency-band partitioning by module and assembly configurations offering from one to three frequency bands. Using a power-management module or tying into an external radar warning receiver, the ALQ-131 becomes a fully automatic, computer-controlled system that matches jamming to the threat environment on a pre-assigned basis that is user-programmable. If an externally mounted pod is not feasible, the modular design permits packaging of the electronics modules inside the aircraft. The computer's operational software can be programmed on the flight line in less than 15 minutes by means of a memory loader/verifier. During normal jamming operations, a continuous central integrated test system monitors the operational and functional status of the equipment. Pod repair and check-out is facilitated by computer-aided fault detection and isolation to the component level with the electronic system test console. The receiver-processor for the ALQ-131 is produced by Loral. It is a plug-in module that provides power management. System is a double-conversion wideband agile super-heterodyne receiver with a crystal video receiver for low band. The module has a self-contained processor that sorts and identifies threats automatically. The sorting process and the exact definition of what is considered a threat is programmable and can be loaded at the flight line with the pod memory loader/verifier. This threat identification is passed to the jammer subsystem via the pod's digibus. The receiver/processor looks through the jamming in a dense, multiple-threat/multiple-jamming environment. It contains a continuous self-test feature to assure operability. The Loral receiver/processor is currently in production. The ALQ-131 is compatible with F-16, F-111, A-7, A-10, F-15 and F-4. In addition, the Netherlands, Egypt, Israel, Pakistan and Japan are under contract to purchase the ECM pods.

MANUFACTURER:
Westinghouse Electronic Warfare Division
Baltimore, Maryland USA

AN/ALQ-133 Quick Look II

Army's airborne non-communications emitter location and identification system (NELISA). The system operates from 500 MHz to 18 GHz and is deployed in an OV-1D Mohawk aircraft as part of a passive battlefield non-communications surveillance network. Location and identification information is passed to the ground data processor and reported to the battlefield commander. Six Quick Look II systems are deployed with each forward-area aviation company. Quick Look II replaces the APQ-142 Quick Look I ESM system. System can be remotely controlled with the AN/USQ-61 digital data link. The ALQ-133 consists of two pods: one contains Elint receiver broadband antennas; the other contains data-processing equipment. The pod antennas will be capable of providing direction-of-arrival measurements over a 90-degree sector abeam of the aircraft to a typical accuracy of 0 degrees to 5 degrees. If some degradation of bearing accuracy is acceptable, the coverage sector can be enlarged to 120 degrees. The data processor is responsible for control of the search receivers, analysis of intercepted signals, and comparison of these against a file of known hostile radar characteristics. Under the RV-1D improvement program, the ALQ-133 will be supplemented with a CW jammer and missile detector system.

MANUFACTURER:
UTL Inc.
Dallas, Texas USA

AN/ALQ-135 Internal Countermeasures Set

Part of the overall F-15 Tactical Electronic Warfare Suite (TEWS). The system is fully integrated with the AN/ALR-56 radar warning receiver and the AN/ALE-45 countermeasures dispenser. The basic ALQ-135 consists of two line replaceable units (LRUs) plus appropriate waveguides and antennas. Each LRU consists of a control oscillator and an RF amplifier that are mission-interchangeable. Over the years the system has continued to evolve along with capabilities of the aircraft and changes in the threat. While maintaining commonality with the original system and support electronics, the AN/ALQ-135 has been updated to include high-band coverage and digital receiver/processor power managed functions.

MANUFACTURER:
Northrop Corp.
Defense Systems Division
Rolling Meadows, Illinois USA

AN/ALQ-136 Helicopter Radar Jammer

Lightweight jammer for Army helos and fixed-wing tactical aircraft. It gives attack helicopters the added dimension of defense that delays acquisition radar's positioning of the aircraft. The system allows the helicopter sufficient time to "pop out" and obtain a favorable firing position. ALQ-136 is an I/J-band jammer designed to operate against anti-aircraft artillery weapons. The entire ALQ-136 weighs 40 pounds and consists of three line-replaceable units: an operational control unit; a pair of spiral antennas, one each for transmit and receive; and the receive/transmit module. When the helicopter is painted by a threat radar, the jammer analyzes the received pulses and responds with appropriate angle and range deception jamming. This microprocessor-based system is software programmable and designed for future expansion requirements.

MANUFACTURER:
ITT Avionics
Nutley, New Jersey USA

AN/ALQ-94/137 ECM System

Self-protection system was developed to protect U.S. Air Force F-111, FB-111 and EF-111 aircraft against surface-to-air missiles, anti-aircraft artillery and airborne interceptors. It is a multiband, deceptive, RF pulse repeater incorporating automatic signal processing and power management to identify, prioritize and jam threat radars in a dense signal environment. The AN/ALQ-94 was developed in the 1960s to meet the requirements of both the Tactical Air Command and the Strategic Air Command. More than 500 systems were built to support this task. When the FB-111 needed greater ECM coverage in the 1970s, the AN/ALQ-137 was developed to include the addition of the aft system and improvements to the forward ALQ-94 system for enhanced processing, techniques generation, frequency coverage and threat change adaptability. A power-managed deception jammer using a wideband crystal-video set-on receiver. Operated in three modes: repeater, CW noise and transponder. The system is broken down into three subsystems: E/F, G/H and I/J bands. Pulse power of more than one kilowatt is provided at four percent to five percent duty cycle, with a power output of 100 watts. Jamming techniques available include RGPO, VGPO, and selective range jamming. The jammer response time is approximately 100 nanoseconds with a memory duration of about 2.5 microseconds. The ALQ-137 system has also been installed on the EF-111 for self-protection. USAF ASD recently awarded a contract for Lot 5 of the ALQ-137, bringing the total production of the ALQ-137 to 120 systems for both SAC and TAC. A combined ALQ-94/137 update program is underway to meet the future self-protection requirements of the F-111 series aircraft. This system will use the most advanced technology and signal processing and will be completely adaptable to changing threat environments through its flight-line reprogrammable capability.

MANUFACTURER:
Sanders Associates Inc.
Nashua, New Hampshire USA

AN/ALQ-142 ESM

Electronic support measures for SH-60 LAMPS Mark III helicopters. Supports the LAMPS ASW, Area Surveillance and Over-the-Horizon targeting missions. Processed information from the AN/ALQ-142 is data linked to a shipboard AN/SLC-32 electronic warfare system and correlated with the on-board data base to enhance system response. The AN/ALQ-142 has a high probability of intercept on a single scan of a hostile radar at extended ranges. Threat bearing is measured on each pulse and emitter identification accomplished by comparing measured parameters with data sorted in the AN/AYK-14 system computer. Emitter identification and bearing are displayed on the AN/SLQ-32(V) operator console. Intercept and direction finding are performed over 360 degrees in azimuth with elevation coverage tailored to LAMPS requirements. Four antenna units, each providing 90 degrees of azimuthal coverage, are located fore and aft on the helicopter with the receiver/processor centrally located inboard. In production and in service.

MANUFACTURER:
Raytheon Co.
Electromagnetic Systems Division
Goleta, California USA

AN/ALQ-144 Infrared Countermeasure Pod

Electrically powered active IRCM system. Primarily designed to provide small and medium sized helicopters with protection against heat-seeking missiles, but has applications on some fixed-wing aircraft. Omnidirectional system consisting of a cylindrical source surrounded by a highly efficient modulation system to confuse incoming missiles. Since commencement of production in 1978, more than 1,300 systems delivered. Installed on AH-1, UH-1, H-3, H-60, AH-64, OH-58 and OV-10 type aircraft. Weighs 28 pounds. An approved export version (ALQ-144(VE)) is available. In production and in service.

MANUFACTURER:
Sanders Associates Inc.
Nashua, New Hampshire USA

AN/ALQ-147 IRCM System

Improved variant of the AN/ALQ-132 infrared countermeasures systems developed by Sanders, the AN/ALQ-147 is designed for aircraft that cannot support the large power generation load associated with ARC Lamp or resistive source systems. Instead, JP fuel is burned in a ram-air filled duct. The AN/ALQ-147 unit for the OV-1 and RV-1 aircraft is mounted on the rear of the fuel tanks. Installation includes a filter, which reduces visible emissions and makes the system suitable for night operations.

MANUFACTURER:
Sanders Associates Inc.
Nashua, New Hampshire USA

AN/ALQ-149 Airborne Communications Jammer

The AN/ALQ-149 will jam hostile communications signals and long-range early warning radars. On-board elements include an acquisition subsystem which has separate communications and radar intercept and processing elements that coordinate their activities with the AN/ALQ-99 radar jammer. The analysis subsystem acquires communications and radar signals and defines appropriate responses. Results are transferred to the central processing subsystem which interfaces with other systems, including the on-board mission computer. The system

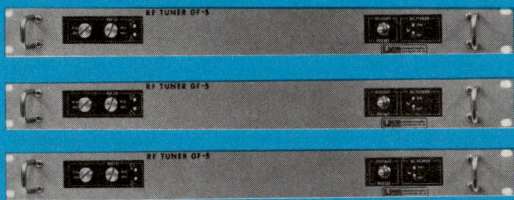

uses the AN/ALQ-99 display and other hardware and software. Transmitters for the AN/ALQ-149 will be housed in a pod.
MANUFACTURER:
Sanders Associates Inc.
Nashua, New Hampshire USA

AN/ALQ-150 Cefire Tiger
Tactical airborne ESM equipment designed for use against multichannel communications. Has electronic support measures capability essential for hand-off to jammers. All user aircraft can target frequency division multiplexed (FDM), frequency modulated (FM), and time division multiplexed (TDM) emitters. System is deployed in four modular subsystems in three RU-21 aircraft of a rear Army aviation company. Frequency ranges per system as follows:
- Aircraft #1: 60 to 115 MHz and 1500 to 9000 MHz
- Aircraft #2: 115 to 480 MHz
- Aircraft #3: 450 to 1500 MHz

System interfaces with the Le Fox Grey Control Processing Center (CPC) and the Forward Control and Analysis Center, thus providing management of ECM missions on a real-time basis. CPC has wideband data links that communicate with up to three Cefire Tiger aircraft simultaneously. Effective radiated power is from 3 to 19 kilowatts, depending on frequency. Flight endurance is up to five hours.
MANUFACTURER:
GTE Government Systems
Mountain View, California USA

AN/ALQ-151 Quick Fix Airborne EW System
Direction-finding, intercept and ECM system operating from 2 to 76 MHz deployed on Army EH-1H and EH-60A helicopters. Designed to identify, locate, listen to, and disrupt enemy command and control communications. Jamming power output is 40 to 150 watts. Quick Fix can interface with all other Army aircraft using a secure air-to-air link. Also interfaces with the division support company operations center using an AN/ARC-164 UHF radio system. Quick Fix, as a communications jammer, will be interoperable with Army aircraft survivability equipment to be deployed on the same platform. Weight about 7,500 pounds.
MANUFACTURER:
ESL Division of TRW
Sunnyvale, California USA

AN/ALQ-153 Tail Warning System
Pulse Doppler radar designed for use in the B-52 bomber. Designed to replace the AN/ALQ-127 tail-warning system. Range gated Doppler that detects approaching missiles and aircraft; can also be fitted into the F-15 and F-111 aircraft and can be integrated with the ALQ-131 for pod applications. Radar provides target information, alerts crew and initiates chaff and flares countermeasures. Built-in test checks the system every 1.5 seconds.
MANUFACTURER:
Westinghouse Electronic Warfare Division
Baltimore, Maryland USA

AN/ALQ-155(V) B-52 Power Management System
Provides integral set-on receivers for each jamming transmitter, plus increased effective radiated power (ERP) density through accurate frequency set-on. System is a power-management evolution for the ALT-28(V) active ECM set providing automated hand-off from the ALR-46 radar warning receiver with nearly instantaneous jammer response and is computer-managed and field-programmable. Containing automatic frequency control in all modes and a wide variety of ECM techniques that are both automated or manual (or semiautomated), the system has a 12-transmitter upload capability. Improvements of ALQ-155(V) are:
- Frequency agility against multiple threats
- Pulse repetition interval trackers
- Cover pulse jamming techniques
- False target generation through pseudorandom noise
- Hybrid IFM receiver and central receiver capability
- Programmable noise optimization
- Compatibility with AN/ALQ-117 deceptive I/J-band jammer
- Downlink jamming
- Increased pulse-up power for CW to pulse operations
- Electronically steerable antenna system-compatibility
- Coherent and incoherent jamming.

MANUFACTURER:
Northrop, Defense Systems Division
Rolling Meadows, Illinois USA

AN/ALQ-156 Missile Detection System
Pulse-Doppler radar detects approaching missile threats with 360 degrees of azimuth coverage and automatically triggers the ejection of IR or RF decoys for the aircraft. Lightweight system with minimal power consumption, solid-state design and reprogrammability. Initial application was for the U.S. Army/Boeing CH-47 helicopters. Adaptable to fixed and rotary wing aircraft. Currently operational on U.S. Army helicopters and fixed-wing special mission aircraft, U.S. Air Force C-130 transports, and U.S. Navy P-3Cs and other tactical aircraft. The system responds to surface-to-air and air-to-air missile threats by triggering the release of a decoy from an associated dispenser system. Infrared missiles are decoyed with use of a flare. May be used to cue active RF countermeasures as well. The system is compatible with the XM-130, ALE-39 and ALE-40 dispenser systems. Pulse-Doppler techniques are used to make the system immune to battlefield clutter down to nap-of-the-earth. Consists of a receiver/transmitter unit, a control unit, and up to four antennas. Weight: 50 pounds.
MANUFACTURER:
Sanders Associates Inc.
Nashua, New Hampshire USA

AN/ALQ-157(V) Infrared Countermeasures System
Protects large helicopter and turboprop transport aircraft against infrared missiles. Consists of four subsystems: two transmitters, a control power supply, EMI filter assembly and a pilot control indicator. Pilots can select one of five jamming codes preprogrammed in the microprocessor. Employs built-in test to automatically perform operational readiness tests. Special maintenance tools and equipment are not required. IR transmitters are mounted on each side of the aircraft, providing complete 360-degree coverage. Weighs 220 pounds and is currently operational on H-46, H-53 and C-130 aircraft.
MANUFACTURER:
Loral Electro-Optical Systems
A division of Loral Corp.
Pasadena, California USA

AN/ALQ-161 Defensive Avionics System

Radio frequency surveillance/electronic countermeasures system for the B-1B bomber. Modifications will probably be added to counter the new look-down, shoot-down Soviet radar systems by adding the 10.4-to-18-GHz frequency band. The new frequencies will also cover some of the newer surface-to-air missiles reported to operate in that band. Monopulse jamming capabilities will also be added, along with other technology such as terrain bounce and scatter jamming. It is a totally integrated EW system capable of providing virtually immediate response to incident threat radar signals. It is configured to automatically optimize jammer frequency set-on, antenna pointing, modulation technique and activation against threat system radars. The incident signals are received in omni-directional and DF antennas. These signals are fed through a receiver in which individual pulse parameters and time of arrival are measured and encoded into a digital word. It is then filled out with the measured direction of arrival quantized in a DF receiver. The digitized pulse descriptors are fed through an active data filter to the computer for pulse train sorting and threat evaluation. The data filter, an associated tracker, is used to remove already processed data to reduce the computer load. The system uses the LC-4516D computer. It is packaged in two line-replaceable units. The LC-4516D is microprogrammable. The central processor functions can grow to a larger instruction set through reprogramming and replacement of microprogrammable, read-only memory (micro/PROM) chips in the control structure. Boeing will integrate the system into the avionics and aircraft, and will develop the defensive management subsystem (DMS) for inputs from all other defensive sensors. The DMS includes controls, displays and blanking. Each threat descriptor in the alarm file is accompanied by a priority rating, which is a function of its operating mode and one or more jamming techniques. Each threat-associated jamming technique is described in terms of its non-time sensitive parameters. The jammer logic also acts as a control switching matrix to tune the frequency sources to the correct frequency, to activate a transmitter and to point the antenna at that threat in the intervals during which it is most sensitive to jamming. Each element can be switched and activated at rapid rates to allow time-sharing among many threat signals as programmed by the jammer logic. The techniques generator is composed of a number of digital waveform devices that can be easily reprogrammed to satisfy changing threat requirements. High-band transmitters are medium-power TWT systems similar to those developed for the F-15 tactical electronic warfare system. The lower-band transmitters were developed by Eaton and include a high-power, solid-state amplifier. The electronically steerable transmitting antennas use variable phase-shift, multi-element arrays. Signals from the transmitter amplifiers feed to a corporate feed network through a waveguide to stripline transition. After division in the corporate feed, the signals are fed to the ferrite phaser networks, which are latched in phase by control signals from the electronic beam-forming network.

MANUFACTURER:
Eaton Corp.
Deer Park, New York USA

AN/ALQ-162 Continuous-Wave Jammer

Airborne continuous-wave radar jammer being developed under joint sponsorship of the U.S. Navy and U.S. Army. The system provides complementary jamming capability on older aircraft currently carrying the AN/ALQ-126 DECM, and is aimed at countering the Soviet SA-6 missile and newer threats. To be integrated with the AN/APR-43 and virtually any RWR in the inventory today. It is proven in threat simulation and flight testing by both developing services. At a weight of 40 pounds and less than 0.5 cubic feet of volume, the ALQ-162 fits almost anywhere on any aircraft. It is also mounted in wing pylons without taking up needed fuel or stores stations, inducing drag, or occupying valuable fuselage electronics bay real estate. The system makes use of advanced jamming techniques, is software programmable to meet new threats, and includes built-in-test devices to increase maintainability. It can operate autonomously using its own receiver/processor, in a stand-alone capacity, or in conjunction with a variety of on-board radar warning receivers. Its reprogrammability provides the flexibility to accommodate a unique or rapidly changing threat environment. Current installation commitments of the AN/ALQ-162 include: the U.S. Navy's A-7E, A-4M, RF-4B, F-4S, and AV-8C; the U.S. Army's EH-1, EH-60, RC-12D, OV-1D, RV-1D, and RU-21; and NATO's F-18, F-16 and F-35.

MANUFACTURER:
Northrop Corp.
Defense Systems Division
Rolling Meadows, Illinois USA

AN/ALQ-164 Electronic Countermeasures Pod

Pod version of the AN/ALQ-126B DECM system with integral Northrop AN/ALQ-162 CW jammer, designed specifically for the AV-8A/B/C Harrier. Two engineering development models were produced and successfully tested on AV-8C aircraft; further OT&E scheduled in mid-1985 lead to production. Potential wide application for fighter, attack, surveillance, transport and ASW aircraft. Pod weight: 350 pounds. Pod dimensions: 85 inches long by 16 inches in diameter.

MANUFACTURER:
Sanders Associates Inc.
Nashua, New Hampshire USA

AN/ALQ-165 Airborne Self-Protection Jammer (ASPJ)

Joint Navy-Air Force development to provide on-board self-protection jamming for the next generation of tactical aircraft. Managed by the Navy, the new jammer takes advantage of previous development efforts on the Westinghouse-Air Force ALQ-131 pod-mounted jammer, the Air Force risk-reduction program and the Navy's wideband dual-mode countermeasures program. It is currently slated for F-14, F/A-18, A-6 and AV-8B USN aircraft installations; it will also be used on the F-16. Advance expected in ASPJ will be improved analysis, sorting and power management. Using newly developed traveling-wave tubes, the ASPJ will be able to respond with high power in both pulsed and CW modes. ASPJ is expected to defeat manual operation of surface-to-air missiles that have been forced into that mode by effective jamming of automatic tracking and guidance modes. The ASPJ will interface with the ALR-67 RWR and ALE-39 chaff dispenser in Navy aircraft. In the Air Force, it will interface with the ALR-74 RWR or ALR-56M, and ALE-40 chaff dispenser. Flight testing of ASPJ on the F-16C and F/A-18 has begun.

MANUFACTURERS:
ITT Avionics
Nutley, New Jersey USA
Westinghouse Electronic Warfare Division
Baltimore, Maryland USA

IRCM
THE ULTIMATE DISTRACTION

Next Generation Countermeasures Are Here Today

Loral EOS is a world leader in design, development and production of Infrared Countermeasures (IRCM) Systems that protect a wide variety of commercial and military aircraft against heat-seeking missiles. Loral EOS develops and manufactures IRCM systems with state-of-the-art technology and microprocessor flexibility that assure performance to the year 2000 and beyond.

LORAL
ELECTRO-OPTICAL SYSTEMS

A SUBSIDIARY OF **LORAL** CORPORATION

AN/ALQ-171 ECM System

Fully automatic electronic countermeasures system. Self-contained receiver/processor/jammer with an advanced power management capability to ensure the most effective use of available jamming power. A conformal configuration was developed to provide protection for Northrop's family of F-5 aircraft. Designed to fit on the underside of the aircraft's fuselage, the conformal countermeasures system allows full use of all weapons and stores stations without affecting the performance and maneuverability of the aircraft. The pod version consists of identical electronics and is available for a number of foreign tactical aircraft. Each system contains multiple transmitters and receivers that share common support electronics such as system controller, signal generator and processors, antenna subsystems, and cooling. Pod and conformal version of the AN/ALQ-171 have common logistics support.
MANUFACTURER:
Northrop Corp.
Defense Systems Division
Rolling Meadows, Illinois USA

AN/ALQ-172 Pave Mint

An update of the ALQ-117 ECM system for the B-52, the system is designed to improve low-level penetration survivability. Also designed to protect B-52 air-launched cruise missile carriers from hostile enemy threats. In production and operational on B-52G aircraft. A follow-on system which uses a phased array antenna is in flight test.
MANUFACTURER:
ITT Avionics
Nutley, New Jersey USA

AN/ALQ-176 ECM Pod

Pod-mounted jammer system designed, developed and tested to fulfill aircraft mission requirements for ECM support, stand-off jamming and combat evaluation and training. The system mounts to standard aircraft stores/munitions stations. Option of using aircraft internal power or a ram air turbine generator. A two- or three-canister arrangement is available: The two-canister configuration (AN/ALQ-176V-1) houses up to three voltage-tuned magnetron (VTM) transmitters; the three-canister configuration (AN/ALQ-176V-2) provides housing for as many as five VTMs. Dependent on frequency band and tube selection, each transmitter produces 150 to 400 watts CW (up to 30 percent bandwidth) at efficiencies of greater than 50 percent. The AN/ALQ-176(V) is controlled (frequency and jamming mode) by standard inventory C-6631 and C-9492A control boxes mounted in the cockpit. The flexibility in ECM pod configurations and standard aircraft stores/munitions stations installation, enables the AN/ALQ-176(V) ECM pod to be used on fighter, attack, transport or training aircraft.
Frequency range: VTM transmitters from 0.8 to 15.5 GHz in 30-percent octave bandwidths.
Power output: 150-400 W (CW) per VTM
Pod length: (V) 1 pod—(78 inches); (V) 2 pod—(102 inches)
Pod max diameter: 25 cm (10 inches)
Pod weight: (V) 1-aircraft power or ram air turbine, 220 pounds
         (V) 2 pod—aircraft power, 310 pounds
Transmitter module weight: 27 pounds
Status: In production.
MANUFACTURER:
Hercules Defense Electronics Systems, Inc.
Clearwater, Florida USA

AN/ALQ-178 RAPPORT

Advanced, fully integrated system designed to be completely internally mounted in a wide variety of tactical aircraft, such as the F-16 and Mirage series. The system is field-programmable and combines radar warning, jamming and flare/chaff management functions in one system. Unique features of the system's distributed architecture permit installation of the ALQ-178 in its entirety or the RWR portion alone, as dictated by user requirements. The system utilizes a central programmable computer for data analysis and system control, with independent microprocessors to direct RWR, display and jamming functions. Detected signals by the RWR are de-interleaved and identified and displayed to the CRT. The power management algorithm matches the countermeasures to the RWR's constantly changing threat picture. Separate forward and aft jammers are used for maximum spatial coverage. It is based on the Loral-developed Rapport-II, which is operational on Belgian air force Mirage V aircraft. Internally mounted ALQ-178 has minimal effect on aircraft aerodynamics and external stores-carrying capabilities. Currently in production.
MANUFACTURER:
Loral Electronic Systems
A division of Loral Corp.
Yonkers, New York USA

AN/ALQ-184(V) ECM Pod

A U.S. Air Force tactical self-protect pod is designed for installation on F-4, F-15, F-16, A-7, A-10 and F-111 combat aircraft, providing countermeasures against surface-to-air missiles, radar-directed gun systems and airborne interceptors. An upgrade of the AN/ALQ-119(V) pod, it functions as a repeater, transponder or noise jammer. Its multibeam architecture based on Rotman lens antenna technology provides reduced countermeasures response time with a tenfold increase in ERP at 100 percent duty cycle. More than 60 successful flight test missions have been flown on U.S. Air Force combat aircraft in multi-threat environments. The pod is in full production and the integrated logistics support system is in place, providing intermediate level support equipment, technical orders, depot spares and repair facilities and maintenance training documentation.
MANUFACTURER:
Raytheon Co.
Electromagnetic Systems Division
Goleta, California USA

AN/ALQ-187 Airborne Self-Protection Suite

Fully automatic electronic countermeasures system integrated with radar warning and chaff/flare systems, internally housed for tactical aircraft self-protection. Primarily designed for F-4, F-16, A-7 and Mirage 2000 aircraft, it detects and defends against surface-to-air missiles, anti-aircraft artillery and air-to-air interceptor weapon systems. It can interface with ALE-39 or ALE-40 chaff/flare dispensers and ALR-69, ALR-66 or ALR-74 RWRs. Fully programmable, automatically detects single or multiple threat radars, selecting ECM programs to counter against pulse, Pulse Doppler or CW ground-based, shipborne or airborne emitters. Flight line reprogrammable, allowing incorporation of latest intelligence and ECM techniques.
MANUFACTURER:
Raytheon Co.
Electromagnetic Systems Division
Goleta, California USA

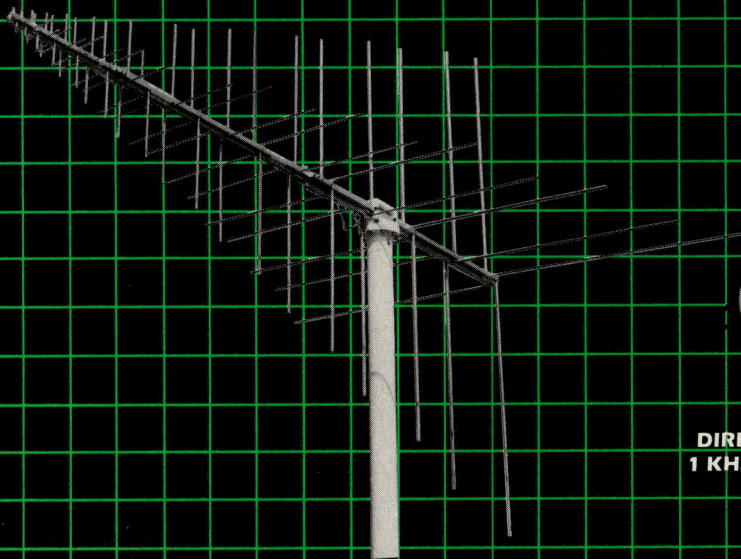

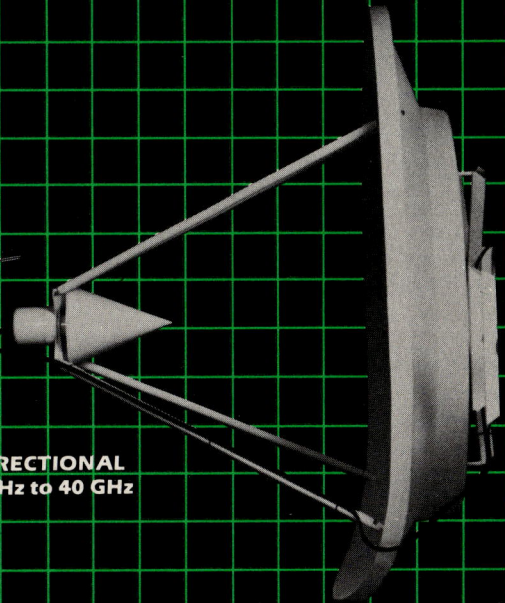

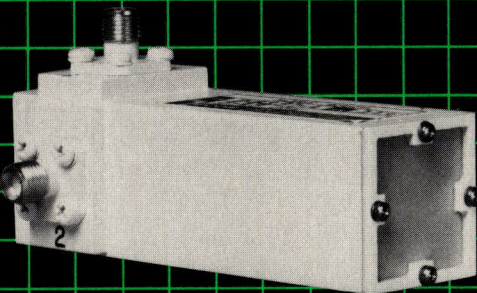

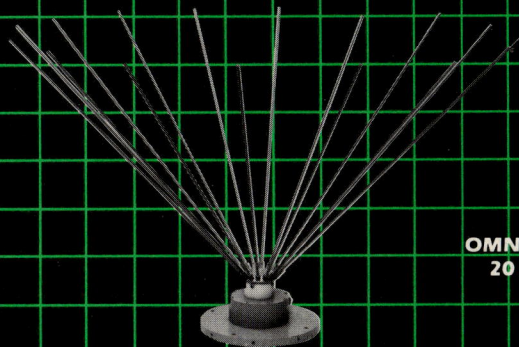

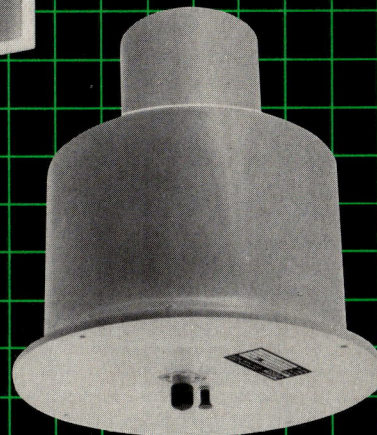

AN/ALR-45 Radar Warning/Control System

ALR-45 is a crystal-video receiver utilizing a hardwired programmable signal processor. The system currently in use by the U.S. Navy is the AN/ALR-45D. Four cavity-backed planar spiral antennas are used with preamplifiers and processor. The AN/ALR-45F (ALR-67 retrofit) replaces the U.S. Navy ALR-45 processor and azimuth indicator with a CP-1293 dual ATAC-16M computer-based threat processor, and the IP-1276, an alphanumeric microprocessor-based, fully contained Mil-Std-1553-compatible display terminal. The CP-1293 features threat software reprogrammability, on-board avionics and EW suite communications and interface control of such systems as HARM missiles, APR-43, ALQ-126A/B, ALE-29/39, ALQ-162, ALQ-164, and the ALQ-165. The CP-1293 and the IP-1276 are standard AN/ALR-67 weapons replaceable assemblies, and are interchangeable in the 45F configuration of F-4J, A-4M, A-7E and the AV-8C aircraft without kit wiring changes.

MANUFACTURER:
Litton Applied Technology
San Jose, California USA

AN/ALR-46(V) Radar Warning Receiver

Employs a wide-open, front-end receiver. Can identify and analyze frequency-agile emitters and can process 16 emitters simultaneously in priority order and feed data to a jammer, anti-missile seeker, or a data collector system. The system uses a signal processor, CM-442/ALR-46, which provides it with in-cockpit threat parameter programming and unambiguous identification. The system is the Air Force's standard radar-warning asset. The (V6) version, which has an analog analyzer, is made completely by Applied Technology.

MANUFACTURERS:
Litton Applied Technology
San Jose, California USA

Singer Co.
Dalmo Victor Division
Belmont, California USA

AN/ALR-47 Surveillance Receiver

Scanning superheterodyne installed in the U.S. Navy's S-3A anti-submarine warfare aircraft, and as the ALR-502, on Canada's CP-140 P-3C aircraft. Receiving frequencies in the E through J bands. The system is designed to search for and detect signals of extremely short duration as might typically be used by submarines. It employs fixed broadband antennas, superheterodyne receivers, automatic digital processing and programmable system control. Its modular construction enables it to extend its frequency and other performance parameters with minimum difficulty. Only spares for repairs are being produced at this time. It is being upgraded to a new system, the ALR-76.

MANUFACTURER:
IBM
Federal Systems Division
Owego, New York USA

AN/ALR-50 Missile Launch Warning Receiver

Designed to intercept and analyze Soviet missile launch and guidance signals associated with SA-2, SA-3 and SA-6 surface-to-air missiles. The determination of hostile firing and guidance activates special visual and aural signal to pilot. It works as a companion with the ALR-45 RWR. When ALR-50 integrated with ALR-56, the combined suite is referred to as "PRIDE." Correlating with radar or RWR signal can automatically initiate defensive measures such as chaff ejection or jamming. Currently in service with F-4, RA-5, A-4, A-6, A-7 and F-14 aircraft. It will be replaced by the APR-43/67.

MANUFACTURER:
Magnavox Co.
Fort Wayne, Indiana USA

AN/ALR-52 IFM Receiver

Multiband IFM (instantaneous frequency measuring) receiver used in the EP-3E Aries aircraft. The ship's variant, nomenclatured AN/WLR-11, is also produced. Typical systems cover 0.5 to 18 GHz using receiver modules that cover octave bandwidths. Additional capabilities include frequency band selection, blanking, separation of interleaved pulse trains, measurement of radar parameters and direction-of-arrival measurement. A modified version has been developed, linking the IFM receiver to a digital computer. The system also employs a 1-to-18-GHz DF antenna system. The computer uses the data to "tag" every received pulse with its frequency, pulse width, time-of-arrival and azimuth.

MANUFACTURER:
ARGOSystems
Sunnyvale, California USA

AN/ALR-56A F-15 Radar Warning Receiver

Used as part of the F-15 tactical electronic warfare system (TEWS) in conjunction with the ALQ-135 internal countermeasures set (ICS), the AN/ALR-56A is the radar warning set in TEWS. Elements include: processor and low-band receiver; high-band receiver countermeasures display; receiver control; power supply; and antenna systems. The processor and low-band receiver contain three major elements: single-channel low-band superheterodyne receiver, dual-channel IF section and a processor. The low-band receiver is electronically tuned under processor control. The dual-channel IF section operates with either high-band dual-channel receiver or the low-band single-channel receiver. The processor controls selection of the receiver to operate with the dual-channel IF section. It contains a preprocessor and a general-purpose digital computer originally developed by Texas Instruments and extensively modified by Loral. The preprocessor contains all video circuits for signal analysis and provides digital outputs to the computer on measured signal parameters, which are subsequently displayed with an audio signal also generated. The computer also controls all other system functions and is software-programmable to accommodate future changes in the threat environment. An all-solid-state, digitally controlled, dual-channel high-band receiver, the ALR-56A can scan from H through J band. Single-conversion superhet with dual YIG-filter preselectors provides additional sensitivity as well as selectivity and high

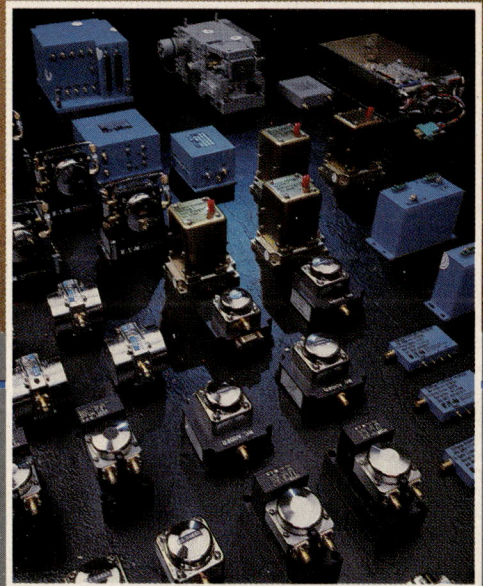

spurious signal rejection. The local oscillator assembly is frequency synthesized. RF input ports are configured to accept two main antenna inputs to each channel and two additional RF inputs that may be designated to either channel. All switching functions are provided internally. Major system functions include:

- Programmed frequency search of threat parameters including low-band priority, look-through and jammer blanking
- Signal acquisition and signal sorting routines
- Signal analysis routines
- Direction finding
- Prioritized threat analysis
- Jammer power management with integral look-through
- Threat data displays for analysis.

System display is an alphanumeric radar-type unit that shows the degree of threat lethality to the platform, along with its range. The overall threat processor is an outgrowth of the MPP series threat processor developments at Loral.

MANUFACTURER:
Loral Electronic Systems
A division of Loral Corp.
Yonkers, New York USA

AN/ALR-56C F-15 Radar Warning Receiver

Digitally controlled, dual-channel, high-band receiver, operating from E through J band. System is a dual-conversion superhet receiver capable of sorting and identifying all known threats to the F-15. RF input ports are configured to accept two main antenna inputs to each channel and two additional RF inputs that may be designated to either channel, with all switching functions provided internally. This arrangement supports DF measurements. Major system functions include:

- Programmed frequency search-of-threat parameters including low-band and designated high-band threat priorities, look-through, and jammer blanking
- Signal acquisition and signal sorting routines
- Signal analysis routines
- Prioritized threat analysis
- Jammer management
- Countermeasures dispenser management
- Threat data visual displays and audible alerts
- Manual and automatic BIT with manual BIT displays.

The system comprises a major modification to the ALR-56A RWR, previously installed on all F-15 aircraft, to allow reception and identification of new threats-independent environment. The ALR-56C is slated for installation on all new production F-15 aircraft starting in mid-1985, and a retrofit program for all Air Force F-15 aircraft starting in mid-1985. A retrofit program for all F-15 C and D models is also programmed.

MANUFACTURER:
Loral Electronic Systems
A division of Loral Corp.
Yonkers, New York USA

AN/ALR-59(V) ESM Set

Broadband scanning super-heterodyne used in the Navy's E-2C airborne early warning aircraft. Direction of arrival is determined by a phase interferometer aided by amplitude comparison to produce system DF accuracy. The system uses four-quadrant antenna arrays. Receiver features four-band simultaneous coverage and is computer-controlled. ALR-59 is

compatible with versatile avionics shop test (VAST) equipment. BITE to the weapon replaceable level. An upgrade of ALR-59 system has resulted in the ALR-73 system.

MANUFACTURER:
Litton Systems, Inc.
Amecom Division
College Park, Maryland USA

AN/ALR-60 (Deep Well) Communications Intercept System

A massive computer-controlled communications intercept and analysis system carried on U.S. Navy EP-3E electronic reconnaissance aircraft. Seven systems were built. Each has multiple operator positions equipped with a CRT terminal that allows the operator to randomly access raw audio data and processed text data. Digitally controlled receivers are operated under computer control on a prioritized search pattern. Computer-formatted CRT pages, temporary-state-recorder techniques and overall computer control of the receiving, recording, prioritized queueing, and operator processing functions assure complete coverage of all signals intercepted.

MANUFACTURER:
GTE Government Systems
Mountain View, California USA

AN/ALR-62(V) Radar Warning System

Standard radar warning set in Air Force F/FB-111 and EF-111A aircraft. A third-generation development that includes a complete antenna set, forward and aft receivers, a digital signal processor and cockpit-mounted threat indicator/countermeasures control units. ALR-62(V)3 is used in the FB-111 aircraft while the ALR-62(V)4 is used in the EF-111A electronic warfare aircraft produced for the Air Force by Grumman Aerospace. ALR-62(V)4 constitutes a major portion of the EF-111A terminal threat warning system. The system monitors and analyzes radar threats and has capacity to "look through" high-power jamming from the ALQ-99 transmitters aboard EF-111A.

MANUFACTURER:
Singer Co., Dalmo Victor Division
Belmont, California USA

AN/ALR-66 Radar Warning System

Fully programmable crystal video radar warning receiver covering E-J frequency bands; digital EW computer to provide rapid, unambiguous emitter identification in complex signal environments. Unique identification of all detected emitters is provided by alphanumeric symbology on a high-brightness CRT display. The system is readily programmable and provisions are included that allow modification of emitter parameters stored in library without special test equipment. More than 1,000 emitters can be stored in the EEPROM emitter library in less than 90 seconds. AN/ALR-66A(V)1 is the standard U.S. Navy ESM system on the LAMPS Mk1 (SH-2F) helicopters and is being updated. It is also in use by international customers for surveillance and ASW in a variety of platforms. AN/ALR-66(V)2 is a high-sensitivity surveillance and OTH targeting system installed in U.S. Navy P3-B ASW aircraft. AN/ALR-66(V)3 is an upgraded (V)2 with an interactive plasma display, and it provides increased capabilities for the ASW mission. The (V) is also installed in P-3B/Cs operated by foreign countries. AN/ALR-66(V)4 is a high-sensitivity version of the (V)1 which includes C/D band, DF, and upgraded software. It is operational in the U.S. Navy. AN/ALR-66(V)6 is an ESM system designed for shipboard applications providing self-protection and OTH targeting capability. Installed in U.S. Navy PHM-class ships. AN/ALR-66(VE) is a high-performance fighter aircraft RWR system suitable for retrofit of older generation aircraft as well as the new generation fighters. It is used by several foreign air forces, including those of NATO countries. AN/ALR-66 is available in an export version as the ALR-606. It is in use in several foreign countries. Major system characteristics:

- 360-degree coverage; 15-degree bearing accuracy
- Designed to interface with jammers, chaff/flare dispensers, and on-board recording systems
- RS-232 or 1553 I/O
- MTBF over 500 hours (demonstrated)
- Continuous over E/J- or C/J-band detection
- CW detection standard.

MANUFACTURER:
General Instrument Corp.
Government Systems Division
Hicksville, New York USA

AN/ALR-67(V) Threat Warning Control System

U.S. Navy's latest tactical threat-warning system. It consists of ALR-45F broadband crystal video receivers, a superheterodyne receiver, an integrated lowband receiver, an antenna array and an azimuth indicator, all under control of the dual ATAC-16M CP-1293 threat processor. It is field-programmable and includes provisions for software updates at the squadron level. The CP-1293 and IP-1276 used in the ALR-67 and ALR-45F configurations are identical and interchangeable. The ALR-67 systems are to be installed in F/A-18, A-6E, F-14, and AV-8B aircraft. System is fully compatible with Mil-Std-1553 data bus requirements and interfaces with systems such as HARM missiles, ALQ-126 A/B, ALE-29/39, ALQ-162, ALQ-164 and ALQ-165.

MANUFACTURER:
Litton Applied Technology
San Jose, California USA

AN/ALR-68 Advanced Radar Warning System

A broadband crystal video receiver radar warning system. Provides radar emitter detection and relative direction over a wide frequency range. Signal identification is made by comparing the received radar emitter characteristics to the ALR-68's emitter database. Emitter status information is then provided to the aircrew and to on-board ECM systems. The system's digital processor is configured with separate operational program and emitter databases, with the database module in the cockpit. The system is an interchangeable upgrade of the AN/ALR-46 and uses some of the same components. A complete set of support equipment, including radar emitter database generation and validation equipment, is available. The system is currently deployed on Federal Republic of Germany Luftwaffe F-4F and RF-4E aircraft.

MANUFACTURERS:
Litton Applied Technology
San Jose California USA

AEG-Telefunken
Ulm, West Germany

AN/ALR-69 Radar Warning Receiver

Outgrowth of the earlier models of ALR-46 warning receivers. Essentially an ALR-46 with a frequency selective receiver system (FSRS) and a low-band launch alert receiver (Compass Sail) added to the basic ALR-46. Compass Sail detects and analyzes SAM guidance beams to warn the pilot when the threat missile system is tracking the target aircraft and simultaneously guiding a missile toward the aircraft. The FSRS is reported to be designed to detect and analyze higher frequency missile guidance radiations that are not associated with the tracking radar and to point to the relative bearing of the missile's approach. The system is designed to automatically activate ECM resources. The CM-479 signal processor provides executive control for the FSRS. It accepts video inputs from five receivers and processed information from the FSRS, sorts and analyzes the data, identifies and labels the radar signal found, tracks the status of the found radar signals and generates signals to provide threat warning to the operator. Functions performed by the FSRS include warning and direction finding on CW signals, accurate frequency measurements on pulse signals for ambiguity resolution, threat antenna scan type and rate analysis, jammer frequency set-on and predictive PRI trackers for jammer power management functions and management of jammer blanking.

MANUFACTURERS:
Litton Applied Technology
San Jose, California USA

Singer Co., Dalmo Victor Division
Belmont, California USA

AN/ALR-73 ESM System

U.S. Navy passive detection system designed to supplement the early-warning radar of Grumman E-2C Hawkeye aircraft. Tailored to role and configuration of the E-2C. Update of ALR-59. Alerts operators to presence of electronic emitters. Intended to augment the AEW and surface, subsurface, command and control role of the E-2C by enhancing the threat detection and identification performed by the aircraft. Completely automatic, computer-controlled, superheterodyne receiver/processing system communicates directly with the E-2C aircraft command and control central processor. Designed for:

(1) Very high probability of intercept in dense environments
(2) Automatic system operation
(3) High reliability
(4) Ease of maintenance.

Features of the ALR-73 related to its intercept probability performance include: four-quadrant antenna coverage, independently controlled receivers, dual processor channels, digital closed-loop, rapid-tuned local oscillator. Others concerned with automated system operation are: low false-alarm report rate, automatic overload logic. AYK-14 computer adaptively controls hardware. Degraded mode operation. Signal preprocessor performs pulse-train separation, DF correlation, frequency tuning/timing and built-in test equipment tasks. All system functions are programmable to accommodate future tactical changes. In production.

MANUFACTURER:
Litton Systems, Inc.
Amecom Division
College Park, Maryland USA

AN/ALR-74 Warning Receiver

Self-protection for F-16 and a broad range of tactical aircraft; redesignated from the ALR-67/69 update. It will combine U.S. Air Force and Navy receivers to provide 2-to-18-GHz warning, a cooperative superheterodyne and an IFM. Under Navy funding, an advanced signal receiver will provide greater sensitivity through a fault-tolerant system capable of operating without diminishing protective capabilities. A Bragg cell/Fourier transform-type signal processor is also in development.

MANUFACTURER:
Litton Applied Technology
San Jose, California USA

AN/ALR-75(V) Receiver Set (SCR-2100)

Solid-state microwave surveillance receiver system tailored for electronic warfare applications. It provides spectrum surveillance, acquisition and analysis capability over the 100-MHz-to-40-GHz spectrum. The system consists of either RETMA rack-mounted or aircraft-style (DZUS) control panels and numerous remote-controlled units in ATR configurations. The use of independent octave band tuners allows simultaneous panoramic reception over the microwave spectrum or a mix of multiple tuners on a given band. Any combination of eight tuners may be used. Options permit simultaneous signal analysis of all eight tuners. Phase lock frequency accuracy is achieved with a frequency synchronizer. Other features include complete digital control, computer compatibility, scratch pad memories, optimized search and acquisition capability, signal identification from memory comparison, digital refreshed displays and other innovative features. System has been proven in both ground and airborne installations. System provides standard video/audio/processing and RF pan CRT displays plus detailed pulse/antenna/scan information (PRI/PRF/PW, pulse frequency, antenna beamwidth, scan type/interval, etc.). Control and/or readout of the receiver/pulse processor can be accomplished via standard IEEE-488 I/O or many special I/O formats.

MANUFACTURER:
Scientific Communications Inc.
Garland, Texas USA

AN/ALR-76 ESM System

Scanning superheterodyne to be installed in the U.S. Navy's S-3B aircraft. Form and fit replacement upgrade to the AN/ALR-47 and has frequency extensions beyond the ALR-47's. Uses VSLI chips that allow a wide range of multimission applications. Salient features are:

• Emitter classification—automatic operation with techniques for ambiguity resolution; radar warning is provided within this feature.

• Emitter location—automatically correlates separate detections of the same emitter and performs automatic location on every emitter.

• Auto self-test—three levels of auto self-test:
(1) continuous monitoring of all functions including the antenna; (2) periodic tests to determine fault locations to the box; and (3) diagnostics that are loadable on aircraft for fault location to the subassembly. Small, lightweight (134 pounds) and modular to allow growth.

MANUFACTURER:
IBM Federal Systems Division
Owego, New York USA

AN/ALR-77 ESM System

ESM for P-3C and P-3G Orion maritime patrol aircraft. Currently undergoing test and evaluation, it replaces the ALQ-78 or ALR-66(V)3 obsolescent systems. ALR-77 designed for detection, classification, tracking and targeting emitting radars. Uses interferometer antennas installed in updated wing tips of Lockheed design. Covers three bands over 360 degrees azimuth. System employs wideband acquisition with auto-switched narrowband analysis and fine DF. Wing tip includes RF front end, which down-converts to common IF. A preprocessor feeds data to operational processor, with the AYK-14 being one postprocessor that may be selected. ESM data is fed to a 1553B data bus. Production system weight is approximately 270 pounds, and power is 900 volt-amps.

MANUFACTURER:
Eaton Corp.
Deer Park, New York USA

AN/ALR-79(V) Radar Warning Receiver

Upgraded AN/ALR-66(V)1 system designed for the SH-2F helicopter. Contains an IFM for measurement of detected frequencies. Operates in the E/J band, providing automatic emitter identification in dense threat environments. Interfaces with active ECM systems. Logistic, maintenance support package available.

MANUFACTURER:
General Instrument Corp.
Government Systems Division
Hicksville, New York USA

AN/ALR-80(V) Radar Warning Receiver

Successor to the AN/ALR-66(VE); operates in the C-J bands, providing automatic emitter identification in complex environments. Designed to interface with active and passive ECM/ESM systems. In-flight recording and playback of all detected radars. Fully reprogrammable and equipped with complete logistics support and independent emitter library. ALR-80 is the direct substitute for APR-36, -37, -39, ALR-45, -67 and -69. Suitable for installation on all aircraft and attack helicopters. Currently in production and service.

MANUFACTURER:
General Instrument Corp.
Government Systems Division
Hicksville, New York USA

AN/ALR-81(V1/2)

Synthesized, microprocessor-controlled scanning superheterodyne system that operates with less than 1.8-degree rms integrated phase noise over the 0.5-to-18.5-GHz range. The receiver combines full band scanning and low phase noise characteristics. It displays both accurate synthesized frequency and calibrated signal amplitude measurements, while distributed microprocessors and independent scan capability provide operator control flexibility. AN/ALR-81 displays high resolution signal presentation with alphanumerics, and incorporates low sidelobe antennas and an 18-to-40-GHz downconverter. It employs digitized HP1345A CRT display. Operating modes include: band scan, sector scan, dwell, multiple sector and manual tune. Band scan allows operator to see full spectrum in sectors (0.5 to 2 GHz, 2 to 4 GHz, 4 to 8 GHz, 8 to 12 GHz and 12 to 18.5 GHz.) Sector scan allows operator to select and view frequency sector with increased frequency and amplitude resolution. Dwell causes momentary halt at marker frequency, writes an IF pan display trace and automatically resumes scan. Multiple sector allows operator to scan and view any five frequency bands. Manual tune receiver tunes on selected frequency for detailed signal analysis using IF pan. Control/display-to-tuner interfaces consists of a fiber-optic control cable and two IF signals. This allows remote location of tuner 200 feet from the control unit. AN/ALR-81 operates remotely via a standard IEEE-488 talker/listener port at the control unit. Other protocols, such as RS-232, RS-432 and Mil-Std-1553, are also available.

MANUFACTURER:
Condor Systems
Campbell, California USA

AN/ALT-40(V)

Noise jammer developed for Navy ERA-3B FEWSG aircraft for training exercises. AEL's I/J-band, high-power microwave amplifier system (HPMAS) is a modularized, liquid-cooled unit housed in a single pressurized assembly and meets Mil-E-5400 requirements. HPMAS provides in excess of 600 watts CW minimum RF output across the 8-to-17-GHz system. Equipment interfaces with external control, test and RF subsystems.

MANUFACTURERS:
Watkins-Johnson Co.
San Jose, California USA

American Electronic Laboratories Inc.
Lansdale, Pennsylvania USA

AN/AM-6988 POET Expendable Jammers

Primed oscillator expendable transponders (POET) in production for the U.S. Navy, it is designed to be dropped by Navy aircraft to confuse enemy radar-guided weapons, the system includes a power source, antenna, receiver and transmitter in a three-cubic-inch package weighing about a pound. Using the latest in MIC technology, the complex transponder, which repeats an incoming signal after suitable modification and uses only about 300 components.

MANUFACTURER:
Sanders Associates Inc.
Nashua, New Hampshire USA

AN/APR-38 Radar Homing and Warning Receiver

Installed on the F-4G Wild Weasel aircraft, it consists of a dual-baseline interferometer, which provides precision direction finding for emitting radars over broad frequency ranges. Receivers perform long-range acquisition sorting and location functions under the control of a general-purpose computer. Emitters over a broad frequency range are reacted to quickly, and the appropriate data is sent to the aircraft's automatic weapons release computer. Fully automatic under software control supervised by the F-4G electronic warfare officer through his control indicator.

MANUFACTURERS:
IBM
Federal Systems Division
Owego, New York USA

Loral Electronic Systems
A division of Loral Corp.
Yonkers, New York USA

AN/APR-39(V)1 Radar Warning Receiver

Radar warning system for helicopters and light to medium fixed-wing fighter, cargo and utility aircraft. Effective against low- to medium-altitude, radar-supported, anti-aircraft guns (AAA) and surface-to-air missile (SAM) systems. Threat recognition is based on frequency, PRI, PW persistence and threshold power level. Indicator has upgrade for night vision goggle compatibility. CW receiver also available for some SAM threats. Nomenclatured support equipment is also produced. Specifics:
 • Warning: Bearing, identity, mode(s) of operation of radar supported AAA and SAM weapons systems.
 • Coverage: Radar bands E, F, G, H, I and most of J. Appropriate portions of bands C and D. 360-degree azimuth.
 • Power: 28 VDC at 2 Amps max.
 • Weight: 8 pounds/3.63 kilograms.
 • Environment: Mil-E-5400, Class 2
 • Status: In production. Applications: OV-1D, OV-10. U-21, UH-1, OH-6, OH-58, UH-60, AH-1, AH- 64, Falcon 20, Merlin, Hawker, 500MD, Gazelle, Lynx, PAH-1, BO-105, CH-47, CH-46; installed on three classes of fast patrol/gun boats.

MANUFACTURER:
E-Systems Inc.
Melpar Division
Falls Church, Virginia USA

AN/APR-43 Compass Sail/Clockwise RWR

Provides C- and D-band CW detection and direction finding associated with certain SAM threats. The system interfaces with the AN/ALR-45 or AN/ALR-45F and the AN/ALQ-162 ECM system used on a variety of U.S. Navy tactical aircraft. It replaces outdated ALR-50 and can operate independently of associated systems sharing only the cockpit display of the primary RWR. The system provides its C/D band CW capabilities to other RWRs lacking such function.

MANUFACTURER:
Loral Electronic Systems
A division of Loral Corp.
Yonkers, New York USA

AN/APR-44 Airborne CW Radar Warning Receiver

Lightweight system designed for Army and Navy reconnaissance helicopters. It provides alert whenever a CW threat emission is received in the 14.5-to-16.5-GHz passband. The system comprises an omnidirectional antenna; a receiver containing an RF filter, detector-limited switch assembly, video amplifier, processing and output circuitry; and a control panel. The receiver is an RF-chopped, crystal video unit. It can detect hostile air defense radars and is used in conjunction with chaff/flare dispenser in AH-1J and AH-1T helicopters.

MANUFACTURER:
American Electronic Laboratories Inc.
Lansdale, Pennsylvania USA

AN/APR-46 (WJ 1840A)

Wideband microwave receiver system; high-performance, ruggedized. Frequency coverage of 0.03 to 18 GHz in eight bands. Max noise figures are 11 dB for 0.03 to 0.5 GHz; 20 dB for 0.5 to 8 GHz; 23 dB for 8 to 18 GHz. Omnidirectional antenna gain 0-3 dBi, slant linear polarization. Single operator. Six- or eight-trace digitally refreshed pan display. Log video and audio displays. The system incorporates WJ-8535-15 omnidirectional antenna, MD-127 demodulator, TN-130 RF tuner, Hewlett-Packard 1304A pan display, C-115 remote receiver control unit, and C-125 remote priority scan control unit.

MANUFACTURER:
Watkins-Johnson Co.
San Jose, California USA

AN/AVR-2 Laser Warning Receiver

Developed to protect U.S. Army and Marine combat helicopters from hostile laser-aided weapons. Integrated with the AN/APR-39 radar warning receiver, the AVR-2 detects, the system locates and identifies laser threats for warning and target cueing. It incorporates optical BITE and provides audible and visual alert through the radar warning receiver cockpit display. Intermediate level of maintenance equipment provides repair capability in the field. Sensor components are installed at optimum field of regard locations. Interface to unit is located adjacent to APR-39.

MANUFACTURER:
Perkin-Elmer Corp., Military Systems Division
Danbury, Connecticut USA

AN/BLD-1 Interferometer Direction-Finding Set

Designed to provide submarine commanders with precision bearing information on RF emitters while the submarine is at periscope depth. It is cued by other on-board electromagnetic sensors. Once the cue is received, the BLD-1 rapidly determines the bearing of the emitter. The bearing information in conjunction with other emitter parameters can be used for target identification, isolation or as support to other fire control sensors on board the submarine.

MANUFACTURER:
Litton Systems, Inc.
Amecom Division
College Park, Maryland USA

AN/GLQ-3A HF/VHF Comjam

Designed to disrupt hostile voice and data ground and airborne communications in the 20-to-230-MHz band. A transportable system, it is housed in an S-380 shelter mounted in 1¼-ton truck. Rapidly erectable coplanar, log-periodic antenna can be used for each of the three segments of the covered frequency spectrum with spot jamming available on any single frequency. The system was modified in 1976 to incorporate all solid-state transmitters and microprocessor controls for look-through, signal-initiated and continuous jamming. It is capable of operating with amplitude modulation, frequency shift keying continuous wave and frequency modulated CW signals. GLQ-3A is being replaced by the MLQ-324 Tacjam system. Jamming

power output from the GLQ-3A is approximately 1,500 watts from a solid-state transmitter controlled by the microprocessor.

MANUFACTURER:
Fairchild Weston Systems
Syosset, New York USA

AN/GLQ-3B Countermeasures System

Designed to disrupt hostile ground and airborne communications, radar and navigational aids operating within the 20-to-230-MHz band. A transportable system, it is housed in an S-250 shelter mounted in a 1¼-ton truck. A rapidly erectable coplaner, log-periodic antenna can be used for each of the three segments of the covered frequency spectrum with spot jamming available on any single frequency. The system was modified in 1976 to incorporate all solid state transmitters and microprocessor controls for look-through, signal-initiated and continuous jamming. It is capable of operating with amplitude modulation, frequency-shift keying, continuous-wave, modulated continuous-wave and frequency-modulated CW signals. Jamming power output varies from 1,450 to 2,300 watts. The GLQ-3B's solid state transmitter is controlled by a microprocessor.

MANUFACTURER:
Fairchild Weston Systems
Syosset, New York USA

AN/GLQ-501 Tactical Signal Simulator (Tass)

Latest in a series of computer-controlled RF and/or video radar signal simulators that can be tailored to meet the user's needs and budget. The system is flexible, versatile and quickly reprogrammable to meet changing threats. It provides realistic simulation of a complete variety of emitters ranging from simple pulsed radars to the exotic signatures of the most advanced systems. The system also provides programmable platform characteristics to simulate threat sources originating from fixed land-based sites through to high-speed missile systems. Tass is user-interactive and can be programmed and operated by users who have no computer experience. Features include: total user-friendly software programmability; frequency range of 0.5 to 18 GHz; up to 1,248 simultaneous emitters with wide selection of characteristics; modular expandable design.

MANUFACTURER:
Canadian Astronautics Ltd.
Ottawa, Canada

AN/MLQ-T4(V) Ground Jammer

Computer-controlled transportable ground jamming system that employs the latest technology to achieve high-power, continuous wave and pulse jamming, using microwave power amplifiers. A trailer, pedestal and shelter comprise the system, providing the capability for tracking aircraft, analyzing received signals and jamming radar emissions.

MANUFACTURER:
American Electronic Laboratories Inc.
Lansdale, Pennsylvania USA

AN/MLQ-34 Tacjam Communications Jamming System

Army's VHF communications jammer replacing the AN/GLQ-3. Tacjam, together with its passive counterpart TACELIS, is part of the Army's overall tactical communications EW (Tacom-

EW) program. The system can independently jam three signals simultaneously and works under remote control from the Tacom-EW control-processing center via command and data links. Independent operation is also possible from the division support company operations center. One CPC can control three Tacjams. Operating in conjunction with TACELIS, Tacjam will provide the supported commander with the means to cause significant disruption, confusion and delay among hostile forces. Psycho-acoustic and deceptive techniques are also included in the jamming modulation program. Remote control and override of jamming is possible. Tacjam has a look-through and read-through capability so as to continue monitoring during the jamming mission. Standard modulations include: RF, FM, and jam by noise, frequency shift key, amplitude modulation, frequency modulation, single sideband (SSB) by voice random CW. Tacjam will have three systems deployed with a forward operations company, three systems with a division support company and two systems with a brigade support company. Key Tacjam components include three solid-state transmitters using VHF power transistors of the latest vintage and three jamming antennas per set. Also included in the overall set is the AN/UYK-19 ruggedized minicomputer and the AN/ULR-17 communications intercept receiver. The latter two subsystems are common to the TSQ-112 TACELIS system. Effective radiated power (ERP) is three to four kilowatts. Tacjam interfaces with TACELIS with operations companies. At a division support or brigade support company level, it will interface with the DSC operations center via the AN/GRC-106 or the AN/VRC-46 tactical radio sets. It is installed in M-548 tracked cargo carrier and is currently in production.

MANUFACTURER:
American Electronic Laboratories Inc.
Lansdale, Pennsylvania USA

AN/MLQ-T6 Communications Data Link Jammer

Mobile, general-purpose, ground-based communications and data link jammer; 960 to 1850 MHz, one kilowatt CW or two kilowatts peak. Subsystems: transmitter, receiver, antennas, pedestal, processor/controller, automatic tracker, control and display console and the shelter.

MANUFACTURER:
Cincinnati Electronics
Cincinnati, Ohio USA

AN/MLQ-33 Close Air Support ECM Set

Army close air support ECM in the 100-to-450-MHz region. It consists of one receive and transmit antenna plus a receiver with a synthesized local oscillator. The system is housed in a standard U.S. Army S-250 shelter. Effective radiated power is 4 kilowatts; designed to jam air-to-ground communications. Scans automatically. On recognition of high-value communications, it will automatically jam it. MLW-33's receiver looks for high value signals and then locks onto one signal. Time to search the entire band is less than or equal to 0.5 second.

MANUFACTURER:
GTE Government Systems
Mountain View, California USA

AN/MSQ-103A Teampack ESM System

500 MHz to 40 GHz non-communications emitter location system mounted in a jeep, shelter or tracked vehicle. It is designed for service at a division level. Built-in computer processing from a ruggedized minicomputer is housed in the shelter. Teampack systems are housed in ballistically protected shelters. Each Teampack system is independently operated with secure voice back to the forward control and analysis center. Antenna rapidly deployable.

MANUFACTURER:
Emerson Electric Co.
St. Louis, Missouri USA

AN/MSQ-T43(V) Modular Threat Emitter

A manned radar with in-band target tracking and a monopulse receiver. Now in production for the U.S. Air Force, Navy and Marine Corps, it is employed worldwide on aircrew tactics/electronic combat training ranges to simulate enemy surface-to-air missile (SAM) radar systems. Three types of SAM radar types can be simulated one at a time. All radar parameters, operating modes and functions are controlled by an imbedded computer. The system consists of two major hardware groups: an operations shelter group and a pedestal group. The shelter group is an S-280 shelter containing all controls for radar operation, video recording and threat programming. The pedestal group contains all transmitters and modulators and a bore-sited video camera. The system can be slaved to another radar. Military standard technical orders and depot support have been acquired by the U.S. Air Force.

MANUFACTURER:
Whittaker Corp.
Simi Valley, California USA

AN/PRD-11 DF/Surveillance System

Combines the WJ-8640-1 receiver, WJ-8975A DF processor, WJ-9180 signal monitor and WJ-9880A (A-1) antenna into a lightweight, ruggedized system. By selecting tuning heads for the receiver and the appropriate antenna configuration, the system can accomplish target acquisition, surveillance and location.

MANUFACTURER:
Watkins-Johnson Co.
Gaithersburg, Maryland USA

AN/PSS-10 Troop Radar Signal Detector

Designed to provide soldiers in the field with the capability to detect and DF the location of troop movement radars. The small detector (24 ounces) mounts two antennas offset by about 90 degrees, allowing reception of signals from a wide are while not picking up emissions from friendly radars from the rear. It operates on a standard military battery for up to 24 hours, and determines bearings of CW or pulse emitters. Bearing accuracies reported to be of about 15 degrees and a coverage in the I, J and K bands, the ones most appropriate for radars capable of spotting soldiers on foot. Current DF capabilities are limited. The detector warns the operator via earphone when the device receives RF signals in the mode (CW or pulse) selected by switch. He then orients the receiver to determine the general direction of the threat, changes the detector to DF mode via toggle switch, and determines approximate bearing.

MANUFACTURER:
General Instrument Corp.
Government Systems Division
Hicksville, New York USA

AN/SLQ-17(V2) EW System for Aircraft Carriers

Designed to perform automatic detection, direction finding and identification of friendly and hostile platforms and missiles. When a threat is detected, the SLQ-17 provides deception techniques tailored for that specific threat. The system consists of a display, AN/UYK-20 computer and a printer located in the carrier's CIC. The signal processor, high power amplifiers and antenna control units are located elsewhere. Equipments are controlled at the CIC display. In the event CIC equipment is destroyed or becomes inoperable, the equipment can be manually set. Operator's screen displays friendly and enemy air, surface and subsurface platforms. System status and faults are automatically displayed to the operator. Once an intercept has been made, the system automatically displays likely platforms along with a probability level of correct identification. As additional information is received, the system automatically improves platform identification ratio based upon new information. The system will also display weapons systems associated with the platform identified in order of decreasing threat to the carrier. Increasing threat levels (fire control or missile homer) are automatically displayed and appropriate deception techniques are generated to prevent launch or lock-on by the seeker. In production.

MANUFACTURER:
Hughes Aircraft Co.
Fullerton, California USA

AN/SLQ-32(V) Shipboard ESM/ECM

Modularly related family of EW suites with varying levels of electronic warfare capability providing surveillance, warning and countermeasures against complex multiple missile attacks. Computerized processing and control provide rapid response, greatly reduce operator workload, allow reprogramming to meet changing threats and aid in maintenance. The system, now in production for more than 20 classes of U.S. Navy ships, has three separate suites:

• (V)1-Basic electronic support measures suite detects and analyzes radar threat emissions, identifying their bearing. Provides data and controls that enable an operator to select and time-launch rapid-bloom chaff rockets. (V)1 is installed on small auxiliary and amphibious warfare ships of the U.S. Navy.

• (V)2-Basic (V)1 system augmented with two receiving subsystems to extend frequency surveillance coverage. Installed in U.S. Navy DDGs, FFGs and *Spruance*-class destroyers.

• (V)3-Includes (V)2 ESM capability and adds active countermeasures with high power jamming capability against missiles and threat platforms. Can perform receive and jam functions simultaneously. In U.S. Navy cruisers, large auxiliary and amphibious warfare ships. The system's primary role is point defense against missile attack, and the system architecture uses crystal video DF detection, combined with IFM receivers for frequency measurement. High effective radiated power is obtained by paralleling and phasing wide bandwidth, medium power TWTs, using a lens technology which provides 100 percent duty cycle ERPs. All three systems use multiple beam antennas for receiving except in the lowest frequency band. The antenna consists of an array of elements fed through coaxial cables by a multiple beam parallel-plate lens. The lens-fed array provides a set of individual, contiguous high-gain beams, all

existing simultaneously, with each beam possessing the full gain of the array aperture. Display format is tactically oriented to provide a scenario of friendly and threat emitters. By designating any assigned symbol, the operator can call parameter information from the computer and the system will automatically display emitter information, based on operator-entered or intelligence-based library. All variants use the UYK-19V computer. The MK-36 super-rapid bloom offboard countermeasures system can be semi-automatically or manually controlled by the SLQ-32(V) operator at the display control console. The system provides the operator with recommendations on launch timing and launcher selection.

MANUFACTURER:
Raytheon Company
Electromagnetic Systems Division
Goleta, California USA

AN/SSQ-72 Classic Outboard

Composed of the HF/VHF direction-finding SRD-19 Diamond system and the SLR-16 communications intercept and analysis receiver. It is a tactical shipboard unit designed to provide over-the-horizon (OTH) detection and identification of surface ships for targeting purposes. The capability is similar to the shore-based Bulls Eye program, and uses hull- and mast-mounted antenna arrays. To be fitted in guided missile frigates, destroyers, and cruisers to provide signals intelligence, early-warning and OTH capabilities to two surface units that accompany each Navy carrier force. In production.

MANUFACTURER:
Sanders Associates Inc.
Nashua, New Hampshire USA

AN/SSQ-81 Radar Target System

System is a radar target for homing missiles. AN/SSQ-81 is an I-band radar target system deployed on a target boat to support testing of missile firings by providing appropriate electromagnetic radiations in space for missile homing devices. It consists of a self-contained gas generator, transmitter, stabilized rotator and antenna. System employs all solid-state design, including a high-power, solid-state magnetic modulator.

MANUFACTURER:
Whittaker, Tasker Systems Division
Simi Valley, California USA

AN/TLQ-15 Mobile Communications Jammer

Army transportable HF communications jammer in the 1.5-to-20-MHz band. Equipment includes: a panoramic indicator unit; frequency counter; modulation generator; RF amplifier; low-pass filter assembly; and an antenna coupler. In addition to the main antenna (a 35-foot telescopic whip with counterpoise) provisions are made for feeding into a dummy load and for connection to an auxiliary antenna. Carried on a 1½-ton truck. Designed to counter HF voice, radio teletype and CW communications. Peak power output is two kilowatts. Three TLQ-15 systems are typically deployed with a forward operations company in standard S-250 shelters and power unit trailer. The system requires one operator.

MANUFACTURER:
American Electronic Laboratories Inc.
Lansdale, Pennsylvania USA

AN/TLQ-17/17A Traffic Jam Ground ECM System

HF/VHF communications intercept and jamming/counter-measures system. It is designed to be carried in jeeps and in EH-1H Quick Fix I helo and EH-60A Quick Fix II helo. Code named "Traffic Jam," it includes a capacity to prevent jamming of friendly voice communications equipment operating in the same area. Limited voice (FM) communications are also possible with the TLQ-17. CW jamming power is 500 watts with a pulse peak of 2.5 kilowatts. TLQ-17A is 1.5 to 80 MHz. Equipment uses a microprocessor-controlled receiver, fully synthesized, digitally tuned, high-stability receiver to allow the operator to monitor and record both voice and digital communication links. The product-improvement program for TLQ-17A includes effective radiated CW jamming power of 550 watts. Also included are new microprocessor-related developments allow-ing complete remote capability. TLQ-17A operates more than 256 preselected frequencies with a receiver tuning time of less than one second; operating modes include search/lock-out, priority/lock-out, monitor/automatic, scan/band selectors and total.

MANUFACTURER:
Fairchild Weston Systems Inc.
Syosset, New York USA

AN/TLQ-17A(V) Communications Countermeasures System

Intercepts and jams communications signals in the 1.5-to-80-MHz band. Modular in design, it can be used in a variety of airborne and ground-based configurations. Presently fielded in three configurations: The (V)1 is jeep-mounted, the (V)2 is carried in the Heliborne "Quick Fix" system, and the (V)3 is mounted in an S-250 shelter in a 1¼-ton truck. The system's microprocessor allows the system to automatically search, find, identify and lock onto hostile transmissions across the entire band, in any selected portion of the band, at up to 256 pre-selected priority frequencies, while locking out designated friendly frequencies. It can be programmed to transmit, auto-matically or manually, a variety of modulations. Internal modulations are AM, FM, CW and SSB. Plug outlets allow use of a tape player, keyer, hand set or other external modulation sources. The receiver is a fully synthesized, digitally controlled, high-stability system with a one-second tuning time. The effec-tive radiated power is 10 to 550 watts.

MANUFACTURER:
Fairchild Weston Systems
Syosset, New York USA

AN/TLQ-501 Tactical Communications Jammer

100W jammer covers between 20 and 80 MHz, in 12.5-kHz steps. Designed to operate in small soft-skinned armored vehicles. Automatic capabilities include search-for and jamming on a priority basis of non-friendly transmissions from a list of up to 16 presettable channels or in a specified range of fre-quencies. System can also be used for intrusion or deception. Includes integral secure communications. Delivers 400W ERP.

MANUFACTURER:
Racal Communications Inc.
Rockville, Maryland USA

AN/TLQ-502 Tactical Communications Jammer

Combined 100W and 500W responsive jammer covering 20 to 80 MHz in 12.5-kHz steps. Selection of the 100-watt or 500-watt amplifier is by manual switch. Designed for use with a selected common LPA, they produce typically 400 watts or 2 kilowatts ERP, respectively. Both modes include automatic search and jamming of target transmissions from a list of 16 pre-programmed channels or in a specified band of frequen-cies. It is also used for intrusion or deception and includes integral secure communications.

MANUFACTURER:
Racal Communications Inc.
Rockville, Maryland USA

AN/TSQ-109 Agtelis Sigint System

Designed to locate transmitters, 500 MHz to 18 GHz. Uses time of arrival and direction of arrival to achieve accuracy in direc-tion finding to one degree rms with a 30-meter circular error probability at a range of 30 kilometers. Complete system con-sists of remote sensor outstations and a two-vehicle control processor set-up. Outstations provide measurements of fre-quency, pulse width, pulse repetition frequency and various other parameters to identify and locate emitters. Control processor controls the outstations by sending search and tuning commands, and receives data; second processor integrates this data with that of airborne systems and transmits the result to a forward control and analysis center. Outstations can be oper-ated independently of the control processors. Three Agtelis systems are normally deployed with each operational company and consist of three outstations and one control processor, all of which are contained in two ½-ton trucks.

MANUFACTURER:
Bunker Ramo
Westlake Village, California USA

AN/TSQ-112 TACELIS ESM System

Tactical automated communications emitter location and identification system (TACELIS) operates from 0.5 to 500 MHz. Performs communications collection, emitter location and processing functions. Mission is to exploit hostile tactical com-munications voice radio by generating threat/early warning, target, and decision information as well as jamming. TACELIS can delay hostile application of decisive combat power at critical junctures in a battle or prebattle situation. This is a major component of the Tacom-EW system and consists of two remote master stations and four remote slave stations deployed with each operations company forward in one 10-ton van and three 6-ton vans plus a truck tractor. Remote master stations are deployed in three 5-ton trucks and each remote slave station is deployed in one 1¼-ton truck. TACELIS' component elements include the various stations plus one AN/UYK-7 computer, and 12 AN/UYK-19 minicomputers as well as AN-ULR-17 receivers. Remote slave stations have a direction-finding capability only, while each remote master station has 14 receivers. Remote slave stations have a direction-finding capa-bility only, while each remote master station has 14 receivers and two search/acquisition receivers. System is interoperable with forward control and analysis centers AN/MLQ-34 Tacjam via the AN/GRC-103 communications link directly out of the various RMSs, and the Cefly Lancer aircraft. The system can be manned by 20 to 30 soldiers and will eventually be controlled from the rear by a control processing center as will its active counterpart, Tacjam. System replaces AN/TRQ-32.

MANUFACTURER:
GTE Government Systems
Mountain View, California USA

AN/TSQ-114 Trailblazer VHF DF System

Sigint system developed for U.S. Army consists of five stations: two serving as master control stations mounted in 1¼-ton trailers and three remote slave stations installed in ¼-ton trucks. The outstations require no operators and the overall system can DF five to six targets in one minute. One complete system is deployed with each brigade support company and operates from 0.5 to 150 MHz for detection and identification and performs direction-finding from 20 to 80 MHz. System includes DF/intercept antennas, radio frequency processors, DF/intercept receivers, line-of-bearing/fix displays, digital tape equipment, audio tape equipment, minicomputers, UHF data links, FM voice transceivers, and communications security (Comsec) equipment. Two intercept/DF control positions are in each control station with two operators per shift in each. The outstations require no operators.

MANUFACTURER:
ESL Division of TRW
Sunnyvale, California USA

AN/ULQ-11 Cefirm Leader Airborne Jamming System

Conducts ESM, jamming, imitative communications deception, and direction finding. System is composed of nine RU-21 aircraft, four of which are designated as "A" aircraft that conduct airborne radio direction finding (ARDF); three, designated as "B" aircraft, conduct intercept and serve as system control; and, the final two "C" aircraft conduct jamming missions. The basic function of the nine aircraft systems is to intercept, locate, and deny hostile forces the use of their communications. Types of communications susceptible to Cefirm Leader ESM and jamming include: amplitude modulation; frequency modulation; frequency shift key; continuous wave (CW, ICW and MCW); single-side-band. Assigned to an aviation company (rear), Cefirm Leader frequency coverage is 2 to 80 MHz with jamming power output at two MHz of 500 watts and 70 watts at 80 MHz operating from on-board aircraft power of 400 Hz.

MANUFACTURER:
Unknown

AN/ULQ-14 Radar Countermeasures System

Ground-based system operates from 8.5 to 17 GHz. It is designed to be the ground element of MULTEWS and provides ECM capabilities required by all Army echelons from division through armored cavalry regiments for countering ground- and air-based threat radar. Primary targets are countermortar/ counterbattery radars, combat surveillance/target acquisition radars, and other radars when line-of-sight propagation is possible. The system is housed in a standard Army S-250 shelter for use with a 1½-ton truck. It has a data interface with AN/ALQ-143 MULTEWS airborne systems, Quick Look II, and the AN/MSQ-103 Teampack mobile ESM system. Capabilities include jamming of four to six signals in a 15-kilometer range.

MANUFACTURER:
Unknown

AN/ULQ-19 (V1) and (V2) Tactical Communications Jammers

A responsive 100-watt jammer system covering the combat net radio band (20 to 80 MHz) in 12.5-kHz steps. It meets requirements for a full automatic, low-cost, highly mobile VHF communications jamming system for use in the forward combat zone. Automatically detects and jams signal activity on any one of 16 preselected target channels within one second, on a priority basis. Because jamming only takes place when targets are active, this can effectively result in the disruption of several nets simultaneously, depending on the individual transmit-to-receive ratios of the target nets. Alternatively, the system can be programmed to scan a specified band of frequencies and disrupt non-friendly transmissions detected above a preset signal level. A manual mode provides operator intrusion or deception of a single, dedicated frequency. The V1 variant is for installation in vehicles, with whip antennas, and can be used on the move. The V2 variant also includes an LPA (manually switched) to provide 400W ERP. Comsec-compatible integral communications are included to provide programmable jamming interrupt on a priority basis. Facilities for full function remote ECM control via digital burst command and control ink. It has compact and rugged construction, and can be mounted in soft-skinned or armored vehicle to provide high mobility.

MANUFACTURER:
Racal Communications, Inc.
Rockville, Maryland USA

AN/USD-9V(2) Guardrail Airborne Comint System

Guardrail is an airborne Sigint system that intercepts, locates and classifies target systems and transmits data to ground processors to provide real-time intelligence information. Single- and multi-channel VHF/UHF communications intelligence collection, direction-finding and reporting system. Composed of a transportable, ground-based facility and C-12 airborne platforms carrying remotely-controlled mission equipment. Secure radio links provide voice and data communications among these facilities and tactical commander terminals in the AN/ TSC-87. There are three Guardrail systems: Guardrail IIA with 14 intercept/DF positions and truck-mounted mobile relay facilities (MRF); Guardrail IV with eight intercept/DF positions which is in service in Korea and interfaces with the Adventurer system in lieu of having mobile relay facilities; and Guardrail V. Deployment is in six aircraft per forward aviation company. Frequency ranges: 20 to 75 MHz; 100 to 150 MHz; and 350 to 450 MHz. Direction finding is performed in the two lower bands. Major components in Guardrail II include six sets of mission equipment and inertial navigation sub-systems. Also part of the suite are integrated processing facilities located in three vans as well as mobile relay facilities. Typical Guardrail operations call for two aircraft launched together to provide an optimum base line for direction finding. Flight endurance is between four and five hours. In Guardrail V two aircraft are launched together to provide an optimum base line for direction finding. Guardrail V product improvement program (PIP) compensates for system shortcomings: no short-field capability in RU-21H aircraft; no weight capacity left for mission system growth; no pressurization in aircraft; limited aircraft survivability equipment needed in a combat environment; and aircraft themselves reaching the limit of the predicted structural flying limits. Mission deficiencies include: low receiver quantity of six per aircraft; slow, manual DF with limited frequency coverage; data links susceptible to enemy ECM; ground processing facilities not interoperable with other system; and, a lack of computerized search. Other PIPs contemplated: to increase Guardrail V on station time and increase its altitude above 10,000 feet, while not raising weight; replacing existing data links with systems from the Army's Corps airborne Sigint collection direction-finding element (Cascade). This improvement makes Guardrail interoperable with TACELIS, the Army's main ground-based communications intelligence asset. Cascade would fly a two-aircraft mission and would equip each aircraft with 16 receivers: two for search, six for VHF single-channel or multichannel intercept, and eight for HF/VHF intercept.

MANUFACTURER:
ESL Division of TRW
Sunnyvale, California USA

AN/VLQ-5 Comfy Sword III Communications Countermeasures System

Currently used for training U.S. Air Force communications operators in ECCM. It is also capable of deployment as a tactical ECM system. The system intercepts, monitors and jams both voice and data links. Modular in design, compact, rugged, highly mobile and can be conveniently installed in virtually any operating platform. In mobile applications, sets up and tears down in less than 10 minutes. Allows for the addition of data recording devices, remote control and other peripherals. Depending on the intended application, Comfy Sword III is supplied with a spectrum of broadband untuned antennas allowing instantaneous frequency agility. A guard antenna provides protection for friendly distress frequencies.

MANUFACTURER:
Fairchild Weston Systems
Syosset, New York USA

AN/WLQ-4 Electronic Support Measures System

A very high-performance electronic support measures (ESM) system deployed on 637-class submarines. The system identifies the nature and source of unknown radar and communications signals, and incorporates a series of minicomputers and microprocessors. Other technical features include: automatic search, acquisition, signal processing, logging and reporting. The WLQ-4 provides semiautomatic fusion of real-time measured data input from external sensors, automated monitoring and fault location performance, 400,000 lines of AN/UYK-20 source code and 50,000 lines of executable code in 40 microprocessors. The system has been designed with growth capability to handle new threats.

MANUFACTURER:
GTE Government Systems, Western Division
Mountain View, California USA

AN/WLR-8(V) Surveillance Receiver

Shipboard, digitally controlled, spectrum scan and analysis receiver, capable of sequential or simultaneous scans over a 50-MHz-to-18-GHz frequency range. It measures signal direction of arrival, frequency, modulation, PRF, pulse width, amplitude and scan interval of received signals and consists of seven superheterodyne RF tuners, all of which are YIG-tuned except for the varactor-tuned lower band. Outputs from RF tuners are served by the processor on a first come, first served basis. A programmable frequency search strategy with parallel tuning of multiple bands is used to maximize the probability of signal intercept to minimize the intercept time. The WLR-8 is built in variable configurations for submarines and large surface ships. There are currently four versions: (V)1 for submarines, (V)2 for the SSN-688 class submarines [frequency range: 550 MHz to 20 GHz], (V)4 for aircraft carriers and other large ships, and (V)5 for Trident submarines. The console contains the control, display and processing unit. Two oscilloscope traces can be switched to any tuner for PAN or time display. Two time-band traces and one PAN trace are used for Amp/Demod analysis display. The system can be expanded in frequency or signal handling capability through hardware additions and software changes. There are WLR-8s in *USS Enterprise* and submarines. Last September a $25 million U.S. Navy contract was awarded for continued production and it includes provisions for spare parts and ancillary equipment.

MANUFACTURER:
GTE Government Systems
Mountain View, California USA

AN/WLR-11 IFM Receiver

An instantaneous frequency measuring receiver to provide rapid analysis of ship missile threats. It features a 100 percent probability of threat signal intercept over the 7-to-18-GHz frequency range. The MX9414-A/ULR automatic processor, used with the WLR-11, can call up emitter parameters from memory or "queue" them in order to improve the overall reaction time. The IFM uses two YIG-filters, a band reject, and a bandpass filter for band rejection and pre-selection. Typical system sensitivity is reported to be –75 to –80 dBm when used in conjunction with the WLR-1. The WLR-11 is used with an omni and a DF antenna capable of scanning at 300 rpm. Also used in conjunction with the MX9414 to analyze PRF and pulse width of received signals, and is being added to all WLR-1 installations. An automatic processing capability is being added to the WLR-1/11 system to speed identification. New configuration is designated WLR-1G.

MANUFACTURER:
ARGOSystems Inc.
Sunnyvale, California USA

INTERNATIONAL HARDWARE

(Domestic and Non-U.S. Equipment
Not Bearing AN/ Designation)

ACS-500 Airborne Comint System
Computerized VHF/UHF data collection system: 20 to 500 MHz.
Tadiran Ltd.
Tel Aviv, Israel

Advanced Optical Countermeasures Pod
Air Force Compass Hammer electro-optical countermeasures program.
Westinghouse Electric Co.
Baltimore, Maryland USA

AEL Type 6040 Universal Jammer Tester
Signal source for use in testing system jamming vulnerability.
American Electronics Laboratories Inc.
Lansdale, Pennsylvania USA

ADVCAP Surveillance Receiver
Advcap receiver/processor group, will have automatic/manual jamming.
Litton Systems Inc.
Amecom Division
College Park, Maryland USA

Airborne Comjam System (ABCJS)
Standoff and escort system jams communications: 100 to 400 MHz.
Tadiran Ltd.
Tel Aviv, Israel

ALEX Shipboard Decoy System
Modular deck-mounted decoy system: anti-ship radar or IR-guided missiles.
Hycor
A Subsidiary of Loral Corp.
Woburn, Massachusetts USA

Apollo Airborne Jammer
Airborne ESM/ECM package combines Guardian series RWR with a jammer.
Marconi Defence Systems Ltd.
Stanmore, Middlesex, UK

AQ 31 Jammer
Aircraft pod mount: E to J bands. Several jamming modes available.
SATT Communications AB
Stockholm, Sweden

AQ 800 Jammer
Pod-mounted, ram-air cooled: four bands, 2 to 4, 4 to 8, 8 to 12, 12 to 18 GHz.
SATT Communications AB
Stockholm, Sweden

AQ 900 Airborne Self-Protection System
Pod-mounted automatic jamming system: Two pods, S and Ku-bands.
SATT Communications AB
Stockholm, Sweden

AR 777 Airborne Microwave Receiver
Computer-controlled superhet signal acquisition and jammer: 1 to 8 GHz.
SATT Communications AB
Stockholm, Sweden

AR 830 Airborne Warning System
Programmable threat warning of radar and laser emissions.
SATT Communications AB
Stockholm, Sweden

AR 861 Radar Warning Receiver
Programmable lightweight RWR system designed for aircraft.
SATT Communications AB
Stockholm, Sweden

ARI 18240/2 Nimrod MR ESM
Acquires, classifies, identifies, localizes and tracks the movements of ships, aircraft, and submarines as well as fixing land based sites.
Loral Electronic Systems
A division of Loral Corp.
Yonkers, New York USA

ARR-81 Comint Receiving Set
One kHz to 500 MHz. An optional receiver covers 500 to 2000 MHz.
Magnavox Government and Industrial Electronics
Fort Wayne, Indiana USA

Artillery-Delivered Expendable Communications Jammer (AD/ExJam)
A standard NATO 1.55-mm cargo round containing five self-contained jammers.
Fairchild Weston Systems
Syosset, New York USA

ASTG EW Simulator
Advanced standard threat generator (ASTG) produces test signals for EW receivers and signal processors: 500 MHz to 18 GHz.
AAI Corp.
Baltimore, Maryland USA

Automatic Countermeasures Dispensing System
Shipboard anti-missile application of chaff and flares.
Elbit Computers Ltd.
Military Systems and Products Division
Haifa, Israel

AT-4910 Comjam Support System
Land-based mobile VHF/UHF jamming and communication support system.
Elisra Electronic Systems Ltd.
Bene-Baraq, Israel

B.AE Infrared Jammer
Protects helicopters from attack by heat-seeking SAM missiles.
British Aerospace Dynamics Group
Bristol Division
Bristol, UK

Barbican EW System
Tactical, mobile battlefield automated equipment; 1 to 18 GHz.
MEL
Crawley, West Sussex UK
Ferranti Computer Systems Ltd.
Bracknell, Berkshire UK

BAREM Self-Protection Jammer
Pod-mounted self-protection jammer for aircraft.
Thomson-CSF, Avionics
Gabriel Peri, Malakoff
France

Barricade Naval Decoy Equipment
A chaff and infrared decoy launcher.
Wallop Industries Ltd.
Andover, Hampshire, UK
Aish and Co. Ltd.
Poole, Dorset, UK

BF Radar Warning Receiver
Provides pilot with alarms, threat type and direction.

Thomson-CSF
Paris, France

BO Series Countermeasures Dispensers
BOZ 100: Pod-mounted advanced dispenser for aircraft. BOZ 3: High capacity chaff pod dispenser for training use. BOP 300: For protection aircraft. BOH 300: Chaff/flare dispenser for helicopters.

Philips Elektronikindustrier AB
Jarfalla, Sweden

Bofors Chaff Rockets
Uses existing ships AA gun mounts: 57-millimeter rocket launchers.

Bofors Ordnance
Bofors, Sweden

Breda Light Chaff Rocket Launching System
Two shipboard launchers and control panel for patrol craft.

Breda Meccanica Bresciana SpA
Brescia, Italy
BFD Difesa-Spazio
Roma, Italy

Breda Multipurpose Naval Rocket Launcher
Launches 105mm multipurpose rockets in its standard version.

Breda Meccanica Bresciana S.p.A.
Brescia, Italy

Broadband Chaff (BBC) Rocket
Compatible with existing launchers and command/control systems.

Plessey Naval Systems Ltd.
Addlestone
Weybridge
Surrey, UK

Buck-Wegmann Decoy System
IR/chaff dispenser decoy system for boats and ships.

Wegmann & Co. GmbH
Federal Republic of Germany (launch system)
Buck Chemisch-Technische Werke GmbH & Co.
Federal Republic of Germany (grenades)

Caiman Offensive GCM Jamming Pod
Two six-KVA jammers that employ barrage jamming.

Thomson-CSF
Paris, France

Canadian Electronic Warfare System (Canews)
Fully automatic, integrated naval system incorporates ESM, ECM and chaff countermeasures: 1 to 18 GHz.

MEL
Crawley, West Sussex, UK
Westinghouse Canada Ltd.
Electronics Systems Division
Hamilton Ontario, Canada

CERES Communications Monitoring System
Computer-enhanced radio emission surveillance provides communications monitoring from 15 kHz to 30 MHz.

Rediffusion Radio Systems Ltd.
Crawley, West Sussex, UK

Challanger IRCM
Lightweight protection for light aircraft. Omnidirectional jammer.

Loral Electro-Optical Systems
Pasadena, California USA

Chief EW Combat System
Mobile integrated EW combat system: ESM, ECM, DF.

HRB-Singer Inc.
State College, Pennsylvania USA

Colibri Heliborne Integrated ESM/ECM System
Integrated ESM/ECM system for helicopter installation.

Elettronica SpA
Rome, Italy

CMR-500B Warning Receiver
Warning receiver detects CW signals: visual and aural output.

American Electronic Laboratories Inc.
Lansdale, Pennsylvania USA

Communication Network Simulator
CNS 8520 Series
Replicates RF signals: 2 MHz to 4 GHz.

Litton Applied Technology
San Jose, California USA

Communications ECM/ESM System
Modular vehicle-mounted communications electronic warfare system (CEWS).

Fairchild Weston Systems Inc.
Syosset, New York USA

Co-NEWS Naval ESM System
Communications naval electronic warfare system: VHF/UHF.

Elettronica SpA
Rome, Italy

Corvus Dispensing Rocket System
Lightweight naval, quick-reaction, chaff-dispensing rocket system.

Vickers Shipbuilding and Engineering Ltd.
Cumbria, UK

CR-2700/CR-2740 Mobile Elint Systems
Automatic search and acquisition of radar signals: High gain antennas and superheterodyne receivers.

Elisra Electronic Systems Ltd.
Bene-Baraq, Israel

CR-2800 Airborne ESM/Elint System
Automatic, computerized system for surveillance/Elint.

Elisra Electronic Systems Ltd.
Bene-Baraq, Israel

C/URD-10(V) UHF Direction Finder
UHF DF from 225 to 399.975 MHz in 7,000 channels: 50 presets.

Probe Systems Inc.
Sunnyvale, California USA

Cutlass Shipboard ESM System
Computerized automated ESM system: 1 to 18 GHz.

Racal Radar Defence Systems Ltd.
Chessington, Surrey, UK

Cygnus ECM System
ECM system with a narrow-beam tracking jammer: 8 to 16 GHz.

Racal Radar Defence Systems Ltd.
Chessington, Surrey, UK

Dagaie Naval Countermeasures Decoy System
Shipborne decoy system: Single- or double-mounted configuration.

Compagnie de Signaux et d'Enterprises Electriques (CSEE)
Paris, France

DALIA ESM System
Associated with the DR 2000 ESM receiver: E to J band, IFM.

Thomson-CSF
Avionics Division
Malakoff, France

DB 3141 and Remora Jammer Pods
Self-protection for aircraft against radar threats in H/I bands.

Thomson-CSF
Paris, France

DF-8000 Tactical DF System
DF system for mobile or fixed site applications.

GTE Government Systems, Western Division
Mountain View, California USA

DR 2000 ESM Receiver
Crystal video radar intercept receiver operating in the E to J band.

Thomson-CSF
Avionics Division
Malakoff, France

DR4000 ESM Receiver
IFM system intercepts signals from C through J bands.

Dalmo Victor Division
The Singer Co.
Fremont, California USA

Dragonfly Airborne C³CM System
Jams in VHF/UHF range.

GTE Government Systems Western Division
Mountain View, California USA

Dynamic Scenario Threat Generator System (DSTGS)
Generates RF environment for development and testing of EW suites.

Flight Systems Inc.
Newport Beach, California USA

E-1700/E-1800 VLF-HF Surveillance Receivers
Ten kHz to 30 MHz. Microprocessor controlled.

AEG-Telefunken
Ulm, West Germany

EB 100 MINIPORT Monitoring System
Portable, battery-operated mini VHF/UHF receiver: 20 MHz to 18 GHz.

Rohde & Schwarz
Muhldorfstrasse, Munchen
Federal Republic of Germany

EF-111A Operational Flight Trainer
Provides simulated combat training: Digital radar landmass simulation.

AAI Corp.
Hunt Valley, Maryland USA

EL/K-1250T Compact VHF/UHF DF Comint Receiver
Dual channel synthesized receiver: 20 to 510 MHz.

IAI Elta Electronics Industries Ltd.
Ashdod, Israel

EL/K-1251, 1252 and 1253 Miniature VHF/UHF Narrowband Tuners
Very compact synthesized tuners: 20 to 90 MHz, 100 to 500 MHz, 500 to 1000 MHz.

IAI Elta Electronics Industries Ltd.
Ashdod, Israel

EL/K-7001 Communications EW System
Compact communications EW system: VHF/UHF.

IAI Elta Electronics Industries Ltd.
Ashdod, Israel

EL/K-7010 Communications Jammer
Modular microprocessor-controlled VHF communication jammer.

IAI Elta Electronics Industries Ltd.
Ashdod, Israel

EL/K-7020 EW System
Modular command and control center.

IAI Elta Electronics Industries Ltd.
Ashdod, Israel

EL/K-7032 Airborne Comint System
Communications surveillance between 20 and 500 MHz.

IAI Elta Electronics Industries Ltd.
Ashdod, Israel

EL/K-7035 All-Platform Comint System
Monitors, locates, analyzes and reports: 20 to 500 MHz.

IAI Elta Electronics Industries Ltd.
Ashdod, Israel

EL/K-7050 Tactical Communications EW System
Combines all passive Comint tasks with high power jamming.

IAI Elta Electronics Industries Ltd.
Ashdod, Israel

EL/L 8202 Advanced Self-Protection Pod
Operates F-J bands, but can be modified to fit specific requirements.

IAI Elta Electronics Industries Ltd.
Ashdod, Israel

EL/L-8231 Airborne ECM System
Internally mounted miniaturized self-protection ECM system.

IAI Elta Electronics Industries Ltd.
Ashod, Israel

EL/L 8240 Internal Self Protection ECM System
Unified radar warning and radar jamming/deception.

IAI Elta Electronics Industries Ltd.
Ashdod, Israel

EL/L-8300 Airborne Sigint System
Sigint, Elint and Comint. Includes command station.

IAI Elta Electronics Industries Ltd.
Ashdod, Israel

EL/L-8303 ESM System
Computerized for shipborne, airborne and ground-based applications.

IAI Elta Electronics Industries Ltd.
Ashdod, Israel

EL/L 8310 Elint Receiver
Superheterodyne receiver with frequency range of 0.5 to 18 GHz.

IAI Elta Electronics Industries Ltd.
Ashdod, Israel

EL/L-8312A Airborne Elint/ESM System
Automatic, sensitive superheterodyne receiving system: 0.5 to 18 GHz.

IAI Elta Electronics Industries Ltd.
Ashdod, Israel

Which One...?

The real target is nearly impossible to find with the RBOC II Rapid Bloom Offboard Countermeasures system from Hycor. Masked by ship-launched infrared decoys and rocket-propelled chaff to confuse IR-seeking missiles and radar, this sophisticated system protects ships of all sizes. An extremely effective countermeasure, RBOC II can be fired either manually or by computerized control interfaced with the ship's detection system at the first hint of a threat.

Since 1972, Hycor has been the **only** company producing complete systems for the U.S. Navy and other friendly countries. Let us put our problem-solving task force to work for you.

For More Information, Contact us at
10 Gill St., Woburn, MA 01801, USA, Tel. (617) 935-5950,
FAX (617) 932-3764, TWX 710-393-6345

LORAL

Hycor
©1987

EL/L-8351 Elint Trainer Simulator
Advanced trainer system, simulating up to 270 emitters simultaneously.
IAI Elta Electronics Industries Ltd.
Ashdod, Israel

Elisa Receiver
Electronic intelligence, search and analysis (Elisa) receiver: C-J Band.
Thomson-CSF Avionics Division
Malakoff, France

ELT/128 Communications Direction Finder
HF, VHF and UHF frequency bands.
Elettronica SpA
Rome, Italy

ELT/156 Radar Warning Receiver
Airborne passive warning for aircraft.
Elettronica SpA
Rome, Italy

ELT Noise Jammer Pods
ELT/457, ELT/458, ELT/459 and ELT/460: airborne noise jammer pods.
Elettronica SpA
Rome, Italy

ELT-Series ESM/ECM
Two to 18 GHz: Four-band IFM receiver performing all ESM functions.
Elettronica SpA
Rome, Italy

ELT-263 ESM System
Integrated system: electronic surveillance in patrol aircraft.
Elettronica SpA
Rome, Italy

ELT/555 Jammer Pod
Pod-housed airborne DECM jammer and warning system.
Elettronica SpA
Rome, Italy

ELT/562 and ELt/566 Deception Jammers
Repeater jammers for internal installation in aircraft.
Elettronica SpA
Rome, Italy

ELT/999 Airborne Comint System
Strategic communications surveillance system for aircraft.
Elettronica SpA
Rome, Italy

Enhanced Radar Warning Equipment (ERWE)
Programmable digital threat processor: C to D and E to J bands.
Litton Applied Technology
San Jose, California USA
AEG Aktiengesellschaft
Ulm, Federal Republic of Germany

EP 1650 Baustein Comint System
Communications intelligence system: Modular, van-mounted.
AEG-Telefunken
Ulm, West Germany

EPO Naval Rocket Launcher
Multibarrel rocket launcher for chaff and infrared flares.
Evershed Power-Optics Ltd.
Surrey, UK

Erijammer 200 Jammer Pod
Automatic, multimode jammer for self-protection of combat aircraft.
Ericsson Radio Systems AB
Stockholm, Sweden

Erijammer A100 Training System
Pod-mounted jammer with built-in antennas and electronics.
Ericsson Radio Systems AB
Stockholm, Sweden

ES-400
Elint system ES-400A; ESM system ES-400U and ECM system ES-400G.
AEG-Telefunken
Ulm, West Germany

ESCORT Airborne Elint System
Electronic support measures equipment for maritime patrol aircraft.
Raytheon Co.
Electromagnetic Systems Division
Goleta, California USA

ESM 500 Series VHF/UHF Receivers
Computerized receivers for detection, monitoring and surveillance.
Rohde & Schwarz
Munich, West Germany

ESM 1000 Receiver Family
Receivers and control unit peripherals for radio reconnaissance.
Rohde & Schwarz
Munich, West Germany

ESM Electronic Warfare Equipment
Shipborne ESM: instantaneous detection, triggers jammer and decoys.
Thomson-CSF
Malakoff, France

ESM System 515R-1
ESM receiver for surveillance, Sigint and DF applications.
Rockwell International
Collins Defense Communications
Cedar Rapids, Iowa USA

ESP Scanning Receiver
Automatic receiver scans at 1000 channels/second: 10 KHz to 1,300 MHz.
Rohde & Schwarz
Munich, West Germany

ESS-2 Electronic Surveillance System
Mobile integrated surveillance and analysis: L to Ku Band.
Selenia, SpA
Rome, Italy

ETTT EW Trainer
Simulates electromagnetic emissions: EA-6B cockpit trainer.
AAI Corp.
Baltimore, Maryland USA

Evade Airborne Decoy System
Four or more dispensers and a cockpit control unit: Chaff or flares.
Wallop Industries Ltd.
Andover, Hampshire, UK

EW1017 Superheterodyne DF Receiver
Four-channel system: wingtip pod-mounted receivers, computer control.
Loral Electronic Systems
A division of Loral Corp.
Yonkers, New York USA

EWS-900 Chaff System
Shipboard chaff system for smaller ships.

Saab Missiles AB
Linkoping, Sweden

EWS-905 Dual-Mode ESM/Warning System
Identifies targets detected by surveillance radar.

Saab Missiles AB
Linkoping, Sweden

FARAD Naval EW System
For small naval ships: passive detection, self-protection.

Elettronica SpA
Rome, Italy

Fast Jam System
Communications ECM system for helicopters: surveillance and autojam.

Elettronica SpA
Rome, Italy

FIX-500 Automatic Position-Fixing System
Shipborne or ground-based mobile VHF/UHF system: 20 to 500 MHz.

Tadiran Ltd.
Tel Aviv, Israel

FIX-3000 Mini Comint/DF System
Ground-based mobile or fixed-site system: 2.0-to-30-MHz range.

Tadiran Ltd.
Tel Aviv, Israel

GE-530M/GE-530S Surveillance Receiver
Modular, sythesized, microprocessor-controlled HF receiver: 5.0 to 30 MHz.

Tadiran Ltd.
Tel Aviv, Israel

Guardian Shipborne EW System
Shipborne active and passive EW system: radar spectrum.

Thorn EMI Electronics Ltd.
Radar Division
Hayes, UK

Guardian Series RWRs
Series of advanced radar warning receivers: crystal video.

Marconi Defence Systems Ltd.
Stanmore, Middlesex, UK

Hand-Emplaced Expendable Jammer
Solid-state jammer for use against communication: HF and VHF bands.

Fairchild Weston Systems Inc.
Syosset, New York USA

HERMES ESM System
Airborne, shipboard or land-based.

Marconi Defence System Ltd.
Stanmore, Middlesex, UK

Helicopter Applique Communications Jammer
HF/VHF/UHF intercept and jamming system:

Racal Communications Inc.
Rockville, Maryland USA

HF Signal Analysis SAT3311
Signal analyzer and demodulator for use with MF/HF receivers.

Racal Communications Ltd.
Berkshire, UK

HOFIN Hostile Fire Indicator
Passive warning system for helicopters.

M.S. Instruments PLC
Bromley, Kent, UK

HUMMEL VHF Jammer
Automatic responding VHF jammer for use against voice and data links.

AEG-Telefunken
Ulm, West Germany

HWR-2 Radar Warning Receiver
Miniature system for helicopters and small patrol craft.

Racal Radar Defence Systems Ltd.
Surrey, UK

IGS-1 Ground EW Suite
Trailer-mounted 1-to-18-GHz microwave collection system.

Selenia SpA
Rome, Italy

IHS-6 Standoff Helicopter Jammer
Integrated ESM/ECM suite for helicopters.

Selenia SpA
Rome, Italy

INS-3 Naval ESM/ECM System
Wide-open crystal video receiver with IFM: 1 to 18 GHz.

Selenia SpA
Rome, Italy

J-3400 Communications Countermeasures System
Mobile passive/active equipment for VHF signal intercept and jamming.

GTE Government Systems
Western Division
Mountain View, California USA

Jamcat Jamming Unit
EW attachment designed for use with any combat net radio (CNR).

Racal-SES Ltd.
Burnham, Berkshire, UK

Jam Pac(V) ECM Jamming Systems
A compact jamming system developed for light aircraft.

Sperry Corp.
Defense Products Group

JANET Shipborne Jammer
Shipborne radar jamming and surveillance system in the H/J bands.

Thomson-CSF
Paris, France

Decca
London, UK

JAS 39 Gripen Integrated ECM System
Automatic and integrated control of all countermeasures functions.

Ericsson Radio Systems AB
Stockholm, Sweden

KATIE Radar Warning Receiver
Airborne radar warning system for light aircraft.

MEL
Crawley, West Sussex, UK

Kestrel Airborne ESM System
Advanced Elint system for surveillance aircraft.

Racal Radar Defence Systems Ltd.
Chessington, Surrey, UK

SYSTEMS
OF THE WEST
INTERNATIONAL HARDWARE

Lacroix Chaff/IR Ammunition
Chaff and infrared flare ammunition for combat aircraft.

Societe E Lacroix, Departement Contre-mesures
Muret, France

Lake 2000 Airborne Jamming Pod
Airborne subsonic jamming pod: 8.5 to 17.5 GHz.

SRA Communications AB (LM Ericsson subsidiary)
Spanga, Sweden

LNR R70 Flexible Microwave Receiving System
Mobile Comint/Elint surveillance system: 0.7 to 18.2 GHz, optional to 40 GHz. Tempest and Abite.

LNR Communications Inc.
Hauppauge, New York USA

Loral Electronic Environment Simulator (LEES)
Computer-based multiple radar-emitter generator, RF and video.

Loral Electronic Systems
A division of Loral Corp.
Yonkers, New York USA

LQ-102 Hand-Emplaced Expendable Jammer
Low-power continuous FM-noise jammer, hand-emplaced.

Cincinnati Electronics
Cincinnati, Ohio USA

LR-5200 Tactical ESM System
Operates between 0.5 and 18 GHz.

Litton Systems, Inc.
Amecom Division
College Park, Maryland USA

M-130 Flare/Chaff Dispenser
Flare/chaff dispenser for Army tactical aircraft.

Tracor Aerospace, Inc.
Austin, Texas USA

M-6880 Series Shipboard ECM Antenna Systems
Operates 8 to 16 GHz. Also produced in a land-mobile configuration.

General Precision Industries, Ltd.
Montreal, Canada

MA1110/1111 Communications DF Processor and Display Units
Covers 1.6 to 1000 MHz with synthesized HF/VHF/UHF receivers.

Racal Communications Ltd.
Berkshire UK

MA1122 Tactical DF Processing and Display Unit
Operates in the 20-to-1000-MHz range.

Racal Communications Ltd.
Berkshire, UK

Magaie Naval Countermeasures Decoy Launcher
For small and medium ships. Derived from the Dagaie system.

Compagnie de Signaux et d'Entreprises Electriques
Paris, France

Magic Mast Antenna Pole
Hydraulically operated antenna pole for armored vehicles, trucks, etc.

GTE Government Systems
Mountain View, California USA

MANTA ESM System
Modular ESM system for submarines: 2 to 18 GHz.

MEL
Crawley, West Sussex, UK

Masquerade Airborne Chaff/IR Decoy System
Chaff and infrared decoy pod for combat aircraft.

Wallop Industries Ltd.
Andover, Hampshire, UK

MATADOR Infrared Countermeasures
Aircraft protection against infrared missiles.

Loral Electro-Optical Systems
A division of Loral Corp.
Pasadena, California USA

MATILDA Missile Alarm System
Microwave analysis, threat indication and launch direction apparatus.

MEL
Crawley, West Sussex, UK

MEL IFM Receivers
From 0.7 to 18 GHz in a variety of octave-plus frequency ranges.

MEL
West Sussex, UK

Mentor Shipborne ESM System
Family of shipboard ESM systems: wide range of antennas.

Marconi Defence Systems Ltd.
Stanmore, Middlesex, UK

MIR-2 ESM System
Compact system for small aircraft and naval vessels: 600 MHz to 18 GHz.

Racal Radar Defence Systems Ltd.
Chessington, Surrey, UK

MIRTS IRCM System
Lightweight, active IR countermeasures system.

Northrop Corp.
Rolling Meadows, Illinois USA

MK-36 Super RBOC
(Rapid Bloom Offboard Countermeasures)
Decoy Launching System
Deck-mounted decoy launcher: self-protection against anti-ship radar or infrared-guided missiles.

Hycor
A division of Loral Corp.
Woburn, Massachusetts USA

Lundy Electronics & Systems Inc.
Lundy Technical Center Division
Pompano Beach, Florida USA

Tracor Aerospace
San Ramon, California USA

Model I-1001 PLO/Synthesizer
Phase-locked oscillator for precision analysis: 1 to 20 MHz.

Ultrasystems Defense & Space Systems Inc.
Sunnyvale, California USA

Model I-1248A High Speed Digitizer
Signals recorder. Two modes: snapshot and continuous.

Ultrasystems Defense & Space
Sunnyvale, California USA

Model 100 Naval DF Set
Shipboard intercept, surveillance and DF: 10 kHz to 180 MHz.

General Precision Industries Ltd.
Montreal, Canada

Model 3600 Reconnaissance/Surveillance Receiver
Digital multiband receiver covering 0.5 GHz to 18 GHz.

American Electronic Laboratories Inc.
Lansdale, Pennsylvania USA

Modular Chaff/Flare Dispenser (MCFD)
Protection for tactical aircraft.

Tracor Aerospace
San Ramon, California USA

Modular Radar Homer
Small, lightweight homer: 1 to 10 GHz.

Fairchild Weston Systems
Syosset, New York USA

Multipurpose Comjam System
HF/VHF/UHF bands: produced in land, sea and airborne versions.

Flight Systems Inc.
Newport Beach, California USA

Multiple Radar Emitter System (MRES)
ECM and C³CM threat generators: mobile, manned or unmanned.

Flight Systems Inc. (Tracor)
Newport Beach, California USA

Naval Antimissile Decoy System—NADS
Two versions: NADS 1-single launcher/NADS 2-double launcher.

Compagnie de Signaux
et d'Enterprises Electriques
Paris, France

Naval ECM Systems-9CM Series
ECM dispensers and jammers integrated with ship's fire control system.

Philips Elektronikindustrier AB
Jarfalla, Sweden

Newton System
Modularized ESM/ECM systems for dense electromagnetic environments.

Elettronica SpA
Rome, Italy

NEWTS
Naval electronic warfare training system (NEWTS): Simulates dense signal environment-full range of intercept, DF, jamming functions.

AAI Corp.
Baltimore, Maryland USA

NS-9000 Naval EW System
Compact shipborne EW self-defense system; automatic and digitized.

Elisra Electronic Systems Ltd.
Bene-Baraq, Israel

NS-9001 Naval Elint System
Automatic analysis and identification of radar emitters: 0.5 to 18 GHz.

Elisra Electronic Systems Ltd.
Bene-Baraq, Israel

NS-9002 Multipurpose EW Simulator
Mobile, self-contained, simulates radar environments: 0.5 to 18 GHz.

Elisra Electronic Systems Ltd.
Bene-Baraq, Israel

NS-9003 Advanced Naval ESM System
Computerized system can be integrated with ECM suite.

Elisra Electronic Systems Ltd.
Bene-Baraq, Israel

NS-9005 Naval ECM System
Multi-threat wideband system: computerized with digital generator.

Elisra Electronic Systems Ltd.
Bene-Baraq, Israel

NS-9009 Compact Naval EW System
For dense environments: 100 percent probability of interception.

Elisra Electronic System Ltd.
Bene-Baraq, Israel

NS-9010 Advanced Shipboard ESM
For use against over-the-horizon radars in the 2-to-18-GHz range.

Elisra Electronic Systems Ltd.
Bene-Baraq, Israel

OG-181/VRC Piranha Applique ECM System
VHF communications jammer: mobile with ERP of up to 1000 watts.

American Electronic Laboratories Inc.
Lansdale, Pennsylvania USA

On-Board Electronic Warfare Simulator (OBEWS)
Airborne simulation system; pod-mounted.

AAI Corp.
Hunt Valley, Maryland USA

PA 010 DF System
HF direction-finding system: 0.5-to-30-MHz range.

Rohde & Schwarz
Muhldorfstrasse, Munich
Federal Republic of Germany

PA 055 Doppler DF System
Land-based, VHF/UHF Doppler DF and monitoring system: 20 MHz to 1 GHz.

Rohde & Schwarz
Muhldorfstrasse, Munich
Federal Republic of Germany

PA 555 DF/Monitoring System
Self-contained, mobile system operating between 20 and 1000 MHz.

Rohde & Schwarz
Muhldorfstrasse, Munich
Federal Republic of Germany

PA 2000 Integrated Signal Interception System
HOPPER TRAP: direction finder for the frequency range 2 to 512 MHz.

Rohde & Schwarz
Muhldorfstrasse, Munich
Federal Republic of Germany

Pinemartin Ground-Based ESM System
Vehicle-mounted tactical ESM surveillance system; 600 MHz to 18 GHz.

Racal Radar Defence Systems Ltd.
Surrey, UK

Plessey HF Monitoring and DF System
Reception, monitoring, direction finding and analysis: 1.5 to 30 MHz.

Plessey Military Communications
Southleigh Park House
Eastleigh Road
Havant, Hampshire, UK

Pod Ka Noise and Deception Jammer
Airborne noise and deception jammer in the X-band.

SRA Communication AB (LM Ericsson subsidiary)
Spanga, Sweden

Porpoise ESM Equipment
Automatic ESM system for submarines: 2 to 18 GHz.

Racal Radar Defence Systems Ltd.
Chessington, Surrey, UK

Portable Dynamic Simulator (PDS) 8601 Series
Computerized land/sea EW threat signal simulator: 0.5 MHz to 18 GHz.

Litton Applied Technology
San Jose, California USA

Prophet Airborne RWR
Radar warning receiver for aircraft: H/I/J bands.

Racal Radar Defence Systems Ltd.
Chessington, Surrey, UK

Protean Shipboard Chaff Launcher
Naval chaff grenade launcher; counters radar guided anti-ship missiles.

MEL
Crawley, West Sussex, UK

PRS 2280 Surveillance Receiver
Modular, remote-controlled HF receiver: 100 memory channels.

Plessey Military Communications
Southleigh Park House
Eastleigh Road
Havant, Hampshire, UK

PRS 3810 Unattended Jammer
Hand-emplaced VHF communication jammers: 20 to 90 MHz.

Plessey Military Communications
Southleigh Park House
Eastleigh Road
Havant, Hampshire, UK

R2111 Frequency Translator
Heterodynes 0.5 to 18 GHz in 2-GHz blocks to 2 to 4 GHz.

Dalmo Victor Division
The Singer Company
Fremont, California USA

R4000 Receiver
Surveillance, reconnaissance and frequency-control: 20 MHz to 40 GHz.

Dalmo Victor Division
The Singer Company
Fremont, California USA

R4700 Wide Bandwidth Tuner
Covers 0.5 to 18 GHz with three inputs: 0.5 to 2 GHz, 2 to 8 GHz, 8 to 18 GHz.

Dalmo Victor Division
The Singer Company
Fremont, California USA

RA1796 HF/VHF/UHF Receiver
Programmable synthesized communications receiver for EW: 20 to 1000 MHz.

Racal Communications Ltd.
Berkshire, UK

RA-6790 HF Series Receivers
HF receivers: synthesized, microprocessor-controlled: 0.5 to 30 MHz.

Racal Communications Inc.
Rockville, Maryland USA

Racal Automated Communications EW System (RACEWS)
Tactical, integrated EW system: computerized.

Racal Communications Ltd.
Berkshire, UK

Radamec Naval Rocket Launcher
Trainable naval rocket launcher: chaff and infrared flares.

Radamec Defence Systems Ltd.
Chertsey Surrey, UK

Rafael Short-Range Chaff Rocket (SRCR)
System for ships: three rockets in a triple-barrel launcher.

Rafael Armament Development Authority
Haifa, Israel

Rafael Long-Range Chaff Rocket (LRCR)
System for ships: tubes mounted on superstructure.

Rafael Armament Development Authority
Haifa, Israel

RAJ101 Ground Radar Jammer
High ERP jammer: fully automatic or semi-automatic modes.

Rafael Armament Development Authority
Haifa, Israel

RAMPART ECM Decoy System
Chaff, smoke and infrared decoys: solar-powered.

Wallop Industries Ltd.
Hampshire, UK

RAMSES Naval ECM
Reprogrammable multimode shipborne ECM system: I/J bands.

Hollandse Signaalapparaten BV
Hengelo, Netherlands

Rapids Naval ESM
Radar passive identification system (Rapids): shipborne ESM system.

Hollandse Signaalapparaten GD Hengelo,
Netherlands
MEL
West Sussex, UK

RAS-1B Airborne Elint System
Modular system: programmable, 0.7 to 18 GHz, optional to 40 GHz.

Tadiran Ltd.
Tel Aviv, Israel

RAS-2A Surveillance and Targeting ESM System
Modular system for aircraft: 0.7-to-18-GHz range: Automatic.

Tadiran Ltd.
Tel Aviv, Israel

Rattler Radar Jammer
Computerized, modular mobile radar jammer: time sharing.

Rafael
Haifa, Israel

RCM Series Countermeasures System
Radar jammers for use against early-warning radars: I/J band.

Racal Radar Defence Systems Ltd.
Chessington, Surrey, UK

RDF 6000 Seeker Direction-Finding (DF) System Mobile
HF/VHF/UHF interception and DF system: 1.6 to 512 MHz.

Racal Communications Inc.
Rockville, Maryland USA
Racal Communications Ltd.
Berkshire, UK

RDF-500 UHF Radar Detector/DF System
For aircraft: Detects UHF radars and determines DOA; 200 to 500 MHz.

Tadiran Ltd.
Tel Aviv, Israel

Today's Technology didn't start this morning...

At Frequency Sources, it began more than twenty years ago with the development of the company's first frequency source. Today, we are a leading supplier of products which generate and control microwave energy.

Today's technology is the result of layered knowledge—knowledge learned over time and adapted to specific situations.

And it is through the application of this knowledge that Frequency Sources has developed the Microwave Monolithic Integrated Circuit (MMIC) shown here in the switch design of a Multiple Output Reference Generator. Such units are typically used in up/down converter LOs in next generation countermeasure systems employing digital RF memories (DRFM).

Fast switching and low losses in the output switch matrix were job requirements best met by silicon monolithic technologies. Designs were fabricated to attain switching speed as low as 25 ns and to ensure full signal coherency of all outputs with a single reference (clock).

Using the building block concept, these units can provide a wide range of frequency spacings, frequency coverage, and power output options. This application of MMIC technology allows significant size reduction and improved reliability and reproducibility.

This is just one example of how Frequency Sources provides cost effective solutions to customer requirements. Solutions made possible by twenty years of developing technology. Solutions that go beyond the specifications. That start with insight and evolve into a technology that's right for the job.

Frequency Sources can put its insight to work for you. Contact us at (617) 256-4113.

today's technology at work

RDL ESM System
Detects radars, provides warning of radar-controlled weapons.

Racal Radar Defence Systems Ltd.
Chessington, Surrey, UK

RG5545A VHF/UHF Receiver
VHF/UHF receiver: 20 to 500 MHz with optional extension to 1000 MHz.

Racal Communications, Inc.
Rockville, Maryland USA

RG Series FDM/VFT Demodulators
Baseband-to-channel or baseband-to-12-channel group signal processing.

Racal Communications Inc.
Rockville, Maryland USA

RG Series Spectrum Monitors
Displays signal activity received by associated ESM receiver.

Racal Communications Inc.
Rockville, Maryland USA

RJS3100/3101 VHF Jammers
Automatic mobile jammer systems in the 20-to-80-GHz band.

Racal Communications Ltd.
Berkshire, UK

RJS3105 Series VHF/UHF Jammer Systems
RJS3106: 110 to 156 MHz. RJS3107: 225 to 400 MHz. RJS3105 covers both.

Racal Communications Ltd.
Berkshire, UK

RJS3140 Hand-Emplaced Unattended Jammer
Low-power barrage jammer: 20-to-90-MHz band.

Racal Communications Ltd.
Berkshire, UK

RQN-1 Radar Intercept Equipment
Wide-open self-protection crystal video; A: 1 to 12 GHz; B: 1 to 18 GHz.

Selenia Industrie Elettroniche Associate SpA
Rome, Italy

RQN-1/3 Direction-Finding ESM
IFM, solid-state microwave amplifiers in the high bands.

Selenia Industrie Elettroniche Associate SpA
Rome, Italy

S2150 Automated Defense Systems
Threat warning system: 2 to 18 GHz. For small surface combatants, subs.

Dalmo Victor Division
The Singer Co.
Fremont, California USA

S3000 Radar Surveillance System
All mobile platforms: 0.5 to 18 GHz.

Dalmo Victor Division
The Singer Co.
Fremont, California USA

S5000 Surveillance System
ESM system for mobile platforms: 0.5 to 40 GHz.

Dalmo Victor Division
The Singer Co.
Fremont, California USA

Sagaie Naval Countermeasures Decoy System
Automatic decoy protection for ships: IR and chaff rockets.

Compagnie de Signaux
et d'Entreprises Electriques
Paris, France

SAPIENS Naval EW System
Surveillance alarm and protection, intelligence EW naval system (SAPIENS).

Thomson-CSF
Avionics Division
Malakaff, France

Compagnie de Signaux
et d'Entreprises Electriques
Paris, France

SARIE-Selective Automatic Radar Identification Equipment
Compares intercepted signals with threat library: radar.

Thorn-EMI Electronics Ltd., Radar Division
Middlesex, UK

SAT 3311 HF Signal Analysis Terminal
Range 100 kHz to 30 MHz: Maximum bandwidth 16 kHz, zoom down to 20 Hz.

Racal Communications Ltd.
Berkshire, UK

Saviour Warning System
Integrated threat warning system: radar/laser sensors for vehicles.

Racal Radar Defence Systems Ltd.
Chessington, Surrey, UK

Sceptre ESM System
Modular shipboard ESM: 2 to 18 GHz, 360 degrees, 100 percent POI.

MEL
Crawley, West Sussex, UK

SCR-2725 Panoramic Display
Digitizes demodulated log video data: 20 to 1200 MHz.

Scientific Communications Inc.
Garland, Texas USA

SCR-2400 ESM System
Integrated 1-to-18-GHz, 360-degree monopulse DF receiver, digitizer.

Scientific Communications Inc.
Garland, Texas USA

SCHALMEI Decoy System
Shipboard IR/RF decoy launcher system: control, launcher and rockets.

AEG-Telefunken
Federal Republic of Germany

SCIMITAR Jammer
On-board ESM system. Variety of jamming modes included.

MEL
Crawley, West Sussex, UK

SCLAR Shipboard Chaff/Flare Rocket System
Multipurpose rocket-launched chaff and flare system.

Breda Meccanica SpA
Brescia, Italy
(Rocket Launcher)

Selenia-Elsag Consortium for Naval Systems
Rome, Italy
(Launcher Control Unit)

SCR-2700
Semiautomatic intercept receiver system for surveillance: 0.1 to 40 GHz.

Scientific Communications Inc.
Garland, Texas USA

Sea Saviour RWR System
Threat warning, chaff or infrared decoy system for ships: 8 to 18 GHz.

Racal Radar Defence Systems Ltd.
Chessington, Surrey, UK

Sea Sentry Naval ESM
Passive electronic surveillance system. Three levels; I/II/III.

Kollmorgen Corp.
Northampton, Massachusetts

Sentry Mobile ESM System
Line of mobile self-contained ESM vehicles: crystal video or superhet.

Marconi Defence Systems Ltd.
Stanmore, Middlesex, UK

Series 2000 Receiver System
VHF/UHF, and peripherals for Sigint: 20 to 500 MHz. Option to 1200 MHz.

Applied Communications
Gaithersburg, Maryland USA

Series-7000 Comjam System
Mobile communications jamming system: 1.5 to 500 MHz.

Tadiran Ltd.
Tel Aviv, Israel

Series-10000 Comint/Comjam System
Modular, tactical HF/VHF/UHF system: mobile or fixed, 1 kW.

Tadiran Ltd.
Tel Aviv, Israel

SERVAL Airborne Radar Warning Receiver
Aural and visual alarms provided: 2 to 18 GHz.

Thomson-CSF
Avionics Division
Malakoff, France

Sherloc Radar Warning Receiver
Aural and visual alarms provided: E to J band, microprocessor.

Thomson-CSF
Avionics Division
Malakoff, France

SHIELD Decoy System
Shipboard system: launchers, control equipment and decoy munitions.

Plessey Naval Systems Ltd
Addlestone
Weybridge
Surrey, UK

Sibyl EW Naval Decoy System
Shipboard decoy system. Multitube trainable rocket launchers.

British Aerospace Dynamics Group
Bracknell Division
Berkshire, UK

BRANDT Armements
Paris, France

Simrad RL1 Laser Warning Device
Detects pulsed laser range finders and target markers: .66 to 1.1 [M]m.

Simrad Optronics A/S
Oslo, Norway

Siren Countermeasures System
Shipboard, intelligent, all-electronic active 130-mm decoy round.

Marconi Defence Systems Ltd.
Stanmore, Middlesex, UK

Sky Shadow (ARI 23246)
ECM aircraft pod: Programmable, smart noise pod; E to J band.

Marconi Defence Systems Ltd.
Stanmore, Middlesex, UK

SLQ-32 Carry-On Device
Connected to AN/SLQ-32 central computer, provides EW simulation.

AAI Corp.
Baltimore, Maryland USA

SL/ALQ-234 Dual-Mode Jamming Pod
Pulsed and CW radar countermeasures jamming pod: G to H bands.

Selenia SpA
Rome, Italy

SLR 600/610 Naval ESM System
E to J bands, wide-open crystal video receiver: C/D bands available.

General Instrument Corp.
Hicksville, New York USA

Smart Guard System
Communications ESM system for helicopters: Automatic, VHF/UHF.

Elettronica, SpA
Rome, Italy

Spectra Elint System
Communications intercept and monitoring system: LF, MF and HF bands.

Hollandse Signaalapparaten BV
Hengelo, Netherlands

Sphinx Naval ESM
Shipboard ESM system for use against pulsed radars: 1 to 18 GHz.

Hollandse Signaalapparaten BV
Hengelo, Holland

SPS-200 Airborne Self-Protection System
Aircraft warning receiver: 0.7 to 18 GHz, CW radar threats.

Elisra Electronic Systems Ltd.
Bene-Baraq, Israel

SR-1A Radar Intercept Receiver
Shipborne ESM radar warning receiver: 2.5 to 18 GHz.

Nera A/S
Bergen, Norway

SR-200 Shipborne ESM System
Modular, mobile system: automatic IFM, 0.5 to 18 GHz, option to 40 GHz.

Sanders Associates
Nashua, New Hampshire USA

SR-2020 Tactical ESM System
Digitally remote-controlled ESM receiver: 0.5 to 500 MHz.

Applied Communications
Gaithersburg, Maryland USA

SR-2152 VHF/UHF Receiver
High-speed Comint receiver: 20 to 500 MHz, option to 1200 MHz.

Applied Communications
Gaithersburg, Maryland USA

SR-2175 Series Surveillance Receivers
Synthesized receiving equipment: VHF/UHF 20 to 512 MHz.

Applied Communications
Frederick, Maryland USA

SRM-2150 Surveillance Receiving System
Signal monitoring: AM, FM, CW, LSB, USB and pulse signals; 20 to 500 MHz.

Applied Communications
Frederick, Maryland USA

Standard Emitter Simulator (Stems) L8320 Series
Range: 0.5 to 40 GHz. Simulates dense EW signal environments.

Litton Applied Technology
San Jose, California USA

Stockade-Seaflash Decoy System
Seaflash remote-controlled boat with chaff launching systems: 20 km.

Wallop Industries Ltd.
Andover, Hampshire, UK

Flight Refuelling Ltd.
Dorset, UK

SYREL Electronic Reconnaissance Pod
Airborne tactical ESM system: C to J band search and analysis.

Thomson-CSF
Avionics Division
Malakoff, France

System 3000 Communication ESM/Comint Systems
Comint search and interception, modular.

Racal Communications Ltd.
Berkshire, UK

Racal Communications Inc.
Rockville, Maryland USA

TACDES Airborne Sigint System
Communication and radar signals: 20 MHz to 18 GHz.

Tadiran Ltd.
Tel Aviv, Israel

TC-586 Portable Direction Finder System
Two standard briefcases: AM/FM signals; 25 to 550 MHz, 800 to 1100 MHz.

Tech Comm Inc.
Sunrise, Florida USA

TC-5100 Direction Finder and Antenna
Fixed or mobile DF: search 110 kHz to 70 MHz, DF 1.5 to 1000 MHz.

Tech Comm Inc.
Sunrise, Florida USA

TCI 800 Series Transportable Comint Systems
Modular, mobile tactical Comint network: computer control.

Technology for Communications International (TCI)
Information Systems Division
Sunnyvale, California USA

TDF-205 Automatic DF/Comint System
Interferometic surveillance system: 20 to 500 MHz. Mobile or fixed.

Tadiran Ltd.
Tel Aviv, Israel

THETIS ESM System
Modular ESM for submarines: IFM threat warning and DF.

Elettronica SpA
Rome, Italy

Timnex Automatic Elint/ESM System
Mobile or fixed: channelled IFM and DF, computer, 2 to 18 GHz.

Elbit Computers Ltd.
Haifa, Israel

TJS-2 Tactical Communications Jammer System
Tactical HF/VHF ECM: ruggedized commercial hardware.

Sanders Associates Inc.
Hudson, New Hampshire USA

TN-123 High Stability Microwave Tuner
Fully synthesized 0.5-to-18-GHz frequency converter.

Watkins-Johnson Co.
Gaithersburg, Maryland USA

TN-1000 Miniceiver
From 0.5 to 18 GHz in a single one-quarter ATR unit: internal synthesizer.

Watkins-Johnson Co.
San Jose, California USA

TRC Communications Intelligence and Jamming Equipment
Receivers and jammers up to 1 GHz: fixed or mobile.

Thomson-CSF Telecommunications Division
Gennevilliers, France

TRC EW Communications Equipment
Receivers: HF/VHF/UHF, microprocessor controlled.

Thomson-CSF Telecommunications Division
Gennevilliers, France

TSS-2 Tactical Communication Surveillance System
Dragon tactical HF/VHF/UHF ESM system. Ruggedized, commercial hardware.

Sanders Associates Inc.
Hudson, New Hampshire USA

Type R505 ECCM Receiver
Can be added to ground radars: pulse length 0.2 to 1.5 microseconds.

Plessy Radar
Chessington, Surrey, UK

Type S373 ECM
Mobile multiband high power jamming system.

Marconi Radar Systems Ltd.
Chelmsford, Essex, UK

Type 405J Mobile Jammer
Mobile ground jammer with ESM: four bands 1 to 2, 2 to 4, 4 to 8 and 18 to 26 GHz.

Plessey Radar
Chessington, Surrey, UK

Type 5000 Chaff Dispensers
Conformal cartridge launcher: chaff or IR flares.

R. Alkan SA
Paris, France

UST-104 Radio Countermeasures System
ESM and ECM/communications system: airborne, ground fixed or mobile.

Rockwell International
Collins Defense Communications
Cedar Rapids, Iowa USA

Vicon 70 Series 33 IRCM Pod
IRCM self-defense, pod-mounted system for fixed-wing combat aircraft.

W. Vinten Ltd.
Bury St. Edmunds
Suffolk, UK

Vicon 77 Chaff/IR Pod
Twelve dispenser modules: 16-inch diameter pod for chaff or flares.
W. Vinten Ltd.
Bury St. Edmunds
Suffolk, UK

Vicon 78 Airborne Decoy Dispensers
Lightweight chaff and IR decoy dispensing systems for aircraft.
W. Vinten Ltd.
Bury St. Edmunds
Suffolk, UK

VKE 3800 HF Multichannel Receiver
Surveillance and analysis: 2 to 30 MHz.
AEG-Telegunken
Ulm, West Germany

VS Digital Voice Storage System
Self-contained system for voice recording and reproduction: digital.
GTE Government Systems
Mountain View, California USA

Wavefinder Passive ESM System
Fixed or mobile: Elint and surveillance of all types of modulators.
Thorn EMI Electronics Ltd.
Computer Systems Division
Wells, Somerset, UK

Weasel
Two-stage Elint-gathering system: 0.5 to 18 GHz, optional to 40 GHz.
Racal Radar Defense Systems Ltd.
Surrey, UK

WJ-1920 Multiparameter Distributed Processing System
Superheterodyne plus IFM over 0.5 to 18 GHz for high POI.
Watkins-Johnson Co.
San Jose, California USA

WJ-1921-1 Pulse Interval Processor (PIP)
High-speed PRF signal sorter: for dense signal environments.
Watkins-Johnson Co.
San Jose, California USA

WJ-1988 Amplitude-Monopulse DF Processor
Direction-finding subsystem: static antenna array, 2 to 18 GHz.
Watkins-Johnson Co.
San Jose, California USA

WJ-1996 Tunable IFM Warning/ESM Receiver
Either an ESM or threat warning receiver: combined IFM and superhet.
Watkins-Johnson Co.
San Jose, California USA

WJ-4810 Communications Jammers
Applique jammer: HF/VHF/UHF, 20 to 1000 watts. Multiple target mode.
Watkins-Johnson Co.
Gaithersburg, Maryland USA

WJ-8599 Active Jammer Antennas
From 0.85 to 17 GHz in four bands: can radiate up to 1 kW of CW RF.
Watkins-Johnson Co.
San Jose, California USA

WJ-8610A-Series Controllers
Central control unit for multiple receiver or demodulator systems.
Watkins-Johnson Co.
Gaithersburg, Maryland USA

WJ-8615D VHF/UHF Compact Receiver
Modular, compact, microprocessor-controlled receiver: 20 to 500 MHz.
Watkins-Johnson Co.
Gaithersburg, Maryland USA

WJ-8955 Mobile ESM System
Mobile signal monitoring/direction finding: 2-to-1100-MHz range.
Watkins-Johnson Co.
Gaithersburg, Maryland USA

WJ-8969 Microwave Receiving System
From 1 to 18 GHz: detects AM, FM, CW and pulse modes.
Watkins-Johnson Co.
San Jose, California USA

WJ-8976 Three-Channel DF System
Provides angle of arrival on monopulse, continuous and spread-spectrum signals.
Watkins-Johnson Co.
Gaithersburg, Maryland USA

WJ-8990 Manpack Tactical Intelligence System (Mantis)
Two-man RF/DF system: 20 to 500 MHz. Options to 500 kHz and 1200 MHz.
Watkins-Johnson Co.
Gaithersburg, Maryland USA

WJ-9040/WJ-8969 Broadband Collection System
Integrated system: 5 kHz to 12.4 GHz, optional for 200 Hz to 18 GHz.
Watkins-Johnson Co.
Gaithersburg, Maryland USA

WJ-9040 Modular Receiving Component System
Rugged mobile or fixed system: 5 kHz to 500 MHz, optional to 1 or 12 GHz.
Watkins-Johnson Co.
Gaithersburg, Maryland USA

WJ-9195C Rapid Acquisition Spectrum Processor
Broadband receiver and spectrum display: can be a system controller.
Watkins-Johnson Co.
Gaithersburg, Maryland USA

WJ-9477-Series Tunable Demodulator
Provides demodulation of AM, FM, and SSB signals from 1 kHz to 30 MHz.
Watkins-Johnson Co.
Gaithersburg, Maryland USA

XR100 Comint Receiving System
VHF/UHF system for AM, FM, CW or pulse signal analysis: 20 to 1000 MHz.
Sintra-Alcatel
Asnieres, France

ZEUS Airborne EW System
Internally mounted, totally integrated airborne defensive EW system.
Marconi Defence Systems Ltd.
Stanmore, Middlesex, UK

SOVIET WEAPON SYSTEMS AND ELECTRONICS

Orders of Battle 155
Aircraft 160
  Bomber/Strike Aircraft 160
  Fighters and Attack Aircraft 163
  Helicopters 170
  Reconnaissance/Electronic Warfare Aircraft 176
  Reconnaissance/Airborne Warning Aircraft 176
  Reconnaissance/Anti-Submarine Warfare Aircraft 177
  Cargo/Transport Aircraft 178

Ships 180
  Aircraft Carriers 180
  Guided Missile Cruisers 182
  Cruisers 186
  Command Cruisers 188
  Guided Missile Destroyers 189
  Destroyers 196
  Frigates/Corvettes 198
  Guided Missile Patrol Boats 209
  Small Combatants 212
  Amphibious Ships 214
  Intelligence Ships 219

Submarines 226
  Ballistic Missile Submarines 226
  Guided Missile Submarines 229
  Attack Submarines 232
  Research Submarines 238
  Auxiliary Submarines 238
  Rescue and Salvage Submarines 239
  Training Submarines 239

Ground Combat Vehicles 240
  Tanks 240

Missiles 248
  Intercontinental Ballistic Missiles 248
  Submarine Launched Ballistic Missiles 252
  Land Attack-Theater Missiles 254
  Anti-Ship Missiles 261
  Anti-Ballistic Missiles 266
  Anti-Submarine Warfare Missiles 267
  Surface-to-Air Missiles 268
  Ground Combat Vehicles/Air Defense 275
  Air-to-Air Missiles 280
  Anti-Tank Missiles 284
  Anti-Radiation Missiles 287

Sensors 288
  Strategic Radars 288
  Tactical Radars 290

Soviet Weapon Systems and Electronics

Orders of Battle

The Soviet listings for this thirteenth edition of *The International Countermeasures Handbook* have been provided by the U.S. Naval Institute Military Database, under the direction of Norman Polmar. It is organized as follows:

> Aircraft
> Ships
> Submarines
> Ground Combat Vehicles
> Missiles
> Sensors

This section is copyrighted© by the Military Data Corporation, a subsidiary of Information Spectrum, Inc.

STRATEGIC ROCKET FORCES

The Strategic Rocket Force is a separate service functioning directly under the General Staff and Ministry of Defense. In 1987 the Strategic Rocket Force was estimated to operate 1,418 ICBMs and 553 IRNF ballistic missiles (previously classified as intermediate-range ballistic missiles/IRBM). ICBMs are based throughout the Soviet Union, with most concentrated in the western and southern areas. The current ICBM strength is listed below.

Intercontinental Ballistic Missiles (1,418 missiles)

SS-11	Sego	440
SS-13	Savage	60
SS-17	Spanker	150
SS-18	Satan	308
SS-19	Stilleto	360
SS-25	Sickle	100

Based on current trends, the number of ICBMs in the Soviet inventory is expected to increase slightly by the mid-1990s with slightly more than one-half of the ICBM force being comprised of mobile SS-24 and SS-25 missiles. By the mid-1990s the other ICBMs in service would be the SS-19 and newer, "heavy," MIRV missiles.

Intermediate-range Nuclear Forces (553 missiles)

SS-4	Sandal	112
SS-20	Saber	441

All SS-4 missiles are deployed west of the Ural mountains; they are being replaced by the SS-20. Two-thirds of the SS-20 missiles are based in European Soviet territory and one-third in the Far East.

SOVIET AIR FORCES

Strategic Aviation

Strategic Aviation is a grouping of the five strategic air armies of the Soviet Air Forces, and is apparently not a single major command organization, as is the U.S. Strategic Air Command. Soviet strategic aviation was previously organized as Long-Range Aviation (Dal'nyaya Aviatsiya) and was in fact such a unified command; however, the recent reorganizations of the Air Forces in 1980 assigned all strategic and theater bombers to the five air armies, each with responsibility for strikes in specific areas.

Long-Range Bombers (165)

Mya-4	Bison	15
Tu-142	Bear-H	50
Tu-20	Bear-G	40
Tu-20	Bear-A/B	60

Medium-Range Bombers (Approx 565)

Tu-22M	Backfire	160
Tu-22	Blinder	135
Tu-16	Badger	270

Strike Aircraft (Approx 450)

Su-24	Fencer	450

Support Aircraft (Approx 300)

MiG-25	Foxbat	25	Reconnaissance
Mya-4	Bison	30	Tanker
Tu-22	Blinder	15	Reconnaissance/ECM
Tu-20	Bear	Few	Reconnaissance
Tu-16	Badger	20	Tanker
Tu-16	Badger	110	Reconnissance/ECM
Yak-28	Brewer	100	Reconnaissance/ECM

Frontal Aviation

Frontal Aviation is the major tactical formation of the Soviet Air Forces providing fighter, attack, and support aircraft for the direct support of ground operations. Frontal Aviation is charged with the gaining air superiority over the combat zones, and striking targets out to "operational depth" behind the front lines.

Counterair/Interceptor Aircraft (Approx 2,850)

MiG-31	Foxhound	10
MiG-29	Fulcrum	115
MiG-25	Foxbat	130
MiG-23	Flogger	1,710
MiG-21	Fishbed	540
Su-27	Flanker	Few
Su-15	Flagon	300
Tu-28	Fiddler	25
Yak-28P	Firebar	20

Ground Attack (Approx 2,850)

MiG-27	Flogger	855
MiG-21	Fishbed	130
Su-25	Frogfoot	200
Su-24	Fencer	700
Su-7/17	Fitter	960

Reconnaissance/Electronic Countermeasures (Approx 600)

MiG-25	Foxbat	170
MiG-21	Fishbed	45
Su-24	Fencer	20
Su-17	Fitter	165
Yak-28	Brewer	200

Helicopters (Approx 4,100)

Mi-26	Halo	20
Mi-24	Hind	1,150
	(includes 830 Hind-D/E gunships)	
Mi-8	Hip	1,760
	(includes 120 Hip-E gunships)	
Mi-6	Hook	465
Mi-4	Hound	25
Mi-2	Hoplite	690

AIR DEFENSE FORCES

The Soviet Air Defense Forces are responsible for the defense of the Soviet Union against attacks by aircraft and missiles.

Fighter Aviation of Air Defense

Fighter-Interceptors (1275)

MiG-23	Flogger-B/G	430
MiG-25	Foxbat-E	300
MiG-29	Fulcrum	Few
MiG-31	Foxhound-A	150
Su-15	Flagon-E/F	200
Yak-28P	Firebar	90
Tu-28P	Fiddler-B	90

Airborne Warning and Control Aircraft (Few)

Il-76	Mainstay	Few
Tu-126	Moss	Few

Zenith Rocket Troops

The PVO rocket troops currently operate an estimated 9,300 strategic surface-to-air missile (SAM) launchers at about 1,200 missile sites. The strategic SAM launchers are reported to be:

Strategic Missile Launchers

SA-1	Guild	2,540
SA-2	Guideline	2,730
SA-3	Goa	1,250
SA-5	Gammon	2,050
SA-10	Grumble	735

Note: SA-10 is replacing SA-1 and, probably, SA-2. At least 20 SA-10 complexes are under construction.

Almost 5,000 tactical SAM launchers are operated by PVO in direct support of the ground forces; these are in addition to hand-held launchers of the SA-9 and SA-13 type.

Tactical Missile Launchers (assignment to Ground Forces indicated)

SA-4	Ganef	1,400	(Army/Front)
SA-6	Gainful	875	(Division)
SA-8	Gecko	764	(Division)
SA-9	Gaskin	545	(Regiment)
SA-11	Gadfly	80	(Division)
SA-13	Gopher	755	(Regiment)

Note: In addition, there are approximately 12,000 anti-aircraft guns deployed for air defense in the Soviet Union.

Radio-Technical Troops

The Radio-Technical Troops operate the massive radar and aircraft control network along the periphery of the Soviet Union, except for those systems that comprise the strategic early warning and ABM systems. The exact number of sites and radars is difficult to determine, since older systems are often retained in service after being replaced by newer units. Western sources estimated that there are more than 2,000 surveillance and target acquisition radars in operation at more than 1,300 sites in the Soviet Union.

Anti-Space and Anti-Rocket Troops

This component of PVO operates the large early-warning radar sites, including the phased-array systems at Krasnoyarsk, Lyaki, Mishelevka, Pechora, and Sary-Shagan. The principal functions of these systems are to detect the launch and to track the flight of ballistic missiles approaching the Soviet Union and to provide strategic warning and guidance for ABM intercept systems. One of the newest radar sites is at Pushkino, near Moscow, to control ABM engagements. In addition, the Soviet Union operates a large network of strategic warning satellites and satellite-intercept systems.

SOVIET NAVY

In numbers of ships the Soviet Union has the world's largest Navy, with four major operating fleets (Northern, Baltic, Black Sea and Pacific) and one flotilla (Caspian). Only active ships are listed below: The + sign indicates additional ships of the class are reported to be under construction.

Ships

Aircraft Carriers (6+)

Leonid Brezhnev Class (2)	Under construction
Kiev Class (4)	VSTOL Carriers
Moskva Class (2)	Helicopter Carriers/ Guided Missile Cruisers

Battle Cruisers (2+)

Kirov Class (2+)	Nuclear-powered/ Multi-purpose

Missile Cruisers (27+)

Kresta /Kynda /Slava Classes (10+)	Anti-ship
Kara /Kresta II Classes (17)	Anti-submarine

Gun Cruisers (9)

Sverdlov Class (Converted) (2)	Command Ships
Sverdlov Class (7)	Light Cruisers

Guided Missile Destroyers (49+)

Udaloy Class (8+)	Anti-submarine
Sovremennyy Class (6+)	Anti-ship
Kanin /Kashin /Kildin / Kotlin (SAM) Classes (35)	Multi-purpose

Gun Destroyers (19)

Kotlin /Skoryy Classes	All-gun type

Frigates (182+)

Krivak Class (32)	Anti-submarine/Additional ships of this class are operated by the KGB Maritime Border Guards
Grisah /Koni / Mirka /Petya (115)	Anti-submarine/Additional ships of this class are operated by the KGB Maritime Border Guards
Riga Class (35)	Patrol

Corvettes (130+)

Nanuchka /Tarantul Classes (35+)	Guided Missile type
Pauk /Poti Classes (75+)	Anti-submarine/Additional ships of this type are operated by the KGB Maritime Border Guards
T-58 (Ex) Class (17)	Patrol, formerly minesweepers/Additional ships of this class are operated by the KGB Maritime Border Guards
T-58 (Ex) Class (3)	Radar Pickets, formerly minesweepers

Missile, Patrol and Torpedo Craft (109)

Matka /Osa /Sarancha Classes (107)	Guided Missile Boats
Babochika /Slepen Types (2)	Patrol/Additional ships of this type are operated by the KGB Maritime Border Guards

Specialized Electronic/Space Support/ Missile Range Ships (75)

Vishnya /Al'pinist / Bal'zam /Primor'ye / Momo /Mayak /Mirnyy / Nikolay Zubov / Okean /Pamir Classes (57+)	Intelligence collection
Marshal Nedelin / Desna /Sibir' Classes (7+)	Missile range instrumentation/One additional nuclear-powered ship is under construction
Kosmonaut Pavel Belyayev /Kosmonaut Yuri Gagarin /Akademik Sergei Korolev / Kosmonaut Vladimir Komarov /Borovichi Classes (11)	Space Event Support/Space Monitoring

Submarines

Ballistic Missile Submarines (78+)

Typhoon/Delta/Yankee Classes (SSBN) (62+)	Nuclear powered, modern submarines

Hotel III (1 SSBN)	Nuclear powered
Golf III/V (2 SSBs)	Diesel-electric
Golf II (13 SSBs)	Diesel-electric

Cruise (Guided) Missile Submarines (63+)

Yankee (Converted) Class (1 SSGN)	Nuclear powered/strategic cruise missiles
Oscar/Papa/Charlie I/II *Echo II* Classes (47+ SSGNs)	Nuclear powered/anti-ship cruise missiles
Juliett Class (15 SSGs)	Diesel-electric/anti-ship cruise missiles

Attack Submarines (227+)

Akula/Alfa/Ex-Echo I/Ex-Hotel II/Mike/ November/Sierra/Victor I/II/III Classes (87+ SSNs)	Nuclear powered/anti-ship/ torpedoes, and land-attack cruise missiles in some newer units
Foxtrot/Kilo/Romeo/ Tango Whiskey/Zulu IV Classes (140+ SS)	Diesel-electric/ anti-ship/torpedoes

Special Purpose and Research Submarines (Approx 25)

Hotel II Converted (1 SSQN)	Nuclear powered/ communications
Golf I Converted (3 SSQ)	Diesel-electric/ communications
Ex-Echo II/Uniform/X-Ray (3 AGSSN)	Nuclear-powered/special purpose and research
India Class (2 AGSS)	Diesel-electric/salvage and rescue
Ex-Foxtrot/Ex-Whiskey/ Ex-Zulu IV/Lima Classes (15 AGSS)	Diesel-electric/special purpose and research

Naval Aviation

Soviet Naval Aviation operates more than 1,600 aircraft, which are integral to the four operating fleets (Northern, Baltic, Black Sea, and Pacific) plus an administration and training establishment. The naval strike aircraft included 115 of the advanced Backfire missile strike aircraft. Backfires and Badger aircraft are also employed in the anti-ship role (the naval Blinder and Bear aircraft do not carry missiles).

Missile Strike/Bomber Aircraft (Approx 355)

Tu-22M	Backfire-B/C	125
Tu-16	Badger-C/G	200
Tu-22	Blinder-A	30

Fighter-Attack Aircraft (Approx 155)

Su-24	Fencer	Few
Su-17	Fitter-C	75
Yak-38	Forger-A/B	70

Reconnaissance/Electronic Aircraft (Approx 190)

Tu-16	Badger-D/E/F/G/H	80
Tu-20	Bear-D	45
Tu-22	Blinder-C	Few
Ka-25	Hormone-B	60

Tanker Aircraft

Tu-16	Badger-A	75

Anti-Submarine Aircraft (Approx 480)

Tu-20	Bear-F	60
Mi-14	Haze-A	105
Ka-27	Helix-A	55
Ka-25	Hormone-A	115
Be-12	Mail	95
Il-38	May	50

Utility/Transport/Training Aircraft

Various Types	400

SOVIET GROUND FORCES

Ground Forces is the term used to designate the Soviet Army. These forces are organized into an extimated 227 divisions plus numerous smaller combat and support units. The divisions, which are at various stages of readiness, are reported as:

Tank divisions	52
Motorized rifle divisions	150
Airborne divisions	7
Defense divisions	2
Artillery divisions	16

There are also 5 inactive, mobilization base divisions, with weapons and equipment assembled and ready for reserve personnel to be called to active duty. (In addition, there are 1 naval infantry division, and 3 naval infantry brigades under control of the Soviet Navy.)

Ground Forces are reported to have the following weapons (totals are rounded):

Main Battle Tanks (53,000)

T-54/55	20,000
T-62	13,600
T-64	9,250
T-72	8,450
T-80	1,400

Light Amphibious Tanks

PT-76	Several Thousand

Self-Propelled Anti-Aircraft Guns

ZSU-23-4   Several Thousand
ZSU-X

Armored Personnel Carriers/Infantry Fighting Vehicles   (59,000)

Surface-to-Surface Missile Launchers

Frog-3/5   155
Frog-7   550

SS-1 Scud-B   635
SS-21 Scarab   100
SS-12 Scaleboard   150
SS-22/SS-23 Spider   Few

Artillery Tubes   (29,000)

Mortars   (11,000)

Multiple Rocket Launchers   (7,000)

In the following listings, the symbol *
indicates the Soviet name or designation.

AIRCRAFT

BOMBER/STRIKE AIRCRAFT

DESIGNATOR: **Tu-(Unnumbered)**
NAME: **Blackjack**

DESCRIPTION: Soviet long-range strategic bomber, under development as potential successor to Bison and Bear bomber/strike aircraft. The Blackjack somewhat resembles the Backfire bomber, but is 25 percent larger and has a different tail configuration. Like the Backfire, it has a variable-geometry wing, but the Blackjack engine intakes are mounted under the fuselage rather than along the fuselage sides, as in the earlier aircraft. It is envisioned that the Blackjack will supplement the Backfire as a long-range naval strike aircraft.

Blackjack

USERS: Soviet Union Air Forces and Navy (planned).

CHARACTERISTICS:

Manufacturer:	Tupolev
Crew:	Approx 5
Weights:	
Empty	260,000 lb (117,950 kg)
Max Takeoff	551,150 lb (250,000 kg)
Dimensions:	
Wing Span	172 ft (52.00 m) extended
	110 ft (33.75 m) swept
Length	175 ft (53.35 m)
Height	45 ft (13.75 m)
Wing Area	2,500 sq ft (232.25 sq m)
Performance:	
Max Speed	1,199 kts (1,380 mph; 2,220 km/h)
Radius	3,942 nm (4,540 mi; 7,300 km) unrefuelled

Armament	Believed capable of carrying up to 36,000 lb (16,330 kg) of bombs and missiles, including the AS-15 strategic cruise missile

DESIGNATOR: **Tu-22M**
NAME: **Backfire**

DESCRIPTION: The Backfire is a long-range, high-performance bomber, the first large bomber with variable-geometry wings to enter into service. The Backfire features two engines buried in the wing roots, a weapons bay which can either carry ordnance internally or air-to-surface missiles semi-recessed, and weapon pylons for external stores.

USERS: Soviet Union Air Forces and Navy.

CHARACTERISTICS:

Manufacturer:	Tupolev
Crew:	4 (Pilot, co-pilot, navigator, defensive systems operator)
Engines:	2 turbofan (Kuznetsov NK-144 derivatives)
Max Power	33,069 lb (15,000 kg) static thrust dry 44,090 lb (20,000 kg) afterburner
Weights:	
Empty	110,000 lb (49,500 kg)
Max Takeoff	270,000 lb (120,500 kg)
Dimensions:	
Wing Span	113 ft 0 in (34.45 m) extended 85 ft 11 in (26.20 m) swept
Length	139 ft 5 in (42.20 m)
Height	33 ft 0 in (10.06 m)
Wing Area	1,830 sq ft (170.0 sq m)
Performance:	
Cruise Speed	432 kts (500 mph; 800 km/h)

Backfire

Max Speed	Low Alt 594 kts (650 mph; 1,050 km/h) High Alt 1,085 kts (1,250 mph; 2,000 km/h)
Ceiling	55,760 ft (17,000 m)
Armament	1 or 2 AS-4 Kitchen air-to-surface missiles carried externally; Also AS-9 anti-radar missile; 26,455 lb (12,000 kg) of bombs internally; 12 to 18 bombs can be carried externally; Two 23-mm remotely fired tail guns
Radar	Down Beat I-band search and targeting for AS-4 Fan Tail fire control for 23-mm guns Bee Hind I-band tail warning radar

DESIGNATOR: Tu-22*
NAME: Blinder

DESCRIPTION: The Blinder was the first supersonic Soviet bomber to enter service. It was built in smaller than expected numbers, apparently because its range fell short of requirements so it did not replace the Tu-16 Badger as originally planned. Soviet strategic aviation operates the Blinder in the missile strike and reconnaissance roles, while those assigned to the Soviet Navy are reconnaissance variants and bombers armed only with free-fall bombs.

USERS: Soviet Union Navy and Air Forces, Iraq and Libya.

CHARACTERISTICS:

Manufacturer:	Tupolev
Crew:	3
Engines:	2 Koliesov VD-7 turbojet
Max Power	30,800 lb (14,000 kg) static thrust each
Weights:	
Empty	88,000 lb (40,000 kg)
Max Takeoff	184,580 lb (83,900 kg)
Dimensions:	
Wing Span	89 ft 3 in (27.70 m)
Length	132 ft 11 in (40.53 m)
Height	35 ft 0 in (10.67 m)

Blinder

Performance:

Cruise Speed	486 kts (560 mph or 900 km/h) at 36,080 ft (11,000 m)
Max Speed	799 kts (920 mph or 1,475 km/h) at 36,080 ft (12,200 m)
Ceiling	60,000 ft (18,300 m)
Combat Radius	1,565 nm (1,800 mi; 2,900 km)
Max Range	3507 nm (4,039 mi; 6,500 km)
Armament	1 23-mm NR-23 cannon in tail
Radar	Bee Hind I-band tail warning

DESIGNATOR: Tu-20/95/142*
NAME: Bear

DESCRIPTION: Long-range strategic bomber, with variants for naval reconnaissance and anti-submarine warfare (ASW). The Bear has been in front-line service and in production for over three decades, and remains the largest aircraft flown by the Soviet Navy. It is the only turboprop-powered strategic bomber in operational service in the world, with four contra-rotating, four-bladed propellers. Comparable in role to the U.S. B-52, the Bear can carry air-to-surface missiles and free fall ordnance. Initially the military versions were designated Tu-20, while the Tupolev design bureau designated the aircraft Tu-95. Bear-F/H variants are designated Tu-142; these variants have a stretched fuselage and enlarged rudder. Most Bears are fitted for in-flight refueling.

Bear

USERS: Soviet Union Navy and Air Forces.

CHARACTERISTICS:

Manufacturer:	Tupolev
Crew:	7-8
Engines:	4 Kuznetsov NK-12MV turboprop
Max Power	12,000 shaft hp each
Weights:	
Max Takeoff	413,600 lb (188,000 kg)
Dimensions:	
Wing Span	167 ft 1 in (51.20 m)
Length	162 ft 4 in (49.50 m)
Height	39 ft 8 in (12.10 m)

Performance:

Cruise	405 kts (465 mph; 750 km/h)
Max Speed	500 kts (575 mph; 925 km/h)
Range	6,750 nm (7,750 mi; 12,500 km)
Armament	4 23-mm cannons in a twin-gun barbette and manned tail turret 2 additional 23-mm cannons in rear ventral turret (Bear-A/B/C/D/E/G) 1 additional 23-mm cannon in nose (Bear-A/B); Bomb bay for 25,000 lb (11,350 kg) of conventional or nuclear ordnance
Radar	Wet Eye search radar Big Bulge I-band radar (Bear-D) Crown Drum I-band radar (Bear-B/C/D/E) Short Horn navigation radar (Bear-D/E) Bee Hind tail warning radar (Bear-A/B) Box Tail tail warning radar (Bear-C) Down Beat bombing/navigation radar (Bear-G)

DESIGNATOR: Tu-16*
NAME: Badger

DESCRIPTION: This is the most widely flown bomber in Soviet service, being employed in the bombing and missile-launch roles as well as several support roles. The Badger is a twin-turbojet medium-range aircraft. Originally designed for free-fall bomb delivery, it has been adapted for use as an air-to-surface missile carrier. The Badger has a swept-wing configuration with large engine nacelles buried in the wing roots. The Tupolev design bureau designation for the Badger was Tu-88.

USERS: Soviet Union Air Forces and Navy, China, Libya, Egypt and Iraq

CHARACTERISTICS:

Manufacturer:	Tupolev
Crew:	6
Engines:	2 Mikulin AM-3 turbojet or 2 Mikulin AM-3M turbojet
Max Power	19,285 lb (8,748 kg) static thrust (AM-3) or 20,950 lb (9,425 kg) static thrust (AM-3M)
Weights:	
Empty	81,840 lb (37,200 kg)
Max Takeoff	158,456 lb (72,000 kg)
Dimensions:	
Wing Span	108 ft 0 in (32.93 m)
Length	118 ft 11¼ in (36.25 m)
Height	35 ft 11¼ in (14.00 m)
Wing Area	1,772.3 sq ft (164.65 sq m)
Performance:	
Max Speed	535 kts (605 mph; 990 km/h)
Ceiling	40,350 ft (12,300 m)
Radius	1,700 nm (1,955 mi; 3,150 km)

DESIGNATOR: Mya-4*
NAME: Bison

DESCRIPTION: The Bison is the first fully operational four-engine jet produced by the Soviet Union. It has swept wings with a small fuel pod at each wingtip. The engines are in paired nacelles buried in the wing roots, and the horizontal stabilizers are mounted at the base of the tail fin. All versions have an inflight refuelling capability, with Bison-B/C models having fixed refuelling probes. Bisons are used as bombers, tankers, and reconnaissance aircraft. At least one has been adapted for use as transporter for the Soviet space shuttle, and most have been modified as tankers.

Badger

Bison

USERS: Soviet Union Air Forces.

CHARACTERISTICS:

Manufacturer:	Myasishchev
Crew:	8
Engines:	4 Mikulin AM-3D turbofan
Max Power	19,180 lb (8,690 kg) static thrust each
Weights:	
Max Takeoff	352,000 lb (159,800 kg)
Dimensions:	
Wing Span	165 ft 7½ in (50.48 m)
Length	154 ft 10 in (47.20 m) less refuel probe
Height	50 ft 0 in (15.25 m)
Performance:	
Max Speed	538 kts (620 mph; 998 km/h)
Ceiling	45,000 ft (13,700 m)
Radius	3,025 nm (3,480 mi; 5,600 km)
Armament	8 23-mm cannon in 4 twin-gun turrets and 10,000 lb (4,540 kg) of ordnance

FIGHTER AND ATTACK AIRCRAFT

DESIGNATOR:	**MiG-31***
NAME:	**Foxhound**

DESCRIPTION: Soviet interceptor based on the design of the MiG-25 Foxbat. Modifications from the Foxbat include the addition of a second (tandem) seat for a weapons operator, Tumansky engines similar to those in the Foxbat-E, a strengthened airframe, significantly improved avionics, and added fuselage mountings to accommodate up to 8 air-to-air missiles (AAM). The Foxhound's range and look-down, shoot-down avionics make it a potent air defense aircraft.

USERS: Soviet Union Air Defense Forces.

CHARACTERISTICS:

Manufacturer:	Mikoyan-Gurevich
Crew:	2 (Pilot, weapons operator)
Engines:	2 Tumansky turbojet
Max Power	30,865 lb (14,002 kg) static thrust each
Weights:	
Empty	48,115 lb (21.825 kg)
Max Takeoff	90,725 lb (41,150 kg)
Dimensions:	
Wing Span	45 ft 11¼ in (14.00 m)
Length	70 ft 6½ in (21.50 m)

Foxhound

Height	18 ft 6 in (5.63 m)
Wing Area	603 sq ft (56.00 sq m)
Performance	
Max Speed	1,375 kts (1,585 mph; 2,550 km/h)
Ceiling	80,000 ft (24,385 m)
Radius	1,135 nm (1,307 mi; 2,103 km)
Armament	Up to 8 AA-9 Amos AAMs
Radar	Pulse-doppler lookdown/shootdown radar in nose

DESIGNATOR:	**MiG-29***
NAME:	**Fulcrum**

DESCRIPTION: The Soviet MiG-29 is a fast, agile fighter approximately the same size as the U.S. F/A-18 Hornet, but similar in configuration and role to the U.S. F-15 Eagle. It has two widely-spaced tail fins, underfuselage engine air intakes, and slightly divergent engine nacelles. The wings are fixed, swept and highly tapered. The Fulcrum also has Leading Edge Root Extensions (LERX) which add lift and hold guns and possibly fuel. It has a minimal fuselage, comprising a cockpit section and radome. Its construction is mostly of aluminum with titanium on parts subject to high thermal stress.

Fulcrum

USERS: Soviet Union Air Defense Forces and Air Forces, India, Iraq (planned), Syria (planned).

CHARACTERISTICS:

Manufacturer:	Mikoyan
Crew:	1
Engines:	2 Tumansky R-33D turbofan
Max Power	11,240 lb (5,098 kg) each, dry 18,300 lb (8,300 kg) each, in afterburner
Weights:	
Empty	17,251 lb (7,825 kg)
Max Takeoff	36,376 lb (16,500 kg)
Dimensions:	
Wing Span	37 ft 9 in (11.50 m)
Length	56 ft 5 in (17.20 m) with nose probe
Height	14 ft 5 in (4.40 m)
Performance:	
Max Speed	
High Alt	1,260 kts (1,451 mph; 2,335 km/h) or Mach 2.2
Sea Level	701 kts (808 mph; 1,300 km/h)

Radius	621 nm (715 mi; 1,150 km)
Armament	1 30-mm or 23-mm cannon in left wing root location; 6 wing pylons for 6 AA-10 Alamo or 2 AA-10 and 4 AA-11 Archer air-to-air missiles
Radar	Flash Dance pulse-doppler

DESIGNATOR: MiG-27*
NAME: Flogger

DESCRIPTION: The variable-sweep wing Flogger was the first modern multi-role Soviet aircraft. Fighter-interceptor variants are designated MiG-23 (see separate entry) and constitute the bulk of the Soviet strike fighter inventory. Several hundred ground attack models (MiG-27) are also in service.

USERS: Soviet Union Air Forces, India.

CHARACTERISTICS:

Manufacturer:	Mikoyan-Gurevich
Crew:	1 (2 in trainer variants)
Engines:	1 Tumansky R-29B turbojet
Max Power	17,635 lb (7,999 kg) static thrust dry 25,350 lb (11,499 kg) static thrust in afterburner
Weights:	
Empty	18,078 lb (8,200 kg)
Normal Takeoff	35,000 lb (15,875 kg)
Max Takeoff	44,996 lb (20,400 kg)
Dimensions:	
Wing Span	
Extended	46 ft 9 in (14.25 m)
Swept	27 ft 6 in (8.38 m)
Length	54 ft 0 in (16.46 m)
Height	18 ft 0 in (5.50 m)
Wing Area	293.4 sq ft (27.26 sq m)
Performance:	
Max Speed	
High Alt	917 kts (1,056 mi; 1,700 km/h) above 36,090 ft (11,000 m)
Sea Level	627 kts (722 mi; 1,162 km/h)
Ceiling	52,493 ft (16,000 m)
Radius	269 nm (310 mi; 499 km) with 4,410 lb (2,000 kg) of external stores
Armament	Six-barrelled 30-mm Gatling cannon, 2 bomb racks and 5 wing pylons for up to 8,818 lb (4,000 kg) of external stores including AS-7 Kerry air-to-surface missile
Radar	Small ranging radar, laser rangefinder and marked target seeker

DESIGNATOR: MiG-25*
NAME: Foxbat

DESCRIPTION: Single-seat twin-engine interceptor designed for high-speed, high-altitude missions. The Foxbat was designed specifically to intercept the U.S. B-

Foxbat

70 high-altitude, Mach 3 strategic bomber, which ironically did not enter operational service. The MiG-25 has fixed-wing, twin-vertical fin configuration, with wedge-inlet rectangular air intake trunks. Structure of arc-welded nickel steel, with titanium on leading edges of wing. Avionics include a ground-control data link for ground-directed interception. In its interceptor configuration, the Foxbat relies on air-to-air missiles.

USERS: Soviet Union Air Defense Forces and Air Forces, Algeria, Iraq, Egypt, Libya, India and Syria.

CHARACTERISTICS:

Manufacturer:	Mikoyan-Gurevich (design)
Crew:	1
Engines:	2 Tumansky R-31 single-shaft turbojet
Max Power	20,500 lb static thrust dry 27,010 lb static thrust in afterburner
Weights:	
Empty	44,100 lb (20,000 kg)
Max Takeoff	82,500 lb (37,425 kg)
Dimensions:	
Wing Span	45 ft 9 in (13.95 m)
Length	78 ft 1¾ in (23.82 m)
Height	20 ft 0¼ in (6.10 m)
Wing area	612 sq ft (56.83 sq m)
Performance:	
Max Speed	1,609 kts (1,850 mph; 2,980 km/h) 50% fuel, 4 missiles
Ceiling	88,580 ft (27,000 m)
Climb Rate	40,950 ft/min (12,480 m/min)
Radius	780 nm (900 mi; 1,450 km)
Armament	4 AA-6 Acrid (2 radar + 2 infrared) or 2 AA-7 Apex + 4 AA-11 Archer or 4 AA-8 Aphid
Radar	Fox Fire fire control radar with 45 nm range (52 mi; 85 km)

DESIGNATOR: MiG-23*
NAME: Flogger

DESCRIPTION: The variable-sweep wing Flogger was the first modern multi-role Soviet aircraft. This fighter-

Flogger

interceptor variant, designated MiG-23, constitutes the bulk of the Soviet strike fighter inventory. Several hundred ground attack models (MiG-27) are also in service.

USERS: Soviet Union Air Defense Forces and Air Forces, Algeria, Angola, Bulgaria, Cuba, Czechoslovakia, China, East Germany, Ethiopia, Hungary, India, Iraq, Libya, North Korea, Poland, Romania, Syria and Vietnam.

CHARACTERISTICS:

Manufacturer:	Mikoyan-Gurevich
Crew:	1 (2 in trainer variants)
Engines:	2 Tumansky R-29B turbojet
Max Power	17,635 lb (7,999 kg) static thrust dry
	23,350 lb (10,591 kg) static thrust in afterburner

Weights:
Empty	18,078 lb (8,200 kg)
Normal Takeoff	34,172 lb (15,500 kg)
Max Takeoff	44,313 lb (20,100 kg)

Dimensions:
Wing Span
Extended	46 ft 9 in (14.25 m)
Swept	27 ft 6 in (8.38 m)
Length	55 ft 1½ in (16.80 m)
Height	18 ft 0 in (5.50 m)
Wing Area	293.4 sq ft (27.26 sq m)

Performance:
Max Speed
High Alt	1,320 kts (1,520 mph; 2,446 km/h) above 36,090 ft (11,000 m)
Sea Level	792 kts (912 mph; 1,468 km/h)
Ceiling	65,617 ft (20,000 m)
Radius	460 nm (530 mi; 853 km) with 4 air-to-air missiles (AAM)
	608 nm 1700 mi; 1,126 km) with 4 AAM and centerline fuel tank
Armament	1 GSh-L twin-barrelled 23-mm cannon, l fuselage pylon and 4 wing pylons for 6 AA-7 Apex or AA-8 Aphid AAM
Radar	High Lark J-band search and track; Sirena 3 radar warning; Doppler navigation radar, and an infrared sensor

DESIGNATOR: MiG-21*
NAME: Fishbed

DESCRIPTION: Single-seat multi-role fighter. Missions include air combat, reconnaissance, electronic countermeasures (ECM), ground attack, and training. Has 4 underwing weapons pylons to carry a mix of weapons and equipment, varying with mission. Capable of carrying up to 3 external fuel tanks.

USERS: Soviet Union Air Forces, Afghanistan, Algeria, Angola, Bangladesh, Bulgaria, China, Cuba, Czechoslovakia, East Germany, Egypt, Ethiopia, Finland, Hungary, India, Indonesia, Iraq, Kampuchea, Laos, Libya, Madagascar, Mali, Mongolia, Mozambique, Nigeria, North Korea, North Yemen, Poland, Romania, Somalia, South Yemen, Sudan, Syria, Uganda, Vietnam, Yugoslavia, Zambia.

CHARACTERISTICS:
Manufacturer:	Mikoyan-Gurevich
Crew:	1
Engine:	1 Tumansky R-13-300 turbojet
Max Power	8,598 lb static thrust dry
	13,668 lb static thrust in afterburner

Weights:
Empty	11,465 lb (5,200 kg)
Max Takeoff	20,940 lb (9,500 kg)

Dimensions:
Wing Span	23 ft 5½ in (7.15 m)
Length	51 ft 8½ in (15.76 m) including pitot boom
Height	14 ft 9 in (4.00 m)
Wing Area	247 sq ft (23 sq m)

Performance:
Max Speed	1,203 kts (1,385 mph; 2,230 km/h)
Ceiling	50,000 ft (15,520 m)
Climb Rate	10,000 ft/min (3,050 m/min)
Radius	400 nm (460 mi; 740 km) with drop tanks
Range	593 nm (683 mi; 1,100 km) internal fuel only
	971 nm (1,108; 1,800 km) with external tanks
Armament	1 twin-barrel 23-mm GSh-23 gun 2 AA-2 Atoll missiles 2 UV-16-57 rocket packs with 16 57-mm rockets each
Radar	Jay Bird search and track radar with 10.8 nm search range (12.5 m; 20 km)

Fishbed

Flanker

DESIGNATOR: Su-27*
NAME: Flanker

DESCRIPTION: Soviet all-weather interceptor/fighter, capable of ground attack missions. The Flanker somewhat resembles the U.S. F-15 Eagle, having twin tail fins, fixed swept wings, and a circular-section fuselage. The Flanker differs from the F-15 by its underfuselage engine air intakes and straight wingtips. It has lookdown/shootdown radar capability and beyond-visual range air-to-air missile armament.

USERS: Soviet Union Air Defense Forces.

CHARACTERISTICS:

Manufacturer:	Sukhoi
Crew:	1
Engines:	2 turbofan
Max Power	20,000 lb (9,070 kg) static thrust dry 30,000 lb (13,610) static thrust in afterburner
Weights:	
Empty	39,000 lb (17,690 kg)
Max Takeoff	63,500 lb (28,805 kg)
Dimensions:	
Wing Span	47 ft 7 in (14.50 m)
Length	69 ft 0 in (21.00 m)
Height	18 ft 0 in (5.50 m)
Performance:	
Max Speed	1,150 kts (1,320 mph; 2,120 km/h)
Climb Rate	60,000 ft/min (304.5 m/min)
Radius	810 nm (930 mi; 1,500 km)
Armament	Up to 6 air-to-air missiles or 13,220 lb (6,000 kg) of other ordnance
Radar	Track-while-scan with 130 nm (150 m; 240 km) earch range, lookdown/shootdown capability

DESIGNATOR: Su-25*
NAME: Frogfoot

DESCRIPTION: The Su-25 is a close air support aircraft designed for use from unimproved airfields. Similar in role to the U.S. A-10 Thunderbolt, the Frogfoot features a heavily armored cockpit and mounts up to 10 weapon pylons for a wide assortment of bombs, rockets, gun pods, and other weapons. The fuselage is flat-sided with the twin turbojet engine nacelles beneath the wing roots (resembling the U.S. prototype A-9 aircraft). The Frogfoot has a single tail fin.

USERS: Soviet Union Air Forces. Czechoslovakia, Iraq, and Hungary.

Frogfoot

CHARACTERISTICS:

Manufacturer:	Sukhoi
Crew:	1
Engines:	2 Tumansky R-13-300 turbojet
Max Power	11,240 lb (5,100 kg) static thrust each
Weights:	
Empty	20,950 lb (9,500 kg)
Max Takeoff	42,330 lb (19,200 kg)
Dimensions:	
Wing Span	46 ft 11 in (14.30 m)
Length	50 ft 6¾ in (15.40 m)
Wing Area	362.75 sq ft (33.70 sq m)
Performance:	
Max Speed	530 kts (608 mph; 980 km/h)
Radius	300 nm (345 mi; 556 km) with external fuel, 4,450 lb (2,000 kg) ordnance load
Armament	1 twin-barrel 30-mm cannon mounted in left bottom front fuselage is standard, and up to 9,900 lbs (4,500 kg) of ordnance is carried on 10 pylons
Radar	Odd Rods identification friend or foe (IFF) system (antennas forward of windscreen and under tail)

DESIGNATOR: Su-24*
NAME: Fencer

DESCRIPTION: The Su-24/Fencer is a high-speed, long-range strike aircraft with variable-sweep wings. The Fencer has a slab-sided rectangular-section fuselage with integral engine air intake trunks, a single swept fin, and two slightly splayed ventral fins. It is described by the U.S. Department of Defense as the best deep-interdiction aircraft in the Soviet inventory. The aircraft is able to conduct precision missions during night or poor weather. It has the same mission as the U.S. F-111, which is similar but somewhat larger.

USERS: Soviet Union Air Forces and Navy.

CHARACTERISTICS:

Manufacturer:	Sukhoi
Crew:	2 (Pilot, weapon systems operator)
Engines:	2 turbojets (probably Lyulka AL-21F variant)
Max Power	Approx 24,250 lb (11,000 kg) each with afterburning
Weights:	
Empty	Approx 41,888 lb (19,000 kg)
Max Takeoff	Approx 87,082 lb (39,500 kg)
Dimensions:	
Wing Span	
Extended	57 ft 5 in (17.50 m)
Swept	34 ft 5 in (10.50 m)
Length	69 ft 10 in (21.29 m)
Height	19 ft 8 in (6.00 m)
Performance:	
Max Speed	
High Alt	1,250 kts (1,440 mph; 2,317 km/h) or Mach 2+ above 36,089 ft (11,000 m)
Sea Level	793 kts (915 mi; 1,470 km/h) or Mach 1.2
Ceiling	54,134 ft (16,500 m)
Combat Radius	Lo-lo-lo more than 174 nm (200 mi; 322 km) hi-lo-lo-hi 701 nm (808 mi; 1,300 km) with 6,614 lb (3,000 kg) external stores, 2 external fuel tanks

Fencer

Armament	1 6-barrel 30-mm Gatling-type cannon and 8 pylons to carry up to 24,250 lb(11,000 kg) of weapons, including conventional bombs, nuclear weapons, and air-to-surface missiles: (AS-10 Karen, AS-11, AS-12 Kegler, AS-13, AS-14 Kedge).
Radar	Pulse-doppler with 50-in (1.27 m) scanner with terrain-avoidance capability, laser rangefinder and marked target seeker, long-range navigation system; accuracy, according to official U.S. sources, is within 180 ft (55 m) of the target under all weather conditions

DESIGNATOR: Su-17/20/22*
NAME: Fitter

DESCRIPTION: The Fitter is a Soviet ground-support fighter with variable-sweep wings. It evolved from the Su-7 fixed-wing strike fighter. The Fitter has an engine air intake in the nose, variable-geometry wings mounted slightly below the center of the fuselage, a single tail fin without tabs, fuselage-mounted horizontal stabilizers, and a single seat canopy. It suffers from limited internal fuel capacity, and is generally seen with 2 large drop tanks. The Fitter has a total of 6 to 8 attachment points for armament and/or fuel. The Soviet designations Su-20 and Su-22 are applied to export versions that have different engines and avionics.

USERS: Soviet Union Air Forces and Navy, Afghanistan, Algeria, Angola, Czechoslovakia, East Germany, Egypt, Hungary, Iraq, Libya, North Korea, North Yemen, Peru, Poland, South Yemen, Syria, Vietnam.

CHARACTERISTICS:

Manufacturer:	Sukhoi
Crew:	1
Engines:	1 Lyulka AL-21F or Tumansky R-29B
Max Power	Lyulka 17,195 lb (7,800 kg) dry Tumansky 17,635 lb (8,000 kg) dry Lyulka 24,200 lb (11,000 kg) afterburn Tumansky 25,350 lb (11,500 kg) in afterburner
Weights:	
Empty	22,000 lb (10,000 kg)
Max Takeoff	38,940 lb (17,700 kg)
Dimensions:	
Wing Span	
Extended	45 ft 11 in (14.00 m)
Swept	34 ft 9 in (10.60 m)
Length	61 ft 6 in (18.75 m) over nose probe
Height	15 ft 7 in (4.75 m)
Wing Area	430 sq ft (40.0 sq m) extended 398 sq ft (37.0 sq m) swept
Performance:	
Max Speed	
Sea Level	Approx 512 kts 1590 mph; 950 km/h) or Mach 0.8 with external stores or 686 kts (790 mph; 1,271 km/h) or Mach 1.05 clean
High Alt	1,150 kts (1,325 mph; 2,133 km/h) or Mach 2.0 at 36,089 ft (11,000 m)
Ceiling	55,574 ft (17,000 m)
Strike Radius	338 nm (390 mi; 628 km) with 4,400 lb (2,000 kg) ext stores, hi-lo-hi mission 194 nm (225 mi; 360 km), lo-lo-lo mission
Armament	2 30-mm NR-30 cannon and up to 7,000 lb (3,150 kg) external stores including bombs; rockets; and AA-2 Atoll air-to-air missiles; AS-9 Kyle anti-radar missiles; and probably AS-7 Kerry air-to-surface missiles

Fitter

Radar	High Fix I-band range-only air intercept Sirena 3 radar warning system

DESIGNATOR: Su-15/Su-21*
NAME: Flagon

DESCRIPTION: The Flagon is a high-speed, all-weather, specialized air defense fighter. It has a delta wing, air intakes on both sides of the fuselage near the cockpit and an elongated nose to house the radar. The first four variants are referred to as Su-15; later changes to the design are thought to have earned the new Soviet designation of Su-21 for the Flagon-E/F/G variants.

USERS: Soviet Union Air Defense Forces and Air Forces.

CHARACTERISTICS:

Manufacturer:	Sukhoi
Crew:	1 (2 in Flagon-C/G)
Engines:	2 Tumansky R-13F2-300 turbojet
Max Power	15,873 lb (7,200 kg) static thrust each with afterburning
Weights:	
Max Takeoff	37,479 lb (17,000 kg)
Dimensions:	
Wing Span	29 ft 6 in (9.00 m)
Length	67 ft 3 in (20.50 m)
Performance:	
Max Speed	
High Alt	1,175 kts (1,353 mph; 2,177 km/h) or Mach 2.05 above 36,000 ft (11,000 m)
Low Alt	728 kts (839 mph; 1,350 km/h) or Mach 1.2 at 3,281 ft (1,000 m)

Flagon

Ceiling	65,616 ft (20,000 m)
Radius	390 nm (450 mi; 725 km) at High Altitude
Armament	4 wing pylons for 2 AA-3 Anab long-range air-to-air missiles and two AA-8 Aphid short-range infrared missiles; fuselage pylons can carry GSh-L 23-mm gun pods
Radar	Skip Spin X-band airborne interception radar in Flagon-A/B/C/D Twin Scan X-band airborne interception radar in Flagon-E/F/G

DESIGNATOR: Tu-28P*
NAME: Fiddler

DESCRIPTION: A two-seat, long-range fighter-interceptor, the Fiddler is the largest aircraft ever designed for that role. It has swept-back wings with large turbojet intakes faired into the fuselage just aft of the cockpit and external fuel tanks faired into the wing trailing edges.

Fiddler

USERS: Soviet Union Air Defense Forces.

CHARACTERISTICS:

Manufacturer:	Tupolev
Crew:	2
Engines:	2 turbojet
Max Power	27,000 lb (12,247 kg) static thrust each
Weights:	
Max Takeoff	100,000 lb (45,000 kg)
Dimensions:	
Wing Span	59 ft 4½ in (18.10 m)
Length	89 ft 3 in (27.20 m)
Performance:	
Max Speed	950 kts (1,090 mph; 1,760 km/h)
Ceiling	65,620 ft (20,000 m)
Radius	810 nm (930 mi; 1,500 km)
Armament	4 AA-5 Ash air-to-air missiles

DESIGNATOR: Yak-38*
NAME: Forger

DESCRIPTION: The Yak-38 is the Soviet Union's only operational fixed-wing VSTOL combat aircraft. Similar in function to the U.S. AV-8 Harrier, it serves in the fighter/attack role on board the *Kiev*-class aircraft carriers. It operates on two lift and one cruise engines, rather than a single engine as in the Harrier.

Forger

USERS: Soviet Union Navy.

CHARACTERISTICS:

Manufacturer:	Yakovlev
Crew:	1 in Forger-A (2 in Forger-B)
Engines:	1 Lyulka AL-21 turbojet cruise engine and 2 Koliesov turbojet lift engines
Max Power	Lyulka approx 17,630 lb (8,000 kg) Koliesov approx 7,710 lb (3,500 kg) each
Weights:	
Empty	16,500 lb 7,485 kg) A version 18,500 lb (8,390 kg) B version
Max Takeoff	25,794 lb (11,700 kg)
Dimensions:	
Wing Span	
Extended	24 ft 0 in (7.32 m)
Swept	16 ft 0 in (4.88 m)
Length	50 ft 10 in (15.50 m) A version 58 ft 0 in (17.68 m) B version
Height	14 ft 4 in (4.37 m)
Wing Area	199 sq ft (18.50 sq m)
Performance:	
Max Speed	
High Alt	545 kts (627 mph; 1,010 km/h) above 36,000 ft (11,000 m)
Sea Level	530 kts (610 mph; 980 km/h)
Ceiling	39,370 ft (12,000 m)
Radius	100 nm (115 mi; 185 km) with 75 min on station Lo-lo-lo 130 nm (150 mi; 240 km) with approx 6,600 lb (2,970 kg) weapons Hi-lo-hi 200 nm (230 mi; 370 km) same payload

Armament	4 pylons for up to 6,600 lb (2,970 kg) of bombs or stores including 2 GSh-23 23-mm gun pods, AA-8 Aphid air-to-air missiles or AS-7 Kerry air-to-surface missiles
Radar	Small ranging radar in nose

DESIGNATOR: Yak-28*
NAME: Firebar

DESCRIPTION: Two-seat all-weather interceptor, available also in reconnaissance, trainer, and ECM versions. Originally designed as a light bomber, but subsequent variants have served primarily in the intercept and reconnaissance roles.

USERS: Soviet Union Air Defense Forces.

CHARACTERISTICS:

Manufacturer:	Yakovlev
Crew:	2
Engines:	2 Tumansky R-11 turbojet
Max Power	13,120 lb (5,951 kg) static thrust each
Weights:	
Max Takeoff	44,000 lb (20,000 kg)
Dimensions:	
Wing Span	42 ft 6 in (12.95 m)
Length	75 ft 5½ in (23.00 m)
Height	12 ft 11½ in (3.95 m)

Firebar

Performance:	
Max Speed	1,080 kts (1,240 mph; 2,000 km/h)
Ceiling	55,000 ft (16,750 m)
Radius	500 nm (575 mi; 925 km)
Armament	2 pylons under each wing for weapons mix
Radar	Short Horn J-band bomb/nav radar

HELICOPTERS

DESIGNATOR: Ka-27*
NAME: Helix

DESCRIPTION: The Helix is a ship-based anti-submarine warfare (ASW) and transport/rescue helicopter developed

Helix

as the successor to the Ka-25 Hormone. Like the Hormone, it features two triple-bladed countra-rotating main rotors. It also has a horizontal stabilizer with endfins which act as rudders and fixed quadricycle landing gear. Fuselage is semi-monocoque pod and boom, with composites used in tailcone and extensive use of titanium for primary components.

USERS: Soviet Union Navy, and India.

CHARACTERISTICS:

Manufacturer:	Kamov
Crew:	2+ (Pilot, navigator) + ASW systems operator or 16 troops
Engines:	2 Isotov TV3-117V turboshaft
Max Power	2,250 shaft hp
Weights:	
Max Takeoff	22,775 lb (12,600 kg)
Dimensions:	
Rotor Diameter	52 ft 2 in (15.9 m)
Length	37 ft 1 in (11.3 m)
Height	17 ft 9 in (5.4 m)
Performance:	
Cruise Speed	124 kts (143 mph; 230 km/h)
Max Speed	135 kts (155 mph; 250 km/h)
Ceiling	19,685 ft (6,000 m)
Range	432 nm (497 mi; 800 km)
Armament	Torpedoes and depth charges
Radar	Search radar in undernose radome

DESIGNATOR:	**Ka-26***
NAME:	**Hoodlum**

DESCRIPTION: This twin-engine, general purpose light helicopter is not currently in service with Soviet military forces but is available for Soviet military use as required. It is now employed in a variety of civilian missions, including cargo/passenger transport, aerial survey, agriculture, fire fighting, search and rescue, air ambulance, and ice clearing/surveillance. It has 2 three-bladed, contra-rotating main rotors and 2 endfins on a horizontal stabilizer.

USERS: Bulgaria and Hungary.

CHARACTERISTICS:

Manufacturer:	Kamov
Crew:	1
Engines:	2 Vedeneyev M-14V-26 piston or 1 TVD-100 turboshaft
Max Power	325 hp each (M-14V-26) or 720 shaft hp (TVD-100)
Weights:	
Empty	4,300 lb (1,950 kg)
Max Takeoff	7,165 lb (3,250 kg)
Dimensions:	
Rotor Diameter	42 ft 8 in (13.00 m)
Length	25 ft 5 in (7.75 m)
Height	13 ft 3½ in (4.05 m)
Performance:	
Cruise Speed	81 kts (93 mph; 150 km/h)
Max Speed	91 kts (105 mph; 170 km/h)
Ceiling	9,840 ft (3,000 m)
Range	647 nm (745 mi; 1,200 km) max fuel

DESIGNATOR:	**Ka-25***
NAME:	**Hormone**

DESCRIPTION: The Hormone is the Soviet Navy's principal anti-submarine warfare (ASW), search and rescue (SAR), and target acquisition helicopter. The fuselage is the semi-monocoque pod and boom type. The two main rotors are three-bladed that counter-rotate, and have automatic folding. The Hormone is distinguished by 3 fins (no tail rotor). Its quadricycle fixed landing gear is fitted with inflatable pontoons for sea landings.

Hormone

USERS: Soviet Union Navy, India, Syria, Vietnam, and Yugoslavia.

CHARACTERISTICS:

Manufacturer:	Kamov
Crew:	2 (Pilot, co-pilot) + 2-3 ASW crew or 12 troops
Engines:	2 Glushenkov GTD-3F turboshaft
Max Power	900 shaft hp each
Weights:	
Empty	10,485 lb (4,765 kg)
Max Takeoff	16,500 lb (7,500 kg)

Dimensions:

Rotor Diameter	51 ft 6 in (15.70 m)
Length	32 ft 0 in (9.75 m)
Height	17 ft 8 in (5.37 m)

Performance:

Cruise Speed	104 kts (120 mph; 193 km/h)
Max Speed	119 kts (136 mph; 220 km/h)
Ceiling	11,500 ft (3,500 m)
Range	135 nm (155 mi; 250 km) Radius
Armament	2 450-mm torpedoes or depth charges
Radar	Puff Ball in Hormone-A Big Bulge in Hormone-B

DESIGNATOR: Mi-28*
NAME: Havoc

DESCRIPTION: The Mi-28 Havoc is an attack helicopter designed to supplement the Mi-24 Hind; however, the Havoc has no troop transport capability. Designed to emphasize the anti-tank, fire support, and air-to-air missions, the Havoc has a tandem cockpit and is a heavily armored aircraft. It has auxiliary wings with weapons pylon and missile rails on the endplate pylon. The fuselage is semi-monocoque, with a much smaller cross-section than the Hind. The main rotor is five-bladed and for landing gear it has two V-strut fixed main wheels and a small tail skid.

USERS: Soviet Union Ground Forces.

CHARACTERISTICS:

Manufacturer:	Mil'
Crew:	2 (Pilot, weapon systems operator)
Engines:	2 pod-mounted turboshaft
Max Power	2,200-2,500 shaft hp (estimated)

Dimensions:

Rotor Diameter	55 ft 9 in (17.00 m)
Length	57 ft 1 in (17.40 m)

Performance:

Max Speed	162 kts (186 mph; 300 km/h)
Radius	130 nm (149 mi; 240 km)
Armament	Large-caliber gun in nose turret standard and various options on wing pylons

Havoc

DESIGNATOR: Mi-26*
NAME: Halo

DESCRIPTION: World's heaviest heavy-lift helicopter in production, the Mi-26 Halo holds several lift world records. It can lift a full rifle company or two airborne infantry combat vehicles. The Halo has semi-monocoque pod and boom fuselage with rear clamshell doors and loading ramp; an eight-bladed, electrically heated main rotor, and a five-bladed tail rotor mounted on starboard of tailboom fin over the horizontal stabilizer.

Halo

USERS: Soviet Union Air Forces, and India.

CHARACTERISTICS:

Manufacturer:	Mil'
Crew:	5 (Pilot, co-pilot, navigator, flight engineer, loadmaster) + 85 troops
Engines:	2 Lotarev D-136 turboshaft
Max Power	11,400 shaft hp each

Weights:

Empty	62,171 lb (28,000 kg)
Max Takeoff	123,450 lb (56,000 kg)

Dimensions:

Rotor Diameter	105 ft (32.00 m)
Length	110 ft 8 in (33.73 m)
Height	26 ft 9 in (8.15 m)

Performance:

Cruise Speed	137 kts (158 mph; 255 km/h)
Max Speed	159 kts (183 mph; 295 km/h)
Ceiling	15,100 ft (4,600 m) service 14,765 ft (4,500 m) In ground effect 5,905 ft (1,800 m) Out of ground effect
Range	432 nm (497 mi; 800 km)
Radar	Doppler/weather radar in nose

DESIGNATOR: Mi-24*
NAME: Hind

DESCRIPTION: The Hind is an assault helicopter used for fire support, escort, anti-tank, and air-to-air combat

Hind

against opposing helicopters. The airframe is all-metal semi-monocoque pod and boom, with two types of crew cabin. All types are heavily armored, and feature two cantilever shoulder wings, each of which has two weapons pylons and an endplate pylon. The Hind has a five-bladed main rotor and three-bladed tail rotor, retractable tricycle landing gear, twin engines mounted above the cabin, and a variable-incidence stabilizer at the base of the tailboom fin.

USERS: Soviet Union Ground Forces, Afghanistan, Algeria, Angola, Bulgaria, Cuba, Czechoslovakia, East Germany, Ethiopia, Hungary, India, Iraq, Libya, Mozambique, Nicaragua, Poland, South Yemen, Syria, Vietnam.

CHARACTERISTICS:

Manufacturer:	Mil'
Crew:	Hind-A/B/C
	3 (Pilot, co-pilot/gunner, and ground engineer) + 8 troops or 4 litters
	Hind-D/E/F
	2 (Pilot in raised rear cockpit, gunner in forward cockpit) + 8 troops or 4 litters
Engines:	2 Isotov TV3-107 turboshaft
Max Power	2,200 shaft hp each
Weights:	
Empty	18,520 lb (8,400 kg)
Max Takeoff	24,250 lb (11,000 kg)
Dimensions:	
Rotor Diameter	55 ft 9 in (17.00 m)
Length	57 ft 5 in (17.50 m)
Height	21 ft 4 in (6.50 m)
Performance:	
Cruise Speed	159 kts (183 mph; 295 km/h)
Max Speed	167 kts (192 mph; 310 km/h)
Ceiling	14,750 ft (4,500 m) service
	7,200 ft (2,200 m) out of ground effect
Radius	86 nm (99 mi; 160 km)
Range	405 nm (466 mi; 750 km)

Armament	
Hind-A	4 UB-32 rocket pods, each with 32 57-mm rockets
	4 AT-2 Swatter missiles
	1 12.7-mm single-barrel DShK machine gun in nose turret
Hind-B	Same as above, minus Swatter
Hind-C	Same as Hind-B, minus nose gun
Hind-D	Same as Hind-A, except nose gun replaced with 12.7-mm Gatling-type gun
Hind-E	Same as Hind-D, except pylons modified to carry up to 12 AT-6 Spiral missiles
Hind-F	Same as Hind-E, except nose gun replaced with twin-barrel 30-mm cannon
Radar	Doppler navigational radar mounted under nose

DESIGNATOR:	**Mi-14***
NAME:	**Haze**

DESCRIPTION: The Mi-14 Haze is a land-based helicopter based on the Mi-8* Hip that is configured for anti-submarine warfare (ASW) and minesweeping operations. Modifications from the Mi-8 include provision of a boat hull, tailboom float, side sponsons, retractable landing gear, port side tail rotor, shortened engine nacelles, weapons bay in the bottom of the hull, and a large undernose radome.

USERS: Soviet Union Navy, Bulgaria, Cuba, East Germany, Libya, Poland, Romania and Syria.

Haze

CHARACTERISTICS:

Manufacturer:	Mil'
Crew:	4 or 5
Engines:	2 Isotov TV3-117 turboshaft
Max Power	2,200 shaft hp each
Weights:	
Empty	15,026 lb (6,816 kg)
Max Takeoff	28,660 lb (13,000 kg)
Dimensions:	
Rotor Diameter	68 ft 10¾ in (21.29 m)

Length	60 ft ¾ in (18.31 m)
Height	22 ft 7¾ in (6.90 m)
Performance:	
Cruise Speed	140 mph (225 km/h)
Max Speed	161 mph (260 km/h)
Ceiling	14,760 ft (4,500 m) (Service)
	6,235 ft (1,900 m) In ground effect
	2,625 ft (800 m) Out of ground effect
Range	251 nm (289 mi; 465 km)
Armament	Torpedo or depth charges
Radar	Doppler radar in box under forward part of tailboom

Hip

DESIGNATOR: Mi-10*
NAME: Harke

DESCRIPTION: This is a flying crane helicopter developed from the Mi-6* Hook. It retains the same dynamics as the Hook, but has a modified fuselage and landing gear. The fuselage was altered to give it an unbroken under-surface the entire length of the aircraft, while the landing gear was modified to a quadricycle arrangement, permitting the aircraft to taxi over the load to be lifted. It also has either a retractable, rear-facing pilot gondola for use in sling-loading operations, or a closed-circuit television system. It is in service in both military and civil roles.

USERS: Soviet Union Air Forces.

CHARACTERISTICS:

Manufacturer:	Mil'
Crew:	2 + 24,250 lb (11,000 kg) payload or 28 troops
Engines:	2 Soloviev D-25V turboshaft
Max Power	5,500 shaft hp each
Weights:	
Empty	60,186 lb (27,300 kg)
Max Takeoff	96,341 lb (43,700 kg)
Dimensions:	
Rotor Diameter	114 ft 10 in (35.00 m)
Length	107 ft 10 in (32.86 m)
Height	38 ft 2 in (9.80 m)
Performance:	
Cruise Speed	109 kts (125 mph; 202 km/h) with sling load
Max Speed	135 kts (155 mph; 250 km/h)
Ceiling	9,850 ft (3,000 m)
Ranges	135 nm (155 mi; 250 km)
	428 nm (494 mi; 795 km) with auxiliary fuel

DESIGNATOR: Mi-8*
NAME: Hip

DESCRIPTION: The Mi-8 Hip is the primary transport helicopter of the Soviet armed forces. It is based on the Mi-4* Hound, with a redesigned five-bladed rotor, twin 1,500 hp-engines, and a larger fuselage. Produced in both military and civilian versions, it features an all-metal, semi-monocoque pod and boom with the tail rotor mounted starboard of a small vertical stabilizer. The tail-boom also has a small horizontal stabilizer. Uses fixed tricycle landing gear.

USERS: Soviet Union Air Forces, Ground Forces and Navy; Afghanistan, Algeria, Angola, Bangladesh, Bulgaria, Cuba, Czechoslovakia, East Germany, Egypt, Ethiopia, Finland, Guinea-Bissau, Hungary, India, Iraq, Kampuchea, Laos, Libya, Mali, Mongolia, Mozambique, Nicaragua, North Korea, Pakistan, Peru, Poland, Romania, Somalia, South Yemen, Sudan, Syria, Vietnam, Yugoslavia and Zambia.

CHARACTERISTICS:

Manufacturer:	Mil'
Crew:	2(Pilot, copilot) + 26 troops or 12 litters and attendant
Engines:	2 Isotov TV2-117 turboshaft
Max Power	2,210 shaft hp each
Weights:	
Empty	15,026 lb (6,816 kg)
Max Takeoff	26,455 lb (12,000 kg)
Dimensions:	
Rotor Diameter	68 ft 10¾ in (21.29 m)
Length	60 ft ¾ in (18.31 m)
Height	18 ft 6 in (5.65 m)
Performance:	
Cruise Speed	139.5 mph (225 km/h)
Max Speed	161 mph (260 km/h)
Ceiling	14,760 ft (4,500 m) service
	6,235 ft (1,900 m) hover (In ground effect)
	2,625 ft (800 m) hover (Out of ground effect)
Range	251 nm (289 mi; 465 km)
Armament	
Hip-C	4 x 57-mm rocket pods, 12.7-mm flexible-mount machine gun
Hip-E	6 x 57-mm rocket pods, 4 AT-2 Swatter missiles
Hip-F	Same as above, except Swatters replaced with 6 AT-3 Sagger missiles
Radar	Doppler radar in box under forward part of tailboom

DESIGNATOR: Mi-6*
NAME: Hook

DESCRIPTION: A heavy transport helicopter. It was the largest helicopter in the world when first flown in 1957. Missions include heavy-lift cargo operations, medical evacuation, troop transport, and civil operations such as fire fighting. It has semi-monocoque pod and boom fuselage, and hydraulically-operated clamshell rear doors for loading and off-loading cargo and personnel. Most models are fitted with two auxiliary wings. It has a five-bladed main rotor, a four-bladed tail rotor, and fixed tricycle landing gear.

USERS: Soviet Union Air Forces and Navy, Algeria, Iraq, Laos, Peru and Vietnam.

Hook

CHARACTERISTICS:

Manufacturer:	Mil'
Crew:	5 (Pilot, co-pilot, navigator, flight engineer, radio operator) + 65 troops or 41 litters and 2 attendants or 26,450 lb (12,000) cargo
Engines:	2 Soloviev D-25V (TV-2BM) turboshaft
Max Power	5,500 shaft hp each
Weights:	
Empty	60,055 lb (27,240 kg)
Max Takeoff	93,700 lb (42,500 kg)
Max Payload	26,450 lb (12,000 kg)
Dimensions:	
Rotor Diameter	114 ft 10 in (35.00 m)
Length	108 ft 11 in (33.18 m)
Height	32 ft 4 in (9.86 m)
Performance:	
Cruise Speed	135 kts (155 mph; 250 km/h)
Max Speed	162 kts (186 mph; 300 km/h)
Ceiling	14,750 ft (4,500 m)
Range	335 nm (385 mi; 620 km) 540 nm (621 mi; 1,000 km) with external tanks
Armament	12.7 mm machine gun in nose in some models

DESIGNATOR: Mi-4*
NAME: Hound

DESCRIPTION: Transport and general purpose helicopter, formerly the principal Soviet transport helicopter. The

Hound

Hound has a four-bladed main rotor. Its three-bladed tail rotor is mounted on a tailboom which extends above the rear fuselage clamshell doors. The Hound has fixed quadricycle landing gear. The Mi-8* Hip, which is now the main Soviet transport helicopter, was derived from the Hound.

STATUS: Initial operational capability in 1953. Most have been retired from Soviet military service, with only an estimated 25 remaining. Hounds are still in active military service in 14 countries.

USERS: Soviet Union Ground Forces, Afghanistan, Albania, Algeria, Bulgaria, China, Cuba, Czechoslovakia, Mali, Mongolia, North Korea, Poland, Romania, Somalia.

CHARACTERISTICS:

Manufacturer:	Mil'
Crew:	2 + 16 troops
Engines:	1 ASh-82V radial piston
Max Power	1,700 hp
Weights:	
Empty	11,614 lb (5,268 kg)
Max Takeoff	17,196 lb (7,800 kg)
Dimensions:	
Rotor Diameter	68 ft 11 in (21.01 m)
Length	55 ft 1 in (16.80 m)
Height	17 ft 0 in (5.19 m)
Performance:	
Cruise Speed	87 kts (100 mph; 161 km/h)
Max Speed	113 kts (130 mph; 209 km/h)
Ceiling	18,000 ft (5,490 m)
Range	135 nm (155 mi; 250 km)

DESIGNATOR: Mi-2*
NAME: Hoplite

DESCRIPTION: The Hoplite Mi-2 is a light utility helicopter, developed in the Soviet Union but manufactured in Poland. Missions include anti-tank, liaison, air ambulance, training, and scout. Produced in both military and civil configurations, its fuselage is built in three sections to facilitate mission tailoring. The fuselage is semi-

Hoplite

monocoque pod and boom, with fixed tricycle landing gear and a tail skid. It has a three-bladed main rotor and a two-bladed tail rotor.

USERS: Soviet Union Ground Forces, Bulgaria, Cuba, Czechoslovakia, East German, Hungary, Libya, Nicaragua, North Korea, Poland, Romania and Syria.

CHARACTERISTICS:

Manufacturer:	WSK Swidnik (designed by Mil')
Crew:	1 + 6-8 passangers
Engines:	2 Isotov GTD-350 turboshaft
Max Power	400 shaft hp each
Weights:	
Empty	5,213 lb (2,365 kg)
Max Takeoff	8,157 lb (3,550 kg)
Dimensions:	
Rotor Diameter	47 ft 7 in (14.50 m)
Length	37 ft 5 in (11.40 m)
Height	12 ft 4 in (3.75 m)
Performance:	
Cruise Speed	108 kts (118 mph; 190 km/h)
Max Speed	113 kts (130 mph; 210 km/h)
Ceiling	13,125 ft (4,000 m) service
Range	237 nm (273 mi; 440 km) internal fuel only
Armament	4 AT-3 Sagger missiles or rocket pods on some military models

RECONNAISSANCE/ELECTRONIC WARFARE AIRCRAFT

DESIGNATOR: Il-20*
NAME: Coot-A

DESCRIPTION: Reconnaisance/electronic warfare (EW) aircraft derived from Il-18* Coot airliner. The recce/EW Coots are designated Il-20 by the Soviet Union and Coot-A by NATO. A cylindrical container for EW equipment, 33 ft 7½ in (10.25 m) long, is carried on the undersurface of the fuselage. On either side of the forward fuselage are smaller containers housing cameras or other sensors. It has 10 other blisters and antennas, eight on the underside and two atop the forward fuselage.

USERS: Soviet Union Navy.

CHARACTERISTICS:

Manufacturer:	Ilyushin
Crew:	5 + 20 EW operators
Engines:	4 Ivchenko AI-20M turboprop
Max Power	4,250 effective hp each
Weights:	
Max Takeoff	141,100 lb (64,000 kg)
Dimensions:	
Wing Span	122 ft 9 in (37.40 m)
Length	117 ft 9 in (35.90 m)
Height	33 ft 4 in (10.17 m)
Wing Area	1,507 sq ft (140 sq m)
Performance:	
Cruise Speed	337 kts (390 mph; 625 km/h)
Max Speed	364 kts (420 mph; 675 km/h)
Range	3,508 nm (4,030 mi; 6,500 km) max fuel
Takeoff Run	4,265 ft (1,300 m)
Landing Run	2,790 ft (850 m)

Coot

RECONNAISSANCE/AIRBORNE EARLY WARNING AIRCRAFT

DESIGNATOR: Tu-126*
NAME: Moss

DESCRIPTION: The Moss is an airborne early-warning (AEW) aircraft derived from Tu-114* Cleat transport, which in turn was a derivative of the Tu-20/95* Bear strategic bomber. It was the first Soviet aircraft equipped with a rotodome (rotating radar antenna). Modifications from the Cleat/Bear configuration included a refueling probe at the nose, several antennas and blisters for elec-

Moss

tronic equipment, horizontal stabilizers mounted on rear mid-fuselage rather than at the tail fin root, a ventral tail fin, and a tail cone instead of a tail gun position.

USERS: Soviet Union Air Defense Forces.

CHARACTERISTICS:

Manufacturer:	Tupolev
Crew:	12-13
Engines:	4 Kuznetsov NK-12MV turboprop
Max Power	14,795 effective hp each
Weights:	
Max Takeoff	374,785 lb (170,000 kg)
Dimensions:	
Wing Span	168 ft 0 in (51.20 m)
Length	181 ft 1 in (55.20 m)
Height	52 ft 8 in (16.05 m)
Wing Area	3,349 sq ft (311.1 sq m)
Performance:	
Cruise Speed	351 kts (404 mph; 650 km/h)
Max Speed	459 kts (528 mph; 850 km/h)
Range	6,775 nm (7,800 mi; 12,550 km) unrefuelled
Radar	Flat Jack AEW system

RECONNAISSANCE/ANTI-SUBMARINE WARFARE AIRCRAFT

DESIGNATOR:	**Il-38***
NAME:	**May**

DESCRIPTION: The May is an anti-submarine warfare (ASW)/patrol aircraft that resembles the U.S. P-3 Orion in appearance and role. The May is derived from the Il-18* airliner with a lengthened fuselage, wings mounted farther forward, a large undernose radome for search radar, a blister fairing for electronic equipment, and a weapons bay. A tail cone houses a magnetic anomaly detector (MAD).

USERS: Soviet Union Navy, and India.

CHARACTERISTICS:

Manufacturer:	Ilyushin
Crew:	12 (Pilot, co-pilot, flight engineer, radar operator, navigator, 3 systems operators, tactical coordinator, 3 observers)
Engines:	4 Ivchenko AI-20M turboprop
Max Power	4,250 shaft hp each
Weights:	
Empty	79,200 lb (36,000 kg)
Max Takeoff	139,700 lb (63,500 kg)

May

Dimensions:	
Wing Span	122 ft 8 in (37.40 m)
Length	129 ft 10½ in (39.60 m)
Height	33 ft 4 in (10.16 m)
Wing Area	1,507 sq ft (140 sq m)
Performance:	
Cruise Speed	216 kts (250 mph; 400 km/h)
Max Speed	324 kts (370 mph; 600 km/h)
Range	3,888 nm (4,474 mi; 7,200 km) max fuel
Takeoff Run	4,265 ft (1,300 m)
Landing Run	2,790 ft (850 m)
Armament	ASW torpedoes, mines, and depth charges
Radar	Wet Eye surface search

DESIGNATOR:	**Be-12***
	M-12
NAME:	**Mail**
	Tchaika*

DESCRIPTION: The Soviet Beriev Be-12 is one of only three remaining amphibian aircraft in active military service; the others being the new Chinese Harbin PS-5 and the Japanese Shin Meiwa PS/US-1. (The PS-1 will be phased out of service by March 1990.) The Be-12 is used by Soviet naval aviators for maritime patrol, anti-submarine warfare (ASW), and search-and-rescue (SAR) missions. The aircraft is characterized by a twin tail design with two turboprop engines mounted above a gull wing. The boat-hull fuselage has a single step. There are fixed wing-tip floats and retractable landing gear. A thimble radome extends from the nose and a magnetic anomaly detector (MAD) boom extends from the tail. In the

Mail

esoteric category of turboprop amphibian, the Be-12 holds all 21 possible international performance records for speed, altitude and weight carrying.

STATUS: Initial operational capability in 1965. First flight approx 1960. Approx 100 built between 1965 and 1972. No longer in production. Approx 95 remain in service with Soviet Naval Aviation.

USER: Soviet Navy.

CHARACTERISTICS:

Manufacturer:	Beriev
Crew:	6-10 (5 flight crew, 1-5 mission crew)
Engines:	2 Ivchenko AI-20D turboprops
Max Power	4,190 shaft hp each
Weights:	
Empty	47,850 lbs (21,700 kg)
Max Takeoff	64,790 lbs (29,450 kg)
Dimensions:	
Wing Span	97 ft 5½ in (29.71 m)
Length	98 ft 11½ in (30.17 m)
Height	22 ft 11½ in (7.00 m) on undercarriage
Wing Area	1,030 sq ft (95.69 sq m)
Performance:	
Cruise Speed	173 kts (200 mph; 320 km/h)
Max Speed	329 kts (380 mph; 610 km/h)
Ceiling	37,000 ft (11,280 m)
Range	2,158 nm (2,485 mi; 4,000 km)
Armamemt	ASW torpedoes on 4 underwing pylons; depth bombs; mines and internal weapons bay
Radar	Surface search radar in nose radome

CARGO/TRANSPORT AIRCRAFT

DESIGNATOR:	An-12*
NAME:	Cub
	Y-8 (Chinese-built designation)

DESCRIPTION: This is a very widely used Soviet cargo and paratroop aircraft, similar in appearance, payload, and role to the U.S. C-130 Hercules. It is flown by the Soviet Navy and possibly several other countries in the electronics intelligence (Elint) roles and in the cargo/transport role by Aeroflot and 9 other nations. At least one Navy Cub has been configured as a testbed for aircraft anti-submarine warfare (ASW) systems. The Cub is a four-engine turboprop aircraft with a high, straight wing, glazed nose, chin radome, large tail fin, and large rear cargo hatch. The Cub lacks an integral cargo loading ramp, instead using an upward-folding, longitudinally hinged cargo door. When the cargo door is folded open, the entire complement of paratroops may be dropped in less than one minute. Military models often have a twin 23-mm cannon in a tail gun position. (The gun position is seen on some civil models).

Cub

USERS: Soviet Union Air Forces and Navy, Algeria, Angola, Bulgaria, China, Czechoslovakia, Ethiopia, India, Iraq, Madagascar, Poland, Vietnam and Yugoslavia.

CHARACTERISTICS:

Manufacturer:	Antonov (Chinese models built by Hanzhong)
Crew:	6 (including tail gunner) + 100 troops or 60 paratroops

Engines:	4 Ivchenko AI-20K turboprop with AV-28 4-blade reversible pitch propellers
Max Power	4,000 shaft hp each
Weights:	
Empty	61,600 lb (28,000 kg)
Max Payload	44,090 lb (20,015 kg)
Max Takeoff	134,200 lb (61,000 kg)
Dimensions:	
Wing Span	124 ft 8 in (38.00 m)
Length	108 ft 7 in (33.10 m)
Height	34 ft 6½ in (10.53 m)
Wing Area	1,310 sq ft (121.70 sq m)
Hatch	25 ft 3 in (7.70 m) x 9 ft 8 in (2.95 m)
Cargo Hold	44 ft 3½ in (13.50 m) x 11 ft 5¾ in (3.50 m) x 8 ft 6¼ in (2.60 m)
Performance:	
Cruise Speed	313 kts (360 mph; 580 km/h)
Max Cruise	361 kts (415 mph; 670 km/h)
Ceiling	33,450 ft (10,200 m)
Range	3,075 nm (3,540 mi; 5,700 km) max fuel 1,942 nm (2,230 mi; 3,600 km) 22,000 lb load
Takeoff Run	2,300 ft (700 m)
Landing Run	1,640 ft (500 m)
Armament	2 x 23-mm NR-23 cannon in tail turret
Radar	Toad Stool navigation

DESIGNATOR: Il-76*
NAME: Candid

DESCRIPTION: Medium/long-range cargo/transport aircraft, produced in both military and civil air versions. Similar in appearance and role to U.S. C-141 Starlifter. Military missions include troop transport, freight, aerial refueling, and Airborne Early Warning (AEW). Cargo versions also capable of dropping paratroops. All versions are capable of extreme cold weather operations.

USERS: Soviet Union Air Forces, India, Iraq, Libya and Syria.

Candid

CHARACTERISTICS:

Manufacturer:	Ilyushin
Crew:	5 + 2 load handlers (military cargo) or 10 AEW operators or 140 troops or 125 paratroops (transport)
Engines:	4 Soloviev D-30KP turbofan
Max Power	26,455 lb (12,000 kg) static thrust
Weights:	
Max Payload	105,820 lb (48,000 kg)
Max Takeoff	418,875 lb (190,000 kg)
Dimensions:	
Wing Span	165 ft 8 in (50.50 m)
Length	152 ft 10 in (46.59 m)
Height	48 ft 5 in (14.76 m)
Wing Area	3,229 sq ft (300.0 sq m)
Hatch	11 ft 1¾ in (3.40 m) x 11 ft 4 in (3.45 m)
Cargo Hold	65 ft 7½ in (20.00 m) x 11 ft 1¾ in (3.4 m) x 11 ft 4¼ in (3.46 m)
Performance:	
Cruise Speed	432 kts (497 mph; 800 km/h)
Max Speed	459 kts (528 mph; 850 km/h)
Ceiling	50,850 ft (15,500 m)
Range	2,700 nm (3,100 mi; 5,000 km) max load
Takeoff Run	2,790 ft (850 m)
Landing Run	1,475 ft (450 m)

SHIPS

AIRCRAFT CARRIERS

NAME: LEONID BREZHNEV* CLASS

NUMBER IN CLASS: 2 NUCLEAR-POWERED AIRCRAFT CARRIERS

DESCRIPTION: These aircraft carriers are the largest warships built in the Soviet Union and have succeeded the *Kiev** class VSTOL carrier in series production. Completion of the *Leonid Brezhnev* can be expected before the end of the 1986-1990 five-year plan. Western intelligence estimates that four of these ships are planned although the number will be directly affected by budgetary considerations. At least the first two ships should be completed. A complement of about 60-75 fixed-wing aircraft is estimated. Aircraft for the *Brezhnev* air wing are still under development, and it is assumed that existing types will be adopted. They are expected to have both air-to-air and ground support capabilities. The ships are expected to be fitted eventually with catapults and arresting wires for the operation of conventional fixed-wing aircraft, although the ultimate flight deck configuration has yet to be confirmed. The lead ship apparently has a STOL ski-ramp forward. Initially they may operate a mixture of VSTOL and STOL aircraft plus helicopters. The ships have an angled flight deck and deck-edge and centerline elevators. It is anticipated that the ships will have a combined nuclear and steam (Conas) propulsion plant, probably similar to that of the missile cruiser *Kirov**.

CHARACTERISTICS:

Displacement	Approx 65,000 tons
Length	Approx 1,000 ft (305.0 m)
Propulsion	4 steam turbines; 4 shafts

COMBAT SYSTEMS:

Aircraft	60 to 75
Missiles	Vertical SA-N-9 launchers
Guns	30-mm close-in (multibarrel)
Radar	Multi-function fixed-array

DESIGNATOR: TAKR*—TAKTICHESKOYE AVIONOSNYY KREYSER (TACTICAL AIRCRAFT-CARRYING CRUISER)

NAME: KIEV CLASS

NUMBER IN CLASS: 4

DESCRIPTION: The *Kiev*-class aircraft carriers have a large island structure on the starboard side and an angled flight deck. Two inboard elevators are provided for handling aircraft, and there are several smaller weapon elevators. No catapults or arresting wires are fitted. Unlike U.S. aircraft carriers, the Kievs have a full missile cruiser armament of anti-air, anti-ship, and anti-submarine weapons. Most of these are fitted forward, depriving the ship of significant forward flight deck area. Portions of the flight deck are covered with blast-resistant, refractory tile for vertical take-off and landing aircraft operations. The hull design features a large underwater bow-mounted sonar; boat stowage is cut into the after hull; and the stern counter has an opening for variable-depth sonar (VDS). The *Baku* appears to have a larger aircraft hangar or aircraft ordnance stowage in place of the SS-N-12 missile reloads in the three other ships. (She has more SS-N-12 launchers to compensate for deletion of the reloads.)

CHARACTERISTICS:

Displacement:	36,000 tons standard
	43,000 tons full load
Dimensions:	
Length	818 ft 4 in (249.5 m) waterline
	895 ft 5 in (273.0 m) overall
Beam	107 ft 3 in (32.7 m)
Deck Width	154 ft 10 in (47.2 m)
Draft	32 ft 10 in (10.0)

Leonid Brezhnev

Kiev

		1 Top Steer
		1 Top Knot
Sonars		Low frequency hull mounted, bistatic edium frequency variable depth
EW Systems		Bell Clout
		Rum Tub
		Side Globe (Kiev and Minsk only)
		Top Hat-A/B

Propulsion:	4 steam turbines; 200,000 shaft hp; 4 shafts
Boilers	8
Speed	32 kts
Range	4,000 nm at 31 kts
	13,500 nm at 18 kts
Manning	Approx 1,200 (including air group)

COMBAT SYSTEMS:

Aircraft	Approx 30
	12 or 13 Yak-38* Forger VSTOL
	14 to 17 Ka-25* Hormone or
	Ka-27* Helix Helicopters
Catapults	None
Elevators	2 in flight deck
Missiles	2 twin SA-N-3 Goblet launchers (72)
	2 twin SA-N-4 launchers (40)
	except the Novorossiysk and Baku
	12 vertical SA-N-9 launchers (96) in place of SA-N-4
	8 SS-N-12 Sandbox surface-to-surface missile tubes (16) except 12 tubes in Baku (no reloads)
Guns	4 76.2-mm/59-caliber AA (2 twin)
	8 30-mm/65-caliber close-in (8 multi-barrel)
ASW Weapons	1 twin SUW-N-1 missile launcher
	2 RBU-6000 rocket launchers
Torpedoes	10 21-in (533-mm) torpedo tubes (2 quintuple launchers)
Radars	Kiev and Minsk
	4 Bass Tilt fire control
	2 Don-2 navigation
	1 Don Kay navigation
	2 Head Lights fire control
	2 Owl Screech fire control
	2 Pop Group fire control
	1 Trap Door fire control
	1 Top Sail 3-D air search
	1 Top Steer 3-D air search
	1 Top Knot air control
	Novorossiysk and Baku
	4 Bass Tilt
	3 Palm Frond navigation
	2 Head Lights
	2 Owl Screech
	1 Trap Door
	1 Top Sail

DESIGNATOR:	**PKR*—PROTIVOLODOCHNYY KREYSER (ANTI-SUBMARINE CRUISER)**
NAME:	**MOSKVA* CLASS**

NUMBER IN CLASS: 2

DESCRIPTION: The Moskva-class carrier is a hybrid helicopter carrier, missile cruiser. The carrier is a missile cruiser forward, and aft of the superstructure is a clear, open flight deck. The superstructure is stepped forward to support missiles launchers and radars, and has a smooth afterface. A small hangar is located between the stack uptakes in the superstructure. Two elevators connect the flight deck to the hangar deck. The Moskva class introduced the SA-N-3 Goblet missile system as well as the Top Sail radar and Head Lights fire control system to Soviet warships. The Moskva was trials ship for the Yak-38* Forger VSTO aircraft, and was modified in the early 1970s for flight tests of the Forger VSTOL aircraft. She has since reverted to her original configuration. Two banks of five 21-in (533-mm) torpedoes originally installed were deleted in the mid-1970s.

Moskva

CHARACTERISTICS:

Displacement:	14,500 tons standard
	17,000-18,000 tons full load
Dimensions:	
Length	619 ft 11 in (189 m) overall
Beam	85 ft 3 in (26 m)

Draft	24 ft 11 in (7.6 m)
Flight Deck	Approx 282 ft x 112 ft (85.95 x 34.14)
Propulsion:	Steam turbines; 100,000 shaft hp; 2 shafts
Boilers	4
Range:	4,500 nm at 29 kts
	14,000 nm at 12 kts
Manning:	Approx 850 (including air group)

COMBAT SYSTEMS:

Helicopters	14 Hormone (Ka-25) helicopters
Catapults	None
Elevators	2 in flight deck
Missiles	2 twin SA-N-3 Goblet launchers (44)
Guns	4 57-mm/70-caliber AA (2 twin)
ASW Weapons	1 twin SUW-N-1 missile launcher
	2 RBU-6000 rocket launchers
Torpedoes	Removed
Radars	3 Don-2 navigation
	2 Head Lights fire control
	1 Head Net-C 3-D search
	2 Muff Cob fire control
	1 Top Sail 3-D search
Sonars	Low frequency hull mounted
	Medium frequency variable depth
EW Systems	Bell Series
	Side Globe
	Top Hat

GUIDED MISSILE CRUISERS

DESIGNATOR: RKR*—RAKETNYY KREYSER (MISSILE CRUISER)

NAME: KIROV* CLASS

NUMBER IN CLASS: 2 + 2 NUCLEAR-POWERED BATTLE CRUISERS

DESCRIPTION: The Kirovs are the largest warships, except for aircraft carriers, built by any nation since World War II; they are significantly larger than their U.S. contemporaries (CGN types). The only larger surface combatants in service with any Navy today are the U.S. battleships of the Iowa (BB 61) class. The *Frunze* has a single 130-mm gun mount and an enlarged deckhouse forward of the bridge and the after superstructure is enlarged, extending farther aft. The two small deckhouses adjacent to *Kirov*'s helicopter deck (each with two Gatling guns) have been removed in Frunze with the ship's Gatling guns moved farther forward, onto the main superstructure. The helicopters are lowered from the large landing deck to the hangar by an elevator just forward of the landing area. The elevator opening is covered by a two-section hatch opening outward. This allows a helicopter to be parked on the lift in the lowered position while the hangar is closed. There is a hull-mounted sonar and a large variable depth sonar fitted in the stern. Active fin stabilizers are

Kirov

provided. The *Kirov* class has an innovative combined nuclear and steam (Conas) power plant, with two reactors that are coupled with oil-fired boilers. The reactors generate an estimated 90,000 shaft horsepower (shp) for 24-25 knots. The boilers can provide an estimated 60,000 shp for a maximum speed of 32 knots. A 1981 U.S. Navy intelligence estimate noted that, "Even if the fuel for the superheater were exhausted, ships of this class would still be able to make an estimated 29 knots using only the nuclear plant." Thus, the ships have a virtually unlimited cruising range.

CHARACTERISTICS:

Displacement:	24,000 tons standard
	28,000 tons full load
Dimensions:	
Length	754 ft 5 in (230.0 m) waterline
	813 ft 5 in (248.0 m) overall
Beam	91 ft 10 in (28.0 m)
Draft	28 ft 11 in (8.8 m)
Propulsion:	Steam turbines; 150,000 shp; 2 shafts
Reactors	2 pressurized-water
Boilers	2
Speed:	32 knots
Range:	Virtually unlimited (See Description)
Manning:	Approx 800

COMBAT SYSTEMS:

Helicopters	3 Ka-25 Hormone or Ka-27 Helix
Missiles	2 twin SA-N-4 launchers [40]
	12 rotary SA-N-6 launchers [96]
	16 vertical SA-N-9 launchers [128] in *Frunze*
	20 SS-N-19 tubes
Guns	2 100-mm/70-cal DP (2 single) in *Kirov*
	2 130-mm/70-cal DP (1 twin) in *Frunze*
	8 30-mm/65-cal close-in (8 multibarrel)
ASW Weapons	2 SS-N-14 Silex launch tubes [8-12] in *Kirov*
	1 RBU-6000 rocket launcher
	2 RBU-1000 rocket launchers
	Torpedoes
Torpedoes	8 21-in (533-mm) torpedo tubes (2 quad)

Radars	4 Bass Tilt (fire control)
	2 Eye Bowl (fire control) in *Kirov*
	1 Kite Screech (fire control)
	3 Palm Frond (navigation)
	2 Pop Group (fire control)
	2 Top Dome (fire control)
	1 Top Pair (3-D air search)
	1 Top Steer (3-D air search)
	2 fire control systems for SA-N-9 in *Frunze*
Sonars	Low-frequency (bow mounted)
	Low-frequency (variable depth)
EW Systems	*Kirov*
	10 Bell-series
	8 Side Globe
	Frunze
	10 Bell-series
	2 Rum Tub

Slava

DESIGNATOR: RKR*—ROKETNYY KREYSER (MISSILE CRUISER)
NAME: SLAVA* CLASS

SHIPS IN CLASS: 3 + 1 GUIDED MISSILE CRUISERS

DESCRIPTION: Slavas are primarily anti-ship cruisers, armed with the same SS-N-12 anti-ship missiles that in are the *Kiev** class aircraft carriers and the modified *Echo II* class submarines. These ships were built simultaneously with the larger *Kirov** class nuclear-propelled battle cruisers.

CHARACTERISTICS:

Displacement:	10,000 tons standard
	12,500 tons full load
Dimensions:	
Length	610 ft 1 in (186.0 m) overall
Beam	66 ft 7 in (20.3 m)
Draft	26 ft 3 in (8.0 m)
Propulsion:	4 gas turbines; 120,000 shaft hp; 2 shafts
Speed	32 knots
Range:	2,000 nm at 30 knots
	8,800 nm at 15 knots
Manning:	Approx 720

COMBAT SYSTEMS:

Helicopters	1 Hormone-B (Ka-25)
Missiles	2 twin SA-N-4 launchers [40]
	8 rotary SA-N-6 launchers [64]
	16 SS-N-12 Sandbox tubes
Guns	2 130 mm/70-cal DP (1 twin)
	6 30-mm/65-cal close-in (6 multibarrel) ASW
Weapons	2 RBU-6000 rocket launchers
Torpedoes	10 21-in (533-mm) torpedo tubes (2 quints)
Radars	3 Bass Tilt (fire control)
	1 Front Door/Front Piece (fire control)
	1 Kite Screech (fire control)
	3 Palm Frond (navigation)
	2 Pop Group (fire control)
	1 Top Dome (fire control)
	1 Top Pair (3-D air search)
	1 Top Steer (3-D air search)
Sonars	Low-frequency (hull mounted)
	Medium-frequency (variable depth)
EW Systems	Bell-series
	4 Rum Tub
	8 Side Globe

DESIGNATOR: BPK*—BOL'SHOY PROTIVOLODOCHNYY KORABOL' (LARGE ANTI-SUBMARINE SHIP)
NAME: KARA CLASS

NUMBER IN CLASS: 7 GUIDED MISSILE CRUISERS

DESCRIPTION: The Karas are large, graceful ships, a refinement of the *Kresta II* design with major anti-air and anti-submarine capabilities. Significantly larger than the *Kresta II* class, they have a heavier complement of guns, are fitted with extensive command and control facilities, and are all gas-turbine propelled. Their large superstructure is dominated by the large, square-topped gas-turbine funnel. The helicopter hangar, just forward of the landing area, is partially recessed below the flight deck. To stow the helicopter the hangar's roof hatch and rear doors open and the helicopter is pushed in and then lowered by elevator to the hangar deck. The *Petropavlovsk* has a higher hangar structure with two Round House tactical air navigation (Tacan) antennas fitted abreast the hangar, in place of RBU-1000 launchers. The *Azov* has been fitted as the trials ship for the SA-N-6 vertical-launch anti-air warfare (AAW) system; a launcher has replaced the after SA-N-3 Goblet system and the Top Dome missile control radar has been fitted in place of the after Head Lights.

SOVIET WEAPON SYSTEMS AND ELECTRONICS
SHIPS

Kara

CHARACTERISTICS:

Displacement:	8,200 tons standard
	9,700 tons full load
Dimensions:	
Length	567 ft 5 in (173.0 m) overall
Beam	61 ft (18.6 m)
Draft	22 ft (6.7 m)
Propulsion:	4 gas turbines; 120,000 shaft hp; 2 shafts
Speed:	34 knots
Range:	3,000 nm at 32 knots
	8,000 nm at 15 knots
Manning:	Approx 525

COMBAT SYSTEMS:

Helicopters	1 Ka-25 Hormone-A
Missiles	2 twin SA-N-3 Goblet launchers [72] except 1 launcher in Azov [36 missiles]
	2 twin SA-N-4 launchers [40]
	Test installation of rotary SA-N-6 launchers in Azov
Guns	4 76.2-mm/59-cal AA (2 twin)
	4.30-mm/65-cal close-in (4 multibarrel)
ASW Weapons	8 SS-N-14 Silex (2 quad)
	2 RBU-6000 rocket launchers
	2 RBU-1000 rocket launchers; removed from Petropavlovsk
Torpedoes	10 21-in (533-mm) torpedo tubes
Radars	2 Bass Tilt (fire control)
	1 Don-2 or Palm Frond (navigation)
	2 Don-Kay (navigation)
	2 Head Lights (fire control) except 1 in Azov plus 1 Top Dome
	1 Head Net-C (3-D air search)
	2 Owl Screech (fire control)
	2 Pop Group (fire control)
	1 Top Sail (3-D air search)
Sonars	Low-frequency (bow mounted)
	Medium-frequency (variable depth)
EW Systems	2 Bell Clout
	2 Bell Slam
	2 Bell Tap except 4 Rum Tub in Kerch and Petropavlovsk

DESIGNATOR: **BPK*—BOL'SHOY PROTIVOLODOCHNYY KORABL' (LARGE ANTI-SUBMARINE SHIP)**

NAME: **KRESTA II CLASS**

NUMBER IN CLASS: 10 GUIDED MISSILE CRUISERS

Kresta II

DESCRIPTION: These are large anti-submarine warfare (ASW)/anti-aircraft warfare (AAW) ships, similar to the *Kresta I* design but with improved surface-to-air missiles and electronics. The SS-N-14 Silex anti-submarine system is provided in place of the earlier ship's 4 SS-N-3 Shaddock anti-ship missiles, and improved helicopter facilities. With essentially the same hull and arrangement as the interim *Kresta I* design, these are much more capable warships. The most prominent feature is the large Top Sail radar antenna surmounting the superstructure pyramid and the Head Lights fire control radars for the SA-N-3 Goblet missile systems. The SA-N-3 has an anti-ship capability, as does the SS-N-14. The helicopter is hangared in the same manner as in the contemporary *Kara* class, with an elevator/hangar arrangement. Unlike the *Kynda*, however, no variable-depth sonar is fitted under the helicopter deck. The last 3 ships have an enlarged superstructure with a two-level deckhouse between the mast tower and funnel. The ships are fitted with fin stabilizers. The first 3 ships do not have the Bass Tilt fire control directors for the 30-mm Gatling guns. Those ships rely only on optical gun directors for those weapons.

CHARACTERISTICS:

Displacement:	6,200 tons standard
	7,700 tons full load
Dimensions:	
Length	521 ft 6 in (159.0 m) overall
Beam	55 ft 9 in (17.0 m)
Draft	19 ft 8 in (6.0 m)
Propulsion:	2 steam turbines; 100,000 shaft hp; 2 shafts
Boilers	4
Speed:	34 knots
Range:	2,400 nm at 32 knots
	10,500 nm at 14 knots
Manning:	Approx 380

COMBAT SYSTEMS:

Helicopters	1 Hormone-A (Ka-25)
Missiles	2 twin SA-N-3 Goblet launchers (72)
Guns	4 57-mm/70-cal AA (2 twin)
	4 30-mm/65-cal close-in (4 multibarrel)
ASW Weapons	8 SS-N-14 Silex (2 quad)
	2 RBU-6000 rocket launchers
	2 RBU-1000 rocket launchers
	Torpedoes
Torpedoes	10 21-in (533-mm) torpedo tubes (2 quints)
Radars	2 Bass Tilt (fire control) in *Admiral Makarov* and later ships
	1 Don-2 (navigation)
	2 Don-Kay (navigation)
	2 Head Lights (fire control)
	1 Head Net-C (3-D air search)
	2 Muff Cob (fire control)
	1 Top Sail (3-D air search)
Sonars	Medium frequency (bow mounted)
EW Systems	1 Bell Clout
	2 Bell Slam
	2 Bell Tap
	8 Side Globe

DESIGNATOR:	**RKR*—ROKETNYY KREYSER (MISSILE CRUISER)**
NAME:	**KRESTA I CLASS**

NUMBER IN CLASS: 4 GUIDED MISSILE CRUISERS

Kresta I

DESCRIPTION: These ships are an interim design, carrying the SS-N-3b Shaddock anti-ship missiles but apparently intended for other weapon systems (see *Kresta II* entry). Considerably larger than the previous *Kynda* series, these cruisers have 2 surface-to-air missile launchers and are the first Soviet surface combatants with a helicopter hangar. No mine rails are fitted, as in the previous post-war cruiser and destroyer classes. Com-pared with the smaller *Kynda* class rocket cruisers (RKR), the *Kresta I* design has only one-half the number of Shaddock launch tubes and one-fourth the total number of missiles. The Kresta missile tubes are mounted under cantilever bridge wings; they cannot be trained but are elevated to about 18 deg for firing. The *Vitse Admiral Drozd* was modified 1973-1975 with a two-deck structure installed between the bridge and radar pyramid with 4 30-mm close-in weapons (Gatling guns) being fitted along with the associated Bass Tilt fire-control radars. The *Sevastopol* received a similar deckhouse in 1980 but the Gatling guns and radars have not been observed.

CHARACTERISTICS:

Displacement:	6,200 tons standard
	7,600 tons full load
Dimensions:	
Length	510 ft (155.5 m) overall
Beam	55 ft 9 in (17.0 m)
Draft	19 ft 8 in (6.0 m)
Propulsion:	Steam turbines; 100,000 shp; 2 shafts
Boilers	4
Speed:	34 kts
Range:	1,600 nm at 34 kts
	7,000 nm at 14 kts
Manning:	Approx 380

COMBAT SYSTEMS:

Helicopters	1 Hormone-B (Ka-25)
Missiles	2 twin SA-N-1 Goa launchers (44)
	4 SS-N-3b Shaddock (2 twin)
Guns	4 57-mm/70-cal AA (2 twin)
	4 30-mm/65-cal close-in (4 multi-barrel) in *Vitse Admiral Drozd*
ASW Weapons	2 RBU-6000 rocket launchers
	2 RBU-1000 rocket launchers
Torpedoes	10 21-in (533-mm) torpedo tubes (2 quints)
Radars	2 Bass Tilt (fire control) in *Vitse Admiral Drozd*
	1 Big Net (air search)
	1 or 2 Don-2 (navigation) except none in *Admiral Zozulya*
	1 Don-Kay (navigation) in ships with 1 Don-2 and *Admiral Zozulya*
	1 Head Net-C (3-D air search)
	2 Muff Cob (fire control)
	2 Palm Frond (navigation) in *Admiral Zozulya*
	2 Peel Group (fire control)
	2 Plinth Net (surface search)
	1 Scoop Pair (fire control)
Sonars	Herkules medium frequency (hull mounted)
EW Systems	1 Bell Clout
	2 Bell-Series
	2 Bell Slam
	2 Bell Tap
	8 Side Globe

DESIGNATOR: RKR*—RAKETNYY KREYSER (MISSILE CRUISER)

NAME: KYNDA CLASS

NUMBER IN CLASS: 4 GUIDED MISSILE CRUISERS

DESCRIPTION: These were among the first of the modern Soviet warships resulting from the defense decisions made in the mid-1950s after the death of Stalin. The *Kynda* class cruisers are only slightly longer than the *Krupnyy* and *Kildin* destroyer classes, but have lines more akin to a conventional cruiser hull and significantly more firepower. These ships introduced the imposing pyramid structures to Soviet ships to support radar and EW antennas. They are the only ships with pyramids and twin funnels. No helicopter hangar or maintenance facilities are provided. Mine rails are installed, and this is the last class of Soviet cruiser-type ships to have them. *Groznyy* was fitted with Gatling guns and Bass Tilt radars in 1980 and the *Varyag* similarly modified in 1981. These ships were built with two Head Net-A radars; some ships subsequently refitted with Head Net-C. The Plinth Net was added after completion.

Kynda

CHARACTERISTICS:

Displacement:	4,400 tons standard
	5,500 tons full load
Dimensions:	
Length	464 ft 9 in (141.7 m) overall
Beam	51 ft 10 in (15.8 m)
Draft	17 ft 5 in 15.3 m)
Propulsion:	Steam turbines; 100,000 shaft hp; 2 shafts
Boilers	4
Speed:	34 knots
Range:	2,000 nm at 32 knots
	7,000 nm at 14 knots
Manning:	Approx 375

COMBAT SYSTEMS:

Helicopters	Landing area only
Missiles	1 twin SA-N-1 Goa launcher [24]
	8 SS-N-3b Shaddock tubes [8 + 8 reloads]
Guns	4 76.2-mm/59-cal AA (2 twin)
	4 30-mm/65-cal close-in (4 multi-barrel) in *Groznyy* and *Varyag*
ASW Weapons	2 RBU-6000 rocket launchers
	Torpedoes
Torpedoes	6 21-in (533-mm) torpedo tubes (2 triple)
Radars	2 Bass Tilt (fire control) in *Groznyy* and *Varyag*
	2 Don-2 (navigation)
	2 Head Net-A in *Admiral Golovko*; 1 in *Admiral Fokin*
	2 Head Net-C in *Groznyy* and *Varyag*; 1 in *Admiral Fokin*
	1 Owl Screech (fire control)
	1 Peel Group (fire control)
	2 Plinth Net (surface search) in *Groznyy* and *Admiral Fokin*
	2 Scoop Pair (fire control)
Sonar	Herkules high-frequency (hull mounted)
EW Systems	1 Bell Clout
	1 Bell Slam
	1 Bell Tap
	2 Guard Dog
	4 Top Hat

CRUISERS

DESIGNATOR: KR*—KREYSER (CRUISER)

NAME: SVERDLOV* CLASS

NUMBER IN CLASS: 7 + 2 LIGHT CRUISERS

Sverdlov

DESCRIPTION: These are the last conventional, all-gun cruisers in commission with any navy. The only ships in commission with larger guns are the U.S. Navy battleships of the *Iowa* (BB 61) class, each with nine 16-inch (400-mm) guns. The Sverdlovs are large, graceful-looking ships, with classic World War II era lines. They reflect an Italian design influence with a prominent, free-standing

conning tower forward, separate funnels, and two tripod masts immediately forward of the funnels. The long forecastle extends to the after gun turrets. Tracks for between 140 and 200 mines are fitted on the after deck. A stern anchor is fitted in addition to the bow anchors. The radar arrangements on these ships have been changed during their long service lives. Three ships have had 30-mm Gatling guns fitted on an enlarged superstructure in the area of their forward mast and funnel with the associated Drum Tilt radars. As built, these ships had ten 21-in (533-mm) torpedo tubes, mounted in two banks on the main deck, outboard of the motor launch stowage. They were removed from all ships by the early 1960s. Although sonar is fitted, no ASW weapons were ever provided except for the (now-deleted) torpedo tubes.

CHARACTERISTICS:

Displacement:	12,900 tons standard
	17,211 tons full load
Dimensions:	
Length	656 ft (200.0 m) waterline
	688 ft 10 in (210.0 m) overall
Beam	70 ft 10 in (21.6 m)
Draft	23 ft 7 in (7.2 m)
Propulsion:	Steam turbines; 110,000 shaft hp; 2 shafts
Boilers	6
Speed:	32 knots
Range:	2,400 nm at 32 knots
	10,000 nm at 13.5 knots
Manning:	Approx 1,000

COMBAT SYSTEMS:

Helicopters	No facilities
Guns	12 152-mm/57-cal SP (4 triple)
	12 100-mm/50-cal DP (6 twin)
	32 37-mm/63-cal AA (16 twin) except 28 guns in ships with 30-mm guns
	16 30-mm/65-cal close-in (8 twin) in *Admiral U.S. Hakov, Alexsandr Suvorov, Oktyabrskaya Revolutsiya*
ASW Weapons	None
Torpodoes	Removed
Mines	Rails for 140 to 200
Radars	1 Big Net or Top Trough (air search)
	1 Don-2 or Neptune (navigation)
	4 Drum Tilt (fire control) in ships with 30-mm guns
	8 Egg Cup(fire control); removed from ships with 30-mm guns
	1 High Sieve or Low Sieve (air search)
	1 Knife Rest (air search) in some ships
	1 Slim Net (air search)
	2 Sun Visor (fire control)
	2 Top Bow (fire control)
Sonars	Tamir-5N
EW Systems	2 Watch Dog

COMMAND CRUISERS

DESIGNATOR:	**KU*—KORABL' UPRAVLENIY (COMMAND SHIP)**
NAME:	**CONVERTED SVERDLOV* CLASS**

NUMBER IN CLASS: 2 COMMAND CRUISERS

DESCRIPTION: These ships were standard *Sverdlov*-class light cruisers that were extensively converted for the command ship role in 1971-1972. Both ships have been fitted with extensive command and communications facilities. A third mast, aft of the second funnel, carries the distinctive Vee Cone high frequency communications antennas. A pair of Big Ball communications satellites were fitted on small deckhouses immediately aft of the second funnel in 1979-1981. The different configurations of these ships probably represent the requirements of the specific fleet commanders for their respective operating areas. Most obvious: the *Senyavin* has a large helicopter hangar aft (requiring removal of all after 152-mm guns). An after superstructure has been built, with the SA-N-4 Gecko missile system and 30-mm guns installed for close-in defense. The *Zhdanov*, without a hangar, does have a helicopter landing area aft and an aircraft control position at the after end of the superstructure. Mine rails and torpedo tubes have been removed from both ships. See entry on *Sverdlov* class for additional notes.

Sverdlov Conversion

CHARACTERISTICS:

Displacement:	12,900 tons standard
	17,200 tons full load
Dimensions:	
Length	656 ft (200.0 m) waterline
	688 ft 10 in (210.0 m) overall
Beam	70 ft 10 in (21.6 m)
Draft	23 ft 7 in (7.2 m)
Propulsion:	Steam turbines; 110,000 shaft hp; 2 shafts
Boilers	6

Speed:	32 knots
Range:	2,400 nm at 32 knots
	10,000 nm at 13.5 knots
Manning:	Approx 1,000

COMBAT SYSTEMS:

Helicopters	1 Ka-25* Hormone-C in *Admiral Senyavin*
Missiles	1 twin SA-N-4 Gecko launcher [20]

Guns

Admiral Senyavin	*Zhdanov*
6 152-mm/57-cal SP (2 triple)	9 152-mm/57-cal SP (3 triple)
12 100-mm/50-cal DP (6 twin)	12 100-mm/50-cal DP (6 twin)
32 37-mm/63-cal AA (16 twin)	16 37-mm/63-cal AA (8 twin)
16 30-mm/65-cal close-in (8 twin)	8 30-mm/65-cal close-in (4 twin)

ASW Weapons	None
Torpodoes	(Removed)
Mines	(Rails removed)
Radars	4 Drum Tilt in *Senyavin*, 2 in *Zhdanov*
	6 Egg Cup (fire control)
	1 Pop Group (fire control)
	2 Sun Visor (fire control)
	2 Top Bow (fire control)
	1 Top Trough (air search)
Sonar	Tamir-5N
EW Systems	None

GUIDED MISSILE DESTROYERS

DESIGNATOR:	**BPK*—BOL'SHOY PROTIVOLODOCHNYY KORABL' (LARGE ANTI-SUBMARINE SHIP)**
NAME:	**UDALOY* CLASS**

NUMBER IN CLASS: 8+ GUIDED MISSILE DESTROYERS

DESCRIPTION: These are large anti-submarine destroyers, similar in concept to the U.S. Navy *Spruance* (DD 963) class. The Soviet ships have a long, low superstructure with their quad SS-N-14 launchers under a cantilevered extension to the bridge structure, as in the *Kresta II* and *Kara* classes. The hangar has separate bays to accommodate two helicopters, which are partially lowered into them by elevators similar to the arrangement in the *Kara* and *Kresta II* classes. There is a control station between the hangar bays; a pair of Round House Tacan structures are provided. A variable-depth sonar is fitted in the stern counter, similar to that of the *Kirov** class cruisers. The early ships were completed with empty positions for the Cross Sword fire control directors (above the bridge and above the hangars). The *Admiral Zhakarov*

Udaloy

was the first ship to have the full SA-N-9/Cross Sword system installed.

CHARACTERISTICS:

Displacement:	6,200–6,700 tons standard
	8,200 tons full load
Dimensions:	
Length	492 ft (150.0 m) waterline
	531 ft 4 in (162.0 m) overall
Beam	63 ft 4 in (19.3 m)
Draft	20 ft 3 in (6.2 m)
Propulsion:	4 gas turbines; 120,000 shaft hp; 2 shafts
Speed:	34 kts
Range:	2,000 nm at 33 kts
	5,000 nm at 20 kts
Manning:	Approx 300

COMBAT SYSTEMS:

Helicopters	2 Helix-A (Ka-27)
Missiles	8 vertical SA-N-9 launchers [64]
Guns	2 100-mm/70-cal DP (2 single)
	4 30-mm/65-cal close-in (4 multibarrel)
ASW Weapons	8 SS-N-14 Silex (2 quad)
	2 RBU-6000 rocket launchers
	Torpedoes
Torpodoes	8 21-in (533-mm) torpedo tubes (2 quad)
Mines	Rails fitted
Radars	2 Bass Tilt (fire control)
	2 Cross Sword (fire control)
	2 Eye Bowl (fire control)
	1 Kite Screech (fire control)
	3 Palm Frond (navigation)
	2 Strut Pair (surface search) in *Udaloy* and *Admiral Kulakov*; 1 in later ships
	1 Top Plate/Top Mesh (air search) in *Marshal Vasil'yevskiy* and later ships
Sonars	Low-frequency (bow mounted)
	Low-frequency (variable depth)
EW Systems	2 Bell Shroud
	2 Bell Squat
	(space provided for additional systems)

SOVIET WEAPON SYSTEMS AND ELECTRONICS
SHIPS

DESIGNATOR: EM*—ESKADRENNYY MINONOSETS (DESTROYER)
NAME: SOVREMENNYY* CLASS

NUMBER IN CLASS: 7+ GUIDED MISSILE DESTROYERS

Sovremennyy

DESCRIPTION: This destroyer class is intended primarily for anti-ship operations. The ships have a major anti-air capability but minimal anti-submarine warfare (ASW) weapons and sensors are provided. With the *Udaloy* class cruisers, the *Sovremenyy* class has succeeded the *Kresta II* cruisers in series production at the Zhdanov yard. This class is similar in size to the *Udaloy*, but with different hull form, propulsion, weapons, and sensors. The basic hull form and propulsion are similar to the *Kresta II*, and both classes are built at the same shipyard. The quad surface-to-surface missile launchers are mounted slightly forward of the bridge structure; the main gun armament is divided fore and aft (the *Udaloy* has guns forward); pressure-fired steam propulsion is provided (vice the *Udaloy's* gas turbines); and there is a telescoping helicopter hangar adjacent to the landing area to accommodate a single Hormone-B (Ka-25*) helicopter for over-the-horizon targeting. This is the first Soviet surface combatant with the helicopter deck amidships instead of at the stern, and the first to use a telescoping hangar. Two small, spherical radomes are located on platforms on both sides of the single stack. They may be associated with over-the-horizon targeting for the SS-N-22 missiles. The use of steam propulsion in the *Sovremennyy* was somewhat surprising in view of the use of gas turbines in the previous *Kara* and the contemporary *Udaloy* classes. (The estimated ranges of the two destroyers classes is approximately the same.) The *Stoikii* is thought to be the first of its class to be fitted with the Top Plaid 3-D search radar.

CHARACTERISTICS:

Displacement:	6,300 tons standard
	7,900 tons full load

Dimensions:		
Length	475 ft 7 in (145.0 m) waterline	
	511 ft 8 in (156.0 m) overall	
Beam	55 ft 9 in (17.0 m)	
Draft	20 ft (6.1 m)	
Propulsion:	Steam turbines; 100,000 shaft hp; 2 shafts	
Boilers	4	
Speed:	34 kts	
Range:	2,400 nm at 32 kts	
	10,500 nm at 14 kts	
Manning:	Approx 380	

COMBAT SYSTEMS:

Helicopters	1 Ka-25 Hormone-B
Missiles	2 SA-N-7 launchers [40]
	8 SS-N-22 (2 quad)
Guns	4 130-mm/70-cal DP guns (2 twin)
	4 30-mm/65-cal close-in (4 multibarrel)
ASW Weapons	2 RBU-1000 rocket launchers Torpedoes
Torpodoes	4 21-in (533-mm) torpedo tubes (2 twin)
Mines	Rails fitted
Radars	1 Band Stand (fire control)
	2 Bass Tilt (fire control)
	6 Front Dome (fire control)
	1 Kite Screech (fire control)
	3 Palm Frond (navigation)
	1 Top Steer (3-D air search) except *Osomotritel'nyy* has Top Steer/Top Plate combination
Sonars	Medium-frequency (bow mounted)
EW Systems	2 Bell Shroud
	2 Bell Squat
	4 unknown

DESIGNATOR: BPK*—BOL'SHOY PROTIVOLODOCHNYY KORABL' (LARGE ANTI-SUBMARINE SHIP)
NAME: KASHIN CLASS

NUMBER IN CLASS: 12 GUIDED MISSILE DESTROYERS (KASHIN CLASS)
6 GUIDED MISSILE DESTROYERS (MODIFIED KASHIN* CLASS)
1 GUIDED MISSILE TRIALS SHIP (CONVERTED KASHIN* CLASS)

DESCRIPTION: The Kashins were the world's first major warships with all-gas turbine propulsion. They are multi-purpose destroyers. These are large, graceful flush-deck destroyers, with a low superstructure topped by 4 large funnels for gas turbine exhaust, 2 radar-topped lattice masts, and 4 smaller radar towers. There is a helicopter landing area aft (with enclosed control station) but no

Kashin

Speed: 38 kts
Range: 1,500 nm at 35 kts
4,000 nm at 20 kts
Manning: Approx 280 unmodified
Approx 300 Modified and Converted

COMBAT SYSTEMS:

Helicopters	Landing area aft
Missiles	2 twin SA-N-1 Goa launcher [36]
	Unmodified and Modified only
	4 SS-N-2c Styx (4 single)
	Modified only
	1 single SA-N-7 launcher [20]
	Converted only
Guns	4 76.2-mm/59-cal AA (2 twin)
	4 30-mm/65-cal close-in (4 multi-barrel)
	Modified only
ASW Weapons	2 RBU-6000 rocket launchers
	2 RBU-1000 rocket launchers
	Unmodified and Converted only
	Torpedoes
Torpodoes	5 21-in (533-mm) torpedo tubes (1 quint)
Mines	Rails fitted in Unmodified ships
	Rails removed in Modified and Converted ships
Radars	Unmodified Kashin
	1 Big Net (3-D air search) except in Odarennyy and Soobrazitel'nyy
	2 or 3 Don-2 or 2 Don-Kay or 2 Palm Frond (navigation)
	1 Head Net-C (3-D air search) except 2 in Odarennyy and Soobrazitel'nyy
	2 Owl Screech (fire control)
	2 Peel Group (fire control)
	Modified Kashin
	2 Don-Kay (navigation)
	1 Big Net (3-D air search)
	1 Head Net-C (3-D air search)
	2 Brass Tilt (fire control)
	2 Owl Screech (fire control)
	2 Peel Group (fire control)
	Converted Kashin
	2 Don-Kay (navigation)
	1 Don-2 (navigation)
	1 Top Steer (air search)
	1 Head Net-C (3-D air search)
	2 Owl Screech (fire control)
	8 Front Dome (fire control)
Sonars	Medium-frequency (hull mounted)
	Medium-frequency (variable depth)
	Modified and Converted only
EW Systems	2 Watch Dog Unmodified only
	2 Bell Shroud Modified only
	2 Bell Squat Modified only
	deleted except for chaff launchers

hangar. Six ships have been extensively modified with improved electronics and installation of four rear-firing, SS-N-2c Styx missiles. The hull was lengthened by approximately 6 ft 7 in (2 m) with a stern variable depth sonar (VDS) installation being provided under the raised helicopter deck. Improved hull-mounted sonar was also fitted. Rapid-fire Gatling guns were provided in place of the two RBU-1000 rocket launchers that were previously mounted. Improved air-search radars were provided. The field of fire of the after 76.2-mm gun mount is severely restricted. In addition to the SS-N-2c missiles, the SA-N-1 Goa system has an anti-ship capability. The Provornyy was converted from a standard Kashin design to a test ship for the SA-N-7 missile system and the associated radars and fire control system. The two SA-N-1 missile systems were removed and an SA-N-7 launcher installed aft; there are spaces for two additional SA-N-7 launchers forward but they have not been installed. Like the cruiser Azov, the trials ship for the SA-N-6/Top Dome system, the Provornyy retains major combat capabilities.

USERS: Soviet Union Navy, and India.

CHARACTERISTICS:

Displacement:	3,750 tons standard
	3,950 tons standard Modified
	4,750 tons full load
	4,950 tons full load Modified
Dimensions:	
Length	472 ft 4 in (144.0 m) overall
	478 ft 11 in (146.0 m) overall Modified
Beam	51 ft 2 in (15.8 m)
Draft	15 ft 9 in (4.8 m)
Propulsion:	4 gas turbines; 96,000 shaft hp; 2 shafts

TOP

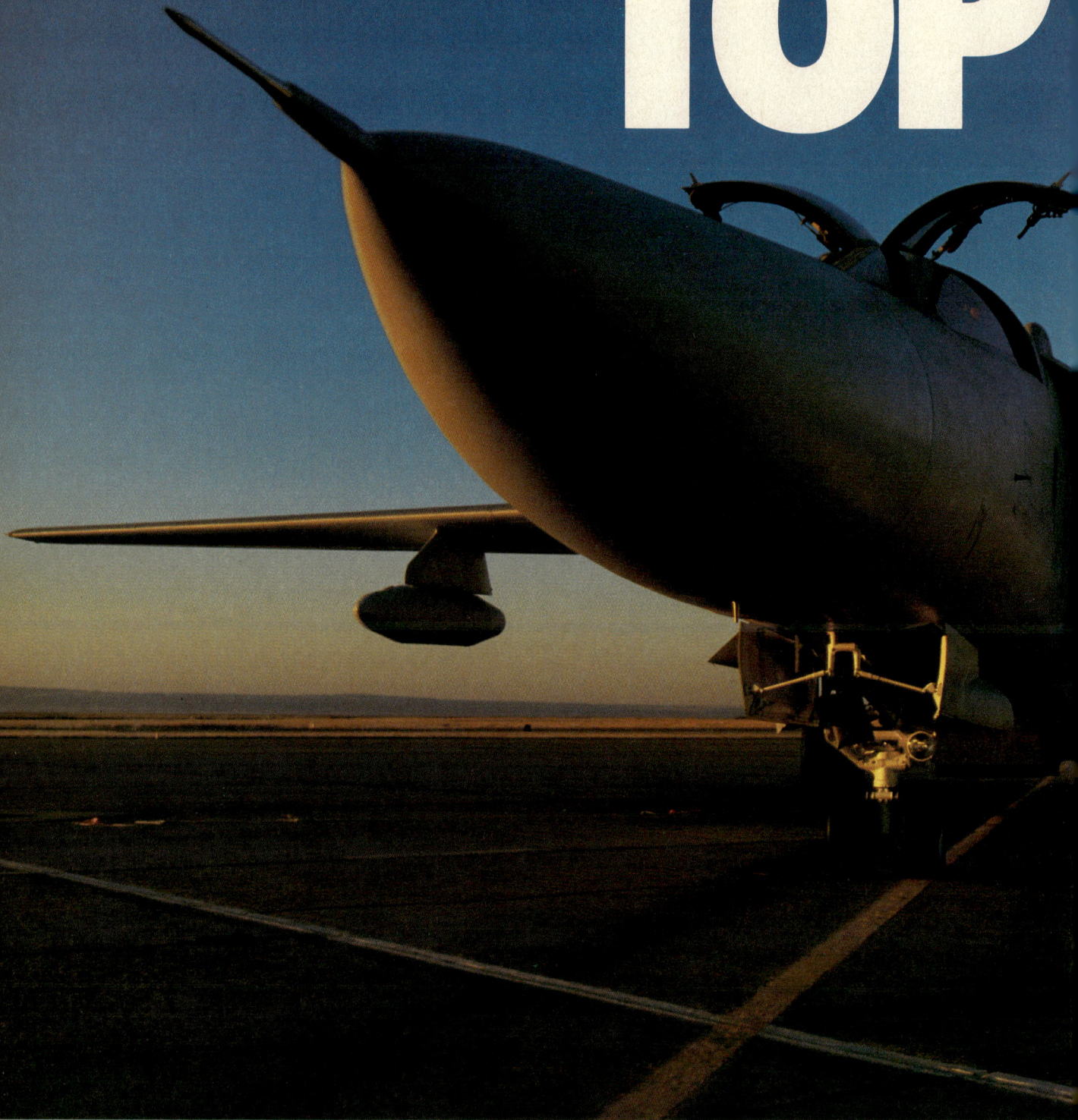

The EF-111A Operational Flight Trainer (OFT) provides invaluable experience for the EF-111A crew. It trains the EW officer for high-density hostile environments and helps develop a coordination between him and the pilot that cannot wait for combat conditions. Designed by AAI in a special way to meet special training needs, this simulator does its job with maximum cost efficiency.

The EF-111A simulator typifies AAI's philosophy in developing high-technology electronic and mechanical systems. Whatever the system, AAI

FLIGHT.

DESIGNATOR: BPK*—BOL'SHOY PROTIVOLODOCHNYY KORABL' (LARGE ANTI-SUBMARINE SHIP)

NAME: KANIN CLASS

NUMBER IN CLASS: 8 GUIDED MISSILE DESTROYERS

DESCRIPTION: These are Soviet anti-air warfare (AAW)/anti-submarine warfare (ASW) destroyers, converted from surface-to-surface missile ships. These ships were built as the *Krupnyy* class missile destroyers, with two SS-N-1 Scrubber launchers; they were modified during construction from *Kotlin* class destroyers. For their present configuration, the forward SS-N-1 launcher was replaced by a second quad 57-mm mount and an RBU-6000; the after SS-N-1 launcher and quad 57-mm gun mount were deleted and the after section of the ship reconfigured for the SA-N-1 Goa system and a larger helicopter deck. The superstructure was modified and additional electronics and twin 30-mm guns fitted, the latter in place of two amidships 57-mm mounts. Torpedo armament increased (from six tubes) and electronics upgraded. The conversions were completed between 1968 and 1977. These ships and the single unmodified *Kildin (Neuderzhimyy*)* have the lightest gun armament in terms of caliber of any destroyers operated by major navies.

Kanin

CHARACTERISTICS:

Displacement: 3,700 tons standard
4,750 tons full load

Dimensions:
Length 455 ft 11 in (139.0 m) overall
Beam 49 ft 2 in (15.0 m)
Draft 16 ft 5 in (5.0 m)
Propulsion: Steam turbines; 80,000 shaft hp; 2 shafts
Boilers 4
Speed: 34 knots

Range: 1,000 nm at 30 knots
4,500 nm at 18 knots
Manning: Approx 300

COMBAT SYSTEMS:

Helicopters	Landing deck
Missiles	1 twin SA-N-1 Goa launcher [16]
Guns	8 57-mm/70-cal AA (2 quad)
	8 30-mm/65-cal close-in (4 twin)
ASW Weapons	3 RBU-6000 rocket launchers
	Torpedoes
Torpodoes	10 21-in (533-mm) torpedo tubes (2 quints)
Radars	2 Don-Kay (navigation)
	2 Drum Tilt (fire control)
	1 Hawk Screech (fire control)
	1 Head Net-C (3-D air search)
	1 Peel Group (fire control)
Sonar	Medium-frequency (bow mounted)
EW Systems	2 Bell-series
	4 Top Hat

DESIGNATOR: BRK*—BOL'SHOY RAKETNYY KORABL' (LARGE MISSILE SHIP)

NAME: MODIFIED KILDIN CLASS

NUMBER IN CLASS: 4 GUIDED MISSILE DESTROYERS (MODIFIED KILDIN CLASS)

DESCRIPTION: The three modified *Kildin* class ships have had their single SS-N-1 Scrubber anti-ship missile launcher replaced by twin 76.2-mm gun mounts and 4 SS-N-2c Styx anti-ship missiles. Four ships were converted during construction from *Kotlin* class destroyers to the world's first missile destroyers. Subsequently, three ships were modified with removal of their SS-N-1 launcher, rejoining the fleet in 1973-1975. The fourth *Kildin*, the *Neuderzhimyy*, was not converted. The hull lines and basic arrangement are common to the *Kildin*, *Kanin*, *Krupnyy*, and *Kotlin* classes. The sole survivor of the four *Kildin*-class destroyers, the *Neuderzhimyy* is in essentially her original configuration with a single SS-N-1 missile launcher aft supplemented by 16 57-mm AA guns. She is the last Soviet warship to carry the SS-N-1. Same basic characteristics as the Modified *Kildin* class.

CHARACTERISTICS:

Displacement: 2,800 tons standard modified
2,600 tons standard unmodified
3,700 tons full load modified
3,500 tons full load unmodified

Dimensions:
Length 414 ft 11 in (126.5 m) overall
Beam 42 ft 8 in (13.0 m)
Draft 15 ft 5 in (4.7 m)
Propulsion: Steam turbines; 72,000 shaft hp; 2 shafts
Boilers: 4

Modified Kildin

Speed:	36 kts
Range:	1,000 nm at 32 kts
	3,600 nm at 18 kts
Manning:	Approx 300

COMBAT SYSTEMS:

Helicopters	No facilities
Missiles	4 SS-N-2c Styx launchers (4 single) (Modified Kildins only)
Guns	4 76.2-mm/59-cal AA (2 twin) (Modified Kildins only)
	16 57-mm/70-cal AA (4 quad) except *Bedovyy* which has 16 45-mm/85-cal AA (4 quad)
ASW Weapons	2 RBU-2500 rocket launchers Torpedoes
Torpodoes	4 21-in (533-mm) torpedo tubes (2 twin)
Radars	Modified *Kildin*
	1 Don-2 (navigation)
	2 Hawk Screech (fire control)
	1 Head Net-C (3-D air search) except Strut Pair in *Bedovyy*
	1 Owl Screech (fire control)
	Unmodified *Kildin*
	1 Knife Rest (surveillance)
	2 Hawk Screech (fire control)
	1 Flat Spin (air search)
	1 Top Bow (fire control)
	1 Slim Net (air and surface search)
Sonars	1 Herkules or Pegasus medium-frequency (hull mounted)
EW Systems	2 Watch Dog

DESIGNATOR:	**EM*—ESKADRENNYY MINONOSETS (DESTROYER)**
NAME:	**KOTLIN SAM CLASS**

NUMBER IN CLASS: 8 GUIDED MISSILE DESTROYERS

DESCRIPTION: These are anti-aircraft ships with a heavier gun batteries than the later *Kildin* class ships. Nine ships were converted from standard *Kotlin* class destroyers to this surface-to-air missile (SAM) configuration. The ninth,

the *Spravedlivyy*, was transferred to Poland in 1970. The *Kotlin*'s after twin 130-mm gun mount and half of the torpedo battery were removed. An SA-N-1 Goa missile system and associated radars were fitted as well as improved electronics and ASW weapons. Details of the ships differ. The *Bravyy* was completed as the trials ship for the SA-N-1. She initially had only four 45-mm AA guns and no torpedo tubes; the armament was increased in the mid-1960s.

Kotlin SAM

USERS: Soviet Union Navy, and Poland.

CHARACTERISTICS:

Displacement:	2,700 tons standard
	3,600 tons full load
Dimensions:	
Length	414 ft 11 in (126.5 m) overall
Beam	42 ft 8 in (13.0 m)
Draft	15 ft 1 in (4.6 m)
Propulsion:	Steam turbines; 72,000 shaft hp; 2 shafts
Boilers	4
Speed:	36 kts
Range:	1,000 nm at 34 kts
	3,600 nm at 18 kts

COMBAT SYSTEMS:

Helicopters	No facilities
Missiles	1 SA-N-1 Goa launcher [16]
Guns	2 130-mm/58-cal DP (1 twin)
	4 45-mm/85-cal AA (1 quad) except *Bravyy*
	12 45-mm (3 quad)
	8 30-mm/65-cal close-in (4 twin) in 4 ships #
ASW Weapons	2 RBU-6000 rocket launchers except *Bravyy*, *Skromnyy* 2 RBU-2500 Torpedoes
Torpodoes	5 21-in (533-mm) torpedo tubes (1 quint)
Mines	Rails removed
Radars	1 or 2 Don-2 (navigation)
	2 Drum Tilt in 4 ships #
	1 Egg Cup (fire control) except

deleted in *Nastochivyy*
1 Hawk Screech (fire control)
1 Head Net-C (3-D air search)
1 Peel Group (fire control)
1 Sun Visor (fire control)

Sonars	Herkules or Pegasus high-frequency (hull mounted)
EW Systems	2 Watch Dog
# Modified Ships	

DESTROYERS

DESIGNATOR:	**EM*—ESKADRENNYY MINONOSETS (DESTROYER)**
NAME:	**KOTLIN CLASS**

Kotlin

NUMBER IN CLASS: 17 DESTROYERS

DESCRIPTION: These are graceful, flush-deck destroyers. Eleven ships were modified and are considered as a sub-class. Other Kotlins have been partially and fully converted to missile configurations. These were the world's last destroyers to be built to classic World War II destroyer lines, mounting heavy dual purpose (DP) and light anti-aircraft (AA) gun batteries, with a large torpedo battery, high speed, and minimal anti-submarine warfare (ASW) weapons and sensors. These were the first of the Soviet Navy's high-speed surface combatants. Eleven ships were modified between 1960 and 1962, having their after bank of torpedo tubes replaced by a deckhouse and improved ASW weapons. The *Moskovskiy Komsomolets* was further modified, being fitted with 2 RBU-6000 launchers forward and, in 1978, a variable-depth sonar was provided. The *Svetlyy* has been fitted with a helicopter platform aft.

CHARACTERISTICS:

Displacement:	2,600 tons standard
	3,500 tons full load
Dimensions:	
Length	414 ft 11 in (126.5 m) overall
Beam	42 ft 8 in (13.0 m)
Draft	15 ft 1 in (4.6 m)
Propulsion:	Steam turbines; 72,000 shaft hp; 2 shafts
Boilers	4
Speed:	36 kts
Range:	1,000 nm at 34 kts
	3,600 nm at 18 kts
Manning:	Approx 335

COMBAT SYSTEMS:

Helicopters	No facilities except landing deck in Svetlyy
Guns	4 130-mm/58-cal Dual Purpose (2 twin)
	16 45-mm/85-cal AA (4 quad)
	4 25-mm/60-cal AA (2 twin) except 8 guns
	(4 twin) in modified ships
ASW weapons	Unmodified ships
	6 BMB-2 depth charge projectors
	2 depth charge racks
	Torpedoes
	Modified ships
	2 RBU-600 rocket launchers except 2 RBU-6000 launchers in Moskovskiy Komsomolets
	Torpedoes
Torpedoes	Unmodified ships
	6 BMB-2 depth charge projectors
	2 depth charge racks
	Torpedoes
	Modified ships
	5 21-inch (533-mm) torpedo tubes
Mines	Rails fitted for approx 70 mines
Radars	2 Don-2 or 1 Neptune (navigation)
	2 Egg Cup (fire control)
	2 Hawk Screech (fire control)
	1 Post Lamp or Top Bow (fire control)
	1 Sun Visor (fire control)
Sonars	Herkules high-frequency (hull mounted)

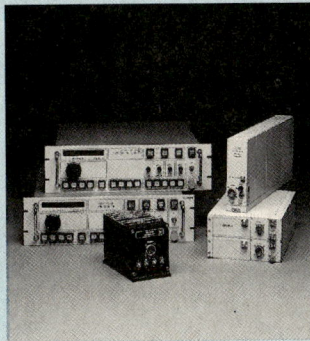

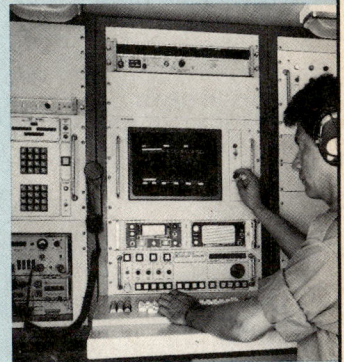

EW Systems	medium-frequency (variable depth) in *Moskovskiy Komsomolets* Watch Dog

DESIGNATOR: EM*—ESKADRENNYY MINONOSETS (DESTROYER)
NAME: SKORYY* CLASS

NUMBER IN CLASS: 20 DESTROYERS

Skoryy

DESCRIPTION: These were the first Soviet destroyers constructed after World War II. Most, if not all are laid up in reserve with some being periodically employed for training duties. This design was derived from the prewar *Ognevoyy** class, the first ship of that class having been laid down in 1939 but not completed until 1943. The *Skoryy* design incorporated some features from German ships, and had improved seakeeping, increased torpedo armament, and more anti-aircraft (AA) guns than the *Ognevoyy* class. In the early 1970s those ships with 8 37-mm weapons were provided with 25-mm AA guns. The main battery lacks sufficient elevation for dual-purpose use.

USERS: Soviet Union Navy, and Indonesia.

CHARACTERISTICS:

Displacement:	2,600 tons standard
	3,130 tons full load
Dimensions:	
Length	397 ft 6 in (121.2 m
Beam	39 ft 4 in (12.0 m)
Draft	14 ft 9 in (4.5 m)
Propulsion:	Steam turbines; 60
Boilers	4
Speed:	34 knots
Range:	850 nm at 30 knots
	3,000 nm at 18 knots
Manning:	Approx 220

COMBAT SYSTEMS:

Helicopters	No facilities
Guns	Unmodified ships
	4 130-mm/50-cal SP (2 twin)
	2 85-mm/-50 cal AA (1 twin)
	7 or 8 37-mm/63-cal AA
	(7 single or 4 twin)
	4 or 6 20-mm AA (2 or 3 twin)
	(see description)
	Modified ships
	4 130-mm/50-cal SP (2 twin)
	5 57-mm/70-cal AA (5 single)
ASW weapons	Unmodified ships
	2 depth charge projectors
	2 depth charge racks
	Torpedoes
	Modified ships
	2 RBU-2500 rocket launchers
	2 depth charge racks
	Torpedoes
Torpodoes	Unmodified ships
	10 21-in (533-mm) torpedo tubes (2 quints)
	Modified ships
	5 21-in (533-mm) torpedo tubes (1 quint)
Mines	Rails fitted for approx 50 mines
Radars	Unmodified ships
	1 Cross Bird (air search)
	1 or 2 Don-2 (navigation)
	1 Half Bow or Post Lamp or Top Bow (fire control)
	1 Slim Net (air search)
	1 High Sieve (air search)
	Modified ships
	1 or 2 Don-2
	2 Hawk Screech (fire control)
Sonars	1 Pegas High-frequency
EW systems	2 Watch Dog

FRIGATES/CORVETTES

DESIGNATOR: SKR*—STOROZHEVOY KORABL' (PATROL SHIP)
NAME: KRIVAK I/II/III CLASSES

NUMBER IN CLASS: 32 ANTI-SUBMARINE FRIGATES

DESCRIPTION: These are anti-submarine warfare (ASW) ships with a relatively short range and lacking the helicopter capability of their Western contemporaries. The *Krivak I/II* classes are similar, the principal difference being their gun caliber. The later *Krivak III* is built for KGB maritime border troops: and has less ASW capability but does embark a utility helicopter. The *Krivaks* represent the largest of surface combatants, frigate and larger,

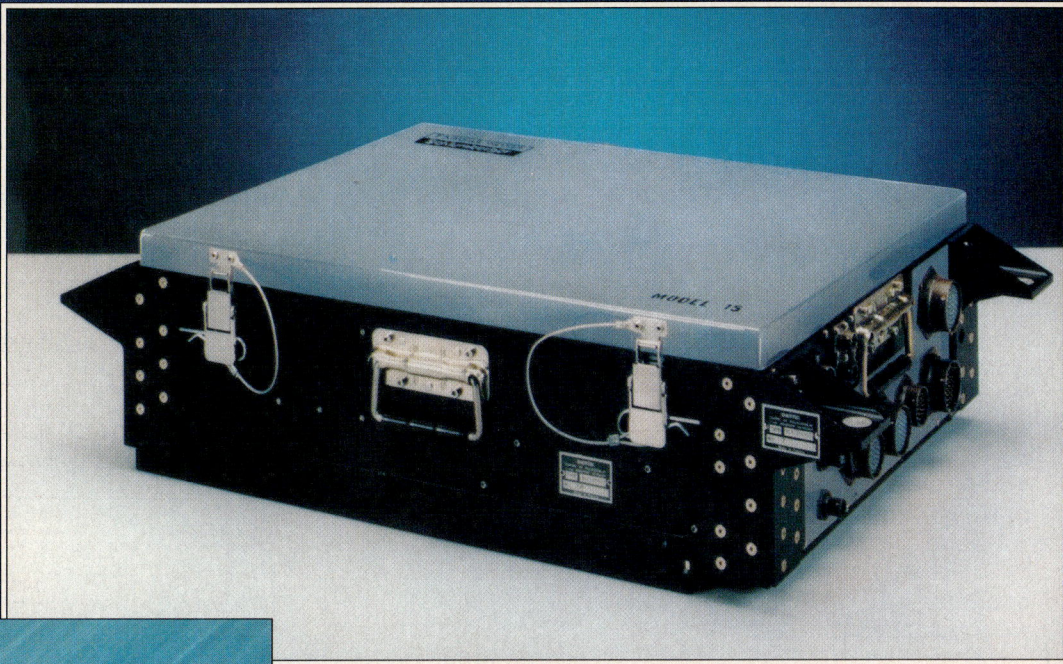

USH - 33
½ the Size
½ the Weight

- Integrated Logistic Support

- Nav Air EW Mission Recorder

- 77 lb., 15 inch reel, less than 2.25 cu. ft.

- Mil Std 1610, Mil–E – 5400, Mil–E – 16400

- Records FM to 500 kHz, Direct to 2 MHz, Serial HDDR to 8 MB/S, 1553 Bus Data, Parallel HDDR to 400 MB/S (special order).

- Available in 14, 28 and 64 channel configurations.

built in the Soviet Union since the Stalin programs of the 1950s. These are among the most heavily armed frigates afloat. However, by Western standards the lack of helicopter facilities on the *Krivak I/II* limits their ASW effectiveness. The *Krivak II* has larger caliber guns than the *Krivak I*, in addition to the improved Owl Screech gun radar/fire control system. Also, the sonar housing is larger in the later ships. The *Krivak III* is the largest class of combat ships operated by the KGB Maritime Border troops. It is also the smallest Soviet combatant ship to regularly embark a helicopter. The principal changes made on the *Krivak III* design for KGB service are a single-barrel 100-mm dual-purpose (DP) gun forward in place of the previous SS-N-14 ASW missile launcher and a raised helicopter platform aft in place of the second SA-N-4 launcher and after 76.2-mm or 100-mm gun mounts. Two 30-mm Gatling guns are mounted on top of the *Krivak III*'s helicopter hanger.

Krivak

CHARACTERISTICS:

Displacement:	*Krivak I* 3,700 tons full load
	Krivak II 3,800 tons full load
	Krivak III 3,900 tons full load
Dimensions:	
Length	405 ft 1 in (123.5 m) overall
Beam	46 ft 3 in (14.1 m)
Draft	15 ft 1 in (4.6 m)
Propulsion:	2 gas turbines 24,200 shaft hp (shp) + 2 boost gas turbines 24,400 shp = 48,600 shp; 2 shafts
Speed:	32 kts
Range:	700 nm at 30 kts
	3,900 nm at 20 kts
Manning:	Approx 200

COMBAT SYSTEMS:

Helicopters	*Krivak I/II* no facilities
	Krivak III 1 Helix (Ka-27)
Missiles	*Krivak I/II* 2 twin SA-N-4 launchers [40]
	Krivak III 3 twin SA-N-4 launchers [20]
Guns	*Krivak I* 4 76.2-mm/59-cal AA (2 twin)

	Krivak II 2 100-mm/70-cal DP (2 single)
	Krivak III 1 100-mm/70-cal DP 2 30-mm close-in (2 multibarrel)
ASW Weapons	*Krivak I/II* 4 SS-N-14 Silex (1 quad)
	2 RBU-6000 rocket launchers
	Torpedoes
Torpodoes	8 21-in (533-mm) torpedo tubes (2 quad)
Mines	Rails for 20 mines
Radars	*Krivak I*
	1 Don-2 or Spin Trough (nav)
	2 Eye Bowl (fire control)
	1 Head Net-C (3-D air search)
	1 Kite Screech(fire control)
	2 Pop Group (fire control)
	Krivak II
	1 Don-Kay or Palm Frond (nav)
	2 Eye Bowl
	1 Head Net-C
	1 Owl Screech (fire control)
	2 Pop Group
	Krivak III
	1 Brass Tilt (fire control)
	1 Don-2 (navigation)
	1 Don-Kay (navigation)
	1 Head Net-C (3-D air search)
	1 Kite Screech (fire control)
	1 Palm Frond (navigation)
	1 Pop Group (fire control)
Sonars	Medium-frequency (bow mounted)
	Medium-frequency (variable depth)
EW Systems	2 Bell Shroud
	2 Bell Squat

DESIGNATOR:	**SKR*—STOROZHEVOY KORABL' (PATROL SHIP)**
NAME:	**KONI CLASS**

NUMBER IN CLASS: 1 LIGHT ANTI-SUBMARINE FRIGATE

DESCRIPTION: The *Koni* is a coastal anti-submarine warfare (ASW) ship that was constructed specifically for foreign transfer. One unit is retained in Soviet service for demonstration and training purposes. The ASW sensors and weapons are both short range and intended for shallow depths (e.g., lack of variable-depth sonar and torpedo tubes which are unsuitable for shallow-water attacks), and the ship is limited in both speed and range. The design follows Soviet small combatant lines, with the *Koni* having similar lines to the smaller *Grisha*. (The *Grisha* has only a two-step bridge structure and no gun mount forward.) The ship has a "split" superstructure, with the space between possibly having been intended for tubes in some roles, as mounted in the *Grisha* classes. There are 2 depth charge racks aft as well as provisions for minelaying and minesweeping equipment. A hull sonar dome is fitted as are fin stabilizers. Deck house arrangements differ in later ships.

USERS: Soviet Union Navy, Algeria, Cuba, East Germany, Libya, and Yugoslavia.

Koni

CHARACTERISTICS:

Displacement:	1,900 tons full load
Dimensions:	
Length	311 ft 7 in (95.0 m) overall
Beam	42 ft (12.8 m)
Draft	13 ft 9 in (4.2 m)
Propulsion:	2 diesels; 15,000 hp +
	1 gas turbine;
	15,000 shp = 30,000 hp; 3 shafts
Speed:	27 kts
Range:	1,800 nm at 14 kts
Manning:	Approx 110

COMBAT SYSTEMS:

Helicopters	No facilities
Missiles	1 twin SA-N-4 launcher [20]
Guns	4 76.2-mm/59-cal AA (2 twin)
	4 30-mm/65-cal close-in (2 twin)
ASW Weapons	2 RBU-6000 rocket launchers
	2 depth charge racks
Torpodoes	None
Mines	Rails for 20 mines
Radars	1 Don-2 (navigation)
	1 Drum Tilt (fire control)
	1 Hawk Screech (fire control)
	1 Pop Group (fire control)
	1 Strut Curve (air search)

Sonars	Medium-frequency (hull mounted)
EW Systems	2 Watch Dog

DESIGNATOR:	**MRK*—MALYY RAKETNYY KORABL' (SMALL MISSILE SHIP)**
NAME:	**TARANTUL I/II/III CLASSES**

NUMBER IN CLASS: 13+ GUIDED MISSILE CORVETTES

DESCRIPTION: The *Tarantul* has the small hull and basic arrangement as the *Pauk* class anti-submarine corvettes, but with a different propulsion system. These missile corvettes are smaller and less capable than the earlier *Nanuchka* class guide missile corvettes, except that the Tarantuls are several knots faster. They are more likely considered as successors to the *Osa* class frigates, but it is unlikely that they can be built in sufficient numbers for even a partial replacement program. The *Tarantul II* differs from the *Tarantul I* only in electronic equipment. *Tarantul III* differs from *Tarantul II* only in that it has 4 more-capable SS-N-22 anti-ship missiles vice the SS-N-2c in the *Tarantul II*.

USERS: Soviet Union Navy, Germany, East and Poland.

Tarantul

CHARACTERISTICS:

Displacement:	480 tons standard
	540 tons full load
Dimensions:	
Length	172 ft 2 in (52.5 m) waterline
	185 ft 4 in (56.5 m) overall
Beam	34 ft 5 in (10.5 m)
Draft	8 ft 3 in (2.5 m)
Propulsion:	Codag 2 gas turbines;
	24 shaft hp + diesels
	Approx 12,000 hp; 2 shafts
Speed:	36 kts
Range:	2,000 nm at 20 kts
	400 nm at 36 kts
Manning:	Approx 40

COMBAT SYSTEMS:

Missiles	4 SS-N-2c Styx anti-ship (2 twin) in *Tarantul I/II*
	4 SS-N-22 anti-ship (2 twin) in *Tarantul III*
Guns	1 76.2-mm/59-cal AA
	2 30-mm/65-cal close-in (2 multibarrel)
Radars	1 Band Stand (fire control) in *Tarantul II/III*
	1 Bass Tilt (fire control)
	1 Kivach (surface search)
	1 Light Bulb (targeting) in *Tarantul II/III*
	1 Plank Shave (targeting)

DESIGNATOR:	**MRK*—MALYY RAKETNYY KORABL' (SMALL MISSILE SHIP)**
NAME:	**NANUCHKA I/III CLASSES**

NUMBER IN CLASS: 24+ GUIDED MISSILE CORVETTES

DESCRIPTION: These are heavily armed coastal missile ships. They provide more gun and anti-air defense capabilities than previous Soviet missile craft. *Nanuchka II* is an export version similar to the *Nanuchka I*, but with 4 SS-N-2c Styx missiles in place of the SS-N-9 Siren system. The foreign *Nanuchkas* have the Square Tie radar within the Band Stand radome. The *Nanuchka III* class has a larger gun battery than the previous units, plus a Gatling gun (also aft); the superstructure is enlarged, with the bridge higher.

USERS: Soviet Union Navy, Algeria, Libya and India.

Nanuchka

CHARACTERISTICS:

Displacement:	770 tons full load
Dimensions:	
Length	194 ft 6 in (59.3 m) overall
Beam	41 ft 4 in (12.6 m)
Draft	7 ft 10 in (2.4 m)
Propulsion:	3 diesels (M504); 30,000 hp; 3 shafts
Speed:	32 kts
Range:	2,500 nm at 12 kts
	900 nm at 30 kts
Manning:	Approx 60

COMBAT SYSTEMS:

Missiles	6 SS-N-9 Siren anti-ship (2 triple)
	1 SA-N-4 Gecko anit-aircraft [20]
Guns	2 57-mm/80-cal AA (1 twin) in *Nanuchka I*
	1 76.2-mm/59-cal AA in *Nanuchka III*
	1 30-mm/65-cal close-in (multibarrel) in *Nanuchka III*
Radars	1 Band Stand (fire control)
	1 Bass Tilt (fire control) in *Nanuchka III*
	1 Muff Cob (fire control) in *Nanuchka I*
	1 Peel Pair (air search)
	1 Pop Group (fire control) in *Nanuchka I*

DESIGNATOR:	**MPK*—MALYY PROTIVOLODOCHNYY KORABL' (SMALL ANTI-SUBMARINE SHIP)**
NAME:	**PAUK CLASS**

NUMBER IN CLASS: 14+ ANTI-SUBMARINE CORVETTES

DESCRIPTION: This class has the same hull as the *Tarantul* missile Corvettes, but with anti-submarine warfare (ASW) weapons and sensors and all-diesel propulsion. A circular housing for a dipping sonar is fitted in the stern on the starboard side. Some later units have the pilot house one deck higher.

USERS: Soviet Union Navy and KGB.

CHARACTERISTICS:

Displacement:	480 tons standard
	580 tons full load
Dimensions:	
Length	191 ft 11 in (58.5 m) overall
Beam	30 ft 10 in (9.4 m)
Draft	8 ft 2 in (2.5 m)
Propulsion:	2 diesels (M504); 20,000 hp; 2 shafts
Speed:	28-32 kts
Manning:	Approx 40

COMBAT SYSTEMS:

Missiles	1 quad SA-N-5 Grail launcher [16]
Guns	1 76.2-mm/59-cal AA
	1 30-mm/65-cal close-in (multibarrel)

Pauk

ASW Weapons	2 RBU-1200 rocket launchers
	2 depth-charge racks
	Torpedoes
Torpodoes	4 16-in (406-mm) torpedo tubes
	(4 single)
Radars	1 Bass Tilt (fire control)
	1 Spin Trough (air search)
	1 Plank Shave (air and surface search)
Sonars	Medium frequency (hull mounted)
	Medium frequency (dipping)
EW Systems	2 Watch Dog

DESIGNATOR: MPK*—MALYY PROTIVOLODOCHNYY KORABL' (SMALL ANTI-SUBMARINE SHIP)

NAME: POTI CLASS

Approximately 60 anti-submarine corvettes were completed during 1961-1967

DESCRIPTION: The *Poti* is a coastal anti-submarine warfare (ASW) corvette, similar to the *Mirka* class light frigate from a technology viewpoint. The early ships were built with an open 57-mm/70-cal twin gun mount, 2 RBU-2500 rocket launchers, and 2 torpedo tubes. Most have been upgraded, with the rocket launchers replaced with RBU 6000s and most ships built with the remote (director)-controlled, automatic 57-mm gun mount. The fixed torpedo tubes are angled out some 15 deg from the centerline. The dipping sonar, installed beginning in the mid-1970s, is the same used in the Ka-25 Hormone-A helicopter.

USERS: Soviet Union Navy, Bulgaria and Romania.

CHARACTERISTICS:

Displacement:	500 tons standard
	580 tons full load
Dimensions:	
Length	194 ft 10 in (59.4 m) overall

Beam	25 ft 11 in (7.9 m)
Draft	6 ft 7 in (2.0 m)
Propulsion:	Codag 2 diesels (M503A)
	8,000 hp + 2 gas
	turbines 40,000 shaft hp; 2 shafts
Speed:	38 kts
Range:	500 nm at 37 kts
	4,500 nm at 10 kts
Manning:	Approx 40

COMBAT SYSTEMS:

Guns	2 57-mm/80-cal AA (1 twin)
ASW Weapons	2 RBU 6000 rocket launchers
Torpodoes	2 or 4 16-in (406-mm)
	torpedo tubes (2 or 4 single)
Radars	1 Don-2 (navigation)
	1 Muff Cob (fire control)
	1 Strut Curve (air search)
Sonars	Herkules high frequency
	(hull mounted)
	High frequency (dipping sonar)
EW Systems	2 Watch Dog

Poti

NAME: GRISHA I/II/III/IV/V CLASSES

NUMBER IN CLASS: 60+ LIGHT ANTI-SUBMARINE FRIGATES

DESCRIPTION: These are small anti-submarine warfare (ASW) frigates, the principal differences being the addition of a 30-mm Gatling gun in the *Grisha III* with the associated Bass Tilt radar. Seven similar ships of the *Grisha II* class were built for the KGB maritime border troops. These ships have a layout similar to the larger *Koni* class, but with only two "steps" forward on the superstructure. The *Grishas* also have a solid mast structure similar to the *Koni* in contrast to the lattice masts of the earlier *Mirkas* and *Petyas*. The *Grisha's* variable depth sonar cannot be used while the ship is underway. Thus, *Grishas* normally perform ASW searches in pairs, alternating in the sprint-and-drift (sonar search) modes. Racks for 12 depth charges can be fitted to the after end of the mine

CRAFTSMANSHIP AND
IT'S OUR *STOCK* ANSWER.

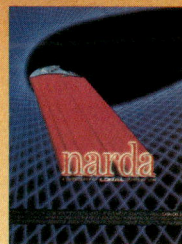

rails. The *Grisha II* was built for the KGB maritime border troops. It differs from the Navy ships in that it has a twin 57-mm gun mount instead of the SA-N-4 missile launcher and associated Pop Group radar/fire control system. Racks for 12 depth charges can be fitted to the after end of the mine rails. Series IV and V are under construction.

CHARACTERISTICS

Displacement:	950 tons standard
	1,200 tons full load
Dimensions:	
Length	234 ft 10 in (71.6 m) overall
Beam	32 ft 2 in (9.8 m)
Draft	12 ft 2 in (3.7 m)
Propulsion:	4 diesels (M503);
	16,000b hp + 1 gas turbine;
	15,000 shp = 31,000 hp;
	3 shafts
Speed:	30 kts
Range:	450 nm at 27 kts
	4,500 nm at 10 kts
Manning:	Approx 60

Grisha

COMBAT SYSTEMS

Helicopters	No facilities
Missiles	1 twin SA-N-4 launcher [20] except in *Grisha II*
Guns	
Grisha I	2 57-mm/80-cal zA (1 twin)
Grisha II	4 57-mm/80 cal AA (2 twin)
Grisha III	2 57-mm/80-cal AA (1 twin)
	1 30-mm/65-cal close-in (multibarrel)
Grisha IV	1 76.2-mm/59-cal AA
	1 30-mm/65-cal close-in (multibarrel)
ASW weapons	2 RBU-6000 rocket launchers except 1 in *Grisha IV*
	2 depth charge racks (see description) torpedoes
Torpedoes	4 21-in (533-mm) torpedo tubes (2 twin)
Mines	Rails for 18 mines
Radars	
Grisha I	1 Don-2 (navigation)
	1 Muff Cob (fire control)
Grisha II	1 Pop Group (fire control)
	1 Strut Curve (air search)
	1 Don-2 (navigation)
	1 Muff Cob (fire control)
Grisha III	1 Strut Curve (air search)
	1 Don-2 (navigation)
	1 Bass Tilt (fire control)
	1 Pop Group (fire control)
Grisha IV	1 Strut Curve (air search)
	1 Don-2 (navigation)
	1 Bass Tilt (fire control)
	1 Pop Group (fire control)
	1 Strut Pair (air search)
Sonars	Medium-frequency (hull mounted)
	High-frequency (dipping)
	except *Grisha II*
EW systems	2 Watch Dog

NAME: **MIRKA I/II CLASSES**

NUMBER IN CLASS: 18 LIGHT ANTI-SUBMARINE FRIGATES

Mirka

DESCRIPTION: The small number of these light frigates were built simultaneously with the larger *Petya* program, being "sandwiched" between the *Petya I* and *II* series at the Kaliningrad yard. Nine ships are of the *Mirka II* configuration with a dipping sonar similar to that in the Hormone-A helicopter installed in a new stern structure. These ships are generally similar to the *Petya* series, but a different propulsion arrangement is used. These ships have 2 shafts, each driven by a gas turbine and diesel compared to the triple-shaft *Petya*. In these ships the turbines were placed all the way aft, preventing the fitting of mine rails or variable depth sonars. The propellers are mounted in tunnels, with the gas turbines powering compressors that inject air into the tunnels to provide a pump-jet action.

CHARACTERISTICS:

Displacement:	950 tons standard
	1,150 tons full load

Dimensions:

Length	270 ft 3 in (82.4 m) overall
Beam	30 ft 2 in (9.2 m)
Draft	9 ft 6 in (2.9 m)
Propulsion:	2 gas turbines 30,000 shaft hp + 2 diesels 12,000 bhp = 42,000 hp; 2 shafts
Speed:	34 kts
Range:	500 nm at 30 kts 4,800 nm at 10 kts
Manning:	Approx 90

COMBAT SYSTEMS:

Helicopters	No facilities
Guns	4 76.2-mm/59-cal AA (2 twin)
ASW weapons	4 RBU-6000 rocket launchers in *Mirka I*
	2 RBU-6000 rocket launchers in *Mirka II*
	1 depth charge rack in *Mirka I* Torpedoes
Torpedoes	5 16-in (400-mm) torpedo tubes (1 quint) in *Mirka I*
	10 16-in (400-mm) torpedo tubes (2 quints) in *Mirka II*
Mines	No rails fitted
Radars	1 Don-2 (navigation)
	1 Hawk Screech (fire control)
	1 Slim Net (air search) except Strut Curve in later *Mirka II* units
Sonars	1 Herkules or Pegas high-frequency (hull mounted)
	1 High-frequency (dipping) in *Mirka II*
EW systems	2 Watch Dog

NAME: PETYA I/II CLASSES

NUMBER IN CLASS: 45 LIGHT ANTI-SUBMARINE FRIGATES

DESCRIPTION: The *Petya* is a light frigate that was produced in larger numbers than the similar *Mirka* class, indicating a more successful design. The *Petyas* have been built and modified into several different configurations. These ships have a large hull-mounted sonar, with a dome that projects almost four feet beneath the keel; the drag created by this structure reduces the ships' potential speed by approximately six knots. The dipping sonar is similar to that fitted in the Ka-25* Hormone-A helicopters. These were the first large Soviet warships to have gas turbine propulsion. The diesel drives the center-line propeller. Several *Petya I*s have been employed in experimental roles. In about 1966, one was fitted with the SUW-N-1 anti-submarine warfare (ASW) rocket launcher in place of the forward 76.2-mm gun mount; in 1967 one ship tested the variable depth sonar for the

Moskva-class helicopter ships; and another ship was modified by 1969 to test towed sonar arrays (a large deckhouse was installed aft). Beginning in 1973, 11 of the *Petya I* type were modified with a large poop deck aft to carry variable depth sonar (VDS). The two after RBU-2500 launchers as well as minelaying capabilities were deleted, as was one depth charge rack. One *Petya II* was modified with a deckhouse aft, the purpose of which is not clear; that modification retains the mine rails. The 11 "modified" *Petya I* ships have a raised stern deck house with a large variable depth sonar. Some ships with this arrangement have their after 76.2-mm twin mount removed as well as their mine rails.

USERS: Soviet Union Navy, Ethiopia, India, Syria and Vietnam.

Petya

CHARACTERISTICS:

Displacement:	950 tons standard 1,150 tons full load
Dimensions:	
Length	268 ft 4 in (81.8 m) overall
Beam	30 ft 2 in (9.2 m)
Draft	9 ft 2 in (2.8 m)
Propulsion:	1 diesel 6,000 bhp + 2 gas turbines 30,000 Shaft hp = 36,000 hp; 3 shafts
Speed:	30 kts
Range:	450 nm at 29 kts 1,800 nm at 16 kts
Manning:	Approx 90

COMBAT SYSTEMS:

Helicopters	No facilities
Guns	4 76.2-mm/59-cal AA (2 twin); reduced to 2 guns in a few modified ships
ASW weapons	4 RBU-2500 rocket launchers in *Petya I*
	2 RBU-6000 rocket launchers in *Petya II*
	2 depth charge racks; 1 in modified *Petya I* and removed from modified *Petya II* Torpedoes
Torpedoes	5 16-in (400-mm) torpedo tubes (1 quint) in *Petya I*

		Propulsion:	Steam turbines; 20,000 shp; 2 shafts
	10 16-in (400-mm) torpedo tubes (2 quints) in *Petya II*	Boilers	2
Mines	Rails for 22 mines; removed from modified *Petya I*	Speed:	30 kts
Radars	1 Don-2 (navigation)	Range:	550 nm at 28 kts
	1 Hawk Screech (fire control)		2,000 nm at 13 kts
	1 Slim Net (air search) in some *Petya I*	Manning:	Approx 170

Radars (continued):
- 1 Strut Curve (air search) in *Petya II* and some *Petya I*

Sonars:
- Herkules high-frequency (hull mounted)
- High-frequency (dipping) in modified *Petya I*
- Medium-frequency (variable depth) in 11 modified *Petya I*

EW systems: 2 Watch Dog

COMBAT SYSTEMS:

Helicopters	No facilities
Guns	3 100-mm/56-cal DP (3 single)
	4 37-mm/63-cal AA (2 twin)
	4 25-mm/60-cal AA (2 twin) in most active ships
ASW weapons	2 RBU-2500 rocket launchers; deleted in 1 ship (with Bell series system)
	2 depth charge racks
	Torpedoes
Torpedoes	2 or 3 21-in (533-mm) torpedo tubes (1 twin or triple)
Mines	Rails for 28 mines
Radars	1 Don-2 or Neptune (navigation)
	1 Hawk Screech (fire control) in 1 ship
	1 Slim Net (air search)
	1 Sun Visor (fire control)
Sonars	Herkules or Pegas high-frequency (hull mounted)
EW systems	2 Watch Dog
	2 Bell series in 1 ship

NAME: RIGA CLASS

NUMBER IN CLASS: 34 FRIGATES

Riga

DESCRIPTION: The *Rigas* are an improved and smaller development of the *Kola* class. One ship has been modified with Bell series electronic warfare (EW) system installed on a short mast fitted aft. Another ship has a Hawk Screech radar/director forward and the Sun Visor radar/director mounted aft. Most, but probably not all ships, have had an anti-submarine warfare (ASW) refit as listed below. Original ASW armament consisted of 1 MBU-600 hedgehog, 4 BMB-2 depth charge mortars, plus depth charge racks, mines, and torpedoes. The original gun armament consisted of the 100-mm and 37-mm guns.

USERS: Soviet Union Navy, Bulgaria and Finland.

CHARACTERISTICS:

Displacement:	1,260 tons standard
	1,480 tons full load
Dimensions:	
Length	298 ft 6 in (91.0 m) overall
Beam	33 ft 6 in (10.2 m)
Draft	10 ft 6 in (3.2 m)

DESIGNATOR: SKR*—STOROZHEVOY KORABL' (PATROL SHIP)

NAME: T-58*

NUMBER IN CLASS: 20+ ex-minesweepers (17 Patrol Corvettes and 3+ Radar Picket Ships)

DESCRIPTION: These ships are two of the many fleet minesweeper classes of steel construction that are used by the Soviet Navy and KGB for patrol duties, and in modified form, by the Navy for radar picket and auxiliary roles, as well as for keeping watch on Western warships while serving in the AGI role. The patrol ships were reclassified as such in 1978, their minesweeping gear removed. In naval service they are called SKR* type (patrol ship) and as used by the KGB Maritime Border Troops they are called PSKR* for Pogranichnyy Storozhevoy Korabl' (border patrol ship). Three of the above patrol ships were further modified to carry the Big Net radar and serve as radar picket ships. They were fitted with a new deck structure aft, and have two twin 30-mm rapid-fire guns installed aft. In this role the Soviet classification KVN* for Korabl' Vozduchnogo Nablyudeniya (radar surveillance ship) is used.

CHARACTERISTICS:

Displacement:	725 tons standard—patrol ship
	760 tons standard—radar picket
	860 tons full load—patrol ship
	880 tons full load—radar picket
Dimensions:	
Length	229 ft 7 in (70.0 m)
Beam	29 ft 10 in (9.1 m)
Draft	8 ft 2 in (2.5 m)
Propulsion:	2 diesels; 4,000 bhp;
	2 shafts
Speed:	18 kts—patrol ship
	17 kts—radar picket
Range:	2,500 mi at 13.5 kts
Manning:	Approx 60—patrol ship
	Approx 100—radar picket

T-58

COMBAT SYSTEMS:

Missiles	None in patrol configuration
	2 quad SA-N-5 Grail launchers [16]
	for radar picket
Guns	4 57-mm/70-cal AA
	(2 twin)—patrol ship
	4 57-mm/70-cal AA
	(1 twin)—radar picket
	4 30-mm/65-cal AA
	(2 twin)—radar picket
ASW weapons	2 RBU-1200 rocket launchers—patrol
	ship 2 depth-charge racks
Mines	Rails for 18 mines—patrol ship
	None-radar picket
Radars	1 Muff Cob (fire control)
	1 Spin Trough (air search)
	1 Don-2 (navigation)—patrol ship
	1 Big Net (air search)—radar picket
	1 Strut Curve (air search)—
	radar picket
Sonars	High frequency (hull-mounted)
EW systems	2 Watch Dog—patrol ship

GUIDED MISSILE PATROL BOATS

DESIGNATOR:	TKA*—TORPEDNYY KATER
	(TORPEDO CUTTER)
NAME:	MATKA CLASS

NUMBER IN CLASS: 16 HYDROFOIL GUIDED MISSILE BOATS

Matka

DESCRIPTION: The *Matka* is a missile-armed version of the *Turya* class hydrofoil torpedo boat, with a larger superstructure to accommodate a more complex missile system and different gun arrangement. Carries two SS-N-2c Styx anti-ship missiles. This class is derived from the *Osa* missile craft, with the same hull and propulsion plant, but with hydrofoils having been fitted forward. At high speeds the craft's stern planes on the water's surface.

CHARACTERISTICS:

Displacement:	225 tons standard
	260 tons full load
Dimensions:	
Length	131 ft 2 in (40.0 m) overall
Beam	24 ft 11 in (7.6 m) hull
	39 ft 4 in (12.0 m) over foils
Draft	6 ft 11 in (2.1 m) hullborne
	10 ft 6 in (3.2 m) foilborne
Propulsion:	3 diesels (M504);
	15,000 hp; 3 shafts
Speed:	40 knots foilborne
Range:	400 nm at 36 knots
	650 nm at 25 knots
Manning:	Approx 30

COMBAT SYSTEMS:

Missiles	2 SS-N-2c Styx anti-ship
	(2 single)

WHEN IT COMES TO MAKING SURE OUR EW SYSTEMS ARE 100%, WE LEAN OVER BACKWARD.

The upside-down plane is a Mirage. The unlikely-looking contraption it's attached to isn't.

It's an advanced test facility for putting our broad line of Thomson-CSF airborne ESM and ECM systems through their paces.

We also bend over backward to develop those systems.

Radar Warning Receiver antenna

We do this by using CAD/CAM techniques and by investing heavily in the brainpower and manhours it takes to prepare the very sophisticated software required to mastermind the systems.

Then we test our ESMS and ECMS as thoroughly as if our own lives depended upon them.

Jamming pod

But first we configure the various sub-systems in a mockup of the plane and put them through exhaustive tests.

Following this we run a more advanced simulation with the plane's other avionics.

When we're satisfied up goes the tower and we lean over every which way to make sure no matter which way up you are, our systems stay in business.

Talking of business, we're one of the few companies that make all their own components which keeps us independent.

And we make both pod-mounted and integrated systems. Our new integrated electronic warfare systems warns, jams, decoys – and processes and manages those various functions.

Elint aircraft

But before it does anything, it's tested. Every which way.

AEROSPACE GROUP

DIRECTION DES VENTES EXPORT MILITAIRES

Tour Vendôme
204, Rond-point du Pont de Sèvres
92516 Boulogne-Billancourt Cedex - FRANCE
Tel.: (1) 46.08.95.79 - Telex: TCSF 204 780 F

THE WILLPOWER

POLARIS AVS/A/86 GE AERO 60

◆ **THOMSON-CSF**

THE BRAINPOWER.   THE WILLPOWER.   THE WINPOWER.

Guns	1 76.2-mm/59-cal AA
	1 30-mm/65-cal close-in (multibarrel)
Radars	1 Bass Tilt (fire control)
	1 Cheese Cake (search)
	1 Plank Shave (targeting)

NAME: **SARANCHA CLASS**

NUMBER IN CLASS: 1 HYDROFOIL GUIDED MISSILE BOAT

DESCRIPTION: This guided missile boat has a stepped hydrofoil hull bottom, with 2 propellers fitted to each of 2 pods mounted on the after foils.

CHARACTERISTICS:

Displacement:	320 tons full load
Dimensions:	
Length	148 ft 3 in (45.2 m) hull
	175 ft 10 in (53.6 m) over foils
	166 ft (50.6 m) foilborne
Beam	36 ft 1 in (11.0 m) hull
	102 ft 8 in (31.3 m) over foils
Draft	8 ft 6 in (2.6 m) hullborne
	23 ft 11 in (7.3 m) foilborne
Propulsion:	2 gas turbines;
	30,000 shaft hp; 4 shafts
Speed:	60 kts

COMBAT SYSTEMS:

Missiles	4 SS-N-9 Siren anti-ship
	(2 twin)
	1 twin SA-N-4 launcher [20]
Guns	1 30-mm/65-cal close-in (multibarrel)
Radars	1 Band Stand (fire control)
	1 Bass Tilt (fire control)
	1 Pop Group (fire control)

DESIGNATOR: **RKA*—ROKETNYY KATER**
                 (MISSILE CUTTER)
NAME: **OSA II CLASS**

NUMBER IN CLASS: Approximately 30 GUIDED MISSILE BOATS

DESCRIPTION: The *Osa II* is an improved version of the basic *Osa I* anti-ship design, capable of carrying the more-capable SS-N-2c Styx missile. These craft can be distinguished from the *Osa I* type by their circular missile tubes with rib-like rings. At least one unit has been observed with a deckhouse between the bridge and Drum Tilt radar. See *Osa I* entry for additional notes.

USERS: Soviet Union Navy, Algeria, Angola, Bulgaria, Cuba, Ethiopia, Finland, India, Iraq, Libya, North Yemen, Somalia, South Yemen, Syria and Vietnam.

Osa I

CHARACTERISTICS:

Displacement:	215 tons standard
	245 tons full load
Dimensions:	
Length	126 ft 7 in (38.6 m) overall
Beam	25 ft (7.6 m)
Draft	6 ft 7 in (2.0 m)
Propulsion:	3 diesels (M504); 15,000 hp;
	3 shafts
Speed:	35 knots
Range:	500 nm at 34 knots
	750 nm at 25 knots
Manning:	Approx 30

COMBAT SYSTEMS:

Missiles	4 SS-N-2b/c Styx anti-ship
	(4 single)
	1 SA-N-5 launcher (hand held)
	in some units
Guns	4 30-mm/65-cal close-in (2 twin)
Radars	1 Drum Tilt (fire control)
	1 Square Tie (targeting)

DESIGNATOR: **RKA*—RAKETNYY KATER**
                 (MISSILE CUTTER)
NAME: **OSA I CLASS**

NUMBER IN CLASS: Approximately 60 GUIDED MISSILE BOATS

DESCRIPTION: These are steel-hulled missile craft, developed to succeed the wood-hulled *Komar* class that carried two SS-N-2 Styx missiles. This class introduced the twin 30-mm rapid-fire, remote-control gun mountings now common to several Soviet ship classes. The *Osa* has an all-welded-steel hull with a superstructure of fabricated steel and aluminum alloy. A "citadel" control station is provided for operation in a nuclear, biological, and chemical (NBC) environment. Its four Styx missile launchers are twice the number in the preceding

Komar class. The launchers are fixed, with the two after launchers, elevated at about 15 deg, firing over the forward launchers, elevated at about 12 deg. The mounting structure of the Drum Tilt gunfire control radar varies. The *Matka, Mol, Stenka,* and *Turya* classes have the basic *Osa* hull and propulsion plant, the Mol being a torpedo boat developed for export.

USERS: Soviet Union Navy, Algeria, Bulgaria, China, Cuba, East Germany, Egypt, India, Iraq, North Korea, Poland, Romania, Syria and Yugoslavia.

CHARACTERISTICS:

Displacement:	185 tons standard
	215 tons full load
Dimensions:	
Length	126 ft 7 in (38.6 m) overall
Beam	25 ft (7.6 m)
Draft	5 ft 11 in (1.8 m)
Propulsion:	3 diesels (M503A*); 12,000 hp;
	3 shafts
Speed:	36 knots
Range:	500 nm at 34 knots
	750 nm at 25 knots
Manning:	Approx 30

COMBAT SYSTEMS:

Missiles	4 SS-N-2a/b Styx anti-ship
	(4 single)
	1 SA-N-5 launcher (hand held)
	in some units
Guns	4 30-mm/65-cal close-in (2 twin)
Radars	1 Drum Tilt (fire control)
	1 Square Tie (targeting)

SMALL COMBATANTS

NAME: **BABOCHKA CLASS**

NUMBER IN CLASS: 1 HYDROFOIL PATROL BOAT

DESCRIPTION: This is a prototype-evaluation anti-submarine warfare (ASW) patrol craft, with fixed, fully submerged foils forward and aft. The torpedo tubes are stacked two above two, in two quad launchers on the bow, angled out to both sides.

CHARACTERISTICS:

Displacement:	400 tons full load
Dimensions:	
Length	164 ft (50.0 m) overall
Beam	27 ft 9 in (8.5 m) hull
	42 ft 8 in (13.0 m) over foils
Propulsion:	2 diesels + 3 gas turbines
	for Approx 30,000 shaft hp;
	3 shafts
Speed:	45+ knots

COMBAT SYSTEMS:

Guns	2 30-mm/65-cal close-in
	(2 multibarrel)
ASW Weapons	Torpedoes
Torpedoes	8 16-in (406-mm)
	torpedo tubes (2 quad)
Radars	1 Bass Tilt (fire control)
	1 Don-2 (navigation)
	1 Peel Cone (search)

NAME: **SLEPEN CLASS**

NUMBER IN CLASS: 1 PATROL GUNBOAT

DESCRIPTION: This is the Soviet Navy's only high-speed gunboat. The craft may be employed as a trials ship for small combatant systems. Similar to the *Matka* design, but without hydrofoils or missiles. As built originally, the ship had a twin 57-mm gun mount forward; replaced in 1975 by the single 76.2-mm gun.

CHARACTERISTICS:

Displacement:	205 tons standard
	230 tons full load
Dimensions:	
Length	126 ft 7 in (38.6 m) overall
Beam	24 ft 11 in (7.6 m)
Draft	6 ft 3 in (1.9 m)
Propulsion:	3 diesels (M504); 15,000 hp;
	3 shafts
Speed:	36 kts
Range:	500 nm at 34 kts
	50 nm at 25 kts
Manning:	Approx 30

COMBAT SYSTEMS:

Guns	1 76.2-mm/59-cal AA
	1 30-mm/65-cal close-in (multibarrel)
Radars	1 Bass Tilt (fire control)
	1 Don-2 (navigation)

DESIGNATOR: **TK***
NAME: **TURYA CLASS**

NUMBER IN CLASS: 31 HYDROFOIL TORPEDO BOATS

DESCRIPTION: These are high-speed coastal anti-submarine warfare (ASW) and torpedo attack craft using a modified *Osa II* hull and propulsion plant. The forward foils are fixed and at high speed, the craft's stern planes on the water. The stern is trimmed by an adjustable flap with twin supports protruding from the stern transom. Later units are reported to have semi-retractable foils to facilitate berthing. The Hormone-A (Ka-25) helicopter dipping sonar is fitted on the starboard quarter. Not provided in units transferred to Cuba.

Turya

CHARACTERISTICS:

Displacement:	215 tons standard
	250 tons full load
Dimensions:	
Length	127 ft 11 in (39.0 m) overall
Beam	24 ft 11 in (7.6 m)
	41 ft (12.5 m) over foils
Draft	6 ft 7 in (2.0 m) hullborne
	13 ft 2 in (4.0 m) over foils
Propulsion:	3 diesels (M504); 15,000 hp;
	3 shafts
Speed:	40 knots
Range:	420 nm at 38 knots
	650 nm at 25 knots
Manning:	Approx 25

COMBAT SYSTEMS:

Guns	2 57-mm/80-cal AA (1 twin)
	2 25-mm/60-cal AA (1 twin)
ASW	Weapons Torpedoes
Torpodoes	4 21-in (533-mm) torpedo tubes
	(4 single)
Radars	1 Muff Cob (fire control)
	1 Pot Drum (fire control)
Sonars	High-frequency (dipping sonar)

NAME: **YAZ CLASS**

NUMBER IN CLASS: 15+ RIVERINE MONITORS

DESCRIPTION: These are low-freeboard riverine monitors employed on the Amur River Flotilla in Siberia. From at least World War II, riverine gunboats built by the Soviets have mounted tank guns.

CHARACTERISTICS:

Displacement:	400 tons full load
Dimensions:	
Length	196 ft 11 in (60.0 m) overall
Propulsion:	Diesels; 2 shafts
Speed:	15 kts

COMBAT SYSTEMS:

Guns	2 100-mm or 120-mm tank guns
	(2 single)

Radars	2 30-mm/65-cal close-in (1 twin)
	1 Bass Tilt (fire control)
	1 Kivach or Spin Trough (navigation)

DESIGNATOR: **AKA*—ARTILLERIYSKIY KATER (ARTILLERY CUTTER)**
NAME: **SHMEL CLASS**

NUMBER IN CLASS: Approximately 80 RIVERINE GUNBOATS

DESCRIPTION: These are heavily armed river craft, similar in concept to the French and U.S. riverine monitors of the Indochina-Vietnam wars. The craft patrol the several Soviet rivers that border on foreign states. These are very shallow-draft craft. The 76.2-mm weapon is mounted in a PT-76* tank turret with one 7.62-mm machine gun coaxially mounted. The other 7.62-mm machine guns are hand held and fired through ports in the open-top deckhouse. The rocket launcher, mounted aft of the deckhouse, is deleted in some units.

CHARACTERISTICS:

Displacement:	60 tons full load
Dimensions:	
Length	92 ft 10 in (28.3 m) overall
Beam	15 ft 1 in (4.6 m)
Draft	3 ft (0.9 m)
Propulsion:	2 diesels (M50F-4); 2,400 hp;
	2 shafts
Speed:	22 knots
Range:	240 nm at 20 knots
	600 nm at 10 knots
Manning:	Approx 15

COMBAT SYSTEMS:

Missiles	1 18-tube 122-mm rocket launcher in some units
Guns	1 76.2-mm/48-cal single-purpose
	2 25-mm/60-cal AA machine guns (1 twin)
	5 7.62-mm machine guns (5 single; See description)
Mines	8 mines can be carried

DESIGNATOR: **SSV*—SUDNO SVYAZZY (COMMUNICATIONS VESSEL)**
NAME: **SSV-10**

NUMBER IN CLASS: 1 RIVERINE FLAGSHIP SSV-10

DESCRIPTION: This ship serves as flagship of the Danube riverine flotilla. Now designated SSV-10 (formerly PS-10).

CHARACTERISTICS:

Displacement:	360 tons full load

Dimension:
Length	160 ft 9 in (49.0 m) overall
Propulsion:	Diesels; 2 shafts
Speed:	12 kts

COMBAT SYSTEMS:
Weapons	None
Radars	Navigation
Sonars	None

AMPHIBIOUS SHIPS

DESIGNATOR:	**BDK*—BOL'SHOY DESANTNYY KORABL' (LARGE LANDING SHIP)**
NAME:	**IVAN ROGOV CLASS**

NUMBER IN CLASS: 2 HELICOPTER/DOCK LANDING SHIPS

Ivan Rogov

DESCRIPTION: These are the largest and most versatile amphibious ships yet constructed for the Soviet Navy. Each ship can embark a Naval Infantry battalion, including vehicles and equipment. They are also the only Soviet amphibious ships with a helicopter facility. These are multi-role amphibious ships, having bow vehicle ramps; a floodable docking well for landing craft and amphibious tractors; and a helicopter hangar with two landing decks forward and aft of the superstructure. The hangar can accommodate four Hormone (Ka-25*) helicopters; they can be moved through the superstructure and down a ramp to the forward landing area. The funnel uptakes are split to provide the helicopter pass through. The float-in docking well can hold three Lebed air-cushion landing craft. The ship has a flat bottom and large tank deck with bow doors to permit the unloading of amphibious vehicles into the water or across the beach. Ten light or medium tanks plus 30 armored personnel carriers can be transported.

CHARACTERISTICS:
Displacement:	11,000 tons standard
	13,000 tons full load

Dimensions:
Length	518 ft 3 in (158.0 m) overall
Beam	78 ft 9 in (24.0 m)
Draft	26 ft 11 in (8.2 m)
Propulsion:	2 gas turbines;
	50,000 shaft horsepower;
	2 shafts
Speed:	23 knots
Range:	8,000 nm at 20 knots
	12,500 nm at 14 knots
Manning:	Approx 200
Troops:	Approx 550

COMBAT SYSTEMS:
Helicopters	4 Ka-25 Hormone-C
Missiles/	1 twin SA-N-4 launcher [20]
Rockets	2 quad SA-N-5 Grail launchers in *Nikolayev*
	1 40-tube 122-mm barrage rocket launcher
Guns	2 76.2-mm/59-cal AA (1 twin)
	4 30-mm close-in (4 multibarrel)
Radars	2 Bass Tilt (fire control)
	2 Don-Kay (navigation) in *Rogov*
	1 Head Net-C (air search)
	1 Owl Screech (fire control)
	1 Palm Frond (navigation) in *Nikolayev*
	1 Pop Group (fire control)
EW Systems	2 Bell Shroud (3 in *Nikolayev*)
	2 Bell Squat

DESIGNATOR:	**BDK*—BOLS'SHOY DESANTNYY KORABL' (LARGE LANDING SHIP)**
NAME:	**ROPUCHA CLASS**

NUMBER IN CLASS: 19 TANK LANDING SHIPS

Ropucha

DESCRIPTION: These Polish-built ships are smaller than the previous, Soviet-built *Alligator* class. The *Ropucha* class has traditional LST (tank landing ship) lines with superstructure aft with bow and stern ramps for unloading vehicles. Cargo capacity is 450 tons with a usable deck space of 600 square meters. Up to 25 armored personnel carriers can be embarked. The superstructure

is large and boxy, with large, side-by-side funnels. The *Ropucha* class, like the *Polnocny* class landing ships, has a long sliding hatch cover above the bow section to permit vehicles and cargo to be lowered into the tank deck by dockside cranes. Several ships have been fitted with SA-N-5 Grail short-range missiles. There are provisions for two barrage rocket launchers on the forecastle, but they have not been installed.

USERS: Soviet Union Navy, South Yemen.

CHARACTERISTICS:

Displacement:	2,200 tons standard
	3,200 tons full load
Dimensions:	
Length	370 ft 8 in (113.0 m) overall
Beam	45 ft 11 in (14.0 m)
Draft	9 ft 6 in (2.9 m)
Propulsion:	2 diesels; 10,000 hp; 2 shafts
Speed:	18 knots
Range:	3,500 nm at 16 knots
	6,000 nm at 12 knots
Manning:	Approx 70
Troops:	Approx 230

COMBAT SYSTEMS:

Helicopters	No facilities
Missiles	4 quad SA-N-5 Grail launchers [32] in some ships
Guns	4 57-mm/80-cal AA (2 twin)
Radars	1 Don-2 (navigation)
	1 Muff Cob (fire control)
	1 Strut Curve (air search)

DESIGNATOR: BDK*—BOL'SHOY DESANTNYY KORABL (LARGE LANDING SHIP)
NAME: ALLIGATOR CLASS

NUMBER IN CLASS: 15 TANK LANDING SHIPS

DESCRIPTION: Built on traditional LST (tank landing ship) lines, the *Alligator*-class ships are less attractive than the

Alligator

later *Ropucha* class but have a significantly larger cargo capacity. The *Alligator* class has a superstructure-aft configuration with bow and stern ramps for unloading vehicles. The arrangement of individual ships differs; early units have three cranes—one 15-ton capacity, two 5-ton capacity; later ships have one crane. There are two to four large hatches above the tank deck to permit vehicles and cargo to be lowered by shipboard or dockside cranes. Later ships have an enclosed bridge and a rocket launcher forward, and the last two ships have 25-mm guns aft. About 25 to 30 tanks or armored personnel carriers or 1,500 tons of cargo can be carried; only about 600 tons can be carried for beaching operations.

CHARACTERISTICS:

Displacement:	3,400 tons standard
	4,700 tons full load
Dimensions:	
Length	370 ft (112.8 m) overall
Beam	50 ft 2 in (15.3 m)
Draft	14 ft 5 in (4.4 m)
Propulsion:	2 diesels; 8,000 hp; 2 shafts
Speed:	18 knots
Range:	9,000 nm at 16 knots
	14,000 nm at 10 knots
Manning:	Approx 75
Troops:	Approx 300

COMBAT SYSTEMS:

Helicopters	No facilities
Missiles	1 40-tube 122-mm barrage rocket launcher in most ships
	3 quad SA-N-5 Grail launchers [24] in some ships
Guns	2 57-mm/70-cal AA (1 twin)
	4 25-mm/60-cal AA (2 twin) in *Fil'chenkov* and *Vikkov*
Radars	2 Don-2 (navigation) and/or Spin Trough (search)

DESIGNATOR: SDK*—SREDNYY DESANTNYY KORABL (MEDIUM LANDING SHIP)
NAME: POLNOCNY A/B/C CLASSES

NUMBER IN CLASS: 39 MEDIUM LANDING SHIPS

DESCRIPTION: This is the largest series of landing ships in service with any navy. The *Polnocny* LSM (medium landing ship) series consists of three principal variants, with the design being enlarged and improved during the construction period. Most of the Soviet units are of the B variant. The *Polnocnys* have a conventional landing ship appearance with bow doors. The A class has a convex bow form while the later series have a concave bow. Superstructure and mast details differ among classes and individual ships. Cargo capacity is about 180 tons in the A and B classes and 250 tons in the C class. The earlier ships can carry 6 to 8 armored personnel carriers and

Polnocny

the C class about 8. There are minor differences within the sub-classes. Some *Polnocny A* ships are fitted with long troughs along the sides and carry 2 self-propelled devices for laying line charges on racks at the stern. These devices are used to clear minefields and obstructions off landing beaches.

USERS: Soviet Union Navy, Angola, Cuba, Egypt, Ethiopia, India, Iraq, Libya, Poland, Somalia, South Yemen, Vietnam.

7 UNITS POLNOCNY A CLASS

CHARACTERISTICS:
Displacement:	770 tons full load
Dimensions:	
Length	239 ft 5 in (73.0 m) overall
Beam	28 ft 2 in (8.6 m)
Draft	6 ft 3 in (1.9 m)
Propulsion:	2 diesels; 5,000 hp; 2 shafts
Speed:	19 knots
Range:	900 nm at 18 knots
	1,500 nm at 14 knots
Manning:	Approx 35
Troops:	Approx 100

COMBAT SYSTEMS:
Helicopters	No facilities
Missiles	2 18-tube 140-mm barrage rocket launchers
	2 or 4 quad SA-N-5 Grail launchers [16-32] in most ships
Guns	2 30-mm/65-cal close-in (1 twin) or 2 14.5-mm machine guns (1 twin)

Radars	1 Spin Trough (search)

23 UNITS POLNOCNY B CLASS

CHARACTERISTICS:
Displacement:	800 tons full load
Dimensions:	
Length	242 ft 9 in (74.0 m) overall
Beam	28 ft 2 in (8.6 m)
Draft	6 ft 7 in (2.0 m)
Propulsion:	2 diesels; 5,000 hp; 2 shafts
Speed:	18 knots
Range:	900 nm at 18 knots
	1,500 nm at 14 knots
Manning:	Approx 40
Troops:	Approx 100

COMBAT SYSTEMS:
Helicopters	No facilities
Missiles	2 18-tube 140-mm barrage rocket launchers
	4 quad SA-N-5 Grail launchers [32]
Guns	2 or 4 30-mm/65-cal close-in (1 or 2 twin)
Radars	1 Drum Tilt (fire control)
	1 Spin Trough (search)

9 UNITS POLNOCNY C CLASS

CHARACTERISTICS:
Displacement:	1,150 tons full load
Dimensions:	
Length	266 ft 8 in (81.3 m) overall
Beam	33 ft 2 in (10.1 m)
Draft	6 ft 11 in (2.1 m)
Propulsion:	2 diesels; 5,000 hp; 2 shafts
Speed:	18 knots
Range:	1,800 nm at 18 knots
	3,000 nm at 14 knots
Manning:	Approx 40
Troops:	Approx 180

COMBAT SYSTEMS:
Helicopters	No facilities
Missiles	2 18-tube 140-mm barrage rocket launchers
	4 quad SA-N-5 Grail launchers [32]
Guns	4 30-mm/65-cal close-in (2 twin)
Radars	1 Drum Tilt (fire control)
	1 Spin Trough (search)

DESIGNATOR:	**DK*—DESANTNYY KORABL' (LANDING SHIP)**
NAME:	**VYDRA CLASS**

NUMBER IN CLASS: 16 UTILITY LANDING SHIPS

DESCRIPTION: These are utility landing ships with an open tank deck. Cargo capacity is approximately 250 tons. No weapons systems are fitted.

USERS: Soviet Union Navy, Bulgaria and Egypt.

CHARACTERISTICS:

Displacement:	425 tons standard
	600 tons full load
Dimensions:	
Length	180 ft 1 in (54.9 m) overall
Beam	24 ft 1 in (7.6 m)
Draft	6 ft 7 in (2.0 m)
Propulsion:	2 diesels; 800 hp; 2 shafts
Speed:	12 knots
Range:	1,900 nm at 12 knots
	2,700 nm at 10 knots
Manning:	Approx 20
Troops:	Approx 100

COMBAT SYSTEMS:

Helicopters	No facilities
Radars	1 Spin Trough (search)

NAME: POMORNIK CLASS

NUMBER IN CLASS: 1+ AIR-CUSHION LANDING CRAFT

DESCRIPTION: The Soviet *Pomornik* is the world's largest air-cushion vehicle. It can travel over 12- to 15-foot obstacles and carry 3 medium tanks or over 200 troops. Unlike U.S. Navy LCAC (Landing Craft Air Cushion), the *Pomornik* has an enclosed cargo hold.

Pomornik

CHARACTERISTICS:

Displacement:	350 tons full load
Dimensions:	
Length	187 ft (57.0 m) overall
Beam	68 ft 11 in (21.0 m)
Propulsion:	Gas turbines; 3 aircraft-type propellers (tractor) + lift fans
Speed:	Approx 55 kts on air cushion
Troops:	200+

COMBAT SYSTEMS:

Missiles	SA-N-5 launchers (?)
Guns	2 30-mm/65-cal close-in (multibarrel)
Radars	1 Band Stand fire control
	1 Bass Tilt fire control
	1 navigation

NAME: UTENOK CLASS

NUMBER IN CLASS: 2+ AIR-CUSHION LANDING CRAFT

DESCRIPTION: The cargo capacity of this air-cushion landing craft is one 45-ton T-72/T-80* tank.

CHARACTERISTICS:

Dimensions:	
Length	86 ft 3 in (26.3 m) overall
Beam	42 ft 8 in (13.0 m)
Propulsion:	Gas turbines

COMBAT SYSTEMS:

Guns	4 30-mm/65-cal close-in (2 twin)

NAME: LEBED CLASS

NUMBER IN CLASS: 20+ AIR-CUSHION LANDING CRAFT

DESCRIPTION: These landing craft are the type seen on board the amphibious ship *Ivan Rogov**. They can carry 2 PT-76* amphibious tanks or 120 troops or approx 45 tons of cargo. The control cabin is offset to starboard with the Gatling gun mounted on top.

Lebed

CHARACTERISTICS:

Displacement:	85 tons full load
Dimensions:	
Length	81 ft 4 in (24.8 m) overall
Beam	35 ft 5 in (10.8 m)
Propulsion:	3 gas turbines; 2 aircraft-type propellers (tractor) + lift fans
Speed:	70 kts on air cushion
Range:	100 nm at 65 kts
	250 nm at 60 kts
Troops:	Approx 120

COMBAT SYSTEMS:

Guns	1 30-mm/65-cal close in (multibarrel)

Aist

DESIGNATOR: DKVP*—DESANTNYY KORABL' NA VOZDUS HNOY PODUSHKE (AIR CUSHION LANDING SHIP)

NAME: AIST CLASS

NUMBER IN CLASS: 17 AIR CUSHION LANDING CRAFT

DESCRIPTION: These are the world's second largest military air-cushion vehicles. They carry four PT-76* amphibious tanks or two medium tanks plus 220 troops or cargo. Bow and stern ramps are fitted to the "drive through" cargo space. The 30-mm gun mounts are forward (only one mount was provided on the prototype).

CHARACTERISTICS:

Displacement:	250 tons full load
Dimensions:	
Length	155 ft 2 in (47.3 m) overall
Beam	57 ft 1 in (17.4 m)
Draft	1 ft (.3 m)
Propulsion:	2 gas turbines (NK-12MV); 4 aircraft-type propellers (2 pusher/2 tractor) + 2 lift fans
Speed:	80 kts on air cushion
Range:	100 nm at 65 kts
	350 nm at 60 kts
Troops:	Approx 220

COMBAT SYSTEMS:

Missiles	2 quad SA-N-5 Grail launchers in later units
Guns	4 30-mm/65-cal close-in (2 twin)
Radars	1 Drum Tilt (fire control)
	1 Spin Trough (search)

NAME: GUS CLASS

NUMBER IN CLASS: 30 AIR CUSHION LANDING CRAFT

DESCRIPTION: The Gus is a naval version of the civilian Skate air-cushion vehicle. They cannot carry vehicles. (The Soviet Union operates the world's largest fleet of commercial air-cushion vehicles.)

CHARACTERISTICS:

Displacement:	27 tons full load
Dimensions:	
Length	69 ft 10 in (21.3 m) overall
Beam	23 ft 3 in (7.1 m)
Draft	8 in (0.2 m)
Propulsion:	3 gas turbines; 2,340 shaft hp; 2 aircraft-type propellers (tractor) + 1 lift fan
Speed:	60 kts
Range:	185 nm at 50 kts
	200 nm at 43 kts
Troops:	Approx 25

COMBAT SYSTEMS:

Radars	Navigation

NAME: ONDATRA CLASS

NUMBER IN CLASS: 16 LANDING CRAFT

DESCRIPTION: These are small personnel/vehicle landing craft. One is normally embarked in each of the Ivan Rogov* class ships for use as a tug for the Lebed air-cushion vehicles. A Rogov could carry six of these craft in the docking well.

CHARACTERISTICS:

Displacement:	90 tons standard
	140 tons full load
Dimensions:	
Length	79 ft 5 in (24.2 m) overall
Beam	19 ft 8 in (6.0 m)
Draft	4 ft 11 in (1.5 m)
Propulsion:	2 diesels; 600 bhp; 2 shafts
Speed:	10 kts
Manning:	4

COMBAT SYSTEMS:

None

NAME: T-4* CLASS

NUMBER IN CLASS: Several landing craft

DESCRIPTION: These are small landing craft, fitted with bow ramps to permit loading of light and medium tanks.

CHARACTERISTICS:

Displacement:	70 tons. Full load
Dimensions:	
Length	62 ft 4 in (19.0 m) overall
Beam	14 ft 1 in (4.3 m)
Draft	3 ft 3 in (1.0 m)
Propulsion:	2 diesels; 600 bhp; 2 shafts
Speed:	10 kts
Manning:	5

COMBAT SYSTEMS:

None

INTELLIGENCE SHIPS

DESIGNATOR: SSV*—SUDNO SVYAZYY
(COMMUNICATIONS VESSEL) AGI
NAME VISHNYA CLASS

NUMBER IN CLASS: 2+ INTELLIGENCE COLLECTION
SHIPS

DESCRIPTION: The hull of this new AGI design bears some resemblance to that of the large *Bal'zam* class intelligence ship. However, the mast and funnel features as well as deck arrangement differ considerably. The *Vishnya* has the heaviest gun armament of any intelligence ship yet seen, with two 30-mm multi-barrel (Gatling) guns fitted forward and another pair aft. Additional electronic antennas are expected to be installed.

Al'pinist

DESIGNATOR: SSV*—SUDNO SVYAZYY
(COMMUNICATIONS VESSEL)
NAME: BAL'ZAM CLASS

NUMBER IN CLASS: 4+ INTELLIGENCE COLLECTION
SHIPS

DESCRIPTION: The *Bal'zam* class appears to have been designed specifically for the intelligence collection role. There are significant at-sea replenishment facilities to permit them to provide supplies and fuel to other ships at sea. No radar is provided for the Gatling gun; only an optical system.

Vishnya

DESIGNATOR: AGI
NAME: AL'PINIST CLASS

NUMBER IN CLASS: 4+ INTELLIGENCE COLLECTION
SHIPS

DESCRIPTION: These are modified stern trawlers employed as intelligence ships (AGI). They are fitted with a bow thruster. The former fish holds contain electronics equipment and additional crew accommodations.

CHARACTERISTICS:

Displacement:	1,200 tons full load
Dimensions:	
Length	176 ft 2 in (53.7 m) overall
Beam	34 ft 5 in (10.5 m)
Draft	14 ft 1 in (4.3 m)
Propulsion:	1 diesel (8NVD48-2U); 1,320 bhp; 1 shaft
Speed:	13 kts
Range:	7,600 nm at 13 kts
Manning:	Approx 50

COMBAT SYSTEMS:

Radars	1 Don-2

Bal'zam

CHARACTERISTICS:

Displacement:	5,000 tons full load
Dimensions:	
Length	346 ft (105.5 m) overall
Beam	50 ft 10 in (15.5 m)
Draft	19 ft (5.8 m)
Propulsion:	2 diesels; 9,000 hp; 2 shafts
Speed:	22 kts
Manning:	Approx 200

SOVIET WEAPON SYSTEMS AND ELECTRONICS
SHIPS

COMBAT SYSTEMS:

Missiles	2 quad SA-N-5 Grail launchers [16]
Guns	1 30-mm close-in (multibarrel)
Radars	2 Don-Kay (navigation)

DESIGNATOR:	**AGI**
NAME:	**PRIMOR'YE* CLASS**

NUMBER IN CLASS: 6 INTELLIGENCE COLLECTION SHIPS

DESCRIPTION: These Soviet AGIs have large, distinctive "box" deckhouses forward and aft on their superstructure to house electronic equipment. They are based on a highly successful, Soviet-built series of stern trawler-factory ships. In their AGI configuration the ships have a distinctive superstructure with three antenna masts while some ships also retain the trawler kingpost aft (i.e., a total of 4 masts).

Primor'ye

CHARACTERISTICS:

Displacement:	2,600 tons standard
	3,700 tons full load
Dimensions:	
Length	277 ft 10 in (84.7 m) overall
Beam	45 ft 11 in (14.0 m)
Draft	18 ft (5.5 m)
Propulsion:	2 diesel (Russkiy);
	2,000 brake hp; 1 shaft
Speed:	13 kts
Range:	12,000 nm at 13 kts
	18,000 nm at 12 kts
Manning:	Approx 160
Radars:	2 Don-Kay (navigation)

DESIGNATOR:	**AGI/GS*—GIDORGRAFICHESKOYE SUDNO (HYDROGRAPHIC VESSEL)**
NAME:	**MAYAK CLASS**

NUMBER IN CLASS: 8 INTELLIGENCE COLLECTION SHIPS

Mayak

DESCRIPTION: These ships are former side trawlers that have been converted to intelligence collection ships (AGI). The ships vary in detail, with the *Girorulevoy* having a flat-topped radome fitted above the bridge, the *Khersones* has a wider main deckhouse, the *Ladoga* has a separate structure forward of the bridge and a third lattice mast, and the *Kurs* a tall deckhouse on the stern. The deckhouse forward of the bridge varies in length and mast configurations vary. One ship was fitted with 2 machine gun twin mounts in 1980.

CHARACTERISTICS:

Displacement:	1,050 tons full load
Dimensions:	
Length	177 ft 9 in (54.2 m) overall
Beam	30 ft 6 in (9.3 m)
Draft	11 ft 10 in (3.6 m)
Propulsion:	1 diesel (8NVD48);
	800 brake hp; 1 shaft
Speed:	11 kts
Range:	9,400 nm 11 kts
	11,000 nm at 7.5 kts
Manning:	Approx 60

COMBAT SYSTEMS:

Guns	4 14.5-mm machine guns (2 twin) in Kursograf
Radars	1 or 2 Don-2 (navigation) and/or Spin Trough (search)

DESIGNATOR:	**AGI/SSV*—SUDNO SVYAZYY (COMMUNICATIONS VESSEL)**
NAME:	**MIRNYY CLASS**

NUMBER IN CLASS: 4 INTELLIGENCE COLLECTION SHIPS

DESCRIPTION: These Soviet intelligence ships are converted whale hunter/catcher ships, easily identified by their high, "notched" bows (for harpoon gun). Details vary, but each has two SA-7 Grail hand-launch positions. New deckhouses were fitted between the superstructure and forward mast in the early 1970s to provide additional working spaces.

Mirnyy

CHARACTERISTICS:

Displacement:	850 tons standard
	1,300 tons full load
Dimensions:	
Length	208 ft 7 in (63.6 m) overall
Beam	31 ft 2 in (9.5 m)
Draft	14 ft 9 in (4.5 m)
Propulsion:	4 diesels; 4,000 hp;
	electric drive; 1 shaft
Speed:	17.5 kts
Range:	8,700 nm at 11 kts
Manning:	Approx 60
Radars	2 Don-2 (navigation)
	1 Spin Trough (search) in some ships

DESIGNATOR: SSV*—SUDNO SVYAZYY (COMMUNICATIONS VESSEL)
NAME: MOMA CLASS

NUMBER IN CLASS: 9 INTELLIGENCE COLLECTION SHIPS

DESCRIPTION: These ships are converted survey ships/buoy tenders, with about 40 ships having been built to this design. The ships vary considerably in detail. Some retain their buoy-handling cranes forward; others have a low deckhouse between the forward mast and superstructure of varying length; forward mast positions (in some ships) vary in height and configuration. There is a deck area aft of the funnel and boat davits for carrying vans with electronic equipment.

Moma

CHARACTERISTICS:

Displacement:	1,260 tons standard
	1,540 tons full load
Dimensions:	
Length	240 ft 5 in (73.3 m) overall
Beam	35 ft 5 in (10.8 m)
Draft	12 ft 6 in (3.8 m)
Propulsion:	2 diesels (Zgoda/Sulzer 6TD48);
	3,600 hp; 2 shafts
Speed:	17 knots
Range:	8,000 nm at 11 knots
Manning:	Approx 100

COMBAT SYSTEMS:

Missiles	2 quad SA-N-5 Grail launchers [16] in some ships
Guns	None
Radars	2 Don-2 (navigation)

DESIGNATOR: AGI/SSV*—SUDNO SVYAZYY (COMMUNICATIONS VESSEL)
NAME: NIKOLAY ZUBOV* CLASS

NUMBER IN CLASS: 3 INTELLIGENCE COLLECTION SHIPS

DESCRIPTION: These are former oceanographic ships converted to intelligence collection ships (AGIs). Positions for 3 quad SA-N-5 Grail launchers [24 missiles] have been provided in all 3 ships. The Gavril Sarychev has been extensively reconstructed with her forecastle deck extended to the stern and an additional level added to her superstructure. The others have a small raised platform aft, which is not a helicopter deck.

CHARACTERISTICS:

Displacement:	2,200 tons standard
	3,100 tons full load

Nikolay Zubov

Dimensions:
Length	297 ft (90.0 m)
Beam	42 ft 11 in (13.0 m)
Draft	15 ft 6 in (4.7 m)
Propulsion:	2 diesels (*Zgoda* 8TD48); 4,800 hp; 2 shafts
Speed:	16.5 kts
Range:	11,000 nm at 14 kts
Manning:	Approx 100

COMBAT SYSTEMS:
Missiles	(See description)
Radars	2 Don-2 (navigation) High Pole-B (IFF)

DESIGNATOR: AGI/GS*—GIDROGRAFICHESKOE SUDNO (COMMUNICATIONS VESSEL)
NAME: OKEAN CLASS
NUMBER IN CLASS: 15 INTELLIGENCE COLLECTION SHIPS

Okean

DESCRIPTION: This is the largest and hence probably most observed class of Soviet AGIs. They are converted side trawlers. Details differ. They retain their trawler arrangement of a tripod mast well forward and a pole mast well aft. There are provisions in most ships for two quad SA-N-5 Grail launchers [16 missiles]. One ship has been fitted with machine guns.

CHARACTERISTICS:
Displacement:	760 tons full load
Dimensions:	
Length	166 ft 8 in (50.8 m) overall
Beam	29 ft 2 in (8.9 m)
Draft	12 ft 2 in (3.7 m)
Propulsion:	1 diesel; 540 hp; 1 shaft
Speed:	11 kts
Range:	7,900 nm at 11 kts
Manning:	Approx 60

COMBAT SYSTEMS:
Missiles	(See description)
Guns	4 14.5-mm machine guns (2 twin) in Barograf
Radars	1 or 2 Don-2 (navigation)

DESIGNATOR: AGI/SSV*—SUDNO SVYAZYY (COMMUNICATIONS VESSEL)
NAME: PAMIR* CLASS
NUMBER IN CLASS: 2 INTELLIGENCE COLLECTION SHIPS

Pamir

DESCRIPTION: These ships are converted salvage tugs. In the AGI configuration their superstructures have been enlarged and antennas fitted. There are positions in these ships for three quad SA-N-5 Grail launchers [24 missiles].

CHARACTERISTICS:
Displacement:	1,443 tons standard 2,300 tons full load
Dimensions:	
Length	255 ft 10 in (78.0 m) overall
Beam	42 ft (12.8 m)
Draft	13 ft 2 in (4.0 m)

Propulsion:	2 diesels (MAN G10V 40/60);
	4,200 hp 2 shafts
Speed:	17.5 kts
Range:	15,200 nm at 17.5 kts
	21,800 nm at 12 kts
Manning:	Approx 120

COMBAT SYSTEMS:

| Missiles | (See description) |
| Radars | 2 Don-2 (navigation) |

AUXILIARY SHIPS

NAME: **NUCLEAR AGM**

NUMBER IN CLASS: 1 MISSILE RANGE
INSTRUMENTATION SHIP (NUCLEAR-POWERED)

DESCRIPTION: This will be the world's largest scientific ship when completed. It probably will be used to support the Soviet test firings of long-range ballistic missiles (ICBMs and SLBMs) and military space activities. This could include monitoring spacecraft outside the control of ground sites in the Soviet Union and providing continuous communications to Soviet manned space flights. Few physical details of the ship are known, although it will be much larger than the missile range instrumentation ship Marshal Nedelin*. Like the Nedelin, this new ship will probably be manned and operated by the Soviet Navy, not a civilian agency.

CHARACTERISTICS:

Length:	Approx 850 ft (259.2 m)
Propulsion:	Steam turbines; 2 shafts
Reactors:	2

NAME: **AKADEMIK SERGEI KOROLEV***

NUMBER IN CLASS: 1 SPACE CONTROL-MONITORING SHIP

DESCRIPTION: This space event support ship (SESS) was built from the keel up for this role, and not converted from another configuration. It is smaller than the Yuri Gagarin*. Tracking and communications equipment include Quad Ring, Ship Bowl radome, Ship Globe, and Vee Tube antennas. The ship contains 80 laboratories.

CHARACTERISTICS:

Displacement:	17,115 tons standard
	21,250 tons full load
Tonnage:	7,067 deadweight tons
Dimensions:	
Length	596 ft 8 in (181.9 m)
Beam	82 ft (25.0 m)
Draft	25 ft 11 in (7.9 m)
Propulsion:	1 diesel (Bryansk/Burmeister
	& Wain) 12,000 bhp; 1 shaft

Akademik Sergei Korolev

Speed:	17.5 kts
Range:	22,500 nm at 16 kts
Manning:	Approx 190 +
	170 scientists-technicians

COMBAT SYSTEMS:

| Helicopters | No facilities |
| Radars | 2 Don-Kay navigation |

NAME: **DESNA CLASS**

NUMBER IN CLASS: 2 MISSILE RANGE
INSTRUMENTATION SHIPS

Desna

DESCRIPTION: These Soviet Navy-manned missile range instrumentation ships (AGMs) were converted from Dzhankoy* class ore/coal carriers. They have a distinctive arrangement, with two island superstructure groupings. Three large missile tracking directors are mounted forward with the Ship Globe radome mounted above the forward island structure. The after structure consists of the funnel, with a faired-in mast (surmounting the short stack) mounting two Vee Cone communication antennas, and a helicopter hangar on each side of the funnel. There is a large helicopter landing area aft. These are the only ships now fitted with the Head Net-B radar.

CHARACTERISTICS:

Displacement:	13,500 tons full load
Dimensions:	
Length	458 ft 11 in (139.9 m)
Beam	59 ft (18.0 m)
Draft	25 ft 11 in (7.9 m)
Propulsion:	1 diesel (MAN); 5,400 bhp; 1 shaft
Speed:	15 kts
Range:	12,000 nm at 13 kts

COMBAT SYSTEMS:

Helicopters	1 or 2 Ka-25 Hormone-C
Radars	2 Don-2 navigation
	1 Head Net-B air search
	1 Ship Globe tracking
EW systems	2 Watch Dog

NAME: **KOSMONAUT PAVEL BELYAYEV* CLASS**

NUMBER IN CLASS: 4 SPACE CONTROL-MONITORING SHIPS

Kosmonaut Pavel Belyayev

DESCRIPTION: These space event support ships (SESS) were converted from Vytegrales*-class cargo/timber carriers. A Quad Spring communication antenna array is fitted amidships in addition to smaller satellite communication antennas.

CHARACTERISTICS:

Displacement:	9,000 tons full load
Tonnage:	2,101 deadweight tons
Dimensions:	
Length	399 ft 6 in (121.8 m) overall

Beam	54 ft 11 in (16.8 m)
Draft	23 ft 11 in (7.3 m)
Propulsion:	1 diesel (Bryansk/Burmeister & Wain 950 VTBF 110); 5,200 bhp; 1 shaft
Speed:	14.5 kts
Manning:	Approx 90
Radar:	1 Don-2 navigation
	1 Kite Screech tracking
	1 Okean navigation

NAME: **KOSMONAUT YURI GAGARIN***

NUMBER IN CLASS: 1 SPACE CONTROL-MONITORING SHIP

Kosmonaut Yuri Gagarin

DESCRIPTION: The space event support ship (SESS) Gagarin is the world's largest ship fitted for scientific activities pending the completion of the Soviet nuclear-propelled missile tracking ship. The Gagarin is also the largest ship in service with turbo-electric propulsion. The ship was built with a Sifiya* class tanker hull and propulsion plant. The ship has a large bulbous bow and is fitted with bow and stern thrusters. Recreation facilities include three swimming pools and a theater seating 300 persons, plus a sports hall. The ship's tracking-communications equipment includes four Quad Ring arrays, and 2 Ship Bowl and 2 Ship Shell stabilized dishes. Two pair of Vee Tube high frequency antennas are rigged outboard of the ship's funnel.

CHARACTERISTICS:

Displacement:	53,500 tons standard
Tonnage:	31,300 deadweight tons
Dimensions:	
Length	760 ft (231.7 m)
Beam	102 ft (31.1 m)
Draft	32 ft 10 in (10.0 m)
Propulsion:	2 steam turbines (Kirov) with electric drive; 19,000 shaft hp; 1 shaft

Boilers: 2
Speed: 17.7 kts
Range: 24,000 nm at 17.7 kts
Manning Approx 160 +
180 scientists-technicians

COMBAT SYSTEMS:
Helicopters No facilities
Radars 1 Don-Kay navigation
1 Okean navigation

NAME: **MARSHAL NEDELIN***

NUMBER IN CLASS: 1 + 1 MISSILE RANGE
INSTRUMENTATION SHIPS

Marshall Nedelin

DESCRIPTION: These are the first of a new class of missile range instrumentation ships to be operated by the Soviet Navy rather than by civilians. They are the world's largest Navy-operated scientific ships, pending completion of a nuclear-propelled, missile range instrumentation ship. These ships can also serves as a sea-based terminal for Soviet spacecraft data and communication links, and could provide naval command ship facilities. These ships have foundations for the installation of 6 30-mm Gatling guns and 3 associated Bass Tilt radar directors. The satellite communications antenna is covered by the large, distinctive Ship Globe radome. Other antennas include 1 Quad Leaf, 3 Quad Wedge, and 4 other tracking antennas, and 6 telemetry reception arrays. A swimming pool is abaft the stack.

CHARACTERISTICS:
Displacement: 24,000 tons full load
Dimensions:
Length 698 ft 8 in (213.0 m)
Beam 88 ft 11 in (27.1 m)
Draft 25 ft 3 in (7.7 m)
Propulsion: Diesel; 2 shafts

COMBAT SYSTEMS:
Helicopters: 2 to 4 Ka-25
Hormone-C/Ka-27 Helix
Guns (See description)

NAME: **SIBIR'* CLASS**

NUMBER IN CLASS: 4 MISSILE RANGE
INSTRUMENTATION SHIPS (Completed 1958)

DESCRIPTION: These ships were converted during construction from *Donbass** class cargo ships. The *Chukota* was the first of the class that was flush-decked; now all are. They have large antenna-bearing kingposts forward and aft of the central superstructure. Two or three missile tracking directors are mounted forward of the bridge. A large helicopter platform is fitted aft. One utility helicopter is normally embarked, although the ships do not have a hangar.

Sibir'

CHARACTERISTICS:
Displacement: 7,800 tons full load
Dimensions:
Length 354 ft 11 in (108.2 m) overall
Beam 47 ft 11 in (14.6 m)
Draft 23 ft 7 in (7.2 m)
Propulsion: Compound piston with low-pressure turbine; 2,500 hp; 1 shaft
Boilers: 2
Speed: 11.5 kts
Range: 11,800 nm at 12 kts

COMBAT SYSTEMS:
Helicopters 1 Ka-25 Hormone-C (no hangar)
Radars 1 Big Net or
Head Net-C 3-D air search
2 Don-2 navigation

SUBMARINES

BALLISTIC MISSILE SUBMARINES (SSBNs)

DESIGNATOR:	PLARB*—PODVODNAYA LODKA ATOMNAYA RAKETNAYA BALLISTICHESKAYA (BALLISTIC MISSILE SUBMARINE NUCLEAR)
NAME:	TYPHOON CLASS

Typhoon

NUMBER IN CLASS: 4+ NUCLEAR-POWERED BALLISTIC MISSILE SUBMARINES

DESCRIPTION: The Typhoon is the largest undersea craft constructed by any nation, being almost half-again as large as the U.S. Navy Trident strategic missile submarines of the *Ohio* (SSBN 726) class. The submarines are approximately the same length, but the Typhoon is almost twice as broad. The Typhoon arrangement differs considerably from previous SSBN designs. There are two large pressure hulls arranged side by side, encased within the outer hull. This arrangement provides a high buoyancy reserve, possibly as much as 40% of total submerged displacement, and increased survivability because of the space between the inner and outer hulls. Placing the missile tubes forward also alleviates complexity of structure, piping and wiring of a large missile compartment between the ship control compartment and the engineering spaces. The two inner, pressure hulls may be modifications of the Delta-class hull. The torpedo tubes appear to be mounted in the bow,indicating a separate pressure hull for the

torpedo tubes/room forward. The 20 large vertical tubes for submarine launched ballistic missiles (SLBMs) are fitted between the hulls, forward of the sail structure. The location of the large sail indicates still another (fourth) pressure hull amidships for the submarine's control spaces. The Typhoon appears to be designed for under-ice operations, that is, it is intended to surface through the Arctic ice pack for launching missiles. The under-ice features appear to include the flat, protected top of the sail structure, retractable bow diving planes, and protected propeller shafts. While the previous Delta and Yankee classes have sail-mounted diving planes, the Typhoon has bow-mounted planes in part because the position of the sail structure aft of the midpoint of the submarine is unsuitable for control planes. Like all Soviet SSBNs, these submarines have twin propeller shafts (all Western SSBNs are single-shaft submarines).

CHARACTERISTICS:

Displacement:	25,000 tons submerged
Length	557 ft 7 in (170.0 m) overall
Beam	82 ft (25.0 m)
Draft	42 ft 8 in (13.0 m)
Propulsion:	Steam turbines; Approx 100,000 shaft hp; 2 shafts/7-bladed propellers
Reactors	2 pressurized water
Speed:	25 knots submerged
Manning:	Approx 150

COMBAT SYSTEMS:

Missiles	20 SS-N-20 SLBM with multiple independently targetable reentry vehicles (MIRV)
ASW Weapons	Torpedoes
Torpedoes	21-inch (533-mm) or 26-inch (650-mm) torpedo tubes (fwd)
Radars	Snoop series
Sonars	Low-frequency

DESIGNATOR:	PLARB*—PODVODNAYA LODKA ATOMNAYA RAKETNAYA BALLISTICHES KAYA (BALLISTIC MISSILE SUBMARINE NUCLEAR)
NAME:	DELTA CLASSES

NUMBER IN CLASS: 38+ NUCLEAR-POWERED BALLISTIC MISSILE SUBMARINES

DESCRIPTION: The Delta is an enlargement of the previous Yankee SSBN design, intended to launch the larger SS-N-8 sub-launched ballistic missile (SLBM). The Delta

I was the world's largest undersea craft when the first unit was completed in 1972. The later Delta designs were further enlarged to accommodate 16 missiles vice the 12 in the Delta I. The Delta I class has "steps" at the after end of the "hump" or "turtle back" deck above the missile compartment. The later Deltas have flat, angled decks. The Delta II class was an interim modification of the Delta I design, lengthened to accommodate 16 tubes but not capable of accommodating the later SLBMs intended for the Delta class. In the Delta III class the "hump" or "turtle back" deck aft of the sail is higher than in previous classes to allow space for the longer SS-N-18 SLBMs. The latest version of the basic Delta design, the Delta IV, was further enlarged to accommodate the advanced SS-N-23 missile, although the design is only 16 ft longer than the Delta III. The submarine has features for under-ice operation and the "turtle back" aft of the sail structure has been further enlarged. The Deltas were the last SSBNs built by the Leninskaya Komsomola shipyard at Komsomol'sk in the Soviet Far East. All subsequent ballistic missile submarines have been constructed at Severodvinsk on the Arctic (White Sea) coast.

Delta

CHARACTERISTICS:

Displacement:
    9,000 tons surfaced Delta I
    10,000 tons surfaced Delta II
    10,500 tons surfaced Delta III
    10,750 tons surfaced Delta IV

    11,750 tons submerged Delta I
    12,750 tons submerged Delta II
    13,250 tons submerged Delta III
    13,550 tons submerged Delta IV

Dimensions:
Length	459 ft 3 in (140.0 m) overall Delta I
	508 ft 5 in (155.0 m) overall Delta II/III/IV
Beam	39 ft 4 in (12.0 m)
Draft	28 ft 6 in (8.7 m)
Propulsion:	2 shafts/5-blade propellers/Steam turbines;
	50,000 shaft hp; 7-blade propellers in Delta IV
Speed:	18 knots surfaced Delta I/II/III
	25 knots submerged Delta I
	24 knots submerged Delta II/III/IV
Manning:	Approx 120

COMBAT SYSTEMS:

Missiles	12 SS-N-8 SLBM Delta I
	16 SS-N-8 SLBM Delta II
	16 SS-N-18 SLBM Delta III
	16 SS-N-23 SLBM Delta IV
ASW Weapons	Torpedoes (18)
Torpedoes	6 21-in (533-mm) torpedo tubes (fwd)
Radars	Snoop Tray
Sonars	Low-frequency/Towed array in Delta IV

DESIGNATOR: PLARB*—PODVODNAYA LODKA ATOMNAYA RAKETNAYA BALLISTICHESKAYA (BALLISTIC MISSILE SUBMARINE, NUCLEAR)

NAME: YANKEE CLASS

NUMBER IN CLASS: 33 NUCLEAR-POWERED SUBMARINES

DESCRIPTION: The Yankee was the first modern Soviet SSBN class, being similar to the earlier U.S. and British Polaris designs. The Yankees were the first Soviet submarines to have sail-mounted diving planes rather than bow planes. They have a smaller rise to the top of the hull aft of the sail, over the missile tubes, than in the later Delta classes. Some have an angled edge at the forward base of their sail for a sonar installation.

Yankee

CHARACTERISTICS:

Displacement:	8,000 tons surfaced/unknown for SSGN
	9,600 tons submerged
	13,650 tons submerged for SSGN
Dimensions:	
Length	429 ft 5 in (130.0 m) overall
	501 ft 10 in (153.0 m) overall for SSGN
Beam	39 ft 4 in (12.0 m)
Draft	28 ft 10 in (8.8 m)
	29 ft 6 in (9.0 m) for SSGN
Propulsion:	Steam turbines; 50,000 shaft hp; 2 shafts/5 bladed propellers
Reactors	2 pressurized water type
Speed:	18 knots surfaced/unknown for SSN/SSGN
	27 knots submerged/23 knots submerged for SSGN
Manning:	Approx 120/unknown for SSGN

COMBAT SYSTEMS:

Missiles	16 SS-N-6 SLBM in Yankee I
	12 SS-N-17 SLBM in Yankee II
	Removed in SSN
	12 SS-N-24 land-attack cruise missiles in SSGN
ASW Weapons	Torpedoes (18)
Torpedoes	6 21 inch (533-mm) torpedo tubes (fwd)
Radars	Snoop Tray/unknown in SSGN
Sonars	Low-frequency/unknown in SSGN

DESIGNATOR:	**PLARB*—PODVODNAYA LODKA ATOMNAYA RAKETNAYA BALLISTICHESKAYA (BALLISTIC MISSILE SUBMARINE NUCLEAR)**
NAME:	**HOTEL CLASSES**

NUMBER IN CLASS: 6 NUCLEAR-POWERED SUBMARINES

DESCRIPTION: The Soviet Navy constructed 8 Hotel-class SSBNs in 1959-1962. Production was halted because of changes in strategic forces policy and the subsequent development of the more-capable Yankee SSBN. The Hotels were converted to the Hotel II configuration in 1962-1967. Subsequently, one was converted in 1969-70 to the Hotel III configuration as a test ship for the SS-N-8 missile, and at least one to a communications submarine. Two others were decommissioned in the early 1980s and apparently stricken. The 4 remaining SSBNs had their 3 SS-N-5 missile tubes disabled in compliance with SALT requirements. Their future use is unknown.

CHARACTERISTICS:

Displacement:	5,500 tons surfaced SSBN
	5,000 tons surfaced SSXN/SSQN
	6,400 tons seubmerged SSBN
	6,000 tons submerged SSXN/SSQN

Hotel

Dimensions:	
Length	426 ft 5 in (130.0 m) overall SSBN
	377 ft 2 in (115.0 m) overall SSXN/SSQN
Beam	29 ft 6 in (9.0 m)
Draft	23 ft (7.0 m)
Propulsion:	Steam turbines; 30,000 shaft hp; 2 shafts/6-bladed propellers
Reactors	2 pressurized-water type
Speed:	20 knots surface
	25 knots submerged
Complement:	Approx 80

COMBAT SYSTEMS:

Missiles	3 SS-N-8; SLBM/SSBN SSXN/SSQN are not configured for missiles
ASW Weapons	Torpedoes
Torpedoes	6 21-in (533-mm) torpedo tubes (fwd)
	4 16-in (406-mm) torpedo tubes (aft)
Radar	Snoop Tray
Sonar	Medium frequency

DESIGNATOR	**PLRB*—PODVODNAYA LODKA RAKETNAYA BALLISTICHESKAYA (BALLISTIC MISSILE SUBMARINE)**
NAME:	**GOLF CLASSES**

NUMBER IN CLASS: 18 GOLF-CLASS SUBMARINES

DESCRIPTION: A total of 23 Golf-class submarines were built, originally armed with the SS-N-4 Sark submarine-launched ballistic missile (SLBM). Of these, 13 were modified to Golf II configuration with installation of underwater-launched SS-N-5 Serb SLBM. This was done from 1961 to 1971. In addition, 3 units were converted to trials ships for new SLBMs; the Golf III in 1973 to test the SS-N-8 for the Delta I/II SSBNs, the Golf IV to test the SS-N-6, and the Golf V in 1974-75 for the SS-N-20. The

Golf

ASW Weapons	Torpedoes
Torpedoes	10 21-in (533-mm) torpedo tubes (6 fwd 4 aft)
Radar	Snoop Tray
Sonar	Medium-frequency

GUIDED MISSILE SUBMARINES (SSGNs)

DESIGNATOR	PLARK*—PODVODNAYA LODKA ATOMNAYA RAKETNAYA KAYLATAYA (CRUISE MISSILE SUBMARINE NUCLEAR)
NAME:	OSCAR CLASS

NUMBER IN CLASS: 3+ NUCLEAR-POWERED GUIDED MISSILE SUBMARINES

DESCRIPTION: The Oscar is a very large cruise missile submarine, carrying three times the number of anti-ship missiles as in the previous Charlie and Echo II SSGNs. The missiles can be launched while the submarine is submerged. The 24 missile tubes are in two rows of 12, in a fixed position at an angle of approximately 40 deg; the tubes are covered by six shutters per side, each covering two missile tubes. The missile tubes are fitted between the inner (pressure) and outer (hydrodynamic) hulls. This provides a "stand-off" distance of some 11½ ft (3.5 m) between the hulls, giving the Oscar significant reserve buoyancy and increased protection against conventional torpedoes. Earlier Western estimates credited the Oscar with a submerged speed of some 30 knots; this has since been revised upward, in further recognition of Soviet emphasis on submarine performance.

Golf IV has been scrapped. Three additional submarines were converted in 1978 to a special communications configuration (SSQ), apparently for use as command ships. Three submarines retaining the basic Golf I SSB configuration have been scrapped. The plans and some components of an additional Golf SSB were given to China and assembled there, with that submarine being launched in 1964 as the test platform for China's SLBM program.

CHARACTERISTICS:

Displacement:	2,300 tons surfaced
	2,900 tons surfaced Golf III
	2,700 tons submerged
	3,300 tons submerged Golf III
Length	328 ft (100.0 m) overall
	360 ft 10 in (110.0 m) overall Golf III
Beam	27 ft 11 in (8.5 m)
Draft	21 ft 8 in (6.6 m)
Propulsion:	3 shafts
	3 diesels; 2,000 hp each
	3 electric motors; 5,300 shaft hp total
Speed:	17 knots surface
	12-14 knots submerged
Range:	9,000 nm at 5 knots
Manning:	Approx 80/Approx 90 Golf III

COMBAT SYSTEMS:

Missiles	Removed from (SSQ) type
	3 SS-N-5 Serb SLBM Golf II
	6 SS-N-8 SLBM Golf III
	1 SS-N-20 SLBM Golf V

Oscar

CHARACTERISTICS:

Displacement:	Approx 11,500 tons surfaced
	Approx 14,500 tons submerged
Dimensions:	
Length	492 ft (150.0 m) overall
Beam	59 ft (18.0 m)
Draft	36 ft (11.0 m)

Propulsion:	Steam turbines; Approx 90,000 shaft hp; 2 shafts/7-bladed propellers
Reactors	2 pressurized water
Speed:	35 knots submerged

COMBAT SYSTEMS:

Missiles	24 SS-N-19 anti-ship
ASW Weapons	Torpedoes (18) SS-N-15 SS-N-16
Torpedoes	21-in (533-mm) and/or 26-in (650-mm) torpedo tubes
Sonars	Shark Gill Low/medium-frequency

DESIGNATOR:	**PLARK*—PODVODNAYA LODKA ATOMNAYA RAKETNAYA KRYLATAYA (CRUISE MISSILE SUBMARINE, NUCLEAR)**
NAME:	**PAPA CLASS**

NUMBER IN CLASS: 1 NUCLEAR-POWERED GUIDED MISSILE SUBMARINE

DESCRIPTION: A single Papa-class SSGN was built as a prototype for advanced SSGN concepts. Earlier Western estimates gave the Papa a speed of 35-40 knots; it has been raised to the upper end of that range. Most Western estimates indicate the Papa carries the SS-N-9 missile; however, this is not certain.

Papa

CHARACTERISTICS:

Displacement:	6,700 tons surfaced/ 8,000 tons submerged
Length	357 ft 6 in (109.0 m) overall
Beam	40 ft (12.2 m)
Draft	31 ft 2 in (9.5 m)
Propulsion:	Steam turbines; 60,000-75,000 shaft hp; 2 shafts/5-bladed propellers

Reactors	2 pressurized water
Speed:	16 kts surfaced/39 kts submerged
Manning:	Approx 85

COMBAT SYSTEMS:

Missiles	10 SS-N-9 anti-ship (see description)
ASW Weapons	Torpedoes SS-N-15
Torpedoes	6 21-in (533-mm) torpedo tubes (fwd)
Sonar	Low-frequency

DESIGNATOR:	**PLARK*—PODVODNAYA LODKA ATOMNAYA RAKETNAYA KRYLATAYA (CRUISE MISSILE SUBMARINE, NUCLEAR)**
NAME:	**CHARLIE CLASS**

NUMBER IN CLASS: 17 NUCLEAR-POWERED GUIDED MISSILE SUBMARINES

DESCRIPTION: The Charlie class has a single reactor, the only Soviet combat nuclear submarines with a single-reactor plant and a single propeller shaft. The Charlie I class was the first submarine capable of launching cruise missiles underwater. They have shorter-range missiles than the earlier Echo SSGN. The Charlie II is an improved version, lengthened forward of the sail structure by 26 1/3 feet to accommodate improved weapons and electronic capabilities. The Charlie II carries the longer-range SS-N-9 anti-ship missiles.

CHARACTERISTICS:

Displacement:	4,000 tons surfaced Charlie I 4,500 tons surfaced Charlie II 5,000 tons submerged Charlie I 5,400 tons submerged Charlie II

Charlie

Length	308 ft 4 in (94.0 m) overall Charlie I
	334 ft 7 in (102.0 m) overall Charlie II
Beam	32 ft 10 in (10.0 m)
Draft	26 ft 3 in (8.0 m)
Propulsion:	1 shaft/5-bladed propeller
	Steam turbines; 15,000 shaft hp
Reactors	1 pressurized water
Speed:	16 kts surfaced/23 kts submerged
Manning:	Approx 100

COMBAT SYSTEMS:

Missiles	8 SS-N-7 anti-ship Charlie I
	8 SS-N-9 anti-ship Charlie II
ASW Weapons	Torpedoes (12)
	SS-N-15
Torpedoes	6 21-in (533-mm) torpedo tubes (fwd)
Radar	Snoop Tray
Sonar	Low-frequency

DESIGNATOR: PLARK*—PODVODNAYA LODKA ATOMNAYA ROKETNAYA KRYLATAYA (CRUISE MISSILE SUBMARINE, NUCLEAR)

NAME: ECHO II CLASS

NUMBER IN CLASS: 29 NUCLEAR-POWERED SUBMARINES

Echo II

DESCRIPTION: The Echo II class was the Soviet Navy's primary anti-aircraft carrier submarine during the 1960s and 1970s. These are the definitive SS-N-3a Shaddock-armed SSGNs, evolving from the Echo I and Juliette SSGN designs. These submarines were originally developed to launch the land-attack version of the Shaddock missile. The Echo II submarines have their large missile tubes mounted in pairs above the pressure hull. While on the surface the forward section of the sail structure rotates 180 deg to reveal the Front-series missile guidance radars. The reactor plant is similar to that of the Hotel SSBN and November SSN classes.

CHARACTERISTICS:

Displacement:	5,000 tons surfaced/
	6,000 tons submerged
Length	377 ft 2 in (115.0 m) overall
Beam	29 ft 6 in (9.0 m)
Draft	24 ft 7 in (7.5 m)
Propulsion:	2 shafts/4-bladed propellers
	Steam turbines; 30,000 shaft hp
Reactors	2 pressurized water
Speed:	20 kts surfaced/23 kts submerged
Manning:	Approx 90

COMBAT SYSTEMS:

Missiles	8 SS-N-3a Shaddock anti-ship or
	8 SS-N-12
	Sandbox anti-ship
ASW Weapons	Torpedoes
Torpedoes	6 21-in (533-mm) torpedo tubes (fwd)
	4 16-in (400-mm) torpedo tubes (aft)
Radars	Front Door
	Front Piece
	Snoop Slab
Sonar	Feniks/Low-frequency

DESIGNATOR: PLRK*—PODVODNAYA LODKA ROKETNAYA KRYLATAYA (CRUISE MISSILE SUBMARINE)

NAME: JULIETT CLASS

NUMBER IN CLASS: 15 GUIDED MISSILE SUBMARINES (SSG)

DESCRIPTION: Sixteen Juliett-Class submarines were originally built. Their SS-N-3a Shaddock missile tubes are paired, above the pressure hull, and elevate in pairs for firing. The Front Door/Front Piece guidance radar is built into the forward edge of the sail structure, which can be opened after rotating 180 deg.

CHARACTERISTICS:

Displacement:	3,000 tons surfaced/
	3,750 tons submerged
Dimensions:	
Length	295 ft 2 in (90.0 m) overall
Beam	32 ft 10 in (10.0 m)
Draft	23 ft (7.0 m)

Juliett

Propulsion:	Diesel-electric; diesel motors 7,000 hp/ Electric motors 5,000 shaft hp; 2 shafts
Speed:	16 knots surfaced/ 14 knots submerged
Range:	9,000 nm at 7 knots
Manning:	Approx 80

COMBAT SYSTEMS:

Missiles	4 SS-N-3a Shaddock anti-ship
ASW Weapons	Torpedoes
Torpedoes	6 21-in (533-mm) torpedo tubes (fwd) 4 16-in (400-mm) torpedo tubes (aft)
Radars	Front Door Front Piece Snoop Slab
Sonar	Medium-frequency

ATTACK SUBMARINES (SSN/SS)

DESIGNATOR: PLA*—PODVODNAYA LODKA ATOMNAYA (SUBMARINE, NUCLEAR)

NAME: AKULA CLASS

NUMBER IN CLASS: 2+ NUCLEAR-POWERED ATTACK SUBMARINES

DESCRIPTION: The Akula is an advanced attack submarine, reportedly using advanced quieting techniques for a significantly reduced noise signature. It is believed to have a titanium hull. The craft is slightly longer than

the contemporary Sierra SSN and is distinguished from the Sierra by the elongated sail structure.

CHARACTERISTICS:

Displacement:	6,800 tons surfaced/ 8,300 tons submerged
Dimensions:	
Length	351 ft (107.0 m) overall
Beam	39 ft 4 in (12.0 m)
Propulsion:	Steam turbines; 1 shaft/ 7-bladed propeller
Reactors	2 pressurized water
Speed:	35+ kts submerged

COMBAT SYSTEMS:

Missiles	Possibly SS-N-21 land-attack cruise
ASW Weapons	Torpedoes SS-N-16
Torpedoes	6 21-inch (533-mm) and/or 26-inch (650-mm) torpedo tubes (fwd)
Sonars	Low-frequency Towed array

Akula

DESIGNATOR: PLA*—PODVODNAYA LODKA ATOMNAYA (SUBMARINE, NUCLEAR)

NAME: MIKE CLASS

NUMBER IN CLASS: 1+ NUCLEAR-POWERED ATTACK SUBMARINES

Mike

DESCRIPTION: The Mike is an apparent follow-on to the Alfa-class SSN, with a titanium hull and an advanced liquid-metal cooled reactor plant. Although slower than the Alfa, the Mike is quieter and has increased weapons and other capabilities compared to most previous attack submarines. This is the world's largest attack submarine built originally for that purpose. Only the converted Yankee SSN is bigger. The Mike is reported to be significantly quieter than previous Soviet attack submarines, correcting the high noise levels experienced with the Alfa's propulsion plant.

CHARACTERISTICS:

Displacement:	7,800 tons surfaced/ 9,700 tons submerged
Length	400 ft 2 in (122.0 m) overall
Beam	39 ft 4 in (12.0 m)
Draft	29 ft 6 in (9.0 m)
Propulsion:	Steam turbines; 60,000 shaft hp 1 shaft/7-bladed propeller
Reactors	2 liquid metal
Speed:	36-38 knots submerged

COMBAT SYSTEMS:

Missiles	Possibly SS-N-21 land-attack cruise missiles
ASW Weapons	Torpedoes SS-N-15 SS-N-16
Torpedoes	21-inch (533-mm) and/or 26-inch (650-mm) Torpedo tubes (fwd)
Sonar	Shark Gill Low-frequency

DESIGNATOR: PLA*—PODVODNAYA LODKA ATOMNAYA (SUBMARINE, NUCLEAR)
NAME: SIERRA CLASS

NUMBER IN CLASS: 1+ NUCLEAR-POWERED ATTACK SUBMARINES

DESCRIPTION: This class is apparently the successor to the Victor III class with improved sonar and quieting, higher speed, and greater operating depth. The Sierra is believed to have a "thin" titanium hull. Note that the Sierra has a greater beam than the Victor III, indicating a different engineering plant and/or additional quieter features.

CHARACTERISTICS:

Displacement:	6,000 tons surfaced/ 7,550 tons submerged
Dimensions:	
Length	360 ft 10 in (110.0 m) overall
Beam	39 ft 4 in (12.0 m)
Propulsion:	Steam turbines; 1 shaft/ 7-bladed propeller
Reactors	2 pressurized water
Speed:	Approx 35 knots submerged

Sierra

COMBAT SYSTEMS:

Missiles	Possibly SS-N-21 land-attack cruise missile
ASW Weapons	Torpedoes SS-N-16
Torpedoes	6 21-inch (533-mm) and/or 26-inch (650-mm) Torpedo tubes (fwd)
Sonars	Shark Gill low-frequency Towed array

DESIGNATOR: PLA*—PODVODNAYA LODKA ATOMNAYA (SUBMARINE, NUCLEAR)
NAME: VICTOR CLASS

NUMBER IN CLASSES: 44+ NUCLEAR-POWERED ATTACK SUBMARINES

DESCRIPTION: The Victor I is an advanced attack submarine, built to the tear-drop hull design for high underwater speeds. Two small, two-bladed propellers are fitted on the stern planes for slow-speed operation. The reactor plant of all Victor class submarines is similar to that fitted in the Yankee and Delta ballistic missile nuclear submarines (SSBNs). The Victor II class is larger to provide additional weapons capability, such as an improved fire-control system and larger weapons. The Victor III is a further improvement, with the additional hull space forward of the sail structure probably for an improved weapons capability. Some or all units have tandem, four-bladed propellers mounted on the same shaft. The propellers rotate in the same direction. Note the reduction in speed from the Victor I, with the same propulsion plant driving a larger submarine resulting in a loss of up to 3 knots. The Victor III was the first Soviet submarine

Victor

fitted with what is probably a towed array sonar, streamed from a distinctive pod fitted atop the vertical stabilizer/rudder.

CHARACTERISTICS:

Displacement:	4,300 tons surfaced Victor I
	4,500 tons surfaced Victor II
	4,800 tons surfaced Victor III
	5,100 tons submerged Victor I
	5,700 tons submerged Victor II
	6,300 tons submerged Victor III
Dimensions:	
Length	311 ft 7 in (95.0 m) overall Victor I
	328 ft (100.0 m) overall Victor I
	347 ft 8 in (106.0 m) overall Victor III
Beam	32 ft 10 in (10.0 m)
Draft	23 ft (7.0 m)
Propulsion:	Steam turbines; 30,000 shaft hp;
	1 shaft/5-bladed propeller Victor I/II
	Steam turbines; 30,000 shaft hp;
	1 shaft/tandem, 4-bladed propeller
	Victor III
Reactors	2 pressurized water
Speed:	20 knots surfaced Victor III
	30-32 knots submerged Victor I
	30 knots submerged Victor II
	29 knots submerged Victor III
Manning:	Approx 80 Victor I/II
	Approx 85 Victor III

COMBAT SYSTEMS:

ASW Weapons	Torpedoes
	SS-N-15
	SS-N-16 Victor II/III
Torpedoes	6 21-in (533-mm) torpedo tubes
	(fwd) Victor I/II
	2 21-in (533-mm) torpedo tubes
	(fwd) Victor III
	4 26-in (650-mm) torpedo tubes
	(fwd) Victor III
Radar	Snoop Tray
Sonars	Low-frequency
	Towed array in Victor III

DESIGNATOR:	PLA*—PODVODNAYA LODKA
	ATOMNAYA
	(SUBMARINE, NUCLEAR)
NAME:	ALFA CLASS

NUMBER IN CLASS: 6 NUCLEAR-POWERED ATTACK SUBMARINES

DESCRIPTION: The Alfa is the world's fastest and deepest-diving combat submarine. The submarine incorporates a number of innovations, including a high power-to-weight reactor plant using liquid metal as the heat exchange medium, a titanium hull, extensive automation, and advanced drag-reduction features. The Alfa was the world's first submarine to have a titanium hull, providing considerable strength while being lighter than a steel-hulled craft. A very high degree of automation is provided in the engineering spaces, which may be unmanned while the submarine is underway. An operating depth of at least 1,970 feet (600 m) with a maximum depth of 2,950 feet (900 m) being credited by Western intelligence. The Alfa has an advanced reactor using a liquid metal such as sodium rather than pressurized-water as the heat exchange medium between the reactor and the steam system. Acoustic data on the first Alfa SSNs indicated that the radiated noise levels at lesser speeds were generally similar to that of other Soviet SSNs, indicating a net improvement in view of the higher power of the Alfa reactor plant. Several drag-reduction features, including probably the use of polymers pumped over the hull, allow higher speeds with low horsepower. The very low manning level in the Alfa indicates a high degree of automation in the propulsion system. Cost was reported very high, leading to the Soviet use of the nickname of "goldfish."

Alfa

CHARACTERISTICS:

Displacement:	2,900 tons surfaced/
	3,700 tons submerged
Dimensions:	
Length	267 ft (81.4 m) overall
Beam	31 ft 2 in (9.5 m)
Draft	23 ft (7.0 m)
Propulsion:	Steam turbines; 25,000 shaft hp;
	1 shaft/7-bladed propeller
Reactors	2 liquid metal
Speed:	43-45 knots submerged
Manning:	Approx 45

COMBAT SYSTEMS:

ASW Weapons	Torpedoes
	SS-N-15
Torpedoes	6 21-in (533-mm) torpedo tubes (fwd)
Radars	Snoop Tray
Sonars	Low-frequency

DESIGNATOR: PLA*—PODVODNAYA LODKA ATOMNAYA (SUBMARINE, NUCLEAR)

NAME: ECHO I CLASS

NUMBER IN CLASS: 5 NUCLEAR-POWERED ATTACK SUBMARINES

DESCRIPTION: The Echo I class is a former cruise-missile submarine that carried 6 SS-N-3c Shaddock missiles. They lack the Front-series radars and other features of the Juliett SSG and Echo II SSGN classes that would permit them to launch the anti-ship versions of the Shaddock (the Echo I was originally designed to launch the land-attack version), so were converted to attack submarine configuration in 1970-1974.

Echo I

CHARACTERISTICS:

Displacement:	4,500 tons surfaced/
	5,500 tons submerged
Dimensions:	
Length	360 ft 10 in (110.0 m) overall
Beam	29 ft 6 in (9.0 m)
Draft	24 ft 7 in (7.5 m)
Propulsion:	Steam turbines; 25,000 shaft hp;
	2 shafts
	5-bladed propellers
Reactors	2 pressurized water
Speed:	20 kts surfaced/25 kts submerged
Manning:	Approx 75

COMBAT SYSTEMS:

ASW Weapons	Torpedoes

Torpedoes	6 21-in (533-mm) torpedo tubes (fwd)
	4 16-in (406-mm) torpedo tubes (aft)
Radar	Snoop Tray
Sonar	Medium frequency

DESIGNATOR: PLA*—PODVODNAYA LODKA ATOMNAYA (SUBMARINE, NUCLEAR)

NAME: NOVEMBER CLASS

NUMBER IN CLASS: 13 NUCLEAR-POWERED ATTACK SUBMARINES

DESCRIPTION: These were the Soviet Navy's first nuclear submarines, the lead unit going to sea 3 years after the *USS Nautilus* (SSN 571), the first U.S. nuclear submarine. The November's reactor plant is similar to that of the Echo II SSGN and Hotel SSBN classes.

November

CHARACTERISTICS:

Displacement:	4,500 tons surfaced/
	5,300 tons submerged
Length	360 ft 10 in (110.0 m) overall
Beam	29 ft 6 in (9.0 m)
Draft	25 ft 3 in (7.7 m)
Propulsion:	Steam turbines; 30,000 shaft hp;
	2 shafts/4- or 6-bladed propellers
Reactors	2 pressurized water
Speed:	30 kts submerged
Manning:	Approx 80

COMBAT SYSTEMS:

ASW Weapons	Torpedoes (32)
Torpedoes	8 21-in (533-mm) torpedo tubes (fwd)
	4 16-in (400-mm) torpedo tubes (aft)
Radar	Snoop Tray
Sonar	Medium-frequency

DESIGNATOR: PL*—PODVODNAYA LODKA (SUBMARINE)

NAME: KILO CLASS

NUMBER IN CLASS: 9+ ATTACK SUBMARINES

Kilo

DESCRIPTION: The Kilo is a medium-range submarine, conventionally powered submarine which is the successor to the Whiskey and Romeo classes and possibly developed for foreign transfer. The Kilo has a hull form somewhat similar to the *U.S. Albacore* (SS-569)/tear-drop design. The Kilo, however, is significantly slower than its foreign contemporaries. This was one of the first Soviet submarine classes to be fitted with an anti-aircraft missile system; installed in the sail structure.

CHARACTERISTICS:

Displacement:	2,500 tons surfaced/ 3,000 tons submerged
Dimensions:	
Length	229 ft 7 in (70.0 m) overall
Beam	32 ft 6 in (9.9 m)
Draft	21 ft 4 in (6.5 m)
Propulsion:	Diesel-electric; 1 shaft/ 6-bladed propeller
Speed:	12 knots surfaced/ 25 knots submerged
Manning:	Approx 60

COMBAT SYSTEMS:

Missiles	SA-N-(undesignated) system
ASW Weapons	Torpedoes (12)
Torpedoes	6 21-in (533 mm) torpedo tubes (fwd)
Radars	Snoop Tray
Sonars	Low-frequency

DESIGNATOR: PL*—PODVODNAYA LODKA (SUBMARINE)
NAME: TANGO CLASS

NUMBER IN CLASS: 20 ATTACK SUBMARINES (SS)

DESCRIPTION: These are long-range torpedo attack submarines, the successor to the Foxtrot-class SS in Soviet service. These submarines have significantly more pressure-hull volume than the Foxtrot class with the increased space providing more battery capacity. Submerged endurance (i.e., on battery) is significantly greater than the Foxtrot class. One or more Tango-class submarines were fitted with a surface-to-air missile system, at least for evaluation and possible for service use.

CHARACTERISTICS:

Displacement:	3,200 tons surfaced/ 3,900 tons submerged
Length	301 ft 9 in (92.0 m) overall
Beam	29 ft 6 in (9.0 m)
Draft	23 ft (7.0 m)
Propulsion:	Diesel-electric; diesel engines 6,000 hp/electric motors 6,000 shaft hp; 3 shafts
Speed:	20 knots surfaced/ 16 knots submerged
Manning:	Approx 70

COMBAT SYSTEMS:

Missiles	SA-N-(undesignated) in one or more units
ASW Weapons	Torpedoes SS-N-15
Torpedoes	10 21-in (533-mm) torpedo tubes (6 fwd 4 aft)
Radar	Snoop Tray
Sonar	Low-frequency

Tango

DESIGNATOR: PL*—PODVODNAYA LODKA (SUBMARINE)
NAME: FOXTROT CLASS

NUMBER IN CLASS: Approx 50 ATTACK SUBMARINES (SS)

DESCRIPTION: These are highly capable, long-range submarines, derived from the Zulu class. This is the second largest class of diesel submarines built by any navy in the post-World War II period. The Soviet Whiskey class was the largest. Submerged (non-snorkel) endurance is estimated at more than 7 days at very slow speed.

CHARACTERISTICS:

Displacement:	1,950 tons surfaced/ 2,400 tons submerged
Dimensions:	
Length	300 ft 2 in (91.5 m) overall
Beam	24 ft 7 in (7.5 m)
Draft	19 ft 8 in (6.0 m)
Propulsion:	Diesel-electric; diesel engines 6,000 hp/electric motors 5,300 shaft hp; 3 shafts

Speed:	16 knots surfaced/
	15.5 knots submerged
Range:	11,000 nm at 8 knots
Manning:	Approx 75-80

COMBAT SYSTEMS:

ASW Weapons	Torpedoes (22)
Torpedoes	10 21-in (533-mm) torpedo tubes
	(6 fwd 4 aft)
Radar	Snoop Tray
Sonar	Medium frequency

Romeo

Foxtrot

DESIGNATOR: PL*—PODVODNAYA LODKA (SUBMARINE)

NAME: ROMEO CLASS

NUMBER IN CLASS: 8 ATTACK SUBMARINES (SS)

DESCRIPTION: The Romeo is a medium-range torpedo attack submarine, a much-improved successor to the previous Whiskey class.

CHARACTERISTICS:

Displacement:	1,330 tons surfaced/
	1,700 tons submerged
Length	252 ft 7 in (77.0 m) overall
Beam	22 ft (6.7 m)
Draft	16 ft 1 in (4.9 m)
Propulsion:	Diesel-electric; diesel engines 4,000 hp/electric motors 3,000 shaft hp; 2 shafts
Speed:	15.5 knots surfaced/
	13 knots submerged
Range:	7,000 nm at 5 knots
Manning:	Approx 55

COMBAT SYSTEMS:

ASW Weapons	Torpedoes (14)
Torpedoes	1 21-in (533-mm) torpedo tubes
	(6 fwd 2 aft)
Radar	Snoop Plate
Sonar	Medium-frequency

DESIGNATOR: PL*—PODVODNAYA LODKA (SUBMARINE)

NAME: ZULU IV CLASS

NUMBER IN CLASS: 5 ATTACK SUBMARINES (SS), 3 OR 4 OCEANOGRAPHIC RESEARCH SUBMARINES (AGSS)

DESCRIPTION: The first Soviet long-range attack or patrol submarines of the postwar era, the Zulu class also provided the platform for the world's first ballistic missile submarine. Only one is reported in service as a pure attack submarine. Three or four are believed to serve as oceanographic research ships. Four additional Zulu-class submarines are in reserve. This class incorporates several German design features. The Zulu I to III configurations of this class had deck guns and no snorkel. All were updated to Zulu IV standards, with guns removed, a snorkel fitted, and other features incorporated. The oceanographic submarines have been observed with the names *Lira, Mars, Veva,* and possibly *Orion.*

Zulu

CHARACTERISTICS:

Displacement:	1,900 tons surfaced/ 2,350 tons submerged
Length	295 ft 2 in (90.0 m) overall
Beam	24 ft 7 in (7.5 m)
Draft	19 ft 8 in (6.0 m)
Propulsion:	Diesel-electric; diesel engines 6,000 hp/electric motors 5,300 shaft hp; 3 shafts
Speed:	18 kts surfaced/ 16 kts submerged
Range:	9,500 nm at 8 kts
Manning:	Approx 70

COMBAT SYSTEMS:

Guns	Removed
ASW Weapons	Torpedoes (22)
Torpedoes	10 21-in (533-mm) torpedo tubes (6 fwd 4 aft)
Radar	Snoop Plate
Sonar	Medium-frequency

DESIGNATOR: PL*—PODVODNAYA LODKE (SUBMARINE)

NAME: WHISKEY CLASS

NUMBER IN CLASS: As many as 70 ATTACK SUBMARINES (SS)

DESCRIPTION: This was the first Soviet postwar submarine, built in larger numbers than any other submarine class in peacetime. Numerous units remain in Soviet service. Of the surviving units, about 50 are in various degrees of active status and about 60 to 70 more are in reserve, in varying degrees of preservation. Some active units are apparently employed in research-experimental roles. Four Soviet shipyards produced 236 Whiskey-class submarines before the program was abruptly halted. Thirteen of these were converted to various missile configurations to carry the SS-N-3 Shaddock missile (SSG); 4 or 5 converted to radar a picket configuration (SSR); and 2 converted to fisheries research ships (renamed *Seveyanka* and *Slavyanka*). All of these specialized boats have been discarded. The class was designed during World War II, but some German features were incorporated during postwar redesign. Units built to the Whiskey I, II, and IV designs had light anti-aircraft guns; most were subsequently modified to the definitive Whiskey V configuration.

USERS: Soviet Union Navy, Albania (4), Bulgaria (2), China (6), Cuba (1), Egypt (7), Indonesia (12), North Korea (2), and Poland (5).

CHARACTERISTICS:

Displacement:	1,050 tons surfaced/ 1,350 tons submerged
Length	246 ft (75.0 m) overall
Beam	20 ft 8 in (6.3 m)
Draft	15 ft 9 in (4.8 m)

Whiskey

Propulsion:	Diesel-electric; diesel engines 4,000 hp/electric motors 2,500 shaft hp; 2 shafts
Speed:	17 knots surfaced/ 13.5 knots submerged
Range:	6,000 nm at 5 knots
Manning:	Approx 50-55

COMBAT SYSTEMS:

Guns	Removed
ASW Weapons	Torpedoes (12)
Torpedoes	6 21-in (533-mm) torpedo tubes (4 fwd 2 aft)
Radar	Snoop Plate
Sonar	Tamir Medium-frequency

RESEARCH SUBMARINE (AGSSN)

NAME: X-RAY CLASS

NUMBER IN CLASS: 1 NUCLEAR-POWERED RESEARCH SUBMARINE

DESCRIPTION: This is a small, one-of-a-kind submarine in some respects similar to the U.S. Navy's nuclear-propelled submersible NR-1.

CHARACTERISTICS:

Length	206 ft 8 in (63.0 m) overall
Reactors	1

COMBAT SYSTEMS:

Torpedoes	(probably unarmed)

AUXILIARY SUBMARINE (AGSSN)

NAME: UNIFORM CLASS

NUMBER IN CLASS: 1 NUCLEAR-POWERED AUXILIARY SUBMARINE

DESCRIPTION: The Uniform is a one-of-a-kind research or special-mission submarine. This is the first Soviet nuclear-propelled submarine of single hull construction.

CHARACTERISTICS:

Displacement:	2,000 tons submerged
Reactors	1

COMBAT SYSTEMS:

Torpedoes	(probably unarmed)

RESCUE AND SALVAGE SUBMARINES

NAME:	**INDIA CLASS**

NUMBER IN CLASS: 2 RESCUE AND SALVAGE SUBMARINES (AGSS)

India

DESCRIPTION: These submarines are specially configured for rescue and salvage operations, being fitted to carrying two 36-foot submersibles. These submarines were designed from the outset for the salvage and rescue roles. Their hulls are configured for high surface speed, probably to permit them to rapidly deploy to operational areas. Two submersibles are carried semi-recessed in tandem deck wells aft of the sail structure. There is direct access from lower hatches in the submersibles into the carrying submarine while submerged; this facilitates clandestine and under-ice operations. It is assumed that they are fitted with bow torpedo tubes.

CHARACTERISTICS:

Displacement:	3,900 tons surfaced/
	4,800 tons submerged
Length	347 ft 8 in (106.0 m) overall
Beam	32 ft 10 in (10.0 m)
Propulsion:	Diesel-electric; 2 shafts
Speed:	15 kts surfaced/
	15 kts submerged

COMBAT SYSTEMS:

Torpedoes	6 21-in (533-mm) torpedo tubes (fwd)
Sonar	Medium-frequency

NAME:	**LIMA CLASS**

NUMBER IN CLASS: 1 AUXILIARY SUBMARINE (AGSS)

DESCRIPTION: The exact purpose and characteristics of this submarine are not publicly known. The hull length-to-beam ratio is relatively small. There is a large, flat deck area; the sail structure is set well amidships, and there is a fixed radar mast. Thrusters appear to be provided; this indicates a research or special-mission role (e.g.) precise maneuvering for handling submersibles.

CHARACTERISTICS:

Displacement:	2,000 tons surfaced/
	2,450 tons submerged
Dimensions:	
Length	282 ft 1 in (86.0 m) overall
Beam	31 ft 2 in (9.5 m)
Draft	24 ft 3 in (7.4 m)
Propulsion:	Diesel-electric
Torpedoes	(believed unarmed)

Bravo

TRAINING SUBMARINES

NAME:	**BRAVO CLASS**

NUMBER IN CLASS: 4 TARGET-TRAINING SUBMARINES

DESCRIPTION: These are specialized target and training submarines for anti-submarine warfare (ASW) forces. They are similar in concept to the *USS Marlin* (SST 2) and *Mackerel* (SST 1) built in the 1950s. The Soviet craft may also have a crew training role and, in wartime, could be employed operationally. The submarines are specially configured to permit their use as "hard" targets for ASW practice torpedoes.

CHARACTERISTICS:

Displacement:	2,400 tons surfaced/
	2,900 tons submerged
Dimensions:	
Length	239 ft 5 in (73.0 m) overall
Beam	32 ft 2 in (9.8 m)
Draft	24 ft 11 in (7.3 m)
Propulsion:	Diesel-electric; 1 shaft
Speed:	14 knots surfaced/
	16 knots submerged
Manning:	Approx 65

COMBAT SYSTEMS:

ASW Weapons	Torpedoes
Torpedoes	6 21-in (533-mm) torpedo tubes (fwd)
Radar	Snoop Tray
Sonar	1 passive array

GROUND COMBAT VEHICLES

TANKS

DESIGNATOR: **T-80***

DESCRIPTION: The T-80 is an updated version of the T-64B* Main Battle Tank (MBT). It has a more reliable suspension design and a turbine powerplant in place of the opposed-piston diesel of the T-64. The T-80 is now thought to be the principal tank in production for Soviet Ground Forces. The 125-mm gun fires fin-stabilized high explosive anti-tank (HEAT-FS) and high-velocity fin-stabilized, discarding sabot armor-piercing (HVAPFSDS) projectiles and the AT-8 Songster* missile.

USERS: Soviet Union Ground Forces, India.

CHARACTERISTICS:

Combat Weight:	92,594-94,799 lb (42,000-43,000 kg)
Ground Pressure	11.80 lb/sq in (0.83 kg/sq cm)
Dimensions:	
Hull Length	24 ft 3 in (7.40 m)
With Gun Forward	32 ft 6 in (9.90 m)
Width	11 ft 2 in (3.40 m) without skirts
Height	7 ft 3 in (2.20 m) without ADMG
Length of Track	
On Ground	14 ft 5 in (4.40 m)
Track Width	23 in (0.58 m
Propulsion:	Multi-fuel Gas Turbine
Max Power	985 hp
Power-to-Weight	22.9-23.4 hp/metric ton
Performance:	
Speed	47 mph (75 km/h)
Range	249 miles (400 km)
Fording	4 ft 7 in (1.40 m) without preparation 16 ft 5 in (5.00 m) with snorkel
Armament:	
Main	1 x 125-mm/56-cal 2A46 Rapira 3 automatic-loading gun/missile launcher firing semi-combustible conventional ammunition and AT-8 (Soviet name, Kobra) anti-tank or anti-helicopter missiles; Forty rounds including 24 in autoloader; AT-8 loaded manually. Fitted with Fume Extractor and Thermal Sleeve.
Elevation	-6/+18 deg
Traverse	360 deg
Initial Muzzle Velocity	
HEAT-FS	2,969 fps (905 mps)
HVAPFSDS	5,299 fps (1,615 mps)

T-80

AT-8 missile	492 fps (150 mps)
Max Speed	1,640 fps (500 mps)
Max range	5,468 yd (5,000 m) (Approx)
Secondary Coaxial	1 x 7.62-mm PKT machine gun (MG) (3,000 rounds)
Other Secondary	1 x 12.7-mm NSVT ADMG
Sensors/Fire Control:	Ballistic Computer; Laser Rangefinder; Radio Command Guidance for AT-8 missile; Infrared (IR) searchlight; Does not seem to have sophisticated night fighting and vision systems
Crew:	3 (Commander, gunner, driver)
Suspension:	Torsion Bar, 6 road wheels, Three Return Rollers, Rear Drive, Front Idler.
Protection:	Hull Armor is a sandwich of steel (3 layers) and glass-fiber, and a lead-impregnated plastic Inner Spall Liner; Reactive Armor Mountings have been added to approx 6,000 T-64B/T-72*/T-80 tanks
Armor	
Upper Glacis	200 mm at 22 deg
Lower Glacis	80-100 mm at 30 deg
Hull	60-80 mm sides, 30-40 mm rear, 100 mm floor
Turret	400-450 mm front, 90-120 mm sides, 60-90 mm rear, 40-50 mm top. May have anti-PGM system;

Fabric skirts may reduce radar glint off moving tracks; Laser-detection system triggers the turret-mounted smoke dischargers; PAZ* collective nuclear, biological, and chemical (NBC) warfare system.

DESIGNATOR: T-72*

DESCRIPTION: The T-72 is a Main Battle Tank (MBT) in use by the Soviet Union and several other countries. Some Western analysts believe that the T-72 is an export-oriented design and was not originally intended for Soviet use.

USERS: Soviet Union Ground Forces, Algeria, Bulgaria, Czechoslovakia, East German, Ethiopia, Finland, Hungary, India, Iran, Iraq, Libya, Poland, Romania, Syria and Yugoslavia.

CHARACTERISTICS:

Combat Weight:	90,389 lb (41,000 kg)
Ground Pressure	11.23 lb/sq in (0.79 kg/sq cm)
Dimensions:	
Hull Length	22 ft 10 in (6.95 m)
With Gun Forward	30 ft 4 in (9.24 m)
Width	11 ft 10 in (3.60 m)
	15 ft 7 in (4.75 m) with skirts
Height	7 ft 9 in (2.37 m)
	9 ft 3 in (2.83 m) to top of cupola
Length of Track	
On Ground	13 ft 11 in (4.25 m)
Track Width	23 in (0.58 m)
Ground Clearance	19 in (0.47 m)
Propulsion:	B-46 (B-55 mod) Supercharged V-12 multifuel diesel
Max Power	780 hp at 2,000 rpm.
Power-to-Weight	19.0 hp/metric ton
Performance:	
Speed	37 mph (60 km/h)
Range	298 miles (480 km) without long-range fuel tanks, 435 miles (700 km) with long-range fuel tanks.
Obstacle	
Vertical	2 ft 9 in (0.85 m)
Trench	9 ft 2 in (2.80 m)
Fording	4 ft 3 in (1.40 m) shallow; 18 ft 0 in (5.50 m) snorkel
Gradient	60%
Armament:	
Main	1 x 125-mm/56-cal Rapira 3 semi-automatic-loading smoothbore gun (39 rounds including 24 in autoloader); Fitted with fume extractor and thermal sleeve
Elevation	–5/+18 deg

T-72

Traverse	360 deg
Secondary Coaxial	1 x 7.62-mm PKT MG (2,500 rounds)
Other Secondary	1 x 12.7-mm NSVT ADMG (500 rounds)
Sensors/Fire Control:	Coincidence Rangefinder; laser rangefinder in later versions (T-74)
Crew:	3 (Commander, gunner, driver)
Suspension:	Torsion bar, six road wheels, three return rollers, rear drive, front idler
Protection:	PAZ nuclear, biological, and chemical (NBC) warfare warning and collective protection; Warsaw Pact T-72s have an inner lead-synthetic lining to protect against neutron weapon radiation and electro-magnetic pulses (EMP) (export versions do not); Fabric track skirts
Armor	
Turret	280 mm conventional cast steel
Glacis	200 mm (possibly laminate)

T-64B

DESIGNATOR: T-64B*

DESCRIPTION: The original T-64* was an unsuccessful Main Battle Tank (MBT) intended as the successor to the T-62. It was produced in limited numbers and distributed only to the Soviet Ground Forces. The T-64B is an improved version which is one of the two principal modern tanks in the Group of Soviet Forces in Germany. The main gun fires fin-stabilized, high-explosive, fragmentation (HE-FRAG-FS), anti-tank (HEAT-FS), as well as high-velocity, armor-piercing, fin-stabilized discarding-sabot (HVAPFSDS) projectiles.

USERS: Soviet Union Ground Forces.

CHARACTERISTICS:

Combat Weight:	83,775 lb (38,000 kg)
Ground Pressure	12.23 lb/sq in (0.86 kg/sq cm)
Dimensions:	
Hull Length	21 ft 4 in (6.50 m)
With Gun Forward	30 ft 2 in (9.20 m)
Width	11 ft 2 in (3.40 m) without skirts
Height	7 ft 3 in (2.20 m)
Length of Track	
On Ground	13 ft 7 in (4.15 m)
Track Width	21 in (0.53 m)
Ground Clearance	15 in (0.38 m)
Propulsion:	Liquid-cooled, five-cylinder opposed-piston diesel
Max Power	750 hp
Power-to-Weight	19.74 hp/metric ton
Performance:	
Speed	37 mph (60 km/h)
Range	311 miles (500 km) without long-range fuel tanks
Obstacle	
Vertical	3 ft 0 in (0.92 m)
Trench	8 ft 11 in (2.72 m)
Fording	4 ft 7 in (1.40 m) without preparation 16 ft 5 in (5.00 m) with snorkel
Gradient	60%
Armament:	
Main	1 x 125-mm/56 cal 2A46 Rapira 3 smoothbore automatic-loading gun/missile launcher, firing semi-combustible conventional ammunition, and AT-8 Songster* (Soviet name Kobra*) anti-tank missiles; Forty rounds — 24 are in autoloader; AT-8 loaded manually; Fitted with fume extractor and thermal sleeve
Elevation	-5/+18 deg
Traverse	360 deg
Initial Muzzle Velocity	
HE-Frag-FS	2,789 fps (850 mps)
HEAT-FS	2,969 fps (905 mps)
HVAPFSDS	5,299 fps (1,615 mps)
Secondary Coaxial	1 x 7.62-mm PKT Machine Gun (3,000 rounds)
Other Secondary	1 x 12.7-mm NSVT Air Defense MG
Sensors/Fire Control:	Laser Designator on T-64B; Gun stabilized in both planes; Infrared (IR) night vision; later models may have forward-looking infrared (FLIR)
Crew:	3 (Commander, gunner, driver)
Suspension:	Hydro-mechanical with 6 dual road wheels, rear drive, front idler, 4 return rollers, 4 hydraulic shock absorbers per side
Protection:	Collective nuclear, biological, and chemical (NBC) warfare protection; Bolt-on reactive armor reportedly added to most

BUSINESS REPLY MAIL

FIRST CLASS   PERMIT NO. 289   PALO ALTO, CA

POSTAGE WILL BE PAID BY ADDRESSEE

EW COMMUNICATIONS, INC.
P.O. BOX 50249
PALO ALTO, CA 94303-9983 USA

BUSINESS REPLY MAIL

FIRST CLASS   PERMIT NO. 289   PALO ALTO, CA

POSTAGE WILL BE PAID BY ADDRESSEE

EW COMMUNICATIONS, INC.
P.O. BOX 50249
PALO ALTO, CA 94303-9983 USA

T-64s; Hull armor is a sandwich of steel (3 layers), and glass-fiber, and a lead-impregnated plastic inner spall liner

Armor
Upper Glacis)	200 mm at 22 deg
Lower Glacis)	80-100 mm at 30 deg
Hull	60-80 mm sides, 30-40 mm rear, 100 mm floor
Turret	400-450 mm front, 90-120 mm sides, 60-90 mm rear, 40-50 mm top

DESIGNATOR: T-62*

DESCRIPTION: As successor to the T-54*/T-55* series, the T-62 was the Soviet Army's principal Main Battle Tank (MBT) until superseded by the T-72* and T-80*. The T-62 has been exported to several other countries. In addition, Israel has captured sufficient numbers to equip some reserve tanks units.

USERS: Soviet Union Ground Forces, Afghanistan, Algeria, Angola, Cuba, Egypt, Ethiopia, Iran, Iraq, Libya, Mongolia, North Korea, North Yemen South Yemen, and Syria.

CHARACTERISTICS:
Combat Weight:	82,673 lb (37,500 kg)
Ground Pressure	11.06 lb/sq in (0.78 kg/sq cm)
Dimensions:	
Hull Length	21 ft 9 in (6.63 m)
With Gun Forward	30 ft 7 in (9.33 m)
Width	10 ft 10 in (3.30 m)
Height	7 ft 10 in (2.40 m)
Length of Track	
On Ground	13 ft 7 in (4.15 m)
Track Width	23 in (0.58 m)
Ground Clearance	17 in (0.43 m)
Propulsion:	V-2-62 water-cooled, V-12 diesel
Max Power	580 hp at 2,000 rpm
Power-to-Weight	15.47 hp/metric ton
Performance:	
Speed	28 mph (45.5 km/h)
Range	
(Basic Fuel)	280 miles (450 km) paved road, 199 miles (320 km) dirt road.
(Long-Range Fuel Tanks)	404 miles (650 km) paved 280 miles (450 kn) dirt
Obstacle	
Vertical	2 ft 8 in (0.80 m)
Trench	9 ft 2 in (2.80 m)
Fording	4 ft 7 in (1.40 m) shallow 18 ft (5.50 m) snorkel
Gradient	60%
Armament:	
Main	1 x 115-mm/52 cal U5TS* (2A20) Rapira 2 semi-automatic smoothbore gun; Forty rounds, typically 12 HVAPFSDS, 6 HEAT, and 22 HE-FRAG); Fitted with fume extractor
Elevation	–4/+17 deg
Traverse	360 deg
Secondary Coaxial	1 x 7.62 mm PKT machine gun (MG) (2,500 rounds)

T-62

Other Secondary	(T-62A) 1 x 12.7 mm DShK MG (500 rounds)
Sensors/Fire Control:	Infrared (IR) night vision equipment; Gun stabilization in both planes; Some may have laser rangefinder and passive night vision system; Some have GPK-48 or -59 directional gyro driving aids; GPK-48 accurate for 15 min without resetting; GPK-59 accurate for 90 min without resetting; Some have map coordinate and direction indicator; Earlier version suitable for 100-km grid square; Later version expands to 1,000-km grid square
Crew:	4 (Commander, gunner, loader, driver)
Suspension:	Torsion bar, five road wheels, rear drive, front idler, two shock absorbers on each side
Protection: Armor	
Glacis	102 mm at 60 degrees
Turret	242 mm front, 153 mm sides, 97 mm rear, 40 mm top. PAZ collective nuclear warning and protection system; No chemical or biological agent filters; Exhaust system smoke generation

DESIGNATOR   **T-54*/T-55***

DESCRIPTION: The Soviet T-54 and T-55 are similar, differing primarily in engine power and gun-laying equipment. This Main Battle Tank (MBT) has served for over 30 years with the Soviet Ground Forces and the armies of many other countries.

USERS: Soviet Union Ground Forces, Afghanistan, Albania, Algeria, Angola, Bangladesh, Bulgaria, Central African Republic, China, Congo, Cuba, Czechoslovakia, Egypt, Equitorial Guinea, East Germany, Ethiopia, Finland, Guinea-Bissau, Hungary, India, Iran, Iraq, Israel, Kampuchea, Laos, Libya, Mali, Mongolia, Morocco, Mozambique, Nicaragua, Nigeria, North Yemen, Pakistan, Peru, Poland, Romania, Somalia, South Yemen, Sudan, Syria, Tanzania, Togo, Uganda, Vietnam, Yugoslavia, Zambia and Zimbabwe.

CHARACTERISTICS:

Combat Weight:	79,366 lb (36,000 kg)
Ground Pressure	11.52 lb/sq in (0.81 kg/sq cm)
Dimensions:	
Hull Length	21 ft 2 in (6.45 m)
With Gun Forward	29 ft 6 in (9.00 m)

T-54/55

Width	10 ft 9 in (3.27 m)
Height	7 ft 10 in (2.40 m)
	5 ft 9 in (1.75 m)
Length of Track On Ground	12 ft 7 in (3.84 m)
Track Width	23 in (0.58 m)
Ground Clearance	17 in (0.43 m)
Propulsion:	V-55 water-cooled, V-12 diesel
Max Power	520 hp at 2,000 rpm (T-54)
	580 hp at 2,000 rpm (T-55)
Power-to-Weight	(T-54) 14.44 hp/metric ton
	(T-55) 16.11 hp/metric ton
Performance:	
Speed	(T-54) 30 mph (48 km/h)
	(T-55) 31 mph (50 km/h)
Range	(T-54) 249 miles (400 km) without long-range fuel tanks.
	(T-55) 311 miles (500 km) same conditions (Both) 373 miles (600 km) with long-range fuel tanks
Obstacle	
Vertical	2 ft 8 in (0.80 m)
Trench	8 ft 10 in (2.70 m)
Fording	4 ft 7 in (1.40 m) shallow
	14 ft 11 in (4.55 m) snorkel
Gradient	60%
Armament:	
Main	1 x 100 mm/56 cal D10T2S rifled gun (34 rounds—T-54) (43 rounds—T-55); Fitted with fume extractor
Elevation	–4/+17 deg
Traverse	360 deg
Secondary Coaxial	1 x 7.62 mm PKT Machine Gun (MG) (3,500 rounds)
Other Secondary	1 x 12.7 mm DshK 38/46 MG (500 rounds) (T-54 had 1 x 7.62 mm bow machine gun, T-55 did not)
Sensors/Fire Control:	No rangefinder or computer on original T-54; some T-55M have laser rangefinder; Some have GPK-48 or -59 directional gyro driving aids; GPK-48 accurate for

15 min without resetting; GPK-59 accurate for 90 min without resetting

Crew:	4 (Commander, gunner, loader, driver)
Suspension:	Torsion bar, 5 road wheels, 2 hydraulic shock absorbers, rear drive, front idler, no return rollers
Protection:	Some have a radiation protection kit
Armor	
Turret	203 mm maximum on the front, 150 mm sides, 64 mm top
Hull	20mm to 99 mm
Glacis	97mm at 58 deg

DESIGNATOR   PT-76*

DESCRIPTION: The PT-76 is a lightly armored amphibious reconnaissance tank. The chassis and power train has been the basis of many light armored vehicles including most other post World War II armored recconaissance vehicles (e.g., BTR50*, BMP*) and the ASU-85* airmobile tank destroyer among others. The main gun fires high velocity armor-piercing (HVAP) and high explosive anti-tank (HEAT) projectiles.

USERS: Soviet Union Ground Forces and Naval Infantry, Afghanistan, Angola, China, Congo, Cuba, East Germany, Egypt, Equatorial Guinea, Finland, Guinea, Guinea-Bissau, Hungary, India, Indonesia, Iraq, Kampuchea, Laos, Madagascar, Mozambique, Nicaragua, North Korea, Pakistan, Poland, Vietnam, Yugoslavia and Zambia.

PT-76

CHARACTERISTICS:

Combat Weight:	30,865 lb (14,000 kg)
Ground Pressure	6.8 lb/sq in (0.48 kg/sq cm)
Dimensions	
Hull Length	22 ft 8 in (6.91 m)
With Gun Forward	25 ft 0 in (7.63 m)
Width	10 ft 4 in (3.14 m)
Height	7 ft 2 in (2.20 m) early model
	7 ft 5 in (2.26 m) late model
Length of Track	
On Ground	13 ft 5 in (4.08 m)
Track Width	15 in (0.36 m)
Ground Clearance	15 in (0.37 m)
Propulsion:	Model V-6 liquid-cooled, in-line six-cylinder diesel
Max Power	240 hp at 1,800 rpm
Power-to-Weight	17.1 hp/metric ton
Performance:	
Speed	27 mph (44 km/h)
Range	174 miles (280 km)
Obstacle	
Vertical	3 ft 7 in (1.10 m)
Trench	9 ft 2 in (2.80 m)
Fording	Amphibious with hydrojet; water speed 6 mph (10 km/h)
Gradient	70%
Armament:	
Main	1 x 76.2-mm D56TM rifled gun (40 rounds); Has muzzle brake and fume extractor
Elevation	–4/+30 deg
Traverse	360 deg
Secondary Coaxial	1 x 7.62 mm SGMT MG (1,000 rounds)
Sensors/Fire Control:	PT-76B gun is stabilized in both planes
Crew:	3 (Commander, gunner, driver)
Suspension:	Torsion bar, 6 road wheels, rear drive, front idler, no return rollers
Protection:	
Armor	
Hull	14 mm maximum
Turret	17 mm maximum on the front
Exhaust-system	Smoke generation

MISSILES

INTERCONTINENTAL BALLISTIC MISSILES (ICBMs)

DESIGNATOR: SS-25
NAME: SICKLE

DESCRIPTION: The SS-25 is a fifth-generation road-mobile solid-fuel ICBM with one reentry vehicle (RV). It has been tested with as many as 4 multiple independently targetable reentry vehicles (MIRVs). The Sickle is replacing SS-11 Segos.

Sickly

USERS: Soviet Union Strategic Rocket Forces.

CHARACTERISTICS:

Missile Weight:	Approx 77,162 lb (35,000 kg)
Throw Weight	Approx 2,600 lb (1,179 kg)
Dimensions:	
Configuration	Resembles U.S. Minuteman II in size and shape with a pointed nose cone and stack of progressively smaller cylinders; missile launching vehicle has seven axles, low sides and divided cab

Length	59-62 ft (18-19 m)
Diameter	5 ft 7 in-5 ft 11 in (1.7-1.8 m)
Propulsion:	Three-stage solid-fuel rocket
Performance:	
Maximum Range	5,666 nm (6,524 mi; 10,500 km)
Warhead:	1 x 550 kiloton RV
Accuracy:	Circular Error Probable (CEP) Approx 0.16 nm (300 m)
Sensors/Fire Control:	Inertial guidance with on-board computer updates

DESIGNATOR: SS-X-24/PL-04
NAME: SCALPEL

DESCRIPTION: The SS-X-24 is a fifth-generation three-stage solid-fuel ICBM with Multiple Independently Targetable Reentry Vehicles (MIRVs) and intended both for fixed silos and for rail car basing.

Scalpel

USERS: Soviet Union Strategic Rocket Forces.

CHARACTERISTICS:

Weight:	
Loaded Railcar	264,554 lb (120,000 kg)
Throw Weight	Approx 8,000 lb (3,600 kg)
Dimensions:	
Configuration	Thick cylinder with tapered nose uses cold launch technique

Length
Railcar	78-85 ft (23.8-26 m)
Missile	69 ft (21 m)
Railcar Width	Approx 9 ft 10 in (3.00 m)
Railcar Height	Approx 16 ft 5 in (5.00 m)
Missile Diameter	Approx 6 ft 7 in (2.00 m)
Propulsion:	Three-stage solid-fuel rocket

Performance:
Maximum Range	5,396 nm (6,213 mi; 10,000 km)
Warheads:	As many as 10 x 100 kiloton MIRVs
Accuracy:	Circular Error Probable (CEP) Approx 0.11 nm (200 m)
Sensors/Fire Control:	Inertial guidance on-board digital computer commands MIRV releases.

DESIGNATOR: SS-19/RS-18*
NAME: STILETTO

DESCRIPTION: One of 3 fourth-generation Soviet ICBMs, the SS-19 differs from the SS-17 and SS-18 in being "hot-launched" from its silo. Of the 3 types, the SS-19 has been deployed in the greatest numbers, apparently as a replacement for the SS-11 Sego.

USERS: Soviet Union Strategic Rocket Forces.

CHARACTERISTICS:
Missile Weight:	Approx 171,960 lb (78,000 kg)
Throw Weight	Mods 1/3: 7,500-8,000 lb (3,402-3,629 kg) Mod 2: 7,500 lb (3,402 kg)

Dimensions:
Configuration	Long, thick cylinder with rounded conical nose; transported in canister which is lowered into silo providing some protection against hot launch damage
Length	Approx 75-80 ft (22.9-24.4 m)
Diameter	8 ft 2 in-9 ft (2.5-2.75 m)
Propulsion:	Two-stage storable-liquid-fuel rocket

Performance:
Maximum range	
Mod 1	5,180 nm (5,965 mi; 9,600 km)
Mod 2	5,450 nm (6,276 mi; 10,100 km)
Mod 3	5,396 nm (6,214 mi; 10,000 km)
Accuracy:	Circular error probable (CEP)
Mod 1	0.19 nm (1,148 ft; 350 m)
Mod 2	Less than 0.16 nm (1,000 ft; 304 m)
Mod 3	Approx 0.13 nm (820 ft; 250 m)

Warhead:
Mod 1	6 x 550 kiloton MIRVs
Mod 2	1 x 4.3 megatons
Mod 3	6 x 550 kiloton MIRVs
Silo hardness:	At least 4,000 lb/sq in (287 kg/sq cm)

DESIGNATOR: SS-18/RS-20*
NAME: SATAN

DESCRIPTION: The SS-18 is the largest operational ICBM in service with any nation. It is a two-stage liquid-fuel cold-launch missile that has been deployed in several mods. The most recent has 10 multiple independently targetable reentry vehicles (MIRVs).

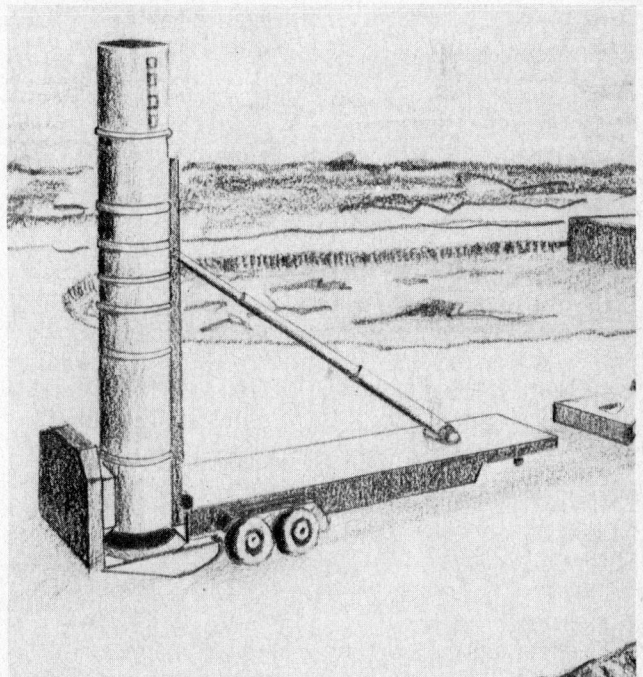

Satan

USERS: Soviet Union Strategic Rocket Forces.

CHARACTERISTICS:
Missile Weight:	Approx 485,000 lb (220,000 kg)
Throw Weight	Approx 16,000-16,700 lb (7,257-7,575 kg)

Dimensions:
Configuration	Long, full cylinder with rounded nose
Length	Estimates vary between 101-121 ft (31-37 m)
Diameter	10-11 ft (3.05-3.35 m)
Propulsion:	Two-stage liquid-fuel rocket; missile is "popped-out" of silo with gas generator after which main engines are ignited.

Performance:

Maximum range

Mod 1	6,475 nm (7,456 mi; 12,000 km)
Mod 3	8,634 nm (9,942 mi; 16,000 km)
Mods 2/4	5,936 nm (6,835 mi; 11,000 km)

Warheads:

Mod 1	1 x 25 megaton RV
Mod 2	8 x 900 - 10 x 500 kiloton MIRVs
Mod 3	1 x 20 megaton RV
Mod 4	10 x 500 kiloton MIRVs

Accuracy:

Circular Error Probable (CEP)

Mods 1/2	0.23 nm (426 m)
Mod 3	0.19 nm (352 m)
Mod 4	0.14 nm (259 m)
Silo Hardness:	At least 4,000 lb/sq in (287 kg/sq cm)
Sensors/Fire Control:	Inertial guidance MIRVs released by on-board digital computer and follow a ballistic path

DESIGNATOR: SS-17/RS-16*
NAME: SPANKER

DESCRIPTION: The SS-17 is a fourth-generation liquid-fueled two-stage ICBM that replaced some of the SS-11 Segos in limited numbers in the Soviet Strategic Rocket Forces. It was the first Soviet ICBM to have a multiple independently targetable reentry vehicle (MIRV) and the first to use a cold launch system to simplify reloading.

USERS: Soviet Union Strategic Rocket Forces.

CHARACTERISTICS:

Missile Weight:	143,300 lb (65,000 kg)
Throw Weight	Approx 6,000 lb (2,722 kg) for all mods

Dimensions:

Configuration	Long cylinder, some mods may have slimmer second stage with conical nose
Length	66-79 ft (20-24 m) may vary according to mod
Diameter	Approx 8 ft 2 in (2.50 m)
Propulsion:	Two-stage storable liquid rocket; missile "popped-out" of silo by gas generator after which the main rocket motors are ignited

Performance:

Maximum range

Mods 1/3	5,396 nm (6,214 mi; 10,000 km)
Mod 2	5,936 nm (6,835 mi; 11,000 km)

Warheads:

Mod 1	4 x kiloton MIRVs
Mod 2	1 x 3.6 megaton RV
Mod 3	4 x 500-750 kiloton MIRVs
Accuracy:	Circular Error Probable (CEP)
Mod 1	Approx 0.24 nm (440 m)

Mod 2	Approx 0.23 nm (430 m)
Mod 3	Approx 0.20 nm (370 m)
Silo hardness:	Est 4,000 lb/sq in (286.8 kg/sq cm)
Sensors/Fire Control:	Inertial guidance. MIRVs released from on-board computer and follow ballistic path to target

DESIGNATOR: SS-13
NAME: SAVAGE

DESCRIPTION: The SS-13 was the first Soviet ICBM with solid-fuel propulsion. Apparently unsuccessful, it was deployed in very small numbers.

Savage

USERS: Soviet Union Strategic Rocket Forces.

CHARACTERISTICS:

Missile Weight:	Approx 77,160 lb (35,000 kg)
Warhead	Approx 1,000-1,500 lb (454-680 kg)

Dimensions:

Configuration	Hot launch from silo; three-stage rocket, each succeeding stage slimmer, stages connected by trusswork, pointed nose cone
Length	65 ft 7 in (20.00 m)

Diameter

First Stage	5 ft 7 in (1.70 m)
Second Stage	4 ft 7.5 in (1.40 m)
Third Stage	3 ft 3 in (1.00 m)
Propulsion:	Three stages, each 4 solid-fuel rocket motors

Performance:

Maximum Range	5,073 nm (5,841 mi; 9,400 km)
Warhead:	1 x 600 kt Reentry Vehicle
Accuracy: Circular Error Probable (CEP)	
Mod 1	Approx 1 nm (1,853 m)
Mod 2	Approx 0.82 nm (1,500 m)
Sensors/Fire Control:	Fly-the-wire pre-launch guidance inertial in-flight guidance
Silo hardness:	Approx 1,200 lb/sq in (86.05 kg/sq cm)

DESIGNATOR: SS-11
NAME: SEGO

DESCRIPTION: The SS-11 is the oldest and least capable ICBM remaining in service with the Soviet Strategic Forces. It is launched from underground silos with limited hardening.

Sego

USERS: Soviet Union Strategic Rocket Forces.

CHARACTERISTICS:

Missile Weight:	Approx 105,000 lb (48,000 kg)
Throw Weight	
Mod 1	2,000 lb (907 kg)
Mod 2/3	2,500 lb (1,134 kg)
Dimensions:	
Configuration	Two-stage rocket, the second slimmer than the first, rounded nose cone, carried in launch canister that protects silo from blast damage during launch
Length	Approx 65 ft 7 in (20.00 m)
Diameter	Approx 7 ft 10 in (2.40 m)
Propulsion:	Two-stage storable liquid rocket, first stage may have four motors
Performance:	
Maximum range	
Mod 1	Approx 5,930 nm (6,835 mi; 11,000 km)
Mod 2	Approx 7,015 nm (8,077 mi; 13,000 km)
Mod 3	Approx 5,720 nm (6,585 mi; 10,600 km)
Warheads:	
Mod 1	1 x 950 kiloton Reentry Vehicle (RV)
Mod 2	1 x 1.10 megaton RV
Mod 3	3 x 350 kiloton Multiple RVs
Accuracy:	Circular Error Probable (CEP)
Mod 1	Approx 0.75 nm (1,400 m)
Mod 2/3	Approx 0.60 nm (1,100 m)
Sensors/Fire Control:	Inertial guidance. Mod 2 has penetration aids. Mod 3 RVs are not independently targetable.

DESIGNATOR: SS–X-10
NAME: NONE

DESCRIPTION: An ICBM apparently designed for the same fractional orbit bombardment system (FOBS) mission as versions of the SS-9 Scarp. The SS-10 differed with its unstorable liquid propulsion system. It was tested in the early 1960s and displayed in a 1965 parade where it was described as a sister to the manned spacecraft launch vehicles.

DESIGNATOR: SS-9
NAME: SCARP

DESCRIPTION: The SS-9 was the world's largest ICBM when it was deployed in 1967. Powered by storable liquid fuel, the Scarp was hot-launched from silos. Most of the 288 deployed carried a single 20-25-megaton reentry vehicle (RV). The Mod 4 variant carried 3 Multiple RVs (MRV) of 5 megatons each but was less accurate and there is doubt that it was ever deployed. The Mod 3 was tested as a fractional orbit bombardment system (FOBS), a depressed trajectory method which may have been intended to approach the United States from the south. Although its testing aroused great concern, it was even less accurate than the Mod 4 and was not deployed.

DESIGNATOR: SS-8
NAME: SASIN

DESCRIPTION: Smaller and more accurate than the SS-6 Sapwood or SS-7 Saddler. The small production probably reflected a return to nonstorable-liquid rockets which, while burning more consistently, limited launch flexibility. The SS-8 carried a 3-megaton nuclear warhead. The Sasins were exchanged for submarine-launched ballistic missiles (SLBMs) under the SALT I agreement.

DESIGNATOR: SS-7
NAME: SADDLER

DESCRIPTION: The SS-7 was the second operational Soviet ICBM and the first deployed in substantial numbers, beginning in 1962. It had a storable-liquid rocket motor and a 3-megaton nuclear warhead. The Mod 3 variant came into service in 1963 and had a 6-megaton warhead. At deployment peak, approximately 130 SS-7s were on soft, above-ground sites and 69 were in soft silos.

DESIGNATOR: SS-6/T-3*
NAME: SAPWOOD

DESCRIPTION: The SS-6 was the world's first operational ICBM with a first flight in August 1957 and an initial operational capability in 1961. It had a 5 megaton nuclear warhead and was propelled by highly unstable liquid fuel.

SUBMARINE LAUNCHED BALLISTIC MISSILES (SLBMs)

DESIGNATOR: SS-N-23
NAME: SKIFF

DESCRIPTION: The SS-N-23 is a submarine-launched ballistic missile (SLBM) developed for the *Delta-IV* class submarine. This liquid-fuel, three-stage missile is believed to have more accuracy, throw weight, and multiple independently targetable re-entry vehicles (MIRVs) than the SS-N-18 and may replace that missile in the *Delta-III*-class submarines.

PLATFORMS: *Delta IV* class

CHARACTERISTICS:

Missile Weight:	Approx 88,185 lb (40,000 kg)
Dimensions:	
Configuration	Tall, pointed cylinder approx the same size as the SS-N-18
Length	Approx 44 ft 7 in (13.60 m)
Propulsion:	Three-stage liquid-fuel rocket
Performance:	
Maximum Range	4,479 nm (5,517 mi; 8,300 km)
Warheads:	10 MIRV
Accuracy:	Circular Error Probable (CEP) 0.48 nm (900 m)

DESIGNATOR: SS-N-20
NAME: STURGEON

DESCRIPTION: The SS-N-20 is the first Soviet solid-propellant submarine-launched ballistic missile (SLBM) to be produced in quantity and is the largest Soviet SLBM yet deployed. The missile was reported to have had significant development problems.

PLATFORMS: *Typhoon* class (20 missiles)

CHARACTERISTICS:

Missile Weight:	Approx 132,277 lb (60,000 kg)
Dimensions:	
Configuration	Tall, pointed cylinder
Length	49 ft 2½ in (15.00 m)
Diameter	Approx 6 ft 7 in (2.00 m)
Propulsion:	Three-stage solid-fuel rocket
Performance:	
Maximum Range	4,479 nm (5,157 mi; 8,300 km)
Accuracy:	Circular Error Probable (CEP) approx 0.53 nm (1.000 m)
Warheads:	6-9 x 100 kiloton multiple independently targetable re-entry vehicles (MIRV)
Sensors/Fire Control:	Inertial with post-boost vehicle independently releasing each RV.

DESIGNATOR: SS-N-18
                 RSM-50*
NAME: STINGRAY

DESCRIPTION: The SS-N-18 is the first Soviet submarine launched ballistic missile (SLBM) to have multiple independently targetable re-entry vehicles (MIRV). It is a two-stage liquid-fueled rocket operational on the *Delta III* class ballistic missile submarines (SSBN).

PLATFORMS: 14 *Delta III* class (16 missiles each).

CHARACTERISTICS:

Missile Weight:	74,957 lb (34,000 kg)
Dimensions:	
Configuration	Tall, pointed cylinder
Length	44 ft 7 in (13.60 m)
Diameter	5 ft 11 in (1.80 m)
Propulsion:	Two-stage liquid-fuel rocket
Performance:	
Maximum Range	
Mods 1/3	3,507 nm (4,039 mi; 6,500 km)
Mod 2	4,317 nm (4,971 mi; 8,000 km)
Accuracy:	Circular Error Probable (CEP) 0.76 nm (l,400 m)
Warheads:	
Mod 1	3 x 200 kiloton MIRVs
Mod 2	1 x 450 kiloton RV
Mod 3	7 x 200 kiloton MIRVs
Sensors/Fire Control:	Inertial with post-boost vehicle providing independent targeting of each RV

DESIGNATOR: SS-N-17
NAME: SNIPE

DESCRIPTION: The first Soviet solid-fueled submarine launched ballistic missile (SLBM), the SS-N-17 has been deployed in only one *Yankee* class ballistic-missile submarine. This missile was also the first with a post boost vehicle (PBV) or "bus" to aim the single re-entry vehicle. Despite its improved accuracy, the SS-N-17 was not deployed beyond the one unit.

PLATFORM:

Submarine	*Yankee II* (12 missiles)

CHARACTERISTICS:

Throw weight	2,500 lb (1,134 kg)
Dimensions:	
Configuration	Bullet-shaped cylinder
Length	34 ft 9 in (10.60 m)
Diameter	5 ft 5 in (1.65 m)
Propulsion:	Two-stage solid-fuel rocket
Performance:	
Maximum Range	At least 2,100 nm (2,423 mi; 3,900 km)
Accuracy:	Circular Error Probable (CEP) 0.75 nm (1,400 m)
Warheads:	1 x approx 1 megaton nuclear RV
Sensors/Fire Control:	Inertial

DESIGNATOR: SS-N-8
NAME: NONE

DESCRIPTION: This submarine launched ballistic missile (SLBM) was developed for the Delta class strategic missile submarines (SSBN). It is the first Soviet two-stage submarine missile and has nearly three times the range of its immediate predecessor, the SS-N-6, allowing Delta SSBNs to remain in Soviet waters and still target the United States.

PLATFORMS:

Delta I class (12 missiles)
Delta II class (16 missiles)
1 *Golf* class (trials)
1 *Hotel* class (trials)

CHARACTERISTICS:

Missile Weight:	66,138 lb (30,000 kg)
Throw Weight	1,500 lb (680 kg)
Dimensions:	
Configuration	Bullet-shaped with smooth taper to nose
Length	42 ft 8 in (13.00 m)
Diameter	5 ft 5 in (1.65 m)
Propulsion:	Two-stage storable-liquid rocket
Performance:	
Maximum range	
Mod 1	4,209 nm (4,847 mi; 7,800 km)
Mod 2	4,910 nm (5,654 mi; 9,100 km)
Accuracy:	Circular Error Probable (CEP) Both Mods 0.86nm (1,600 m)
Warheads:	Both Mods 1 x 800 kiloton re-entry vehicle
Sensors/Fire Control:	Stellar inertial guidance

DESIGNATOR:	**SS-N-6**
NAME:	**NONE**

DESCRIPTION: This submarine-launched ballistic missile (SLBM) was deployed in the Soviet *Yankee* class nuclear-powered ballistic missile submarines (SSBNs) beginning in 1967. The missile is a single-stage, liquid-fuel, underwater-launch missile with a greater range than the earlier SS-N-5 Serb.

PLATFORMS: *Yankee I* class (16 missiles each)

CHARACTERISTICS:

Missile Weight:	41,667 lb (18,900 kg)
Throw Weight	1,500 lb (680 kg)
Dimensions:	
Configuration	Thick cylinder with smaller rounded re-entry vehicle (RV) on Mod 1, fuller section RV on Mod 3
Length	32 ft 10 in (10.00 m)
Diameter	5 ft 11 in (1.80 m)
Propulsion:	Single-stage liquid-fuel rocket
Performance:	
Maximum range	
Mod 1	1,300 nm (1,497 mi; 2,409 km)
Mods 2/3	1,600 nm (1,842 mi; 2,965 km)

SS-N-6

Accuracy:	Circular Error Probable (CEP) All Mods 1 nm (1,900 m)
Warheads:	
Mod 1	1 x 700 kiloton RV
Mod 2	1 x 650 kiloton RV
Mod 3	2 x 350 kiloton multiple RVs
Sensors/Fire Control:	Inertial guidance

DESIGNATOR:	**SS-N-5/R-21*(Soviet service number)/D-4* (Soviet design bureau number)**
NAME:	**SERB**

DESCRIPTION: The SS-N-5 was the Soviet Navy's second-generation submarine-launched ballistic missile (SLBM), featuring underwater launch and a greater range compared to the earlier SS-N-4 Sark. The SS-N-5 replaced the SS-N-4 in 13 of the *Golf* class diesel-electric submarines and all 8 of the nuclear-propelled *Hotel* class submarines.

PLATFORMS:

Submarines	*Golf*-class (3 missiles each)

CHARACTERISTICS:

Missile Weight:	36,376 lb (16,500 kg)
Throw Weight	2,000 lb (907 kg)
Dimensions:	
Configuration	Stacked cylinders, upper slimmer than lower, rounded warhead
Length	35 ft 1 in (10.70 m)
Diameter	3 ft 11 in (1.20 m)
Propulsion:	Liquid-fuel rocket; cold-gas launched through 18 electrically triggered nozzles in a short section which is jettisoned when the main engine ignites
Performance:	
Maximum Range	Early version 700 nm (806 mi; 1,300 km) Improved version 900 nm (1,036 mi; 1,668 km)
Accuracy:	Circular Error Probable (CEP) 1.5 nm (1.7 mi; 2.8 km)

Warheads: 1 x 800 kiloton re-entry vehicle
Sensors/Fire Control: Inertial guidance

DESIGNATOR: SS-N-4
NAME: SARK

DESCRIPTION: The SS-N-4 was the Soviet Navy's first operational submarine-launched ballistic missile (SLBM). It was a surface-launched weapon with storable liquid fuel, a range of approx 350 nm (404 mi; 650 km), and a Circular Error Probable (CEP) of 1.6 nm (1.9 mi; 3 km). Diesel-propelled Zulu V (2 tubes) and Golf (3 tubes) submarines and nuclear-propelled *Hotel*-class (3 tubes) submarines carried the SS-N-4 from 1958 to about 1980. The Zulus were scrapped while 13 Golfs and all 8 Hotels were converted to fire the SS-N-5 Serb SLBM.

LAND ATTACK-THEATER MISSILES

DESIGNATOR: AS-15
NAME: KENT

DESCRIPTION: The AS-15 is a transonic air-launched long-range cruise missile similar in concept and mission to the U.S. BGM-109 Tomahawk. Similar Soviet missiles include the submarine-launched SS-N-21 and the coastal defense version known as the SSC-X-4. All three may be variants of the same airframe.

PLATFORMS: Tu-142/95* Bear-H

CHARACTERISTICS:

Dimensions:
Configuration   Similar to the U.S. Tomahawk with a torpedo-shaped body, pop-out unswept wings and tail group, and probably a dorsal engine intake
Length   23 ft 0 in (7.00 m)
Propulsion: One turbofan
Performance:
Speed   Approx Mach 0.7
Maximum Range   About 1,850 mi (3,000 km)
Warhead: 1 x 250 kt
Accuracy: Circular Error Probable (CEP); Approx 45.7 m
Sensors/Fire Control: Inertial at launch followed by terrain correlation and matching (TERCOM) until impact.

DESIGNATOR   AS-10
NAME: KAREN

DESCRIPTION: The AS-10 is a Soviet short-range, laser-guided, air-to-surface missile (ASM) recently introduced

into Soviet service. It may be a development of the AS-7 Kerry.

PLATFORMS:
Attack aircraft   MiG-27* Flogger
                       Su-17* Fitter
                       Su-24* Fencer

CHARACTERISTICS:
Missile Weight: Approx 660 lb (300 kg)
  Warhead   220 lb (100 kg)
Dimensions:
  Configuration   Reported to resemble AS-7
  Length   9 ft 10 in (3.00 m)
Diameter   Approx 12 in (305 mm)
Propulsion:
  Solid-Fuel Rocket
Performance:
  Speed   Mach 0.8-Mach 1
  Maximum Range   6 nm (6.8 mi; 11 km)
Warhead: Conventional high-explosive
Sensors/Fire Control: Semi-active laser homing using electro-optical seeker

DESIGNATOR: AS-7
NAME: KERRY

DESCRIPTION: The AS-7 was the first Soviet air-to-surface missile (ASM) belonging to the current generation of small, high-performance ASMs.

PLATFORMS:
Fighters   MiG-23BN* Flogger-F/H
                       MiG-27* Flogger-D/J
                       Su-17* Fitter
                       Su-24* Fence
Attack   Yak-38* Forger

CHARACTERISTICS:
Missile Weight: About 880 lb (400 kg)
  Warhead   220 lb (100 kg)
Dimensions:
  Configuration   Resembles U.S. Bullpup (AGM-12); thick cylinder with pointed nose, four small swept foreplanes, "chopped delta" mainplanes near tail and booster ring at tail
  Length   11 ft 6 in (3.50 m)
  Diameter   Approx 12 in (305 mm)
Propulsion: Solid-fuel Rocket
Performance:
  Speed   Mach 0.6 to Mach 1
  Maximum Range   6.8 mi (11 km)
Warhead: Conventional high-explosive
Sensors/Fire Control: Radio command guidance

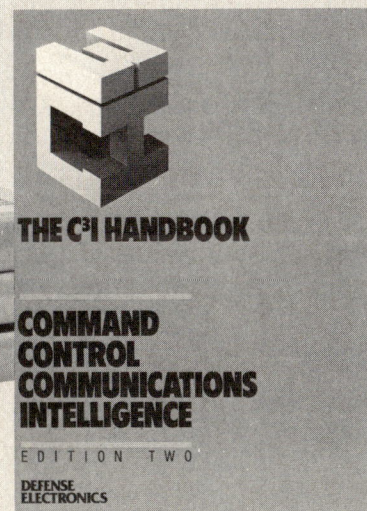

DESIGNATOR: AS-3
NAME: KANGAROO

DESCRIPTION: The largest air-to-surface missile in the Soviet inventory, the aging AS-3 Kangaroo is still available in declining numbers. It was designed to attack soft strategic targets.

Kangaroo

PLATFORMS: Tu-20/95* Bear-B/C

CHARACTERISTICS:

Missile Weight:	24,250 lb (11,000 kg) (figure may include 8,157 lb (3,700 kg) of fuel)
Warhead:	5,070 lb (2,300 kg)
Dimensions:	
Configuration	Essentially a swept-wing fighter aircraft without cockpit or flight controls
Length	49 ft 1 in (14.96 m)
Wing Span	30 ft 0 in (9.15 m)
Propulsion:	Turbojet engine
Performance:	
Speed	Approx Mach 1.8
Maximum Range	404 mi (650 km)
Warhead:	1 megaton nuclear warhead
Accuracy:	Circular error probable (CEP) 0.54 nm (l,000 m)
Sensors/Fire Control:	Beam-riding at launch, preprogrammed autopilot with command guidance override for navigational fixes from then on to impact; No terminal homing

DESIGNATOR: SS-NX-24
NAME: NONE

DESCRIPTION: The SS-NX-24 is a an advanced, submarine-launched, land-attack cruise missile, of much larger dimensions than the SS-NX-21 torpedo-tube launched cruise missile.

PLATFORMS: Converted *Yankee*-class SSBN

CHARACTERISTICS:

Dimensions:	
Configuration	Pointed cylinder with swept wings, smaller swept tail group with nozzle extending further aft
Length	Approx 41 ft 0 in (12.50 m)
Diameter	Approx 4 ft (1.20 m)
Wing-Span	Approx 18 ft (5.50 m)
Propulsion:	Probably turbojet or turbofan sustainer
Performance:	
Speed	Approx Mach 2
Warhead:	Large nuclear
Sensors/Fire Control:	Probably inertial with terminal homing

DESIGNATOR: SS-23
NAME: SPIDER

DESCRIPTION: The SS-23 Spider is a Soviet mobile, short-to-medium range, nuclear-capable ballistic missile. It is replacing the SS-1 Scud B at the Front and Army levels of the Soviet Ground Forces.

Spider

USERS: Soviet Union Ground Forces.

CHARACTERISTICS:

Dimensions:	
Configuration	Only the 8-wheeled carrier has been seen
Propulsion:	Probably single-stage solid-fuel rocket
Performance:	
Range	Approx 270 nm (311 mi; 500 km)
Warhead:	Approx 100-kiloton nuclear, or conventional high explosive or

Accuracy:	chemical Circular Error Probable (CEP) Approx 0.18 nm (350 m)
Guidance:	Inertial

DESIGNATOR: SS-21
NAME: SCARAB/TOCHKA*

DESCRIPTION: The SS-21 Scarab is a Soviet mobile, short-range artillery missile that has nearly replaced the older FROG-7.

Scarab

USERS: Soviet Union Ground Forces, Czechoslovakia, East Germany, Syria.

CHARACTERISTICS:

Dimensions:	
Configuration	Fat cylinder with pointed nose and 4 stabilizing fins at tail
Length	31 ft 0 in (9.44 m)
Diameter	18 in (460 mm)
Propulsion:	Single-stage solid-fuel rocket
Performance:	
Range	54 nm (62 mi; 100 km)
Warhead:	100-kiloton nuclear
Accuracy:	Circular Error Probable (CEP) 0.02 nm (50 m)
Guidance:	Mid-course guidance but type unknown
Carrier:	6 x 6 wheeled vehicle similar to SA-8 Gecko transporter; Has collective Nuclear, Biological, and Chemical (NBC) warfare protection, and is amphibious

DESIGNATOR: SS-N-21
NAME: NONE

DESCRIPTION: The SS-N-21 is an advanced land-attack cruise missile believed to be launched from standard Soviet submarine torpedo tubes, much like the U.S. Navy Tomahawk land-attack missile (TLAM). The air-launched version of this missile is the AS-15 Kent and the ground-launched version is the SSC-X-4.

PLATFORMS:

Submarines	Nuclear-powered attack submarines

CHARACTERISTICS:

Dimensions:	
Configuration	Similar to the U.S. Tomahawk, with a torpedo-shaped body, pop-out wings and tail group, and probably a ventral intake for the engine
Length	21 ft 0 in (6.40 m)
Diameter	Designed for launch from 21 in (533 mm) tubes
Propulsion:	Turbofan engine
Performance:	
Speed	Approx Mach 0.7
Maximum Range	Approx 1,200 nm (1,382 mi; 2,224 km)
Warhead:	Nuclear
Sensors/Fire Control:	Inertial; probably with Terrain Contour Matching (Tercom) homing

DESIGNATOR: SS-20
NAME: SABER/PIONER*

DESCRIPTION: The SS-20 Saber is a mobile Intermediate-range Nuclear Forces (INF) missile. It is a solid-fuel rocket with three multiple independently targetable reentry vehicles (MIRVs).

USERS: Soviet Union Strategic Rocket Forces.

CHARACTERISTICS:

Weight:	With vehicle 110,231 lb (50,000 kg)
Dimensions:	
Configuration	Two-stage rocket, the second stage slimmer than the first, with a smaller post-boost vehicle on top missile launching vehicle, 12 x 12 wheeled MAZ-type heavy truck variants include launcher, transloaders carrying 1-2 reloads, a command and control vehicle for the missile battery
Length	54 ft 0 in (16.46 m)
Diameter	6 ft 7 in (2.00 m)
Propulsion:	Two-stage solid-fuel rocket

Performance:
Maximum range 2,698 nm (3,107 mi; 5,000 km)
Warheads: 3 x 150 kiloton MIRV, some may have had single RV;
Accuracy: Circular Error Probable (CEP) 0.22 nm (400 m) from surveyed site
Sensors/Fire Control: Inertial guidance

Saber

DESIGNATOR: SS-15
NAME: SCROOGE

DESCRIPTION: The SS-15 was a 3-stage solid-fuel, intermediate-range ballistic missile (IRBM) developed at the same time as the SS-14 Scamp/Scapegoat. Its range was approx 3,040 nm (3,500 mi; 5,630 km) carrying one 500 kiloton to 1 megaton warhead. The Scrooge, first seen in May 1965, was transported and launched from a tracked carrier based on the JS-III heavy tank.

DESIGNATOR: SS-14
NAME: SCAMP/SCAPEGOAT

DESCRIPTION: The SS-14 was an ICBM derived from the solid-fueled SS-13 Savage that used the top two stages of the latter as an intermediate-range ballistic missile (IRBM). It had a range of approx 2,170 nm 12,500 mi; 4,000 km) carrying a 500 kiloton to 1 megaton warhead. The SS-14 was first seen in May 1965 as a tracked vehicle derived from the JS-III heavy tank and carrying a new missile container. The vehicle was code named Scamp; the missile inside the container was later seen separately and code named Scapegoat. Later film analysis showed that Scapegoat was carried by Scamp.

DESIGNATOR: SS-12 Mod 2 (formerly SS-22)
NAME: SCALEBOARD

DESCRIPTION: Once considered to be a new missile and designated SS-22, the SS-12 Mod 2 is a mobile, short-to-medium range ballistic missile deployed at the front level of Soviet Ground Forces.

Scaleboard

USERS: Soviet Union Ground Forces.

CHARACTERISTICS:
Combat Weight: Probably similar to SS-12 Mod 1
  Warhead Heavier than SS-12 Mod 1
Dimensions:
  Length Similar to SS-12 Mod 1
  Diameter Similar to SS-12 Mod 1
Propulsion: Solid-fuel Rocket
Performance:
  Range 559 miles (900 km)
Warhead: 500 kt — 1 megaton nuclear
Accuracy: Circular Error Probable (CEP) 0.20 nm (370 m)
Guidance: Inertial
Carrier: 8 x 8 MAZ-543 missile launching vehicles

DESIGNATOR: SS-5
NAME: SKEAN

DESCRIPTION: The SS-5 was an intermediate-range ballistic missile (IRBM) with a one-megaton nuclear warhead. It was a storable-liquid fueled rocket.

DESIGNATOR: SS-4
NAME: SANDAL

DESCRIPTION: The SS-4 is the oldest missile still in service with the Soviet Strategic Rocket Forces, as part of the Intermediate-range Nuclear Forces (INF) in the western Soviet Union.

USERS: Soviet Union Strategic Rocket Forces.

CHARACTERISTICS:
Missile Weight: 59,525 lb (27,000 kg)
  Throw Weight 3,000 lb (1,361 kg)
Dimensions:
  Configuration Thick cylinder with pointed

Sandal

	nose, small steering vanes at tail
Length	68 ft 0 in (20.73 m)
Diameter	5 ft 3 in (1.60 m)
Propulsion:	One RD-214 four-chamber storable liquid-fueled (Nitric acid/Kerosene) rocket with a static thrust of 163,142 lb (74,000 kg)
Performance:	
Maximum Range	1,079 nm (1,243 mi; 2,000 km)
Maximum Speed	Mach 6-7
Warhead:	1 x 1 megaton nuclear Reentry Vehicle may also have conventional high-explosive warhead
Accuracy:	Circular Error Probable (CEP) 1.24 nm (2,300 m)
Silo Hardness:	Most deployed in soft sites; a few were emplaced in hardened silos
Sensors/Fire Control:	Early SS-4s had radio command guidance all converted to inertial guidance by 1962 steering by efflux influence on small tail vanes

DESIGNATOR:	SSC-X-4
NAME:	NONE

DESCRIPTION: The SSC-X-4 is a transonic, ground-launched, long-range cruise missile similar in concept and mission to the U.S. BGM-109G Gryphon/Tomahawk Ground Launched Cruise Missile (GLCM). Similar Soviet missiles include the submarine-launched SS-N-X-21 and the air-launched AS-15 Kent.

CHARACTERISTICS:
Dimensions:

Configuration	If similar to Tomahawk, then the missile will have a torpedo shape

	with pop-out unswept wings, tail group, and ventral engine intake
Length	23 ft 0 in (7.00 m)
Propulsion:	One turbofan
Performance:	
Speed	Approx Mach 0.7
Maximum Range	About 1,850 mi (3,000 km)
Warhead:	Nuclear
Accuracy:	Circular Error Probable (CEP) approx 150 ft (45.7 m)
Sensors/Fire Control:	Inertial at launch followed by terrain correlation and matching (Tercom) until impact

DESIGNATOR:	SS-3/T-1*
NAME:	SHYSTER

DESCRIPTION: The SS-3 was an early Intermediate-Range Ballistic Missile (IRBM) deployed in limited numbers from 1956 to 1968. Its range was 648 nm (746 mi; 1,200 km) and it carried a 300-kiloton nuclear warhead. The missile required fueling before launch and had a Circular Error Probable (CEP) estimated at 2 nm (2.3 mi; 3.7 km).

DESIGNATOR:	SS-1c/8K11*/R-I7*
NAME:	SCUD B

Scud B

DESCRIPTION: The Scud series are Soviet mobile, nuclear-capable, short-range ballistic missiles for battlefield support of Front and Army levels within the Soviet Ground Forces.

USERS: Soviet Union Ground Forces, Bulgaria, Czechoslovakia, East Germany, Egypt, Hungary, Iraq, Romania and Syria.

CHARACTERISTICS:

Combat Weight:	13,889 lb (6,300 kg)
Warhead:	1,697-1,896 lb (770-860 kg)
Dimensions:	
Configuration	Long cylinder with blunted conical nose, four small tail fins
Length	36 ft 11 in (11.25 m)
Diameter	2 ft 9 in (0.85 m)
Propulsion:	One-stage storable liquid propellant rocket
Performance:	
Range	162 nm (186 mi; 300 km)
Time into action	Approx 1 hour from pre-surveyed site
Warhead:	1-10 kiloton nuclear
Accuracy:	Circular Error Probable (CEP) 0.5 nm (900 m)
Guidance:	Simplified inertial guidance rocket steered by small vanes on each fin extending into efflux
Carrier:	8 x 8 MAZ-S43 wheeled missile launch vehicle

DESIGNATOR: FROG-7/9K21*/R-75*
NAME: LUNA M*

DESCRIPTION: The latest in a series of Soviet unguided, nuclear-capable, short-range artillery rockets, the FROG-7 is the only one mounted on a wheeled carrier. (The previous versions had tracked carriers.)

USERS: (FROG series) Soviet Union Ground Forces, Algeria, Bulgaria, Cuba, Czechoslovakia, East Germany, Egypt, Hungary, Iraq, Kuwait, Poland, Romania, Syria and Yugoslavia.

CHARACTERISTICS:

Combat Weight:	4,409-5,511 lb (2,000-2,500 kg)
Warhead:	992 lb (450 kg)
Dimensions:	
Configuration	Cylindrical body with four fins on tail, conical nose cone.
Length	29 ft 6 in (9.0 m)
Diameter	22 in (550 mm)
Propulsion:	1 large solid-fuel rocket surrounded by a ring of 18 much smaller nozzles

FROG-7

Performance:	
Range	
Maximum	37.8 nm (43.5 mi; 70 km)
Minimum	6.5 nm (7.5 mil; 12 km)
Time into action	15-30 minutes after arriving at pre-surveyed sites
Reload time	20 minutes
Warhead:	Approx 200-kiloton nuclear or chemical or high explosive
Accuracy:	Circular Error Probable (CEP) 0.20-0.21 nm (380-400 m)
Guidance:	Unguided, spin-stabilized
Carrier:	8 x 8 (8-wheel drive) ZIL-135 missile launch vehicle uses firing jacks for stability and on-board crane for reloading. Another ZIL-135 variant carries 3 reloads.

DESIGNATOR: FROG-3/T5D*

DESCRIPTION: FROGs comprise a series of mobile, short-range artillery rockets intended for nuclear battlefield support. FROG-3 was the most widely used of the first 5 variants. FROG-7 is described in a separate database entry.

USERS: Soviet Union Ground Forces, Algeria, Bulgaria, Cuba, Czechoslovakia, East Germany, Egypt, Hungary, Iraq, Kuwait, Poland, Romania, Syria and Yugoslavia.

CHARACTERISTICS:

Combat Weight:	2,250 lb (4,960 kg)
Warhead	992 lb (450 kg)
Dimensions:	
Configuration	Long, two-part tube with four small fins at the tail; warhead

section has a larger diameter than rocket

Length	34 ft 5 in (10.5 m)
Diameter	
Body	16 in (400 mm)
Warhead	22 in (550 mm)
Propulsion:	Two-stage solid-fuel rocket
Performance:	
Range	24.8 miles (40 km)
Warhead:	One 20-kiloton nuclear or high explosive
Accuracy:	Circular Error Probable (CEP) 0.35 nm (650 m)
Guidance:	Spin-stabilized through ballistic flight, no mid-course guidance; Bread Bin radar mounted on trailer
Carrier:	Fully-tracked chassis based on PT-76 light tank with two return rollers added to suspension; carries FROG-3 on erectable launcher.
Crew:	6 (Commander, 5-man firing crew)

ANTI-SHIP MISSILES

DESIGNATOR: AS-6
NAME: KINGFISH

DESCRIPTION: The Soviet AS-6 Kingfish is a high-speed air-to-surface (ASM) anti-ship missile. It is launched at 36,000 ft (11,000 m), climbs to cruise altitude—which may be as high as 59,000 ft (18,000 m) and, in the terminal phase, dives steeply into the target. It is reported to be relatively accurate.

Kingfish

PLATFORMS:
Bombers   Tu-16 Badger-C/G (2 missiles)

CHARACTERISTICS:

Missile Weight:	11,000 lb (4,990 kg)
Warhead	1,102-2,205 lb (500-1,000 kg)
Dimensions:	
Configuration	Long cylinder with long-chord delta wings, small tail group, ogival nose
Length	34 ft 5 in (10.50 m)
Diameter	2 ft 11 in (0.90 m)
Wing Span	8 ft 8 in (2.50 m)
Propulsion:	Probably 1 liquid-fuel rocket
Performance:	
Speed	Mach 2.5-3.5
Maximum Range	
Low Altitude	155-186 mi (250-300 km)
High Altitude	348-435 mi (560-700 km)
Warhead:	Conventional high-explosive or 200-350 kiloton nuclear warhead
Sensors/Fire Control:	Inertial guidance in first part of flight active J-band radar homing in terminal phase

DESIGNATOR: AS-5
NAME: KELT

DESCRIPTION: The AS-5 Kelt is a Soviet second-generation anti-ship missile that generally resembles the missile it replaced, the AS-1 Kennel.

PLATFORMS:
Bombers   Tu-16 Badger-C/G

CHARACTERISTICS:

Combat Weight:	10,360 lb (4,700 kg)
Warhead	2,200 lb (1,000 kg)
Dimensions:	
Configuration	Portly airframe with rounded nose and thickened keel, swept wings with stall fences on top surface, swept tail group with aerial at top
Length	28 ft 2 in (8.59 m)
Wing Span	15 ft 1 in (4.60 m)
Propulsion:	Liquid-fuel rocket
Performance:	
Speed	
Low Altitude	Mach 0.9
High Altitude	Mach 1.2
Maximum Range	
Low Altitude	100 mi (161 km)
High Altitude	200 mi (322 km)
Warhead:	Conventional high-explosive or nuclear warhead
Sensors/Fire Control:	Autopilot terminal homing is either active J-band or passive radar in the missile.

DESIGNATOR: AS-4
NAME: KITCHEN
BURYA*

DESCRIPTION: The AS-4 is the primary Soviet air-launched anti-ship missile. The flight path includes launch at approximately 20,000 ft (6,100 m), a rapid climb to high altitude, cruise at Mach 2.5-3.5, and a steep dive onto the target. In 1985, U.S. Secretary of Defense Weinberger stated that, in combination with the Tu-22M Backfire-B bomber, the AS-4 was "the greatest menace" existing to U.S. naval forces.

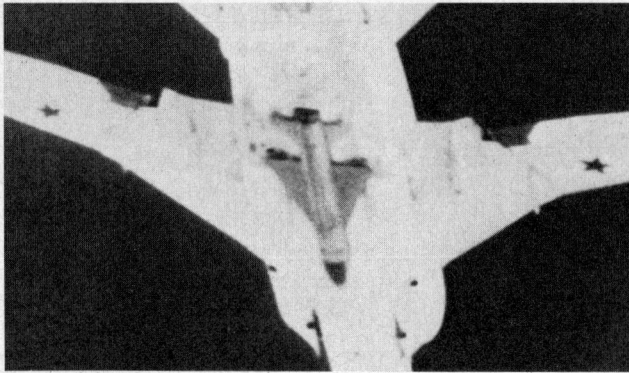

Kitchen

PLATFORMS:
Bombers   Tu-22* Blinder-B
Tu-22M Backfire-B
Tu-20/9S Bear-G

CHARACTERISTICS:

Missile Weight:	14,300 lb (6,500 kg)
Warhead	2,200 lb (1,000 kg)
Dimensions:	
Configuration	Delta wings and cruciform "chopped delta" fins on fuselage, ventral fin folded during carriage
Length	37 ft 1 in (11.30 m)
Wing Span	9 ft 10 in-11 ft (3-3.35 m)
Propulsion:	Storable-liquid rocket
Performance:	
Speed	Mach 2.5-3.5
Maximum Range	
High Altitude	285 mi (459 km)
Low Altitude	170-185 (274-298 km)
Warhead:	Conventional high-explosive or 350 kiloton nuclear warhead
Sensors/Fire Control:	Inertial guidance plus terminal homing with J-band radar anti-radar version uses passive radar homing; Nuclear land attack version uses inertial with estimated circular error probable (CEP) of 0.54 nm (1,000 m)

DESIGNATOR: AS-2
NAME: KIPPER

DESCRIPTION: The Soviet AS-2 is an obsolescent, air-launched, anti-ship missile. It is being replaced by the AS-5 Kelt.

PLATFORM
Bombers   Tupolev Tu-16 Badger-C/G (2 missiles)

CHARACTERISTICS:

Missile Weight:	9,260 lb (4,200 kg)
Warhead	2,205 lb (1,000 kg)
Dimensions:	
Configuration	Pilotless airplane shape with swept wings and tailplane, stubby fin and rudder and turbojet slung under after half of fuselage
Length	31 ft 2 in (9.50 m)
Diameter	3 ft 0 in (0.91 m)
Wing Span	16 ft 1 in (4.90 m)
Propulsion:	One turbojet (perhaps Lyulka AL-5 11,023 lb (5,000 kg; 49 kN)
Performance:	
Speed	Mach 1.2
Maximum Range	
High Altitude	130 mi (210 km)
Low Altitude	62 mi (100 km)
Warhead:	Conventional high-explosive 2,205 lb (1,000 kg) or 400 kiloton nuclear warhead
Sensors/Fire Control:	Launched on pre-programmed autopilot that can be overridden enroute by launch aircraft; Terminal active radar homing

DESIGNATOR: AS-1
NAME: KENNEL
Komet*

DESCRIPTION: The AS-1 is believed to have been the first operational Soviet air-to-surface missile. Development started in 1946 or possibly earlier. It was developed initially for use with the Tu-4* Bull, a Soviet copy of the Boeing B-29 Superfortress. The missile was developed for use on surface ships, but engine limitations dictated that it could be used only from aircraft. The coastal defense variant was designated SSC-2b Samlet.

DESIGNATOR: SS-N-22
NAME: NONE

DESCRIPTION: The Soviet SS-N-22 is an improved version of the SS-N-9 anti-ship cruise missile.

PLATFORMS:

Destroyers	*Sovremennyy** class
Corvettes	*Tarantul III** class

CHARACTERISTICS:

Dimensions:	
Length	Approx 30 ft 0 in (9.15 m)
Propulsion:	Solid-fuel Rocket
Performance:	
Speed	Approx Mach 2.5
Maximum Range	Approx 60 nm (68 mi; 110 km)
Warhead:	Conventional high explosive or 200 kiloton nuclear
Sensors/Fire Control:	Band Stand G/H-band search and missile tracking and control radar mid-course guidance from Ka-2S Hormone-B helicopter

DESIGNATOR: SS-N-19
NAME: NONE

DESCRIPTION: The SS-N-19 is a supersonic long-range, anti-ship cruise missile that evolved from the SS-N-3/12. It is deployed in both surface ships and submarines.

PLATFORMS:

Submarines	*Oscar* class (24 tubes)
Battle cruisers	*Kirov** class (20 tubes)

CHARACTERISTICS:

Warhead Weight	2,000 lbs (907 kg)
Dimensions:	
Configuration	Long cylinder with swept wings and small tailplanes, no vertical tail, ventral angled turbojet intake, two small detachable boosters
Propulsion:	Two solid-fuel boosters; One turbojet sustainer engine
Performance:	
Speed	Mach 1+
Maximum Range	299 nm (342 mi; 550 km)
Warhead:	Conventional high explosive or 500 kiloton nuclear warhead Sensors/Fire Control: inertial with anti-radar homing

DESIGNATOR: SS-NX-13
NAME: NONE

DESCRIPTION: The SS-NX-13 was a submarine-launched, tactical ballistic missile intended for attacking aircraft carriers. Flight tested from 1970 to 1973, this two-stage, liquid-fueled rocket had a range of 370 nm (426 mi; 686 km). The SS-NX-13 apparently was developed for launching from the *Yankee*-class submarines, probably with satellite targeting at launch and a terminal radar homing system capable of maneuvering 30 nm (35 mi; 56 km). A nuclear warhead would have been employed with the operational missile. Some Western analysts believe a later version may have been planned; this variant would have been used in an anti-ballistic-missile submarine role.

DESIGNATOR: SS-N-12
NAME: SANDBOX

DESCRIPTION: The SS-N-12 followed the SS-N-3 Shaddock anti-ship missile, being capable of supersonic speed. Despite the higher speed, the Sandbox probably requires mid-course guidance updates from surface ship radars or, in the case of the submarine launch, from aircraft.

Sandbox

PLATFORMS:

Submarines	Modified *Echo II* (8 missiles)
Aircraft Carriers	*Kiev** class (16 missiles)
Cruisers	*Slava** class (16 missiles)

CHARACTERISTICS:

Warhead Weight	2,205 lb (1,000 kg)
Dimensions:	
Configuration	Two-section tailless airframe, cylindrical forebody with pointed nose, larger aft section with ventral turbojet intake, swept wings with trailing edge control surfaces, small tail fins
Length	38 ft 4½ in (11.70 m)
Propulsion:	Turbojet
Performance:	
Speed	Mach 2.5
Maximum Range	300 nm (345 mi; 556 km)
Warhead:	Conventional high-explosive or 350 kiloton nuclear

SOVIET WEAPON SYSTEMS AND ELECTRONICS
MISSILES

Sensors/Fire Control:
Radar

Trap Door mid-course missile guidance on *Kiev*
Front Door/Front Piece on *Slava* and *Echo-II*
Terminal homing (active or passive)

DESIGNATOR: SS-N-9
NAME: NONE

DESCRIPTION: The SS-N-9 is a supersonic anti-ship missile, launched initially from surface ships and subsequently fitted in the *Charlie-II* cruise-missile submarines for underwater launch.

PLATFORMS:

Submarines	*Charlie-II* class (8 missiles)
	Papa (10 missiles)
Small combatants	*Nanuchka I/III** class (6 missiles)
	*Sarancha** (4 missiles)

CHARACTERISTICS:

Missile Weight:	7,275 lb (3,300 kg)
Warhead	1,102 lb (500 kg)
Dimensions:	
Configuration	Tailless, winged airframe with rounded nose, shoulder-mounted "cropped delta" wings with trailing edge control surfaces, small stabilizers at tail, ventral detachable booster
Length	29 ft 0 in (8.84 m)
Propulsion:	Solid-fuel rocket booster; Solid-fuel rocket sustainer
Performance:	
Speed	Mach 1.4
Maximum Range	60 nm (69 mi; 111 km); without mid-course guidance = 27 nm (31 mi; 50 km)
Warhead:	Conventional high-explosive or 200 kiloton nuclear warhead
Sensors/Fire Control:	
Radar	Surface ships have Band Stand G/H search and missile tracking and control radar terminal homing (passive or active); Inertial autopilot

DESIGNATOR: SS-N-7
NAME: SIREN

DESCRIPTION: The SS-N-7 was the Soviet Navy's first underwater-launched, anti-ship cruise missile. It was deployed only in 12 *Charlie-I* class nuclear-propelled guided missile submarines (SSGN). The SS-N-7 was succeeded in later *Charlie-II* class SSGNs by the SS-N-9, which has about double the earlier missile's range.

PLATFORMS:

Submarines	*Charlie I* class (8 missiles)

CHARACTERISTICS:

Missile Weight:	6,393 lb (2,900 kg)
Warhead	1,102 lb (500 kg)
Dimensions:	
Configuration	Stubby winged rocket, shoulder-mounted "cropped delta" wings with trailing edge controls, triform tail surfaces.
Length	22 ft 11½ in (7.00 m)
Propulsion:	Solid-fuel Rocket
Performance:	
Speed	Mach 0.9
Maximum Range	35 nm (40 mi; 64.9 km)
Practical Range	27 nm (31 mi; 50 km)
Warhead:	Conventional high-explosive or 200 kiloton nuclear warhead
Sensors/Fire Control:	Autopilot over most of flight; low-altitude flight profile but not sea-skimming; terminal homing is either active J-band radar homing or passive radar homing

DESIGNATOR: SS-N-3
NAME: SHADDOCK

DESCRIPTION: The Shaddock is a large, air-breathing cruise missile originally developed for the strategic attack role in the SS-N-3c variant. The SS-N-3c was followed by the anti-ship -3a and -3b variants. Deployed on several Soviet submarines and cruisers.

Shaddock

PLATFORMS:

Submarines	*Juliett* class (4 missiles)
	Echo II class (8 missiles)
Cruisers	*Kynda** class (16 missiles)

Discarded from	*Kresta I** class (4 missiles) *Echo I* and *Whiskey* submarine classes

CHARACTERISTICS:

Missile Weight:	11,905 lb (5,400 kg)
Warhead	2,205 lb (1,000 kg)
Dimensions:	
Configuration	Tailless "airplane" with pointed nose, "cropped delta" anhedral mainplanes, tapered turbojet nozzle, ventral jet intake, two detachable boosters with angled nozzles
Length	
SS-N-3a/b	33 ft 7 in (10.20 m)
SS-N-3c	38 ft 6 in (11.75 m)
Diameter	3 ft 2 in (0.98 m)
Wing Span	16 ft 5 in (5.00 m)
Propulsion:	Two solid-fuel boosters; one turbo-jet sustainer engine
Performance:	
Speed	Mach 0.9
Maximum range	
SS-N-3a/b	250 nm (288 mi; 463 km)
SS-N-3c	400+ nm (461 mi; 741 km)
Warhead:	Conventional high-explosive or 350 kiloton nuclear warhead
Sensors/Fire Control:	Front Door or Front Piece fire control radars on submarines; Scoop Pair I-band fire control radar for mid-course guidance on *Kynda, Kresta I*; inertial with mid-course guidance through video data link from targeting ship or aircraft to launching ship to missile in form of radar picture with target indicated; submarine must remain on surface for up to 25 minutes after launch

DESIGNATOR:	**SS-N-2**
NAME:	**STYX**

DESCRIPTION: The Styx is an anti-ship missile for coastal defense developed for small combat craft. This surface-to-surface missile (SSM) has undergone several upgrades, with the designation SS-N-11 briefly used to indicate the SS-N-2c version. It is a subsonic missile that was first deployed in Soviet *Komar**-class missile boats.

PLATFORMS:

Destroyers	Modified *Kashin** class
	Modified *Kildin** class
Small combatants	*Komar* class
	*Osa I/II** classes

	*Matka** class
	*Tarantul** class
	*Nanuchka** class (foreign transfers)

USERS: Soviet Union Navy, Algeria, Bulgaria, Cuba, East Germany, Egypt, Ethiopia, Finland, India, Iraq, Libya, North Korea, Poland, Romania, Somalia, South Yemen, Syria, Vietnam and Yugoslavia.

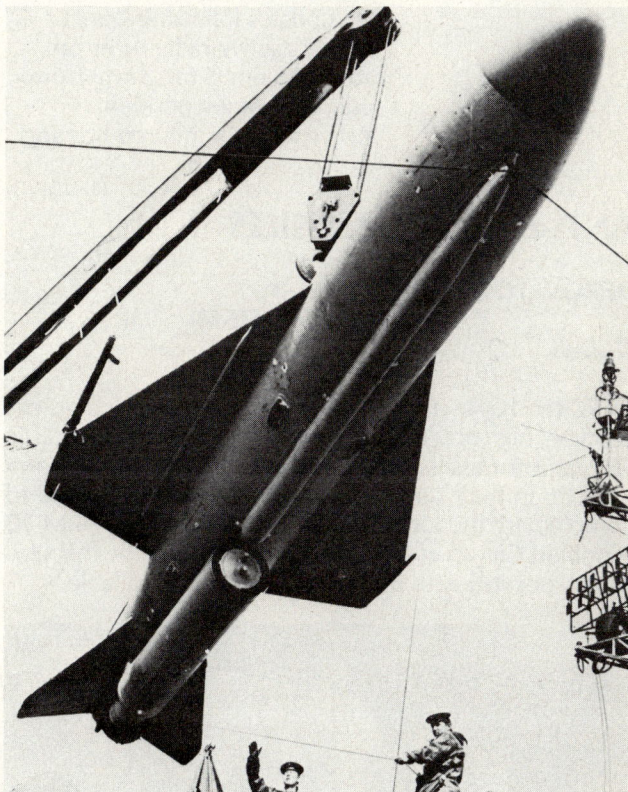

Styx

CHARACTERISTICS:

Missile weight	
SS-N-2a/b	Approx 5,500 lb (2,500 kg)
Warhead	1,102 lb (500 kg)
Dimensions:	
Configuration	Resembles stubby jet aircraft with "cropped delta" mainplanes with guidance aerials at tips; tri-form tail group; rocket booster under fuselage, rounded nose cone
Length	
SS-N-2a/b	19 ft 0 in (5.80 m)
SS-N-2c	21 ft 4 in (6.50 m)
Diameter	29.5 in (750 mm)
Wing Span	9 ft 2 in (2.80 m)
Propulsion:	Solid-fuel Rocket booster; turbojet sustainer

Performance:
Speed	Mach 0.9
Maximum range	
SS-N-2a 25 nm	(28.8 mi; 46.3 km)
SS-N-2b 27 nm	(31.1 mi; 50 km)
SS-N-2c 45 nm	(51.8 mi; 83.4 km)

Warhead: Conventional high explosives

Sensors/Fire Control:

SS-N-2a/b   Square Tie search and target acquisition radar.
No updates following launch; missile's active radar turns on about 4.3 nm (5 mi; 8 km) from estimated target position

SS-N-2c   Active radar or infrared homing

ANTI-BALLISTIC MISSILES

DESIGNATOR: **ABM-X-3**
(Formerly SH-08)

NAME: **GAZELLE**

DESCRIPTION: The Gazelle is a hypersonic, quick-reaction, high acceleration endo-atmospheric anti-ballistic interceptor missile (ABM) that would intercept re-entry vehicles in their terminal phase. It is being deployed to complement the longer-range exo-atmospheric ABM-1B modified Galosh at ABM sites around Moscow. It is silo-based, possibly with underground reloads available.

Gazelle

CHARACTERISTICS:

Performance:

Range	Greater than 43 nm (50 mi; 80 km) est.
Warhead:	Nuclear, possibly in the low kiloton range
Sensors/ Fire Control:	Flat Twin terminal intercept control radar, modular to permit storage and rapid assembly; Pawn Shop missile guidance

radar, also modular; Pill Box 4-sided phased-array radar; Cat House target tracking radar working with Dog House A-band phased-array re-entry vehicle acquisition and tracking radar with 1,476 nm (1,700 mi; 2,736 km) range against small targets; Hen House BMD detection and tracking radar; 11 radars at 6 points along the Soviet periphery; Large Phased-Array Radars (LPAR) — 6 located along the Soviet periphery, 3 more under construction including the controversial Krasnoyarsk installation

DESIGNATOR: **ABM-1B UR-96***
NAME: **GALOSH**

DESCRIPTION: The ABM-1 is the oldest Soviet anti-ballistic interceptor Missile (ABM) still in service. Originally deployed at four sites to the north and west of Moscow, the early version is being replaced by a modified variant of the Galosh as well as later ABM designs.

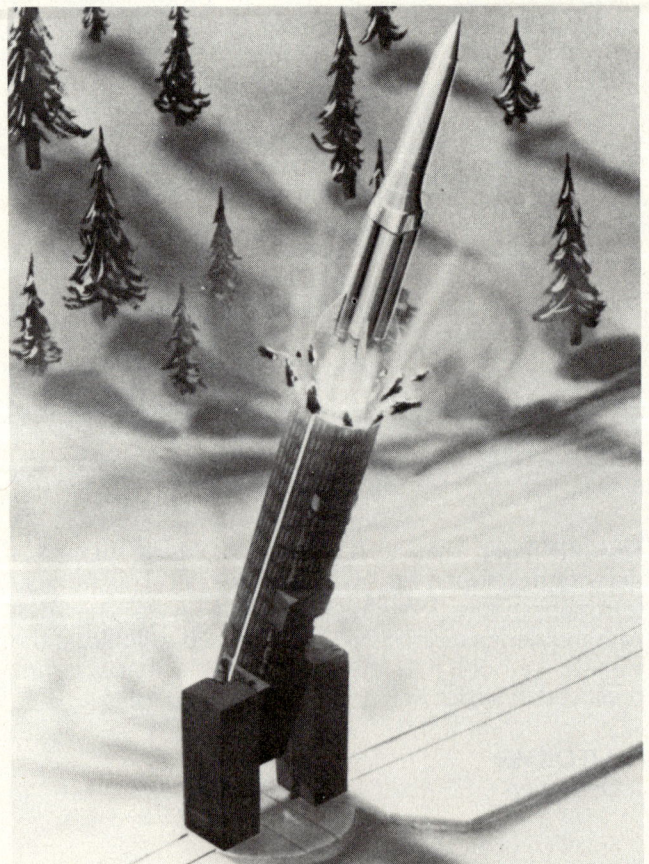

Galosh

USERS: Soviet Union Air Defense Forces.

CHARACTERISTICS:

Missile Weight:	Approx 72,000 lb (32,700 kg)
Dimensions:	
Configuration	Artist's conception shows short, pointed cylinder atop cluster of four boosters, fins at rear of boosters
Length	Approx 65 ft 0 in (19.80 m)
Diameter	Approx 8 ft 6 in (2.60 m)
Propulsion:	Three-stage rocket
Performance:	
Maximum Range	Approx 174 nm (200 mi; 322 km)
Maximum Altitude	Exo-atmospheric
Warhead:	3 megaton nuclear
Sensors/ Fire Control:	Try Add Ballistic Missile Defense (BMD) guidance and engagement radar (6 installations for each 16 missiles); Cat House target tracking radar working with Dog House A-band phased-array re-entry vehicle acquisition and tracking radar with 1,476 nm (1,700 mi; 2,736 km) range against small targets; Hen House BMD detection and tracking radar (11 in total) at six points along the Soviet periphery

ANTI-SUBMARINE WARFARE MISSILES

DESIGNATOR:	**SS-N-16**
NAME:	**NONE**

DESCRIPTION: A further development of the SS-N-16, this submarine-launched missile substitutes an anti-submarine homing torpedo for the SS-N-16's nuclear warhead. In concept, the SS-N-16 is most similar to the U.S. Navy's Sea Lance anti-submarine warfare (ASW) missile.

PLATFORMS:

Attack submarines	*Victor II/III* class; possibly also in *Akula* class, *Mike* class, *Oscar* class, *Sierra* class

CHARACTERISTICS:

Dimensions	
Diameter	26 in (650 mm)
Propulsion:	Solid-fuel rocket, ballistic flight torpedo propulsion unknown but probably either steam, electric (battery), or closed-cycle thermal drive
Performance:	
Maximum Range	30-50 nm (55-92 km)

Warhead:	Conventional high-explosive (homing torpedo)
Sensors/Fire Control:	
Guidance	Submarine sonar provides initial targeting information search pattern after water re-entry and descent to prescribed depth, a programmed maneuver begins until the torpedo's own homing system detects a target

DESIGNATOR:	**SS-N-15**
NAME:	**NONE**

DESCRIPTION: The SS-N-15 is a submarine-tube launched ASW rocket with a nuclear warhead which is similar in concept to the U.S. Navy's Subroc (submarine rocket). It is reportedly carried on most Soviet nuclear-powered attack submarines.

PLATFORMS:

Guided Missile Submarines	*Oscar* class, *Papa* class and *Charlie I/II* classes
Attack submarines	*Victor I/II/III* classes, *Alfa* class, *Mike* class, *Sierra* class, *Akula* class
Diesel attack subs	*Tango* class

CHARACTERISTICS:

Dimensions:	
Diameter	21 in (533 mm) maximum
Propulsion:	Solid-fuel rocket
Performance:	
Maximum Range	20 nm (23 mi; 37 km)
Warhead:	Nuclear
Sensors/Fire Control:	
Guidance	Inertial; initial targeting data from the submarine's passive sonar; probably bearing/range for firing solution from active sonar

DESIGNATOR:	**SS-N-14**
NAME:	**SILEX**

DESCRIPTION: Similar in concept to the French Navy's Malafon and the Australian Navy's Ikara, the SS-N-14 Silex is a surface-launched, rocket-propelled ASW weapon that carries a homing torpedo out to the first sonar convergence zone.

PLATFORMS:

Nuclear Missile Cruiser	*Kirov* (1 launcher and reloads)
Missile Cruisers	*Kresta I/II* classes (8 missiles), *Kara* class (8 missiles)

Silex

Destroyers	*Udaloy* class (8 missiles)
Frigates	*Krivak I/II* class (4 missiles)

CHARACTERISTICS:

Dimensions:	
Length	Approx 25 ft (7.60 m)
Propulsion:	Glider powered by solid-fuel rocket, water entry slowed by parachute
Performance:	
Maximum Range	Approx 30 nm (55.6 km)
Minimum Range	4 nm (7.4 km)
Warhead:	Conventional high-explosive (ASW homing torpedo)
Sensors/Fire Control:	
Guidance	Command radio and inertial; acoustic terminal homing; *Kresta II* class has Head Light D/F/G/H-band fire control radar; *Kirov, Udaloy,* and *Krivak* classes have Eye Bowl F-band fire control radar

DESIGNATOR:	**SUW-N-1 (launcher)/FRAS-1 (Free Rocket Anti-Submarine)**
NAME:	**NONE**

DESCRIPTION: The SUW-N-1 is a short-range ASW missile system found on two classes of Soviet aircraft carriers, apparently provided in place of the SS-N-14 found in other modern Soviet warships. It is an adaptation of the FROG-7 unguided artillery rocket.

PLATFORMS:

Aircraft carriers	*Moskva** class
	*Kiev** class

CHARACTERISTICS:

Dimensions:	
Length	Approx 20 ft 4 in (6.20 m)

SUW-N-1

Diameter	Approx 22 in (550 mm)
Propulsion:	Solid-fuel Rocket, ballistic flight
Performance:	
Maximum Range	16 nm (30 km)
Warhead:	Nuclear
Sensors/Fire Control:	
Guidance	Inertial or unguided

SURFACE-TO-AIR MISSILES

DESIGNATOR:	**SA-13**
NAME:	**GOPHER**

DESCRIPTION: The SA-13 Gopher, mounted in 4-box launcher on an MT-LB chassis, is the successor to the SA-9 Gaskin as the short-range surface-to-air missile (SAM) for Soviet Ground Forces.

USERS: Soviet Union Ground Forces and Navy (Naval Infantry), Angola, Bulgaria, Czechoslovakia, Libya and Syria.

CHARACTERISTICS:

Combat Weight:	121 lb (55 kg)
Warhead	9 lb (4 kg)
Dimensions:	
Configuration:	May resemble a lengthened SA-9
Length	7 ft 2 in (2.20 m)
Diameter	4.7 in (120 mm)
Propulsion:	Solid-fuel rocket
Performance:	
Speed	Mach 2+
Maximum Range	5.4 nm (6.2 mi; 10 km)
Minimum Range	1,640 ft (500 m)
Effective Altitude	165-16,500 ft (50-5,000 m)
Warhead:	Conventional high-explosive
Sensors/ Fire Control:	Range-only radar mounted on missile vehicle; Hat Box system (either passive radar warning or radar link with ZSU-23-4 Gun Dish radar); Dual-wavelength infrared seeker

DESIGNATOR: SA-10a
NAME: GRUMBLE

DESCRIPTION: Considered a greatly improved successor to the SA-2 Guideline, the SA-10 surface-to-air missile (SAM) system has been fielded in both fixed (SA-10a) and mobile (SA-10b) versions. The mobile version is described in a separate database entry. The SA-10 is credited with the ability to intercept not only small radar cross-section cruise missiles flying at low altitudes, but also tactical ballistic missiles (e.g, Pershing II) and possibly some strategic ballistic missiles (i.e. re-entry vehicles).

Grumble

USERS: Soviet Union Air Defense Forces.

CHARACTERISTICS:
Dimensions:

Configuration	Slender cylinder with ogival nose, two sets of cruciform long-chord delta wings, the larger at mid-body, the smaller (presumably steerable) at the tail
Length	23 ft 6 in (7.16 m)
Diameter	17.7 in (450 mm)
Propulsion:	Solid-fuel rocket
Performance:	
Speed	Mach 6 (acceleration to this speed is said to be 100g)
Maximum Range	54 nm (62 mi; 100 km)
Maximum Altitude	Tested to 100,000 ft (30,480 m)
Minimum Altitude	Less than 1,000 ft (305 m)
Warhead:	Conventional high-explosive some could have nuclear warheads
Sensors/ Fire Control:	Uses three different ground-based radars including a planar array type; Missile has active radar terminal homing; System can engage several targets at once

DESIGNATOR: SA-9
NAME: GASKIN

DESCRIPTION: Mounted on BRDM-2 armored reconnaissance vehicles, the SA-9 is a short-range mobile surface-to-air missile (SAM) intended to support ground combat troops.

Gaskin

USERS: Soviet Union Ground Forces, Algeria, Angola, Cuba, Czechoslovakia, East Germany, Egypt, Hungary, India, Iraq, North Yemen, Poland, South Yemen, Syria, Vietnam, and Yugoslavia.

CHARACTERISTICS:

Combat Weight:	66 lb (29.9 kg)
Dimensions:	
Configuration	Probably thin cylinder with rounded nose, four steerable foreplanes, and four larger stabilizers at tail
Length	5 ft 9 in-6 ft 7 in (1.75-2.00 m)
Diameter	4.3-4.7 in (110-120 mm)
Propulsion:	Solid-fuel rocket
Performance:	
Speed	Mach 1.5+
Maximum Range	4.3 nm (5 mi; 8 km)
Maximum Altitude	16,404 ft (5,000 m)
Minimum Altitude	150 ft (45.7 m)
Warhead:	Conventional high-explosive fragmentation
Sensors/ Fire Control:	Gun Dish 3-band fire control radar optical tracking by operator infrared homing seeker

SOVIET WEAPON SYSTEMS AND ELECTRONICS
MISSILES

DESIGNATOR: SA-7
NAME: GRAIL

DESCRIPTION: Soviet small, portable short-range shoulder-fired surface-to-air missile (SAM). It is also fired as the SA-N-5 from various small combatant and auxiliary ships. In very widespread use by nations, guerrilla groups, and terrorist organizations. Also formerly known as Strella* (Soviet Strela = Arrow).

Grail

USERS: Soviet Union Ground Forces and Naval Infantry, Algeria, Angola, Botswana, Bulgaria, Cuba, Czechoslovakia, East Germany, Egypt, Ethiopia, Finland, Guinea-Bissau, Hungary, India, Iran, Iraq, Kuwait, Laos, Lebanon, Libya, Mauritania, Morocco, Nicaragua, North Korea, Poland, Romania, Seychelles, South Yemen, Sudan, Syria, Tanzania, Uganda, Vietnam, Yugoslavia, and Zambia.

CHARACTERISTICS:

Combat Weight:	20 lb (9.07-9.2 kg)
Warhead	5.5 lb (2.5 kg)
Dimensions:	
Configuration	Launch tube has simple sights; power supply is under tube missile has long pipe-like body with very small foreplanes and angled steering vanes extending to the rear of the missile
Length	4 ft 3 in (1.30 m)
Diameter	2.75 in (70 mm)
Propulsion:	Solid-fuel rocket
Performance:	
Speed	Mach 1.5
Maximum Range	4.3-5.2 nm (5-6 mi; 8-9.65 km)
Effective Altitude	5,000 ft (1,524 m)
Warhead:	High-explosive fragmentation
Sensors/ Fire Control:	Rear-aspect infrared (IR) homing; missile is launched when tracker light on launch tube indicates lock-on; Later versions have filter in the IR seeker to reduce effect of flare countermeasures

DESIGNATOR: SA-5/S-200*
NAME: GAMMON

DESCRIPTION: The SA-5 Gammon is a Soviet long-range, medium-to-high altitude strategic surface-to-air missile (SAM). It is a fixed-site, point defense SAM that has been upgraded to operate in conjunction with the SA-10a Grumble. The SA-5 designation was first given to a large SAM seen in 1963 and called Griffon which was not deployed in large numbers.

USERS: Soviet Union Air Defense Forces, Libya and Syria.

CHARACTERISTICS:

Launch Weight	Approx 22,046 lb (10,000 kg)
Dimensions:	
Configuration	Long cylindrical body with conical nose four long-chord delta wings in cruciform layout and four small rectangular control surfaces at extreme rear; Four jettisonable boosters with canted nozzles
Length	34 ft 9 in (10.59 m)
Diameter	34 in (0.86 m)
Wing Span	9 ft 6 in (2.90 m)
Propulsion:	Four solid-fuel boosters; Solid-fuel second stage
Performance:	
Speed	Mach 3.5 +
Maximum Range	162 nm (186 mi; 300 km)
Minimum Range	32-43 nm (37-50 mi; 60-80 km)
Maximum Altitude	Over 100,000 ft (30,480 m)
Effective Altitude	Approx 95,000 ft (29,000 m)

Gammon

Warhead:	Conventional high-explosive; some may have had nuclear warheads
Sensors/ Fire Control:	Radio command from Square Pair H-band fire control radar, and terminal semi-active radar homing in warhead; SA-5 sites also have two other associated radars: —Back Net E-band early warning (EW) and ground control intercept (GCI) radar (perhaps being replaced by Bar Lock E/F-band target acquisition and GCI radar); —Side Net (Soviet PRV-ll) E-band height-finding radar

DESIGNATOR: SA-3/S-125*/5B24*/5V27U*
NAME: GOA/PECHORA*

DESCRIPTION: The SA-3 is a low-altitude surface-to-air missile (SAM) system in widespread use. It is considered to be the Soviet equivalent of the U.S. Hawk SAM system.

Goa

USERS: Soviet Union Air Defense Forces, Afghanistan, Algeria, Angola, Bulgaria, Cuba, Czechoslovakia, East Germany, Egypt, Ethiopia, Hungary, India, Iraq, Libya, Mali, North Korea, Poland, Somalia, South Yemen, Syria, Tanzania, Vietnam, Yugoslavia and Zambia.

CHARACTERISTICS:

Combat Weight:	1,402 lb (636 kg)
Warhead	Approx 132 lb (60 kg)
Dimensions:	
Configuration	Cylindrical booster with 4 square stabilizers in cruciform layout; Second stage tapers to point, mounts 4 small steerable delta foreplanes, 4 larger "cropped delta" mainplanes with aerials at the wing tips

Length	22 ft 0 in (6.70 m)
Diameter	
Booster	24 in (600 mm)
Second Stage	10-18 in (250-450 mm)
Wing Span	
Booster	4 ft 11 in (1.50 m)
Second Stage	4 ft 0 in (1.20 m)
Propulsion:	Solid-fuel booster; Solid-fuel sustainer
Performance:	
Speed	Mach 3+
Maximum Slant Range	15-18.5 mi (25-30 km)
Maximum Altitude	Over 43,000 ft (13,100 m)
Minimum Altitude	150 ft (45.7 m)
Warhead:	Conventional high-explosive
Sensors/ Fire Control:	Radio command guidance with radar terminal homing; Flat Face (Soviet designation is P-15) C-band target acquisition radar; Squat Eye (P-15M) C-band target acquisition radar used instead of Flat Face when low-altitude defense is stressed; Low Blow I-band missile control radar; Spoon Rest (P-12) early warning radar.

DESIGNATOR: SA-2/V750VK/S-75*/V75SM*
NAME: GUIDELINE/DVINA*

DESCRIPTION: The SA-2 is a trailer-mounted mobile surface-to-air missile (SAM) in widespread use in the Soviet Air Defense Forces and in many other countries. Many updated and exported versions exist.

USERS: Soviet Union Air Defense Forces, Albania, Algeria, Angola, Bulgaria, Cuba, Czechoslovakia, East Germany, Egypt, Ethiopia, Hungary, India, Iraq, Libya, North Korea, North Yemen, Poland, Romania, Somalia, South Yemen, Sudan, Syria, Vietnam, and Yugoslavia.

CHARACTERISTICS:

Combat Weight:	5,070 lb (2,300 kg)
Warhead	286 lb (130 kg)
Dimensions:	
Configuration	Two sections; the booster has a larger diameter with 4 large guidance "cropped delta" fins in a cruciform layout; Top of booster tapers to meet flight section; second stage has smaller diameter, 4 blunted delta fins near the rear and smaller fins still further back; Long conical nose
Length	34 ft 9 in (10.59 m)
Diameter	28 in (700 mm) booster

Guideline

Propulsion:	20 in (500 mm) second stage Solid-fuel booster; (nitric acid/hydrocarbon propellant) sustainer

Performance:
Speed	Mach 3.5
Slant Range	18.8-27 nm (21.7-31 mi; 35-50 km)
Minimum Range	3.8-4.9 nm (4.3-5.6 mi; 7-9 km)
Maximum Altitude	90,000 ft (27,432 m)
Minimum Altitude	300 ft (91.4 m)
Reload Time	10 min
Warhead:	High explosive with proximity fuze; some may have nuclear warheads
Sensors/ Fire Control:	Radio command guidance from Fan Song-A/B E/F-band or Fan Song D/E G-band missile control radar; many installations have Spoon Rest early warning radar

DESIGNATOR:	SA-1/R-113*
NAME:	GUILD

DESCRIPTION: The SA-1 Guild is a Soviet strategic surface-to-air missile (SAM) that entered service over 30 years ago. It is a fixed-site, medium-altitude missile thought to be ineffective against targets with a small radar cross-section. The SA-1 was deployed in two concentric rings around Moscow. A typical SA-1 site included 60 missiles and one Yo-Yo radar. The SA-1 is being replaced by the SA-10 Grumble SAM.

USERS: Soviet Union Air Defense Forces.

CHARACTERISTICS:
Missile Weight:	6,614 lb (3,000 kg)
Dimensions:	
Configuration	Long cylinder with small cruciform "cropped delta" steerable foreplanes; Larger "cropped delta" mainplanes and a bulged nozzle at the rear
Length	39 ft 4 in (12.0 m)
Diameter	2 ft 3 in (0.70 m)

Wing Span	Approx 8 ft 10 in (2.70 m)
Propulsion:	Single-stage liquid-fuel rocket
Performance:	
Speed	Mach 2.5
Maximum Range	27 nm (31 mi; 50 km)
Maximum Altitude	Approx 60,000 ft (18,290 m)
Warhead:	Possibly nuclear and conventional high-explosive
Sensors/ Fire Control:	Radio command with terminal homing Yo-Yo E/F-band target acquisition and missile tracking radar

Guild

DESIGNATOR:	SA-N-9
NAME:	NONE

DESCRIPTION: The SA-N-9 is a shipboard short-range surface-to-air missile (SAM) fitted in several Soviet ships. It has a capability against anti-ship cruise missiles as well as aircraft. The SA-N-9 is launched from an 8-round rotary vertical launcher. There are 16 launchers in the *Kirov** class (128 missiles), 8 launchers in the *Kiev** class (64 missiles), and 8 launchers in the *Udaloy** class (64 missiles).

SA-N-9

PLATFORMS:

Aircraft carriers	Kiev* class (Novorossiysk*, Baku* only)
Battle cruiser	Kirov* (Frunze* only)
Destroyers	Udaloy* class

CHARACTERISTICS:

Propulsion:	Probably solid-fuel rocket
Performance:	
Speed	Supersonic
Maximum Range	Approx 8 nm (9.3 mi; 15 km)
Effective Altitude	40,000-60,000 ft (12,192-18,288 m)
Warhead:	Conventional high-explosive
Sensors/ Fire Control:	Cross Sword fire control radar in Udaloy class; the Frunze*, Novorossiysk*, and Baku* may use other radars for missile guidance

DESIGNATOR: SA-N-7
NAME: GADFLY

DESCRIPTION: The SA-N-7 missile is reported to be a naval version of the SA-11 land-based surface-to-air missile (SAM). As with the longer-range SA-N-6 Grumble, the SA-N-7 system is a sophisticated weapon system, being capable of multiple engagements. The SA-N-7 launcher is a single-rail device, loaded with the launching rail in a vertical position from a below-decks magazine that holds 20 missiles. The Sovremennyy* class destroyers have 2 launchers and a total of 40 missiles.

PLATFORMS:

Destroyers	Kashin class (Provornyy* only) Sovremennyy* class

CHARACTERISTICS:

Missile Weight:	Unknown; SA-11 is 1,433 lb (650 kg)
Warhead	Unknown; SA-11 is 198 lb (90 kg)
Dimensions:	
Length	16 ft 5 in (5.00 m)
Diameter	Unknown; SA-11 is 16 in (400 mm)
Wing Span	Unknown; SA-11 is 3 ft 11 in (1.20 m)
Propulsion:	Solid-fuel Rocket
Performance:	
Speed	Mach 3
Maximum Range	Approx 15 nm (17 mi; 28 km)
Effective Altitude	
Maximum	46,000 ft (14,021 m)
Minimum	100 ft (30.5 m)
Warhead:	Conventional high-explosive
Sensors/ Fire Control:	Sovremennyys have six Front Dome H/I-band radars, each of which is thought capable of

illuminating an individual target once it is designated by the Top Steer D/E/F 3-dimensional air search radar and associated digital computers; missile guidance is by radio command and semi-active radar homing with infrared or electro-optical backup for operations under heavy attack or in a heavy jamming environment

DESIGNATOR: SA-N-6
NAME: GRUMBLE

DESCRIPTION: Thought to be a naval version of the SA-10 land-based surface-to-air missile, the SA-N-6 is a long-range SAM deployed in large Soviet surface ships. It may also have an anti-ship capability. The SA-N-6 is the world's first major vertical launch system to have become operational. Each vertical launcher has a rotary 8-missile magazine below deck. The launcher ejects the missile, which rises to approx 50 ft (15 m) before the rocket motor ignites.

Grumble (Naval)

PLATFORMS:

Cruisers	Kara class (Azov* only) Kirov* class Slava* class

CHARACTERISTICS:

Weight, warhead:	198 lb (90 kg)
Dimensions:	
Length	Approx 23 ft (7.00 m)
Propulsion:	Solid-fuel rocket
Performance:	
Speed	Approx Mach 6
Maximum Range	More than 30 nm (35 mi; 55 km)
Maximum Altitude	100,000 ft (30,480 m)
Warhead:	Conventional high-explosive; may have nuclear warhead

Sensors/Fire Control: Top Dome I- or J-band radar system missile guidance is thought to be track-via-missile (i.e., the missile provides radar updates to the launching ship)

DESIGNATOR: SA-N-4
NAME: GECKO

DESCRIPTION: The SA-N-4 is a widely deployed Soviet short-range, surface-to-air missile (SAM) system for point defense of combatant, amphibious, and auxiliary ships. (It is believed to be a naval SA-8 Gecko.)

PLATFORMS:

Aircraft Carriers	Kiev* (first 2 ships only)
Cruisers	Kara class
	Kirov* class
	Slava* class
Command Ships	Mod Sverdlov* class
Frigates	Krivak class
	Koni class (for export)
Missile Corvettes	Nanuchka class
Missile Boats	Sarancha
Amphibious Ships	Ivan Rogov* class
Auxiliary Ships	Berezinea*

USERS: Soviet Union Navy, Algeria, Cuba, East Germany, India, Libya, Romania, and Yugoslavia.

CHARACTERISTICS:

Missile Weight:	419 lb (190 kg)
Warhead	110 lb (50 kg)
Dimensions:	
Configuration	Long cylinder, ogival nose, small steerable cruciform foreplanes, cruciform "cropped delta" stabilizers at tail
Length	10 ft 6 in (3.20 m)
Diameter	8.2 in (210 mm)
Wing Span	2 ft 1 in (0.64 m)
Propulsion:	Solid-fuel Rocket
Performance:	
Speed	Approx Mach 2.5
Maximum Range	9.2 mi (14.8 km)
Effective Altitude	10,000 ft (3,048 m)
Warhead:	Conventional high-explosive
Sensors/Fire Control:	Pop Group fire control radar system (may be derived from the SA-8 Land Roll system) consisting of three radars that may operate as a monopole, frequency hopping system; Acquisition probably in the G-or H-bands and tracking in the I-band missile has a semi-active radar homing seeker

Mount: Twin-rail launcher raises up out of a 9-ft 6-in (2.80-m) diameter silo. The silo is thought to be 16 ft 5 in (5.00 m) deep and contains the 20-missile magazine. The launcher retracts to reload.

DESIGNATOR: SA-N-3
NAME: GOBLET

DESCRIPTION: The SA-N-3 is a shipboard long-range, surface-to-air missile (SAM) analogous in concept and purpose to the U.S. Terrier and Standard SAMs. It is a low-to-medium-altitude weapon with some anti-ship capabililty. It succeeded the SA-N-1 in Soviet service.

Goblet

PLATFORMS:

Aircraft carriers	Moskva* class
Helicopter carriers	Kiev* class
Cruisers	Kresta II class
	Kara class

CHARACTERISTICS:

Missile Weight:	1,212 lb (550 kg)
Warhead	176 lb (80 kg)
Dimensions:	
Configuration	Thick cylinder with cruciform "cropped delta" mainplanes, small cruciform tail fins, and pointed nose
Length	
Booster Cap On	20 ft 4 in (6.20 m)
Booster Cap Off	19 ft 8 in (6.00 m)
Diameter	13.2 in (335 mm)
Wing Span	4 ft 11 in (1.50 m)
Propulsion:	Solid-fuel booster; after burnout, nozzle drops off and empty booster section becomes combustion chamber for ramjet sustainer engine

Performance:

Speed	Mach 2.5
Maximum Range	
Early Version	16.2 nm (18.6 mi; 30 km)
Later Version	30 nm (34.5 mi; 55.5 km)
Maximum Altitude	82,000 ft (25,000 m)
Warhead:	Conventional high-explosive
Sensors/	Head Lights fire control radar
Fire Control:	acquisition and tracking bands are D/F/G/H; missile has semi-active radar homing seeker
Mount:	Twin-rail launcher launcher elevates to 90 deg for reloading from 2 or 4 hatches all ships have one 36-round magazine per launcher except for those of the *Moskvas,* which hold 22 missiles

DESIGNATOR:	SA-N-1
NAME:	GOA

DESCRIPTION: Similar in many respects to the land-based SA-3, the SA-N-1 is a shipboard surface-to-air missile (SAM) system first deployed in the early 1960s.

Goa (Naval)

PLATFORMS:

Cruisers	*Kynda* class
	Kresta I class
Destroyers	*Kotlin SAM* class
	Kanin class
	Kashin class

CHARACTERISTICS:

Missile Weight:	882 lb (400 kg)
Warhead	132 lb (60 kg)
Dimensions:	
Configuration	Cylindrical booster section with 4 large pop-out rectangular fins;

Second stage is tapered with long pointed nose, cruciform foreplanes; cruciform "cropped delta" mainplanes near second stage tail

Length	22 ft 0 in (6.70 m)
Diameter	
Booster	27.5 in (701 mm)
Second Stage	18 in (460 mm)
Wing Span	4 ft 11 in (1.50 m)
Propulsion:	Tandem solid-fuel boosters; solid-fuel sustainer rocket

Performance:

Speed	Mach 2
Maximum Range	19.6 mi (31.5 km)
Altitude	
Maximum	50,000 ft (15,240 m)
Minimum	300 ft (91.4 m)
Warhead:	Conventional high-explosive
Sensors/	Peel Group system includes 4
Fire Control:	radars for target tracking in the I-band; missile guidance in the E-band is radio command
Mount:	Twin-rail roll-stabilized launcher missiles are reloaded with rails at 90 deg elevation from magazine below deck, magazine capacity ranges from 16 to 24 missiles.

GROUND COMBAT VEHICLES/ AIR DEFENSE

DESIGNATOR:	SA-12A
NAME:	GLADIATOR

DESCRIPTION: Tracked missile launching vehicle for long-range surface-to-air missiles.

USERS: Soviet Union Ground Forces.

CHARACTERISTICS:

Combat Weight:	66,138 lb (30,000 kg)
Dimensions:	
Hull Length	41 ft 0 in (12.5 m)
Width	11 ft 6 in (3.5 m)
Height	12 ft 6 in (3.8 m)
Propulsion:	D12A diesel engine
Maximum Power	525 hp
Power-to-weight	17.5 hp/metric ton
Performance:	
Speed	31 mph (50 km/h)
Range	279 miles (450 km)
Armament:	
Main	2 x SA-12 SAMs
Missile Weight	4,409 lb (2,000 kg)
Warhead	331 lb (150 kg)
Missile length	23 ft 7 in (7.20 m)

Gladiator

Diameter	20 in (0.50 m)
Wing Span	4 ft 11 in (1.50 m)
Elevation	+90 deg
Maximum Speed	Mach 3
Maximum Range	Approx 54 nm (62.1 mi; 100 km)
Minimum Range	Approx 3 nm (3.4 mi; 5.5 km)
Maximum Altitude	Approx 100,000 ft (more than 30,000 m)
Minimum Altitude	Approx 295 ft (90 m)
Sensors/ Fire Control:	Missile guidance radar on telescoping mast; Phased array search and fire control radars on separate vehicle
Crew:	4
Suspension:	Six road wheels, no return rollers
Protection:	Probably has collective nuclear, biological, and chemical (NBC) warfare protection

DESIGNATOR:	**SA-11**
NAME:	**GADFLY**

DESCRIPTION: This tracked surface-to-air missile (SAM) launcher is joining the SA-6 Gainful in providing low-to-medium altitude air defense for the Soviet Ground Forces.

USERS: Soviet Union Air Defense Forces.

CHARACTERISTICS: (Estimated)

Combat Weight:	35,274 lb (16,000 kg)
Dimensions:	
Hull Length	30 ft 10 in (9.40 m)

Width	15 ft 2 in (3.10 m)
Height	12 ft 2 in (3.70 m)
Propulsion:	Diesel engine
Performance:	
Speed	31 mph (50 km/h)
Range	186 miles (300 km)
Armament:	
Main	4 x SA-11 SAMs
Missile Weight	1,433 lb (650 kg)
Warhead	198 lb (90 kg)
Missile Length	18 ft 4 in (5.60 m)
Diameter	16 in (400 mm)
Wing Span	3 ft 11 in (1.20 m)
Configuration	Long, pointed cylinder with narrow main planes along mid body and four trapezoidal control surfaces at tail
Propulsion	Solid-fuel rocket
Maximum Speed	Mach 3
Maximum Altitude	49,213 ft (15,000 m)
Maximum Range	18.6 mi (30 km)
Minimum Altitude	82-98 ft (25-30 m)
Minimum Range	1.9 mi (3 km)
Sensors/ Fire Control:	On-board Flap Lid monopulse tracking and guidance radar (thought to be variation of the Front Dome system found in Sovremennyy* class destroyers); One early-warning and acquisition radar vehicle (SSNR) with Clam Shell I-band 3-D radar deployed in each 4-launcher battery
Crew:	4
Suspension:	Torsion bar, six road wheels, three shock absorber points (two double, one single), rear drive, front idler, four return rollers

DESIGNATOR:	**SA-10b**
NAME:	**GRUMBLE**

DESCRIPTION: Advanced Soviet mobile air-defense system consisting of an 8-wheeled missile launching vehicle with 4 SA-10 Mach 6 active radar-homing, surface-to-air missiles (SAM). The carrier is based on the MAZ-7910* tractor.

USERS: Soviet Union Ground Forces Air Defense Forces.

CHARACTERISTICS:

Combat Weight:	44,092 lb (20,000 kg)
Dimensions:	
Hull Length	30 ft 10 in (9.40 m)
Width	10 ft 2 in (3.10 m)
Height	12 ft 2 in (3.70 m)
Propulsion:	D12A diesel engine

Maximum Power	525 hp
Power-to-weight	26.25 hp/metric ton
Performance:	
Speed	53 mph (85 km/h)
Range	404 miles (650 km)
Armament:	
Main	4 x SA-10 Grumble vertically launched SAM (reloads carried in separate transloader vehicle)
Configurations	Slender cylinder with ogival nose, two sets of cruciform long-chord delta wings, the larger at mid-body, the smaller (presumably steerable) at the tail
Length	23 ft 6 in (7.16 m)
Diameter	17.7 in (450 mm)
Propulsion	Solid-fuel rocket
Maximum Speed	Mach 6 (acceleration to this speed is said to be 100g)
Maximum Range	54 nm (62 mi; 100 km)
Altitude	
Maximum	Up to 100,000 ft (30,480 m)
Minimum	Less than 1,000 ft (305 m)
Warhead:	Conventional high-explosive some may have nuclear warheads
Sensors/Fire Control:	Separate radar vehicle carries target tracking and fire control planar array radar (continuous wave and pulse-doppler)
Crew:	4
Suspension:	8-wheeled (4 steering)

DESIGNATOR: SA-8/ZRK-( )*
NAME: GECKO/ROMB*

DESCRIPTION: The SA-8 Gecko is a short-ranged surface-to-air missile (SAM) based on a 6-wheel missile launching vehicle (Soviet designation, Transporter 5937). The system is designed to defend ground forces against low-altitude air attack.

USERS: Soviet Union Ground Forces, Algeria, Angola, East Germany, Guinea, Hungary, India, Iraq, Jordan, Kuwait, Libya, Mozambique, Poland, Syria, and Yugoslavia.

CHARACTERISTICS:

Combat Weight:	19,841 lb (9,000 kg)
Dimensions:	
Hull Length	30 ft 0 in (9.14 m)
Width	9 ft 6 in (2.90 m)
Height	
Radar Down	13 ft 9 in (4.20 m)
Hull Top	6 ft 1 in (1.85 m)
Spacing between	
Axles 1-2	10 ft 1 in (3.08 m)

Gecko

Axles 2-3	9 ft 2 in (2.79 m)
Propulsion:	Zil-37S diesel engine
Maximum Power	175 hp at 2,000 rpm
Power-to-weight	19.44 hp/metric ton
Performance:	
Speed	37 mph (60 km/h)
Range	311 mi (500 km)
Fording	Amphibious; 2 hydrojets
Armament:	
Main	4 x SA-8 SAMs (16 reloads in transloader)
Missile Weight	419 lb (190 kg)
Warhead	90-110 lb (40.8-SO kg)
Missile Length	10 ft 6 in (3.20 m)
Diameter	8.25 in (210 mm)
Wing Span	2 ft 0 in (0.60 m)
Configuration	Long cylinder with pointed nose, small steerable foreplanes, larger fixed stabilizers at tail
Propulsion	Solid-fuel rocket, probably dual thrust range
Range	
Slant	6.5 nm (7.5 mi; 12 km)
Minimum	0.9 nm (1 mi; 1.6 km)
Altitude	
Maximum	39,370 ft (12,000 m)
Effective	20,000 ft (6,096 m)
Minimum	33 ft (10 m)
Traverse	360 degrees
Sensors/ Fire Control:	Infrared night vision systems for commander and driver; missile guidance is radio command with semi-active radar or infrared terminal homing; on-board radars are Land Roll H-band search and early warning (range 18.6 mi (30 km)), J-band monopulse target tracking, (range 15.5 mi (25 km)), 2 sets of

I-band monopulse missile guidance transmitters; electro-optical system with low-light-level TV

Crew: 3 (Commander, missile operator, driver)

Suspension: 6 x 6 (4-wheel steering on first and third axles); probable torsion bar; central tire-pressure regulation

Protection: Collective overpressure nuclear, biological, and chemical (NBC) warfare protection with air filtration

Gainful

DESIGNATOR: SA-6/ZRK-SD 9M9*
NAME: GAINFUL/KUB*

DESCRIPTION: Tracked, armored carrier armed with a medium-range surface-to-air missile (SAM).

USERS: Soviet Union Ground Forces, Algeria, Angola, Bulgaria, Cuba, Czechoslovakia, East Germany, Egypt, Ethiopia, Guinea, Guinea-Bissau, Hungary, India, Iraq, Kuwait, Libya, Mozambique, North Yemen, Poland, Romania, Somalia, Syria, Tanzania, Vietnam, and Yugoslavia.

CHARACTERISTICS:

Combat Weight:	30,864 lb (14,000 kg)
Ground Pressure	6.83 lb/sq in (0.48 kg/sq cm)
Dimensions:	
Hull Length	22 ft 3 in (6.79 m)
With Missiles	24 ft 3 in (7.39 m)
Width	10 ft 5 in (3.18 m)
Height, Hull Top	5 ft 11 in (1.80 m)
Missiles	11 ft 4 in (3.45 m)
Length Of Track	
On Ground	12 ft 6 in (3.80 m)
Track Width	14 in (0.36 m)
Ground Clearance	16 in (0.40 m)
Propulsion:	Model V-6R water-cooled, in-line 6-cylinder diesel engine; Maximum power 240 hp at 1,800 rpm
Power-to-weight	17.14 hp/metric ton
Performance:	
Speed	27 mph (44 km/h)
Range	161 miles (260 km)
Obstacle	
Vertical	3 ft 3 in (1.00 m)
Trench	8 ft 2 in (2.50 m)
Fording	3 ft 7 in (1.10 m)
Gradient	35%
Armament:	
Main	3 x 9M9 (SA-6) SAMs
Missile Weight	1,212 lb (550 kg)
Warhead	176 lb (79.8 kg)

Missile Length	20 ft 4 in (6.20 m)
Diameter	13.2 in (335 mm)
Configuration	Thin tapered cylinder with four ramjet intakes and four trapezoidal mainplanes around the mid-body; Four more guidance fins with aerials at tail
Propulsion	Solid-fuel booster, the casing of which becomes a ramjet combustion chamber for sustained flight
Maximum Speed	Mach 2.8
Maximum Range	18.5 mi (30 km)
Minimum Range	2.3 mi (3.7 km)
Maximum Altitude	59,000 ft (17,983 m)
Effective	43,000 ft (13,106 m)
Minimum Altitude	300 ft (91.4 m)
Elevation	+85 deg
Traverse	360 deg
Sensors/ Fire Control:	Infrared night vision equipment on vehicle; electro-optical sighting system on vehicle; Gainful is associated with: —Long Track E-band early warning radar (93 mi/150 km) mounted on AT-T tractor, —Thin Skin H-band height-finding radar (149 mi/240 km) on truck or trailer. —Straight Flush combined I-band low-altitude and G/H-band medium-range radars on SA-6 transporter-launcher
Crew:	3
Suspension:	Torsion bar, six road wheels, two hydraulic shock absorbers, rear drive, front idler, no return rollers
Protection:	Collective overpressure and filtration protection against nuclear, biological, and chemical warfare
Armor/Maximum	9.4 mm.

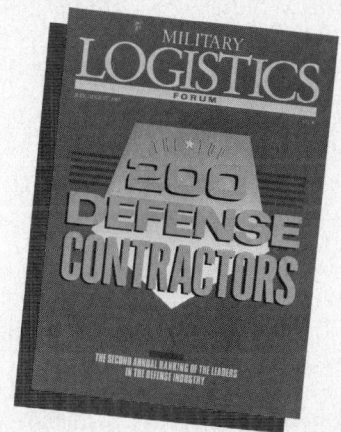

DESIGNATOR: SA-4/ZRD-SD 9M8*
NAME: GANEF/KRUG*

DESCRIPTION: This a a tracked, self-propelled vehicle that carries the medium-range Ganef missile in support of Soviet ground forces.

Ganef

USERS: Soviet Union Ground Forces, Bulgaria, Czechoslovakia, East Germany, Hungary and Poland.

CHARACTERISTICS:

Combat Weight:	66,138 lb (30,000 kg)
Dimensions:	
Hull Length	25 ft 7 in (7.80 m)
With Missiles	31 ft 0 in (9.46 m)
Width	10 ft 6 in (3.20 m)
Height	14 ft 8 in (4.47 m)
Length Of Track	
On Ground	16 ft 5 in (5.00 m)
Track Width	21 in (0.54 m)
Ground Clearance	17 in (0.44 m)
Propulsion:	Water-cooled diesel
Maximum Power	600 hp
Power-to-weight	20.0 hp/metric ton
Performance:	
Speed	31 mph (50 km/h)
Range	300 miles (500 km)
Obstacle	
Vertical	3 ft 3 in (1.00 m)
Trench	8 ft 2 in (2.50 m)
Fording	3 ft 3 in (1.00 m)
Gradient	30%
Armament	
Main	2 x 9M8M1* (SA-4a) or 9M8M2* (SA-4b) SAMs (no reloads carried on launch vehicle)
Missile Weight	5,511 lb (2,500 kg)
Warhead	220-300 lb (100-136 kg)
Missile Length	28 ft 10.5 in (8.80 m)
Diameter	2 ft 8 in-2 ft 11 in (0.81 0.90 m)

Wing Span	7 ft 7 in (2.30 m) body
	8 ft 6 in (2.60 m) booster
Configuration	Thin warhead and guidance section mounted on much larger propulsion section; latter has double-taper foreplanes with aerials and larger mainplanes at tail; four small boosters with canted nozzles
Propulsion:	Solid-fuel boosters; ramjet sustainer with intake surrounding warhead

Other Factors:	9M8M1	9M8M2
Maximum Altitude	16.77 mi (27 km)	14.90 mi (24 km)
Maximum Range	44.71 mi (72 km)	31.05 mi (50 km)
Minimum Range	5.78 mi (9.3 km)	.68 mi (1.1 km)
Maximum Speed	Mach 2.5	
Launcher		
Elevation	+45 deg	
Traverse	360 deg	

Sensors/ Fire Control:	Infrared night vision system; electro-optical fire control system on launcher; missile guidance is radio command with semi-active radar terminal homing; SA-4 Ganef launcher associated with; —Long Track E-band early warning radar (93 mi/150 km) on modified AT-T artillery tractor; —Thin Skin H-band height-finding radar (149 mi/240 km) on truck or trailer; —Pat Hand H-band fire-control and continuous-wave guidance radar on Ganef vehicle chassis.
Crew:	5 (Commander, driver, 3 missile crew)
Suspension:	Torsion bar, seven road wheels, two hydraulic shock absorbers, four return rollers, front drive, rear idler
Protection:	Collective overpressure protection against nuclear, biological, and chemical warfare
Armor/Maximum	15 mm

AIR-AIR MISSILES

DESIGNATOR: AA-10
NAME: ALAMO

DESCRIPTION: The AA-10 Alamo is an advanced Soviet air-to-air missile carried by the Su-27 Flanker and MiG-29 Fulcrum fighter aircraft. When fitted with an infrared seeker, Alamo is longer and heavier than its U.S. counterpart, the AIM-9 Sidewinder. The semi-active radar homing version of the AA-10 is approximately the same size as the U.S. AIM-7 Sparrow III and is described by official U.S. sources as a beyond-visual-range missile. Although photographic analysis has yielded much about the missile's dimensions and probable weight, little information on the weapon's performance has been made public.

USERS: Soviet Union Air Forces.

CHARACTERISTICS: (Estimated)

Weight:	
Infrared	342 lbs (155 kg)
SAR	441 lbs (200 kg)
Dimensions:	
Configuration	Infrared: Cylindrical body, blunt nose, tapered nozzle, cruciform antenna, "ears" forward of fixed cruciform inverse-tapered foreplanes, "cropped delta" tail fins SAR: Longer afterbody, pointed nose
Length	
Infrared	10 ft 6 in (3.20 m)
SAR	13 ft 2 in (4.00 m)
Diameter	7¼ in (185 mm)
Propulsion:	
Infrared	Single-stage boost motor
SAR	Single-stage boost motor and sustainer engine
Performance:	
Speed	Supersonic
Range	BVR for SAR
Warhead	Conventional high explosive
Sensors/Fire Control	
Guidance	
Infrared	Cooled aeeker, probably with all-aspect capability, flare countermeasures filters
SAR	I/J-band monopulse seeker

DESIGNATOR: AA-9
NAME: AMOS

DESCRIPTION: The Soviet AA-9 is a long-range, radar-homing missile that has reportedly demonstrated a lookdown, shoot-down capability against small, low-flying targets.

PLATFORMS:

Fighters	MiG-31* Foxhound MiG-29* Fulcrum

CHARACTERISTICS:

Dimensions:	
Configuration	Cylindrical body of perhaps two diameters; the larger forward section (the warhead and guidance), and a slightly thinner rear section (motor); rear-mounted cruciform wings are long narrow rectangles with swept leading edges and movable control surfaces on trailing edges
Length	Approx 13 ft (4 m)
Propulsion:	Solid-fuel rocket
Performance:	
Maximum Range	
High-Altitude	25-28 mi (40.2-45.1 km)
Low-Altitude	12.5 mi (20.1 km)
	Effective head-on range against a 10.76 sq ft (1 sq m) target 10 mi (16 km)
Warhead:	Conventional high-explosive
Sensors/ Fire Control:	MiG-29 has long-range track-while-scan radar and pulse Doppler look-down/shoot-down fire control system; MiG-31's radar is also a Pulse-Doppler type; search range 190 mi (305 km); Track range 167 mi (270 km)

DESIGNATOR: AA-8/R-60*
NAME: APHID

DESCRIPTION: The AA-8 Aphid is a highly maneuverable, short-range infrared homing "dogfighting" missile mounted on several types of Soviet aircraft.

PLATFORMS:

Fighters	MiG-21* Fishbed MiG-23* Flogger MiG-25* Foxbat MiG-29* Fulcrum MiG-31* Foxhound Su-15/21* Flagon
Attack	Su-25* Frogfoot Yak-38* Forger

USERS: Soviet Union Air Forces and Air Defense Troops, Cuba, Libya and Iraq.

CHARACTERISTICS:

Combat Weight:	121 lb (54.9 kg)
Warhead	13.2 lb (6.5 kg)
Dimensions:	
Configuration	Cylindrical body with two sets of small foreplanes (one rectangular and the other delta-shaped) and one set of "cropped delta" mainplanes at the rear, all in-line cruciform in layout
Length	7 ft 2.5 in (2.20 m)
Diameter	4.75 in (120 mm)

Wing Span	1 ft 3.75 in (0.40 m)
Propulsion:	Solid-fuel Rocket
Performance:	
Speed	Mach 2.5+
Maximum Range	3-4.3 mi (4.8-6.9 km)
Minimum Range	1,650 ft (503 m)
Warhead:	Conventional high-explosive
Sensors/	IR seeker is thought to have
Fire Control:	better low-level target discrimination than earlier Soviet IR missiles and some off boresight seeking capability; often deployed with AA-2 Atoll semi-active radar homing missile and fired in mixed pairs

DESIGNATOR: AA-7/R-23T* (Infrared version)
R-23R* (semi-active
radar homing)

NAME: APEX

DESCRIPTION: The AA-7 Apex is one of the new generation of Soviet aircraft missiles deployed in the mid-1970s. It is medium-range and has both infrared and semi-active radar (SAR) homing variants, with one or more of each type mounted on the same aircraft.

PLATFORM:

Fighters	MiG-23* Flogger
	MiG-25* Foxbat

USERS: Soviet Union Air Forces and Air Defense Forces, Bulgaria, Iraq, Libya and Syria.

CHARACTERISTICS:

Combat Weight:	705 lb (320 kg)
Warhead	88 lb (40 kg)
Dimensions:	
Configuration	Cylindrical body with small double-tapered foreplanes; "Cropped delta" mainplanes with control surfaces on the trailing edges, and trapezoidal control surfaces at the tail; all cruciform in layout; IR nose is rounded with seeker window, SAR nose is pointed
Length	15 ft 1 in (4.60 m) SAR; IR shorter
Diameter	8.75 in (222 mm)
Wing Span	3 ft 5.5 in (1.05 m)
Propulsion:	Solid-fuel rocket
Performance:	
Speed	Approx Mach 3+
Maximum Range	9.3 mi (15 km) IR
	21.7 mi (35 km) SAR
	13.8 mi (22.3 km) SAR in look-down, shoot-down attack
Warhead:	Conventional high-explosive

Sensors/	AA-7s are often launched in
Fire Control:	pairs, the IR version launching first, the SAR version following less than a second later; SAR version can be used in a look-down, shoot-down mode. Associated MiG-23 radar is High Lark I/J-band

DESIGNATOR: AA-6
NAME: ACRID

DESCRIPTION: The long-range AA-6 is the largest air-to-air missile (AAM) in Soviet service and is the principal missile armament on the MiG-25* Foxbat. As with most Soviet AAMs, the Acrid comes in both infrared and semi-active radar (SAR) homing versions.

USERS: Soviet Union Air Defense Forces, Iraq, Libya and Syria.

CHARACTERISTICS:

Combat Weight:	1,650 lb (750 kg) estimated
Warhead	220 lb (100 kg)
Dimensions:	
Configuration	Long cylindrical body with 4 small delta foreplanes and 4 large "cropped delta" wings; IR nose is a blunted ogive; SAR nose is pointed.
Length	
IR	19 ft 0 in (5.79 m)
SAR	20 ft 7 in (6.29 m)
Diameter	12 in (300 mm) estimated
Wing Span	7 ft 5 in (2.25 m)
Propulsion:	Solid-fuel rocket
Performance:	
Speed	Mach 2.2 cruise
Maximum Range	
IR	Approx 23 mi (37 km)
SAR	Approx 29 mi (47 km)
Warhead:	Conventional high-explosive
Sensors/	SAR version (outboard pylon)
Fire Control:	tracks illumination provided by wingtip-mounted, continuous-wave device; associated radar is Fox Fire I/J-band

DESIGNATOR: AA-5
NAME: ASH

DESCRIPTION: The Ash medium-range air-to-air missile (AAM), in both infrared and semi-active radar (SAR) homing versions, is the principal armament of the Tu-28P* Fiddler air defense fighter.

PLATFORMS:

Fighters	Tu-28P* Fiddler
	Yak-28* Firebar

CHARACTERISTICS:

Combat Weight:	441 lb (200 kg)
Warhead	Approx 150 lb (68 kg)
Dimensions:	
Configuration	Cylindrical body with 4 large delta wings along middle third and 4 smaller trapezoidal control surfaces on the rear, both in cruciform arrangement; IR nose is blunter than SAR nose.
Length	17 ft 4.5 in (5.30 m) IR
	17 ft 0 in (5.18 m) SAR
Diameter	12 in (300 mm)
Wing Span	4 ft 3 in (1.30 m)
Propulsion:	Solid-fuel rocket
Performance:	
Speed	Mach 2.5 to Mach 3
Maximum Range	18.5 mi (29.8 km)
Warhead:	Conventional high-explosive
Sensors/	
Fire Control:	Radar-Big Nose I/J-band

DESIGNATOR: AA-4
NAME: AWL

DESCRIPTION: The AA-4 Awl was an unsuccessful Soviet air-to-air missile (AAM) that entered service in 1960. It was approx 16 ft 5 in (5.00 m) long and resembled the U.S. Sparrow III AAM. The Awl had a short Soviet operational career and was not transferred to any other country.

DESIGNATOR: AA-3
NAME: ANAB

DESCRIPTION: Deployed in both infrared and semi-active radar (SAR) homing versions, the Anab is the Soviet counterpart in range and ubiquity to the U.S. Sparrow air-to-air missile.

PLATFORMS:

Fighters	Su-9/11*Fishpot
	Su-15/21* Flagon
	Yak-28* Firebar

USERS: Soviet Union Air Forces and Air Defense Forces, Bulgaria, Czechoslovakia, East Germany, Hungary, Poland and Romania.

CHARACTERISTICS:

Combat Weight:	606 lb (275 kg)
Warhead	Approx 80 lb (36.3 kg)
Dimensions:	
Configuration	Cylindrical body with 4 small control fins ahead of 4 large trapezoidal wings, both in a cruciform arrangement; IR nose is ogival; SAR is pointed
Length	Estimates vary from 11 ft 10 in

	(3.60 m) to 13 ft 5 in (4.09 m)
Diameter	11 in (279 mm)
Wing Span	4 ft 3 in (1.30 m)
Propulsion:	Solid-fuel rocket
Performance:	
Speed	Mach 2-2.5
Maximum Range	10-12 mi (16.1-19.3 km)
Warhead:	Conventional high-explosive
Sensors/	IR and SAR versions usually
Fire Control:	mounted on same aircraft; associated aircraft radars for SAR version include Skip Spin (Soviet name: Uragan 5B*) I/J-band, Twin Scan I/J-band (replaces Skip Spin)

DESIGNATOR: AA-2/K13A*/RS-3S*/5V06*
   (probable Soviet designations)
NAME: ATOLL

DESCRIPTION: The oldest air-to-air missile (AAM) still in Soviet service, the AA-2 has a infrared seeker in the early versions and a semi-active radar (SAR) homing in the Advanced Atoll variant.

PLATFORMS:

Fighters	MiG-21* Fishbed
	MiG-23* Flogger (export version)
	Su-22* Fitter (Peruvian version)

USERS: Soviet Union Air Forces and Air Defense Forces, Algeria, Angola, Bangladesh, Bulgaria, Cuba, Czechoslovakia, East Germany, Egypt, Finland, Hungary, India, Iraq, Laos, Libya, Nigeria, North Yemen, Poland, Romania, Somalia, South Yemen, Syria, Vietnam and Yugoslavia.

CHARACTERISTICS:

Combat Weight:	154 lb (70 kg)
Warhead	13 lb (6 kg)
Dimensions:	
Configuration	Strongly resembles early U.S. Sidewinder AAM; cylindrical body with cruciform triangular foreplanes and cruciform trapezoidal rear control surfaces; Atoll has blunt nose; Advanced Atoll has pointed SAR nose
Length	
Atoll	9 ft 2 in (2.80 m)
Advanced Atoll	9 ft 15 in-10 ft 2 in (3 3.10 m)
Diameter	4.72 in (120 mm)
Propulsion:	Solid-fuel rocket
Performance:	
Speed	Mach 2.5
Maximum Range	3-4 mi (4.8-6.4 km)
Warhead:	Conventional high-explosive
Sensors/	Infrared rear-aspect, bore-
Fire Control:	sighted seeker on Atoll; Semi-

active radar homing on Advanced Atoll; associated aircraft radars include:
—Jay Bird I/J-band
—High Lark I/J-band
—Spin Scan I-band
Steered by both sets of control surfaces; Foreplanes 180 deg apart work in unison; rear surfaces have gyroscopically stabilized tabs

DESIGNATOR: AA-1/K-S* (Soviet design bureau designation)
RS-2U* (Soviet service designation)
NAME: ALKALI

DESCRIPTION: The AA-1 Alkali was the first Soviet air-to-air missile (AAM). It was a short-range, radar-guided weapon able to undertake only tail-aspect attacks.

PLATFORMS:
Fighters   MiG-17* Fresco
MiG-19* Farmer
Su-9* Fishpot

USERS: Bulgaria, Cuba and Poland.

CHARACTERISTICS:
Missile Weight:	198 lb (90 kg)
Warhead	66 lb (30 kg)
Dimensions:	
Configuration	Cylindrical body with pointed nose and tail fairing, cruciform delta foreplanes and mainplanes, and prominent antenna in nose
Length	6 ft 2 in (1.88 m)
Diameter	7 in (178 mm)
Wing Span	
Foreplanes	12.5 in (320 mm)
Mainplanes	22.8 in (580 mm)
Propulsion:	Single-stage solid-fuel rocket
Performance:	
Speed	Mach 2
Maximum Range	3.2-4.3 nm (3.7-5 mi; 6-8 km)
Warhead:	Conventional high-explosive
Sensors/	Radar-guided by:
Fire Control:	—Scan Fix I-band radar in MiG-17, MiG-19
	—Scan Odd I-band radar in MiG-19
	—Spin Scan I-band radar in Su-9

ANTI-TANK MISSILES

DESIGNATOR: AT-8
NAME: SONGSTER

DESCRIPTION: The AT-8 Songster is a Soviet guided missile launched from 125-mm guns fitted in T-64B and T-80 Main Battle Tanks. The use of the AT-8 is uncertain. Some western sources contend that it is an anti-tank weapon, possibly muzzle loaded, which is deployed in some tanks to make up for the poor quality control of conventional projectile production. If muzzle loaded, the AT-8 would seem to be an opening shot weapon only as reloading the gun under combat conditions would be hazardous at best. More recent analysis suggests that the AT-8 is a two-piece missile compatible with the auto-loader in Soviet tanks. In this view, Songster's primary role is as an anti-helicopter weapon or as a counterfire weapon against enemy anti-tank missile positions or vehicles. It is not clear how widely the AT-8 is deployed. The concensus suggests that not all tanks in a battalion would be equipped with the 4-5 missiles thought to make up a basic load.

USERS: Soviet Union Ground Forces.

CHARACTERISTICS:
Weight:	55 lbs (25 kg)
Dimensions:	
Configuration	Probably stowed in two parts, the booster and a guided round
Length	3 ft 11 in (1.2 m) when assembled for firing
Diameter	4.7 in (120 mm)
Propulsion:	Low-velocity booster motor
	Solid-fuel sustainer motor
Performance:	
Speed	
At Launch	492 fps (150 mps)
Max	1,640 fps (500 mps)
Maximum Range	4,374 yds (4,000 m)
Flight Times	7 sec to 3,281 yds (3,000 m)
	9 sec to 4,374 yds (4,000 m)
Warhead:	Conventional high explosive, possible either fragmentation or armor-piercing
Sensors/	Radio command guidance in I-band (8 Ghz)
Fire Control:	Missile probably tracked through gunner's sight

DESIGNATOR: AT-6
NAME: SPIRAL

DESCRIPTION: The AT-6 Spiral is a Soviet helicopter-carried anti-tank missile. It is larger and heavier than previous Soviet anti-tank missiles and, unlike its predecessors, does not appear to have a ground-based variant.

PLATFORMS:

Helicopters	Mi-24* Hind-E
	Mi-8* Hip-E

USERS: Soviet Union Ground Forces, East Germany and Poland.

CHARACTERISTICS:

Warhead Weight:	Approx 22 lb (10 kg)
Dimensions:	
Diameter	Approx 5.5 in (140 mm)
Propulsion:	Solid-fuel rocket
Performance:	
Range	
Maximum	5,470-8,750 yd (5,000-8,000 m)
Flight Time	
Maximum range	11 sec to 5,470 yd (5,000 m)
Warhead:	High-explosive anti-tank (HEAT)
Sensors/	Guidance—Probably radio
Fire Control:	command, although some sources report infrared homing

DESIGNATOR: AT-5
NAME: SPANDREL

DESCRIPTION: Relatively heavy, long-range anti-tank missile mounted on the BRDM-2 amphibious armored reconnaissance vehicle. Some East German BRDMs have a mixed battery of three AT-4 Spigots and two AT-5 Spandrels.

USERS: Soviet Union Ground Forces, Czechoslovakia, East Germany, and Syria.

CHARACTERISTICS:

Combat Weight:	26-40 lb (12-18 kg)
Warhead	15.5 lb 17 kg)

Spandrel

Dimensions:

Configuration	Only the launch tube has been publicly seen; it is cylindrical with a flared rear, a long slot on top, and a blow-out cap on the front
Length	Launcher 4 ft 7 in (1.40 m)
Diameter	7 in (180 mm)
Propulsion:	One solid-fuel rocket, possibly with a gas-generator for launch
Performance:	
Speed	492-820 fps (150-250 mps)
Minimum Range	109 yds (100 m)
Maximum Range	4,374 yds (4,000 m)
Warhead:	High-explosive anti-tank (HEAT)
Penetration	20-24 in (500-600 mm)
Sensors/	Guidance—semi-automatic
Fire Control:	command to line of sight (SACLOS)

DESIGNATOR: AT-4
NAME: SPIGOT/FAGOT*

DESCRIPTION: Portable, wire-guided anti-tank missile intended to replace the AT-3 Sagger in Soviet Ground Forces, the AT-4 is thought to bear a strong resemblance to the Western Milan ATGW.

Spigot

USERS: Soviet Union Ground Forces, Czechoslovakia, East Germany, Finland, Hungary, Poland and Syria.

CHARACTERISTICS:

Combat Weight:	22-26.5 lb (10-12 kg)

Warhead	5.5 lb (2.5 kg)
Dimensions:	
Configuration	Only the launch tube has been seen, mounted on a ground tripod
Launcher Length	3 ft 11 in (1.20 m)
Diameter	5 in (136 mm)
Propulsion:	2-stage solid fuel rocket; First stage is gas generator for launch; Second state is sustainer motor
Performance:	
Speed	492-820 fps (150-250 mps)
Range	
Maximum	2,187 yd (2,000 m)
Minimum	76 yd (70 m)
Warhead:	High-explosive anti-tank (HEAT)
Penetration	19.7 in (500 mm)
Sensors/ Fire Control:	Guidance—semi-automatic command to line of sight (SACLOS); System has two separate optical sensors, one for the operator to lay on target, the other for automatically correcting the missile
Crew:	1 or 2

DESIGNATOR:	**AT-3/PUR-63*** (Soviet army designation) **9M14M*** (Soviet industrial designation)
NAME:	**SAGGER/MALYUTKA*** (Soviet name = Little Child)

DESCRIPTION: Medium wire-guided anti-tank missile in widespread Soviet use with ground launchers and on armored fighting vehicles and helicopters.

Sagger

USERS: Soviet Union Ground Forces, Algeria, Angola, Bulgaria, Cuba, Czechoslovakia, East Germany, Egypt, Ethiopia, Hungary, India, Iraq, Libya, Mozambique, North Korea, Poland, Romania, Syria, Uganda and Zambia.

CHARACTERISTICS:

Combat Weight:	25 lb (11.3 kg)
Warhead	6 lb (2.7 kg)
Dimensions:	
Configuration	Cylindrical body with conical nose, folding compound-sweep cruciform; Rear-mounted wings
Length	2 ft 10 in (0.86 m)
Diameter	4.6 in (119 mm)
Wing Span	1 ft 6.5 in (0.47 m)
Propulsion:	2-stage solid rocket; 1st stage is a booster, 2nd stage is a sustainer
Performance:	
Speed	394 fps (120 mps)
Minimum Range	
Vehicle Launch	547 yd (500 m)
Ground Launch	1,094 yd (1,000 m)
Arming Range	77-219 yds (70-200 m)
Guidance Capture Window	547-875 yd (500-800 m)
Maximum Range	3,281 yd (3,000 m) in 27 seconds
Rate of Fire	2 rounds/minute in ground based form
Warhead:	High-explosive anti-tank (HEAT)
Penetration	15.75-16 in (400-410 mm)
Sensors/ Fire Control:	Guidance—manual command to line of sight (MCLOS); operator may be 15 meters from the launcher on the ground and 80 meters from a BRDM-2 launcher; Sagger C uses semi-automatic CLOS (SACLOS)
Crew:	3-man firing team (two operators and one site-defense soldier with RPG-7 anti-tank grenades) operate a 4-missile "battery."

DESIGNATOR:	**AT-2/PUR-62*** (Soviet army designation)
NAME:	**SWATTER/FALANGA*** (Soviet name = Phalanx)

DESCRIPTION: The AT-2 is a Soviet radio-guided anti-tank missile which may be deployed on ground launchers, on the BRDM-2 armored reconnaissance vehicle, and on helicopters.

USERS: Soviet Union Ground Forces, Egypt, Iraq and Syria.

CHARACTERISTICS:

Combat weight:	
Swatter A/B	59.5 lb (27 kg)
Swatter C	64 lb (29 kg)
Warhead:	Approx 11.5 lb (5.25 kg)

Dimensions:

Configuration — Cylindrical with rounded nose, two small foreplanes, rear-mounted cruciform wings with elevons

Length — 3 ft 10 in (1.16 m)

Diameter — 5.8 in (148 mm)

Wing Span — 2 ft 2 in (0.66 m)

Propulsion: — 1 x solid fuel rocket

Performance:

Speed — 492 fps (150 mps)

Minimum Range — 547 yd (500 m)

Maximum Range

Swatter A — 2,734 yd (2,500 m) in 17 sec

Swatter B — 3,827 yd (3,500 m) in 23 sec

Swatter C — 4,374 yd (4,000 m) in 26-27 sec

Warhead: — High-explosive anti-tank (HEAT)

Penetration — 19.7 in (500 mm)

Sensors/Fire Control:

Swatter A/B — Radio command to line of sight (CLOS) on three frequencies to counter jamming;

Swatter C — Semi-automatic infrared/radio CLOS (SACLOS) with IR terminal homing in the missile

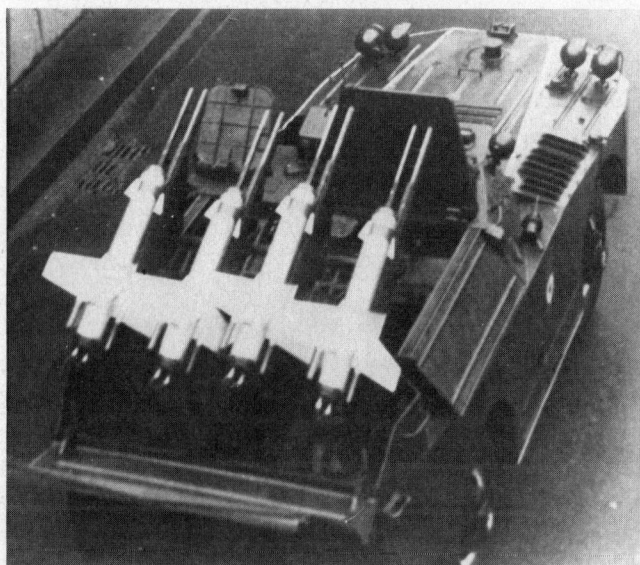

Swatter

DESIGNATOR: AT-1/PUR-61* (Soviet designation) 3M6 (Soviet industrial designation)

NAME: SNAPPER/SHMEL* (Soviet name = Bumblebee)

DESCRIPTION: First-generation wire-guided Soviet anti-tank missile. Now obsolete and no longer in service with Soviet forces.

USERS: Cuba, Egypt, Poland and Romania.

CHARACTERISTICS:

Combat Weight: — 53.5 lb (24.26 kg)

Warhead — 11.5 lb (5.2 kg)

Dimensions:

Configuration — Conical nose, cylindrical body, broad wings in cruciform layout with vibrating trailing-edge spoilers

Length — 3 ft 9 in (1.15 m)

Diameter — 5.3 in (134 mm)

Wing Span — 2 ft 5.5 in (0.75 m)

Propulsion: — 1 solid-fuel rocket

Performance:

Speed — 344 fps (105 mps)

Minimum Range — 547 yd (500 m)

Maximum Range — 2,734 yd (2,500 m) in 19 seconds

Warhead: — High-explosive anti-tank (HEAT)

Penetration — 14-15 in (356-380 mm)

Sensors/Fire Control: — Guidance—command to line of sight (CLOS); operator may be 50 meters from the launcher

ANTI-RADIATION MISSILE

DESIGNATOR: AS-9

NAME: KYLE

DESCRIPTION: The AS-9 Kyle is a Soviet long-range anti-radar missile. It is considerably larger than its U.S. counterparts.

PLATFORMS:

Fighters — Su-24* Fencer

Bombers — Tu-16* Badger

Tu-22M* Backfire

CHARACTERISTICS:

Warhead Weight: — 331-441 lb (150-200 kg)

Dimensions:

Length — 19 ft 9 in (6.03 m)

Diameter — 19 in (488 mm)

Propulsion: — Turbojet engine

Performance:

Speed — Estimated at Mach 3

Maximum Range — 50-62 mi (80-100 km)

Warhead: — Conventional high-explosive

Sensors/Fire Control: — Passive radar homing

SENSORS

STRATEGIC RADARS

DESIGNATOR: **ABM Radars**

DESCRIPTION: The Soviet ballistic missile defense (BMD) radar network is a nationwide system encompassing a variety of radar types. The system has been deployed over the past 30 years with earlier radars now being supplemented or replaced by newer radars with similar functions. The BMD radar network has three main parts: (1) Those radars which are colocated with the 100-missile ABM system which surrounds Moscow, (2) The early warning radar system established along the periphery of the Soviet Union, and (3) The new centrally located radar at Krasnoyarsk.

Radars associated with the Moscow ABM system fall into two generations, the first coming into service in the late 1960s and early 1970s and the second becoming operational in the mid-to-late 1980s. The first generation includes:

Try Add: BMD guidance and engagement radar system similar in function to the U.S. missile site radar (MSR) now cancelled. Try Add, as the code name implies, is a three-radar system consisting of one large target tracking and two smaller interception radars. Two Try Add sites were established for each 16-missile ABM-1 Galosh complex.

Dog House: A-band re-entry vehicle acquisition and tracking phased array radar (PAR) located at Naro-Fominsk, about 19 nm (22 mi/35 km) south of Moscow. It went into operation in 1969. The radar has a peak power output of 20 megawatts, a pulse repetition frequency of 50 pulses/second, and a range of 1,476 nm (1,700 mi/2,736 km) against small targets. Dog House derives its name from the gabled shape of the two radar faces.

Cat House: Target-tracking PAR located at Chekov, about 35 nm (40 mi/65 km) southwest of Moscow. This radar is operated in conjunction with the Dog House radar. It went into operation in 1974-75.

All these radars are fixed-site radars, and represent obsolete technology that may not be able to handle the multiple tracks of even the older multiple-warhead missiles. They have been updated regularly and remain in service.

A second generation of radars is now coming into service as new ABMs are fielded. The original Galosh exoatmospheric ABM is being replaced by the ABM-10 (SH-04) modified Galosh and the endoatmospheric ABM-X-3 Gazelle is also being deployed. Thirty-two Galoshes and 68 Gazelles are being deployed around Moscow. This new generation of ABM radars included two radars apparently intended to assume the Try Add role in support of the Gazelle.

Flat Twin: Terminal intercept control radar, modular to permit storage and rapid assembly. The radar is transportable. The time to disassemble, transport and reassemble at a new, prepared site is given as 6 to 8 weeks. This relative mobility is considered important to a surprise Soviet "break-out" of the ABM Treaty.

Pawn Shop: Missile guidance radar is also modular. This radar is housed in a van-like structure.

Another new generation ABM radar is the Pill Box battle engagement large PAR, located at Pushkino, about 17 nm (20 mi/32 km) northeast of Moscow. Unlike Dog House and Cat House, Pill Box has four large faces, each 500 ft wide and 120 ft high in which the PAR is set, giving the radar full 360-degree coverage.

Along the periphery of the Soviet homeland are several long-range early warning systems. Three major radar systems are involved:

Steel Work: Over-the-horizon backscatter (OTH-B) radars are located at three sites:
—Nikolayev in the Caucasus mountains,
—Gomel, about 152 nm (175 mi/282 km) southeast of

Dog House

Pill Box

Minsk, and

—Nikolayevsk-ba-Amur in the Soviet Far East.

A fourth has been reported to be under construction. Steel Work is intended to provide up to 30 minutes early warning of a U.S. intercontinental ballistic missile strike and a shorter warning time against a Chinese attack. Peak power is at least 20-40 megawatts and possibly more.

Hen House: These radars are located at six points along the Soviet periphery. These BMD detection and target acquisition radars provide confirmation and initial classification of the scale, timing and characteristics of an attack. Eleven Hen House radars are located at six sites:

—Olenogorsk, near Murmansk,

—Skrunda, Latvia,

—Sevastopol, on the Crimean peninsula,

—Nikolayev, in the Caucasus mountains,

—Sary Shagan, and

—Angarsk, near Irkutsk.

Each Hen House has a central building and two long wings, the latter carrying the antenna arrays. The phased array system is about 1,000 ft (305 m) long and 50 ft (15 m) high; the arrays are inclined about 45 degrees. The antenna operates in the A-band at a peak power of over 10 MW, PRFs are between 25 and 100 pulses/sec. Maxi-

mum range is about 3,240 nm (3,730 mi/6,000 km). The radar scans 5 beams in a complex pattern of two in elevation, two in azimuth, and one in a circular pattern.

Hen House is being supplemented or replaced by 8-9 LPARs sometimes identified as Hen Roost. Five are located near Hen House installations at Olenogorsk, Skrunda, Nikolayev, Sary Shagan, and Angarsk. New sites along the Soviet perimeter are:

—Pechora, east of Archangelsk,

—Baranovichi, 74 nm (85 mi/137 km) southwest of Minsk, and

—Mukachevo, in the Western Ukraine.

Of those 8, those in the northwest quadrant, Skrunda, Baranovichi, and Mukachevo, are still under construction. Hen Roost LPARs consist of two large buildings, one a transmitter, the other a receiver, which have one approximately 300-foot (91 m)-long phased array radar face each. The radar is reported to operate in the B-band, and has a peak power of 5 MW. The beamwidth is 2 degrees by 2 degrees. The LPARs have much greater ability to track and classify multiple targets but they are said to be limited in their discrimination and in their resistance to electromagnetic pulse (EMP) blackout.

The ninth LPAR is located at Krasnoyarsk and is the centerpiece of U.S. allegations of Soviet infringements

Hen House

on the 1972 ABM Treaty. It appears to be similar to the other 8 LPARs, but is positioned not to look out from the Soviet periphery but rather northeast across Siberia. Such an orientation would enable the radar to act as a BMD battle management radar similar to the Pill Box. Under the ABM Treaty, BMD radars are permitted only near the Moscow ABM site.

The Soviet response has come in two parts: First, they have contended that the radar is for tracking of space vehicles and to verify U.S. compliance with the ABM treaty. Second, the Soviets have offered to dismantle the Krasnoyarsk LPAR if the U.S. stops modernization of the Thule, Greenland and Flyingdales, England, ballistic missile early warning system (BMEWS) radars. The U.S. program is alleged by the Soviets to transfer BMD capability outside of its national territory, which is also a violation of the ABM Treaty. Whatever its actual purpose and capability, the Krasnoyarsk radar will close the remaining gap in the perimeter early warning system and significantly upgrade its performance as well.

Krasnoyarsk

CHARACTERISTICS:

Band	E

NAME: **BALL END**

DESCRIPTION: Ball End is a Soviet navigation radar found in several Soviet ships. Ball Gun, a similar radar, has a solid truncated parabolic reflector.

CHARACTERISTICS:

Band	I

PLATFORM:

Minesweeper	T-43

TACTICAL RADARS

NAME: **BACK NET**

DESCRIPTION: This is a Soviet ground-based, early-warning and Ground Control Intercept (GCI) radar. It has a 6-feed antenna system which consists of 2 back-to-back parabolic sections. It is being replaced by the Bar Lock radar.

LPA

NAME: **BAND STAND**

DESCRIPTION: Band Stand is a Soviet naval search and missile tracking and control radar used with the SS-N-9 and SS-N-22 anti-ship missiles. The antenna is housed in a tall cylindrical radome. The *Nanuchka II*-class missile corvettes transferred to Third World navies have the SS-N-2 Styx missile instead of the SS-N-9 and their Band Stand radome contains a Square Tie radar.

CHARACTERISTICS:
Band	G/H
Range	25-35 nm (46.3-64.9 km)

PLATFORMS/USERS:
Destroyers	*Sovremennyy** class
Corvettes	*Nanuchka I/III* class
	*Tarantul** class
Small Combatants	*Sarancha*
Landing Craft	*Pomornik* class

Band Stand

NAME: **BAR LOCK/P-50***

DESCRIPTION: This is a Soviet ground-based target acquisition and Ground Control Intercept (GCI) radar in use with Soviet and Eastern Bloc forces. It is mounted on a trailer, its antenna consisting of two truncated parabolic reflectors with clipped corners. Each reflector has 3 feeds to generate an aggregate of 6 stacked beams. The trailer body rotates in azimuth at up to 12 rpm. Known to be in use since at least the mid-1960s.

CHARACTERISTICS:
Band	E/F
Range	Approx 125 nm (230 km)
Power	Approx 1 MW each beam
Pulse Repetition Frequency	375 pulses per second

Bar Lock

NAME: **BASS TILT**

DESCRIPTION: Bass Tilt is a Soviet naval fire control radar. On larger ships it directs the 30-mm multi-barrel, close-in defense gun systems. On smaller combat craft Bass Tilt also can be used to control the 76.2-mm guns and in the *Grisha III* the 57-mm guns. The earlier Drum Tilt looks quite similar but is smaller; Muff Cob also has a drum-like radome.

CHARACTERISTICS:
Band	H
Range	10-12 nm (18.5-22.2 km)

PLATFORMS/USERS:
Aircraft Carriers	*Kiev** class
Battle Cruisers	*Kirov** class
Cruisers	*Slava** class
	Kara class
	Kresta I/II classes
	Kynda (*Groznyy, Varyag* only)
Destroyers	*Udaloy** class
	*Sovremennyy** class
	Modified *Kashin* class
Frigates	*Krivak III* class
	Grisha III/IV class
Corvettes	*Pauk* class
	Nanuchka III class
	Tarantul class
Small Combatants	*Babochka* class
	Matka class
	Sarancha
	Slepen
River monitor	*Yaz* class
Amphibious Ships	*Ivan Rogov** class
Landing Craft	*Pomornik* class
Auxiliary Ships	Various classes

NAME: BEE HIND

DESCRIPTION: This is a Soviet tail-warning aircraft and gunfire control radar fitted in bomber-type aircraft.

CHARACTERISTICS:
Band   I

PLATFORMS:
Bombers   Tu-16* Badger
              Tu-20/95/142* Bear
              Tu-22* Blinder
              Mya-4* Bison

NAME: BIG BAR/BIG MESH

DESCRIPTION: These are Soviet ground-based search and ground control intercept (GCI) radars with a configuration similar to the Bar Lock GCI radar. These radars, however, have the upper, rear antenna angled some 25 deg from the horizontal, thus providing a limited height-finding capability.

CHARACTERISTICS:
Band   E
Range   90 nm (167 km)

NAME: BIG BULGE

DESCRIPTION: This is a Soviet aircraft surface search and missile targeting radar. It is used with a video data link to transmit radar pictures of target ships to missile-launching aircraft or submarines. The Big Bulge-A is fitted in a large, under-fuselage radome bulge on Bear-D long-range naval reconnaissance aircraft and Big Bulge-B is in a chin mounting on the Hormone-B ship-based helicopter.

CHARACTERISTICS:
Band   I/J
Range   230 nm (426 km)

PLATFORMS:
Bombers   Tu-20* Bear-D
Helicopters   Ka-25* Hormone-B

DESIGNATOR: MT-SON*
NAME: BIG FRED

DESCRIPTION: Big Fred is a Soviet artillery and mortar locating radar mounted on the MT-LB armored personnel carrier. The rectangular antenna is mounted on a rotating turret and folds forward for travelling. There is usually one Big Fred per tank and motorized rifle division.

CHARACTERISTICS:
Band   J
Range   Approx 6 nm (11 km)

PLATFORMS: MT-LB* armored personnel carrier

Big Bulge

NAME: BIG NET

DESCRIPTION: Big Net is a Soviet naval long-range, three-dimensional air surveillance radar. First seen at sea in the missile range ship *Sibir** in the early 1960s, Big Net is generally associated with the SA-N-1 Goa surface-to-air (SAM) system. (It was also fitted in the cruiser *Dzerzhinsky** with the SA-N-2 Guideline SAM). The open lattice antenna has an elliptical parabolic form with an offset-fed reflector providing a narrow azimuth beam; the feedhorn is curved and underslung, with twin balancing vanes. The *Kirov** class battle cruisers have a Big Net mounted back-to-back with a Top Sail to form the Top Pair radar.

CHARACTERISTICS:

Band	C
Range	
High Altitude	200 nm (370 km)
Low Altitude	85 nm (157.5 km) (estimated)

PLATFORMS/USERS:

Battle Cruisers	*Kirov** class (as Top Pair)
Cruisers	*Kresta I* class
	*Sverdlov** class (some ships)
Destroyers	Modified *Kashin* class
	Kashin class
Radar Pickets	*T-58** class
Auxiliary Ships	*Sibir**

NAME: BIG NOSE

DESCRIPTION: This is a Soviet airborne air-intercept radar in the Tu-28P* Fiddler interceptor. It is associated with the radar-guided AA-5 Ash air-to-air missile.

CHARACTERISTICS:

Band	I/J
Range	
Search	35-60 nm (64.9-111.2 km)
Tracking	28 nm (51.9 km)

PLATFORMS:

Fighters	Tu-28P* Fiddler

NAME: BOAT SAIL

DESCRIPTION: Boat Sail was a Soviet naval long-range search radar that was fitted in Whiskey-class submarines converted to the Canvas Bag radar pickets configurations. The radar could be folded and often had a canvas covering over the antenna. All Soviet radar picket submarines have been discarded.

CHARACTERISTICS:

Band	E/F

NAME: BOW Series

DESCRIPTION: The Bow series of Soviet naval radars are target designators found on older surface warships. Half Bow is associated with torpedos; Top Bow is used for 152-mm gun control.

CHARACTERISTICS:

Band	I/J
Range	
Long Bow	10 nm (18.5 km)

PLATFORMS/USERS:

Cruisers	*Sverdlov** class
Command Ships	Modified *Sverdlov** class
Destroyers	*Kildin* class
	*Skoryy** class

NAME: BOX TAIL

DESCRIPTION: This is a Soviet aircraft tail warning radar.

CHARACTERISTICS:

Band	I

PLATFORMS:

Bombers	Tu-20* Bear-D/F

Big Net

Box Tail

Cheese Cake

NAME: **BREAD BIN**

DESCRIPTION: Bread Bin is a Soviet meteorlogical radar mounted on a trailer and used with early versions of the FROG (Free Rocket Over Ground) series of artillery rockets. It also is used with other Soviet artillery weapons.

CHARACTERISTICS:

Band	I

NAME: **CAKE SERIES—**
                                  PATTY CAKE
                                  ROCK CAKE
                                  SPONGE CAKE
                                  STONE CAKE

DESCRIPTION: The Cake series of Soviet height-finding radars have large, peel-shaped nodding antennas with horizontal beam widths of 3 to 5 deg and vertical beam widths of 1 to 5 deg. The antenna is "nodded" mechanically at the rate of 30-40 cycles per minute. Patty Cake has back-to-back antennas, Rock Cake is transportable, and Sponge Cake and Stone Cake are both trailer-mounted.

CHARACTERISTICS:

Band	E
Range	108 nm (200 km)
Sponge Cake	162 nm (300 km)

NAME: **CHEESE CAKE**

DESCRIPTION: Cheese Cake is a Soviet naval missile targeting radar for SS-N-2c Styx anti-ship missiles. Installed in limited numbers from the late l970s.

CHARACTERISTICS:

Band	I
Range	17-25 nm (31.5-46.3 km)

PLATFORMS/USERS:

Small Combatants	*Matka* class

NAME: **CLAM SHELL**

DESCRIPTION: Clam Shell is a Soviet 3-dimensional target acquisition radar mounted on a tracked vehicle (known as an SSNR). The radar is associated with the SA-11 Gadfly surface-to-air missile (SAM), the SSNR accompanying 4 tracked Gadfly launchers in each battery.

CHARACTERISTICS:

Band	I

DESIGNATOR: **GUIS-2***
NAME: **CROSS BIRD**

DESCRIPTION: Cross Bird is a Soviet naval long-range, air-search radar developed from the British Type 291 World War II-era P-band radar. It was fitted in early postwar Soviet warships and will soon be discarded. Sea Gull was an improved version with the Soviet designation of GUIS-2M.

CHARACTERISTICS:

Band	A

PLATFORMS/USERS:

Destroyers	*Skoryy** class

NAME: CROSS SWORD

DESCRIPTION: Cross Sword is a Soviet naval fire control radar for the SA-N-9 surface-to-air missile (SAM) system. The radar system reportedly has suffered from development problems and was delayed in being fitted in the *Udaloy*-class ASW destroyers. Ships which had deployed with one or two Cross Swords installed have later been seen missing one or both sets. The radar system sits on a conical base and probably rotates a full 360 deg. The upper part of the assembly has back-to-back mesh antenna. The lower part has a circular flat plate reflector with a tripod feed. Other ships carrying the SA-N-9 SAM system—the later *Kirov** class battle cruisers and *Kiev** class aircraft carriers—have not been seen with Cross Sword. It is thought that other embarked radars are employed for missile guidance.

PLATFORMS/USERS:

Destroyers	*Udaloy** class

NAME: CROWN DRUM

DESCRIPTION: This is a Soviet aircraft long-range, high-density surface-search radar associated with the AS-3 Kangaroo air-to-surface missile. It is faired into the chin position of Bear-B/C bombers that are being phased out of service, with some being converted to the Bear-G configuration.

CHARACTERISTICS:

Band	E/I

PLATFORMS:

Bombers	Tu-20* Bear-B/C

NAME: DON

DESCRIPTION: The Don series of Soviet naval radars are primarily for surface ship navigation, but probably have a target-designation function as well for certain warships. They are horizontally polarized radars.

CHARACTERISTICS:

Band	I
Range	Approx 25 nm (46 km)

PLATFORMS/USERS:

Don-2

Aircraft Carriers	*Kiev** class (except *Novorssyisk**)
	Moskva* class
Cruisers	*Kara* class
	Kresta I/II classes
	Kynda class
	Sverdlov* class (some ships)
Destroyers	*Kashin* class
	Modified *Kildin* class

	Kotlin SAM class
	Kotlin class (some ships)
	Skoryy* class
Frigates	*Krivak I/II/III* classes
	Koni
	Grisha I/II/III/IV classes
	Petya I/II classes
	Mirka I/II classes
	Riga class
Corvettes	*Poti* class
	T-58* class
Small Combatants	*Babochka*
	Slepen
Minesweepers	*Alesha** class
	Natya* class
	Yurka* class
	T-43* class
	Vanya* class
Amphibious Ships	*Ropucha* class
	Alligator class
Auxiliary Ships	Various

Don-Kay

Aircraft Carriers	*Kiev** class (except *Novorssyisk**)
Cruisers	*Kara* class
	Kresta I/II classes
Destroyers	Modified *Kashin* class
	Kashin class
	Kanin class
Frigates	*Krivak III* class
Minesweepers	Modified *Vanya** class
Amphibious Ships	*Ropucha* class
	Alligator class
Auxiliary	Various

NAME: DOWN BEAT

DESCRIPTION: This is a Soviet aircraft bombing and navigation radar employed to target the AS-4 Kitchen anti-ship and later air-to-surface missiles.

CHARACTERISTICS:

Band	I
Range	Approx 175 nm (324 km)

PLATFORMS:

Bombers	Tu-22M* Backfire-B
	Tu-22* Blinder

NAME: DRUM TILT

DESCRIPTION: The first of a series of Soviet naval drum-shaped weapons control radars, Drum Tilt is the acquisition and tracking radar for twin 25-mm and 30-mm multi-barrel (not Gatling) anti-aircraft guns. The plastic radome is tilted about 25 deg above the horizontal and houses a circular parabolic dish.

CHARACTERISTICS:

Band	H/I
Range	Approx 22 nm (40.8 km)

PLATFORMS/USERS:

Cruisers	Modified *Sverdlov** class
	*Sverdlov** class (3 ships)
Destroyers	*Kanin* class
	Kotlin SAM class (4 ships)
Frigates	*Koni* class
Small Combatants	*Osa I/II* classes
	*Stenka** class
Minesweepers	*Natya** class
	*Yurka** class
Amphibious Ships	*Polnocny B/C* classes
Landing Craft	*Aist* class

Drum Tilt

NAME: EGG CUP

DESCRIPTION: Egg Cup is a Soviet naval gunfire spotting radar for 152-mm and older 130-mm and 100-mm guns. It is a modified Skin Head radar and provides range-only data.

CHARACTERISTICS:

Band	E
Range	12-15 nm (22.2-27.8 km)

PLATFORMS/USERS:

Cruisers	*Sverdlov** class (some ships)
Command Ships	Modified *Sverdlov** class
Destroyers	*Kotlin SAM* class
	Kotlin class

DESIGNATOR: RMS-1*
NAME: END TRAY

DESCRIPTION: End Tray is a Soviet trailer-mounted meteorological radar associated with FROG-7 battlefield support missiles and artillery batteries. End Tray has a parabolic dish tracker.

CHARACTERISTICS:

Band	D

NAME: EYE BOWL

DESCRIPTION: Eye Bowl is a Soviet fire control radar for use with the SS-N-14 Silex anti-submarine warfare (ASW) missile in ships not fitted with the larger Head Lights radar. Eye Bowls are fitted in pairs, with each set showing a circular parabolic dish and tripod-mounted feed horn.

CHARACTERISTICS:

Band	F
Range	10-12 nm (18.5-22.2 km)

PLATFORMS/USERS:

Battle Cruiser	*Kirov** (one ship only)
Destroyers	*Udaloy** class
Frigates	*Krivak I/II* classes

NAME: FAN SONG

DESCRIPTION: The Fan Song series are Soviet trailer-mounted missile control radars for the SA-2 Guideline surface-to-air missile (SAM). Two orthogonal antennas—one horizontal and one vertical—use an electromechanical "flapping" motion (Lewis scanner) to radiate beams

Fan Song E

which scan designated sectors in a track-while-scan mode. The SA-2 fire control system can track six missiles simultaneously while guiding three. Fan Song E has two more dishes which confer a Lobe On Receive Only electronic counter-countermeasures capability on the radar. Seven versions of the Fan Song have been identified, with the two major variants being the Fan Song-A/B and the Fan Song E.

CHARACTERISTICS:

Band
Fan Song-A/B	E/F
Fan Song-E	G

Beam Width
Fan Song-A/B	10 deg x 2 deg
Fan Song-E	7.5 deg x 1.5 deg

Peak Power
Fan Song-A/B	600 kW
Fan Song-E	1.5 MW

Range
Fan Song-A/B	32-65 nm (60-120 km)
Fan Song-E	38-90 nm (70-145 km)

Fan Song F

NAME: FAN TAIL

DESCRIPTION: This is a Soviet aircraft gunfire control radar for the remote-controlled tail gun turret in the Backfire-B bomber.

PLATFORMS:
Bombers	Tu-22M* Backfire-B

DESIGNATOR: SON-9/9A*
NAME: FIRE CAN

DESCRIPTION: Fire Can is a Soviet trailer-mounted gun control radar based on the U.S. World War II-era SCR-584. It is used with 57-mm (S-60), 85-mm (M1939/KS-12), and 100-mm (KS-IQ) anti-aircraft guns. The radar has a perforated parabolic dish which is fed by a rotating dipole. The Fire Dish, Fire Wheel, and Whiff are radars derived from Fire Can.

CHARACTERISTICS:
Band	E/F
Peak Power	300 kW
Pulse Repetition	
Frequency	1840-1900 pulses/sec

Range
Maximum	
Acquisition	43 nm (80 km)
Track	19 nm (35 km)

Fire Can

NAME: FLAP WHEEL

DESCRIPTION: Flap Wheel is a Soviet van-mounted fire control radar used in conjunction with 57-mm (5-60) and 130-mm (KS-30) towed anti-aircraft guns. The antenna is a parabolic dish, which is conically scanned, and a Yagi array. Flap Wheel is believed to be similar to the Gun Dish fire control radar mounted on the ZSU-23-4* air defense vehicle.

CHARACTERISTICS:
Band	I/J
Range	Approx 19 nm (35 km)

PLATFORMS/USERS:
Towed guns	57 mm (5-60)
	130 mm (KS-30)

DESIGNATOR: P-15*

NAME: FLAT FACE

DESCRIPTION: Flat Face is a Soviet target acquisition radar used with mobile surface-to-air missile (SAM) systems. It is a truck-mounted system with 2 stacked elliptical paraboloid antennas, each measuring approx 36 ft x 18 ft (11 m x 5.5 m).

CHARACTERISTICS:

Band	C
Peak Power	500 kW
Beam Width	
Vertical	5 deg
Horizontal	2 deg
Range 135 nm	(250 km)

PLATFORMS/USERS: SA-3 Goa
SA-6 Gainful
SA-8 Gecko

Flat Face

NAME: FLAT JACK

DESCRIPTION: This is a Soviet Airborne Warning and Control System (AWACS) radar mounted on the Tu-126 Moss aircraft. The version installed in the Moss has limited capability and cannot distinguish targets over land from background "clutter." The antenna is fitted in a rotodome (rotating radome) about 36 to 38 ft in diameter, mounted above the after fuselage.

PLATFORMS:

Early warning aircraft	Tu-126* Moss
	Il-76TD* Mainstay

NAME: FRONT DOME

DESCRIPTION: Front Dome is a Soviet naval target illumination and missile tracking radar system for the SA-N-7 naval surface-to-air missile (SAM). Each of the radomes resembles a Bass Tilt gunfire control radar but is fixed in elevation. The Sovremennyy*-class destroyers are fitted with 6 radomes, one on either side of the mainmast, and 2 on each side of the foremast. The SA-N-7 trials ship Provornyy* has 8 radomes.

CHARACTERISTICS:

Band	H/I

PLATFORMS/USERS:

Destroyers	Provornyy*
	(Kashin class trials ship)
	Sovremennyy* class

Front Dome

NAME: FRONT PIECE
FRONT DOOR
TRAP DOOR

DESCRIPTION: Front Door/Front Piece are Soviet submarine sail-mounted radars for mid-course guidance of the SS-N-3a Shaddock and SS-N-12 Sandbox anti-ship missiles. The antenna is a truncated parabolic mesh structure standing on end with a large feed horn mounted on the upper edge. Submarines house their Front series radars in the forward edge of the sail which swings open for operation. The Slava's* Front Door/ Front Piece set is mounted on the pyramid mast; the Kiev's* similar Trap Door retracts into the ship's forecastle when not in use.

CHARACTERISTICS:

Band	F

PLATFORMS/USERS:
Front Door/Front Piece

Submarines	Echo-II class
	Juliett class

Cruisers	*Slava** class
Trap Door	
Aircraft Carriers	*Kiev** class

NAME: **GAGE**

DESCRIPTION: The Gage series of Soviet target acquisition radars serve the SA-1 Guild surface-to-air missile (SAM) in conjunction with the Yo-Yo missile control radar.

CHARACTERISTICS:

Band	E/F
Peak power	2 MW

DESIGNATOR: **GS Series**

DESCRIPTION: The GS series are battlefield surveillance radars, some being carried in manpacks and others in tactical vehicles.

CHARACTERISTICS:

Band	J
Peak power	
GS-11	10 kW
GS-12	25 kW
GS-13	50 kW
Range	
GS-11	0.6 nm (1.2 km)
GS-12	1.9 nm (3.5 km)
GS-13	6.5.nm (12 km)

PLATFORMS:

GS-11	Manpack
GS-12	Manpack or vehicle
GS-13	Vehicle-mounted only

NAME: **GUN DISH**

DESCRIPTION: Gun Disk is the fire control radar mounted on the ZSU-23-4* quad 23-mm cannon air defense vehicle. It is also associated with the SA-9 Gaskin surface-to-air missile (SAM) system mounted on BRDM-2A* armored reconnaissance vehicles. Optical and infrared sighting are included in the fire control system. The Gun Dish is mounted on the gun turret and rotates with it, but it can be elevated independently of the guns. The drum-like radar can be folded down when not in use.

CHARACTERISTICS:

Band	J
Range	
Search	27 nm (50 km)
Acquisition	3.5-4.5 nm (6.5-8.3 km)

PLATFORMS/USERS: Soviet Union-Ground Forces, ZSU-23-4 vehicle BRDM-2A with SA-9 Gaskin.

Gun Dish

NAME: **HAWK SCREECH**

DESCRIPTION: Hawk Screech is a Soviet naval gunfire control radar for 45-mm, 57-mm, 76.2-mm, and 100-mm anti-aircraft guns. The pedestal-mounted radar has a circular parabolic dish with a four-legged feed horn mounting. The transmitter and receiver are mounted in a large box behind the antenna. Hawk Screech is always accompanied by an optical back-up sensor. A similar radar of later design is the Owl Screech. See separate database entry.

CHARACTERISTICS:

Band	I
Range	8-12 nm (14.8-22.2 km)

PLATFORMS/USERS:

Destroyers	*Kanin* class
	Modified *Kildin* class
	Kotlin SAM class

Hawk Screech

Frigates	Kotlin class
	Modified Skoryy* class
	Koni class
	Petya I/II classes
	Mirka I/II classes
	Riga class
Auxiliary Ships	Various

NAME: HEAD LIGHTS

DESCRIPTION: The Head Lights radar system is a Soviet naval fire control radar for the SA-N-3 Goblet naval surface-to-air missile (SAM) and SS-N-14 Silex anti-submarine warfare (ASW) missile. Head Lights has two nearly 13-ft (4-m) diameter open mesh dishes, two smaller antenna of similar construction with tripod feed horns under the latter pair. The main dishes are balanced by vanes at the rear. A fifth, still smaller dish may provide command signals to the missile. The entire array can elevate and can traverse 360 deg. Head Lights-A is fitted in aircraft carriers and guides the SA-N-3 SAMs. Head Lights-B is fitted in cruisers and additionally supports the SS-N-14 ASW missiles. A Head Lights-C variant has also been identified.

CHARACTERISTICS:

Band	D/F/G/H
Range	40 nm (74.1 km)

PLATFORMS/USERS:

Head Lights-A

Aircraft Carriers	Kiev* class
	Moskva* class

Head Lights-B

Cruisers	Kara class
	Kresta II class

Head Lights

NAME: HEAD NET

DESCRIPTION: The Head Net series are naval air search radars fitted in a variety of Soviet ships.

Head Net-A: Fitted in a few older cruisers and destroyers. Head Net-B/C are two different versions of back-to-back Head Net-As. The back-to-back arrangement in Head B/C provides the advantages of reducing structural interference to radar emissions, improving data rates, and possibly making track transfer from two- to three-dimensional radars faster and more accurate through common stabilization. Head Net-A has a large, open work elliptical parabolic reflector illuminated by a feed horn carried on a boom projecting form under the scanner's lower edge. The antenna is fitted with larger balancing vanes.

Head Net-B: Consists of back-to-back Head Net-A antennas with one angled 15 deg in elevation to provide both high- and low-altitude coverage. Fitted only in two missile range instrumentation ships.

Head Net-C: Also consists of back-to-back Head Net-As with one angled 30 deg from the horizontal for simultaneous use as air-search and height-finding radar. This version is the most widely deployed, often accompanied by Top Sail in the larger units.

Head Net C

CHARACTERISTICS:

Band	
Head Net-A	C
Head Net-B	E/F
Head Net-C	E/F

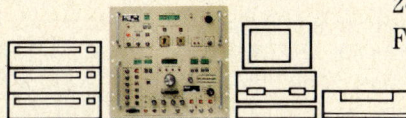

Range
Head Net-A
  High Altitude   120 nm (222 km)
  Medium Altitude 70 nm (130 km)
Head Net-C   60-70 nm (111-130 km)

PLATFORMS/USERS:
Head Net-A
Cruisers   Kynda class (Admiral Golovko*,
                    Admiral Fokin* only)
Head Net-B
Missile Range   Modified Desna* class (2 ships)
Instrumentation
Ships
Head Net-C
Aircraft Carriers   Moskva* class
Cruisers   Kara class
                    Kresta I/II classes
                    Kynda class
                    (except Admiral Golovko)
Destroyers   Modified Kashin class
                    Kashin class
                    Kanin class
                    Kotlin SAM class
Frigates   Krivak I/II/III classes
Amphibious Ships   Ivan Rogov class

NAME: **HIGH FIX**

DESCRIPTION: This is a Soviet aircraft range-only inter-cept radar in fighter-type aircraft.

CHARACTERISTICS:
Band   I
Range   5 nm (9.2 km)

PLATFORMS:
Fighters   Su-20* Fitter-C/D

NAME: **HIGH LARK**

DESCRIPTION: This is a Soviet aircraft air-intercept radar fitted in several fighters. It is associated with the AA-2/7/8 Atoll/Apex/Aphid air-to-air missiles. The later MiG-29* Fulcrum and Su-27* Flanker may have an up-graded version of the High Lark, or more probably, a newer track-while-scan radar.

CHARACTERISTICS:
Band   I/J
Range
  Search   Approx 50 nm (92.7 km)
  Tracking   Approx 35 nm (64.9 km)

PLATFORMS:
Fighters   MiG-23* Flogger-B/G

NAME: **HIGH SIEVE**

DESCRIPTION: High Sieve is an older Soviet naval air- and surface-search radar.

CHARACTERISTICS:
Band   I
Range   25 nm (46.3 km)

PLATFORMS/USERS:
Cruisers   Sverdlov* class (some ships)
Destroyers   Skoryy* class
Patrol Boats   Purga* (KGB Maritime Troops)

NAME: **JAY BIRD**

DESCRIPTION: This is a Soviet aircraft range-only air intercept radar fitted in several fighter aircraft. It is associated with the AA-2/3/4/7/8 Atoll/Anab/Awl/Apex/Aphid air-to-air and the AS-7 Kerry air-to-surface missiles. Its range is relatively short.

CHARACTERISTICS:
Band   I/J
Range
  Search   Approx 16.2 nm (30 km)
  Tracking   Approx 10.8 nm (20 km)
Peak power   Approx 100 kW
Beamwidth   Approx 3.5 deg square

PLATFORMS:
Fighters   Su-24* Fencer
                    MiG-21* Fishbed-J
                    Su-20/22* Fitter
                    MiG-23* Flogger-E

NAME: **KITE SCREECH**

DESCRIPTION: Kite Screech is a Soviet naval fire control radar for 100-mm and newer 130-mm guns. It resembles the older Hawk Screech and Owl Screech radars but its illuminating feed supports are more closely grouped around the center of the circular paraboloid reflector. While the other two radars use conical scanning, Kite Screech may use monopulse tracking. The Kite Screech dish has a hole in the upper left (facing the antenna) in line with a tube mounted behind the dish; this is probably an optical sensor.

CHARACTERISTICS:
Band   I/J
Range   10-12 nm (18.5-22 km)

PLATFORMS/USERS:
Battle Cruisers   Kirov* class
Cruisers   Slava* class
Destroyers   Udaloy* class
                    Sovremennyy* class
Frigates   Krivak II/III classes

Kite Screech

NAME: KIVACH

DESCRIPTION: Kivach is a Soviet late-model navigation radar in use on some small combatants.

PLATFORMS/USERS:

Corvettes	*Tarantul II/III* classes
River Monitors	*Yaz* class

DESIGNATOR: P-8*

NAME: KNIFE REST

DESCRIPTION: The Knife Rest series of radars were used for early warning in conjunction with the SA-2 Guideline surface-to-air missile (SAM) system. The antenna consists of twin Yagi array. Knife Rest-A is ground-mounted; Knife Rest-B/C are truck-mounted. A naval variant is fitted in some *Sverdlov**-class cruisers and T-43*-class minesweepers.

CHARACTERISTICS:

Band	A
Peak Power	100 kW (Knife Rest A)
Beam width	
Vertical	20-25 deg
Horizontal	20-25 deg
Range	
Knife Rest-A	Approx 200 nm (370 km)
Knife Rest-B	Approx 49 nm (90 km)

NAME: LAND ROLL

DESCRIPTION: Land Roll is the target acquisition radar mounted on the SA-8 Gecko mobile surface-to-air missile (SAM) vehicle. It consists of one rotating fan-shaped surveillance radar mounted above the SA-8 launchers, one large tracking antenna and two smaller missile guidance antennas. The tracking antenna is a parabolic dish with the sides squared off; the smaller missile guidance radars (also cropped parabolic reflectors) are located one to each side.

CHARACTERISTICS:

Band		
Surveillance	G/H	
Tracking radar	J	
Guidance radars	I	
Range		
Surveillance	16 nm (30 km)	

NAME: LIGHT BULB

DESCRIPTION: Light Bulb is a Soviet naval search radar fitted at the top of the mast in *Tarantul* class corvettes. (The radar's name describes the shape of the radome and mounting quite well.)

PLATFORMS/USERS:

Corvettes	*Tarantul II/III* classes

NAME: LONG EYE

DESCRIPTION: A Soviet trailer-mounted battlefield surveillance radar. Dish antennas are fitted back-to-back.

CHARACTERISTICS:

Range	11.5-14 nm (21.3-26 km)

NAME: LONG TALK

DESCRIPTION: Long Talk is a Soviet air search and surveillance radar used with the Two Spot precision approach radar (PAR) to make up the standard ground controlled approach (GCA) system. Long Talk has a trailer-mounted truncated parabolic antenna with a smaller upright antenna at each end of the reflector.

OPERATIONAL NOTES: The Soviets have erected several Long Talk/Two Spot systems in Afghanistan.

NAME: LONG TRACK

DESCRIPTION: Long Track is a Soviet battlefield surveillance and target acquisition radar used with several mobile surface-to-air missile (SAM) systems. Long Track's large elliptical parabolic antenna detects the target first and relays it to Pat Hand, Straight Flush, or Land Roll missile guidance radars. The radar is mounted on a modified AT-T* tracked carrier.

CHARACTERISTICS:

Band	Lower E
Range	70-90 nm (130-167 km)

PLATFORMS/USERS:

Soviet Union	Air Defense Vehicles
Ground Forces	SA-4 Ganef
	SA-6 Gainful
	SA-8 Gecko

Long Track

NAME: **LONG TROUGH**

DESCRIPTION: Long Trough is a Soviet air search radar mounted on a half-track vehicle. Its antenna consists of a horizontal array of feeds in a parabolic cylindrical section which is said to resemble a trough.

CHARACTERISTICS:

Band	E
Range	90-115 nm (167-213 km)

NAME: **LOW BLOW**

DESCRIPTION: Low Blow is a missile control radar for the SA-3 surface-to-air missile (SAM) system. The radar has an unusual arrangement of antennas with two scanning parabolic dishes, each with a prominent feed horn, one set above the other. The upper dish is supported by two mechanically scanned trough-like antennas set at right angles to each other and oriented 45 deg from the horizontal. The trough position improves the radar's ability to see through ground clutter. Low Blow is com-

Low Blow

monly used with Flat Face or Squint Eye C-band target acquisition radars. Low Blow is usually sited as the handle of a "T" with four SA-3 launchers comprising the crossbar.

CHARACTERISTICS:

Band	E
Peak Power	250 kW
Range	20-28 nm (37-52 km)

NAME: **MESH BRICK**

DESCRIPTION: Mesh Brick is possibly a Soviet mortar-locating radar mounted on an open, 4-wheel trailer. The antenna has an offset feed parabolic section with 3 spiral feeds.

CHARACTERISTICS:

Range	4.5-6 nm (8.3-11.1 km)

NAME: **MUFF COB**

DESCRIPTION: Muff Cob is a Soviet naval fire control radar for twin 57-mm gun mounts. Unlike Drum Tilt, which is fixed in elevation, Muff Cob can be elevated. Attached to the mounting is a cylinder housing a television camera.

CHARACTERISTICS:

Band	H

PLATFORMS/USERS:

Helicopter carriers	Moskva* class
Cruisers	Kresta I/II classes
Frigates	Grisha I/II classes
Corvettes	Nanuchka I class
	Poti class
	T-58* class
Radar Pickets	Modified T-58* class
Small Combatants	Turya class
Minesweepers	Alesha* class
Amphibious Ships	Ropucha class
Auxiliaries ships	Various

SOVIET WEAPON SYSTEMS AND ELECTRONICS
SENSORS

NAME: MUSHROOM

DESCRIPTION: This is a Soviet aircraft bombing and navigation radar used in several bomber and EW/ECM aircraft. The radome is located under the forward fuselage.

CHARACTERISTICS:

Band	I
Range	Approx 175 nm (324 km)

PLATFORMS:

Bombers	Tu-16* Badger-A
	Tu-20* Bear-A
	Mya-4* Bison
Reconnaissance	Tu-20* Bear-D

NAME: NEPTUNE

DESCRIPTION: Neptune is a Soviet naval navigation radar fitted in older surface warships and a few naval auxiliaries.

CHARACTERISTICS:

Band	I
Range	20-25 nm (37-46.3 km)

PLATFORMS/USERS:

Cruisers	*Sverdlov** class (some ships)
Destroyers	*Kotlin* class (some ships)
Frigates	*Riga* class
Auxiliaries	Various

NAME: NYSA SERIES

DESCRIPTION: The Nysa series radars are long-range acquisition radars for surface-to-air missile (SAM) systems. The Nysa B is an elliptical parabaloid "nodding" height-finder antenna mounted on a 4-wheel trailer. The Nysa C is composed of 2 horizontal parabolic reflectors (each with a 26-dipole array) stacked one over the other and mounted on an 8-wheel trailer fitted with outriggers.

CHARACTERISTICS:

Band	Probably A

NAME: ODD LOT

DESCRIPTION: Odd Lot is a Soviet van-mounted, early-warning radar with height-finding and ground-controlled intercept (GCI) capabilities. The radar has 2 antennas, one vertical parabolic and the other a horizontal orange-peel parabolic.

CHARACTERISTICS:

Band	E

NAME: ODD PAIR

DESCRIPTION: Odd Pair is a Soviet van-mounted, height-finding radar. Vertical orange-peel parabolic antennas of two different sizes are mounted side-by-side.

CHARACTERISTICS:

Band	E

NAME: OWL SCREECH

DESCRIPTION: Owl Screech is a Soviet naval fire control radar for 76.2-mm guns. Presumably an improved Hawk Screech, Owl Screech has an angled feed nested within four supports that, while grouped more closely than Hawk Screech, are not as perpendicular as with Kite Screech.

CHARACTERISTICS:

Band	I
Range	15-18 nm (27.8-33.4 km)

PLATFORMS/USERS:

Aircraft Carriers	*Kiev** class
Cruisers	*Kara* class
	Kynda class
Destroyers	Modified *Kashin* class
	Kashin class
	Modified *Kildin* class
Frigates	*Krivak I* class
Amphibious Ships	*Ivan Rogov** class
Icebreaker	*Ivan Susanin** class (KGB)

NAME: PALM FROND

DESCRIPTION: Palm Frond is a Soviet search and navigation radar being fitted in newer ships replacing the Don series in older ships. The antenna is a truncated elliptical paraboloid with a prominent boom for the illuminating feed.

CHARACTERISTICS:

Band	A
Range	25 nm (46.3 km)

PLATFORMS/USERS:

Aircraft Carriers	*Kiev** class (*Novorossyisk** only)
Battle Cruisers	*Kirov** class
Cruisers	*Slava** class
	Kara class
	Kresta I class
Destroyers	*Udaloy** class
	*Sovremennyy** class
	Kashin class (some ships)
Frigates	*Krivak II/III* classes
Amphibious Ships	*Ivan Rogov** class
Auxiliary Ships	Various

only from EEV
2·5 to 40GHz

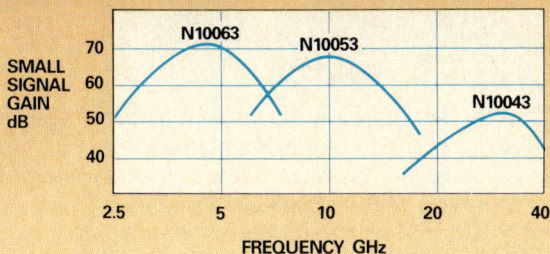

SOVIET WEAPON SYSTEMS AND ELECTRONICS
SENSORS

NAME: SSNR KRUG*/PAT HAND

DESCRIPTION: Pat Hand is a Soviet missile fire control radar mounted on a variant of the SA-4 Ganef tracked surface-to-air missile (SAM) launch vehicle. It is used to track targets and guide 1 or 2 Ganefs to intercept. The vehicle carries a large circular antenna for target tracking and a smaller antenna for missile guidance. SSNR is the Soviet acronym for radar vehicle.

Pat Hand

STATUS: Approx 560 Pat Hand vehicles built before production ended in the 1970s. In each Front or Army SAM brigade, there are 3 battalions each with 9 SA-4 launch vehicles and 1 Pat Hand SSNR. In service with Soviet Ground Forces and several Warsaw Pact countries.

CHARACTERISTICS:
Band	H
Range	Approx 35-45 nm (65-83 km)

NAME: PEEL CONE

DESCRIPTION: Peel Cone is a Soviet naval search radar fitted only on the single Babochka hydrofoil patrol combatant.

NAME: PEEL GROUP

DESCRIPTION: Peel Group is a Soviet naval missile fire control radar for the SA-N-1 Goa surface-to-air missile (SAM) system. The antenna array is complex, consisting of two large and two small solid reflectors. One of the large, orange-peel paraboloid reflectors is arranged horizontally, the other vertically. The two smaller antenna are of a similar shape and disposed similarly. All have prominent lattice-work booms supporting the illuminating feeds. The entire array is sited on a tall, presumably stabilized pedestal mounting, and is counterbalanced by a single, thick vane to the rear.

CHARACTERISTICS:
Band		
Tracking	I	
Guidance	E	

PLATFORMS/USERS:
Cruisers	Kresta 1 class
	Kynda class
Destroyers	Modified Kashin class
	Kashin class
	Kanin class
	Kotlin SAM class

NAME: PEEL PAIR

DESCRIPTION: Peel Pair is a Soviet naval surface search and navigation seen only on the Nanuchka I class missile corvettes. The radar has two orange-peel paraboloid antennas mounted back-to-back on a pedestal.

Peel Group

NAME: PLANK SHEAVE

DESCRIPTION: Plank Sheave is a relatively new Soviet naval air-and surface-ship radar for smaller ships. It is thought to be an improvement over the earlier Strut Pair. The antenna is a solid, truncated paraboloid with an open-work upper edge and an underslung feed horn boom.

PLATFORMS/USERS:
Corvettes	Pauk class

NAME: PLINTH NET

DESCRIPTION: Plinth Net is a Soviet medium-range search radar, apparently associated with the SS-N-3b Shaddock anti-ship missile systems Some western sources speculate that Plinth Net may be used for the video data-link antenna for the Shaddocks rather than a radar.

The cropped truncated paraboloid mesh antenna is mounted half-way up the mast in both classes of cruisers in which it is fitted. *Kynda* class cruisers have one antenna on a centerline platform forward of the mast; the *Kresta I* has two, one to port and one starboard.

CHARACTERISTICS:

Band	E
Range	
Air Search	Approx 80 nm (148 km)
Surface Search	Approx 20 nm (37 km)

NAME: POP GROUP

DESCRIPTION: Pop Group is a Soviet naval fire control radar for the SA-N-4 Gecko naval surface-to-air missile (SAM) system. It is similar to the ground-based Land Roll radar used with the SA-8 Gecko air defense vehicle. The assembly has three antennas. The top-mounted, truncated paraboloid antenna may rotate independently of the lower assembly and is thought to be for target search. Two flat, circular plates of different sizes are mounted on the front below the target search radar. One is a target tracker, the other a missile guidance radar and both are electronically scanned. Frequency-hopping may be employed by the system.

CHARACTERISTICS:

Band	
Target Search	F
Target Track and Missile Guidance	H/I
Range	Approx 35-40 nm (65-74 km)

PLATFORMS/USERS:

Aircraft Carriers	*Kiev** class (except *Novorossyisk**)
Battle Cruisers	*Kirov** class
Cruisers	*Slava** class
Command Ships	Modified *Sverdlov** class
Frigates	*Krivak I/II/III* classes
	Koni class
	Grisha I/III/IV classes
Corvettes	*Nanuchka I* class
	Sarancha
Amphibious Ships	*Ivan Rogov** class
Replenishment Oiler	*Berezina**

DESIGNATOR: SNAR-2*/SNAR-6*
NAME: PORK TROUGH

DESCRIPTION: Pork Trough is a Soviet counter-battery radar using a rotating horizontal parabolic cylinder antenna mounted on the fully tracked AT-LM artillery tractor. Pork Trough-2 (SNAR-6) is a variant which operates at a higher frequency.

CHARACTERISTICS:

Band	
Pork Trough (SNAR-2)	I
Pork Trough-2 (SNAR-6)	J

NAME: POST LAMP

DESCRIPTION: Post Lamp is a Soviet naval fire-control radar used in older destroyers for both guns and torpedoes.

CHARACTERISTICS:

Band	I
Range	15-20 nm (27.8-37 km)

PLATFORMS/USERS:

Destroyers	*Kotlin* class
	*Skoryy** class

NAME: POT DRUM/BAKLAN*

DESCRIPTION: Pot Drum is a Soviet naval surface search radar fitted in some small combatants. The radar is identical to the Square Tie installed in *Osa I/II* missile boats but it is housed in a drum-like radome measuring approx 5 ft (1.5 m) in diameter. (Pot Head is a similar but smaller (4 ft; 1.2 m) radar with a flat top. It is on Soviet and East German torpedo boats, few of which are in service.)

CHARACTERISTICS:

Band	I
Range	18-20 nm (33.4-37 km)

PLATFORMS/USERS:

Small Combatants	*Turya** class
	*Stenka** class

DESIGNATOR: PSNR-1*

DESCRIPTION: PSNR-1 is a portable Soviet battlefield surveillance radar mounted on a tripod. The radar is used by reconnaissance companies in Soviet tank and motorized rifle divisions as well as by divisional anti-tank battalions.

CHARACTERISTICS:

Bands	I (9.6 GHz)
	J
Range	2.7-5.4 nm (5-10 km)

NAME: PUFF BALL

DESCRIPTION: This is a Soviet aircraft bombing and navigation radar used in Badger bombers for targeting AS-2/5/6 Kipper/Kelt/Kingfish air-to-surface missiles (although some aircraft carrying the AS-5 may have the Short Horn radar). Also fitted in Badger and Bison reconnaissance/EW aircraft, and employed as surface-search radar for the Hormone-A ASW helicopter.

CHARACTERISTICS:

Band	I
Range	Approx 175 nm (324 km)

PLATFORM

Bombers	Tu-16* Badger-C and later models
	Mya-4* Bison-C/D
Helicopters	Ka-25* Hormone-A

NAME: ROUND HOUSE

DESCRIPTION: Round House is a Soviet naval tactical aircraft control and navigation (TACAN) radar. It provides all-weather homing and control of aircraft. Fitted in ships with multi-helicopter capability.

PLATFORMS/USERS:

Battle Cruisers	Kirov* class
Cruisers	Kara class (1 ship)
Destroyers	Udaloy* class

NAME: SCOOP PAIR

DESCRIPTION: Scoop Pair is a Soviet naval mid-course missile guidance radar in surface ships for use with the SS-N-3b Shaddock anti-ship missile system. The radar consists of "mirror-image" antennas (each approx 13 ft 9 in or 4.2 m span) above and below a spherical housing. Each antenna has a heavy boom extending out in front of the truncated parabolic open lattice reflector; the boom holds stacked feed horns. Each antenna also has prominent balancing wind vanes extending to the rear.

CHARACTERISTICS:

Band	E

PLATFORMS/USERS:

Cruisers	Kresta I class
	Kynda class

Scoop Pair

NAME: SCOREBOARD

DESCRIPTION: Scoreboard is a Soviet ground-based identification friend or foe (IFF) interrogator. The radar has a rectangular rotating antenna array of dipole radiators. Reportedly of very dated design.

NAME: SHIP GLOBE

DESCRIPTION: Ship Globe is a large Soviet tracking radar used to monitor long-range missile tests and satellites activity. The parabolic reflector is approx 52 ft (16 m) in diameter. It is in use on missile range and space event support ships (SESS).

PLATFORMS/USERS:

Missile Range Ships	Marshal Nedelin* class
	Desna* class
SESS ships	Akademic Sergei Korolev*
	Kosmonaut Vladimir Komarov*

NAME: SHORT HORN

DESCRIPTION: This is a Soviet aircraft bombing and navigation radar used in earlier bomber-type aircraft and in some updated aircraft. It is reportedly associated with AS-5/6 air-to-surface missiles. It has frequency agility and diversity capabilities.

CHARACTERISTICS:

Band	J
Range	Approx 115 nm (213 km)

PLATFORMS:

Bombers	Tu-16* Badger-A/C/G
	Tu-22* Blinder-A/D
Reconnaissance	Yak-28* Brewer-B/C/D/E

NAME: SIDE NET

DESCRIPTION: Side Net is a Soviet height-finding radar with a vertical orange-peel parabolic antenna and prominent feed horn. It is used in a ground control intercept (GCI) mode with early-warning radars such as Bar Lock, Back Net, and Tall King. Side Net is also used with the SA-2 Guideline, SA-3 Goa, and SA-5 Gammon surface-to-air missile (SAM) systems.

CHARACTERISTICS:

Antenna Size	
Height	27 ft 11 in (8.50 m)
Width	11 ft 6 in (3.50 m)
Band	E
Range	97 nm (180 km)

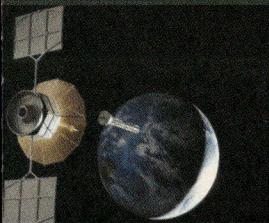

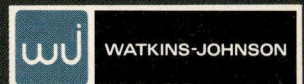

Side Net

NAME: SKIP SPIN/URAGON 5B*

DESCRIPTION: This is a Soviet airborne intercept radar fitted in older Soviet air defense fighters. It is associated with AA-2/3 Atoll/Anab air-to-air missiles and possibly with AA-8 Aphid missiles.

CHARACTERISTICS:
Band	I/J
Range	21.6 nm (40 km)

PLATFORMS:
Fighters	Yak-28P* Firebar
	Su-15* Flagon-E/F

NAME: SLIM NET

DESCRIPTION: Slim Net is a Soviet naval high-definition air-and surface-search radar. It replaced the Hair Net radar on cruisers and destroyers. The antenna is an open, lattice-type tapered rectangle with a four-legged feed horn mounting and two large balancing vanes. Span of the antenna is Approx 18 ft (5.5 m) and the maximum depth Approx 6 ft (1.8 m).

CHARACTERISTICS:
Band	E
Range	
Maximum	More than 175 nm (324 km)

PLATFORMS/USERS:
Destroyers	Kildin class (one ship)
	Kotlin class
	Modified Skoryy* class
Frigates	Petya I class
	Mirka I/II class
	Riga class
Auxiliary Ships	Various

Slim Net

NAME: SMALL FRED

DESCRIPTION: Small Fred is a Soviet battlefield surveillance radar fitted on the fully tracked BMP M1975 (Soviet designation PRP-3 or BRM-1) mobile reconnaissance post. The antenna is a rectangular array mounted on the rear of the turret; it folds forward when not in use.

CHARACTERISTICS:
Band	J
Range	
Detection	10.8 nm (20 km)
Tracking	3.8 nm (7 km)

DESIGNATOR: ARSOM-2P*
NAME: SMALL YAWN

DESCRIPTION: Small Yawn is a Soviet counter-battery radar mounted on the AT-L* tracked carrier. The antenna is a small rotating dish (possibly conically scanned).

CHARACTERISTICS:
Band	I
Range	9-14 nm (16.7-26 km)

NAME: SNOOP SERIES

DESCRIPTION: The Snoop series of radars are Soviet mast-mounted submarine radars. The earliest in the series was Snoop Plate, fitted in early post-World War II-era diesel submarines. Snoop Slab is a larger radar fitted in

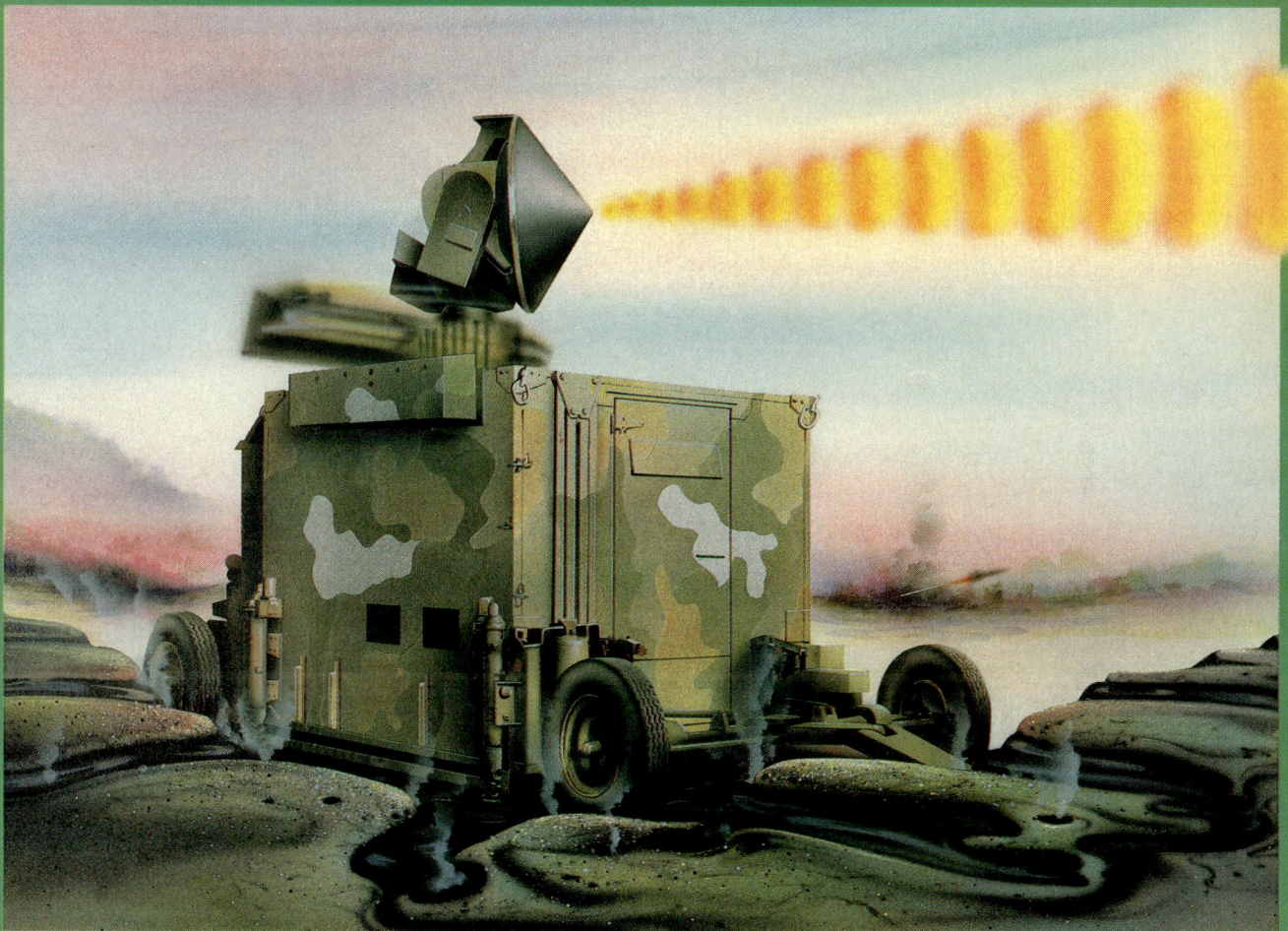

Echo and *Juliett*; it is thought to have a long range and to able to back up the Front Door missile guidance radar on these submarines. Snoop Tray is the most widely used submarine surface search radar.

CHARACTERISTICS:

Band	I
Range	Against aircraft 25 nm (46 km)
(Snoop Plate)	Against ships 12 nm (22.2 km)

PLATFORMS/USERS:

Submarines	Snoop Plate
	Romeo class
	Whiskey class
	Zulu class
	Snoop Slab
	Echo class
	Juliett class
	Snoop Tray
	Alfa class
	Bravo class
	Charlie I/II classes
	Delta I/II/III/IV classes
	Echo I/II classes
	Foxtrot class
	Golf II/III classes
	Hotel II/III classes
	Kilo class
	November class
	Tango class
	Victor I/II/III classes
	Yankee class

NAME: SPIN SCAN

DESCRIPTION: This is a Soviet airborne intercept radar fitted in early versions of the Fishbed and the Fishpot fighters. It is housed in the center of the nose air intake. It is associated with the AA-1/7/8 Alkali/Apex/Aphid air-to-air missiles.

CHARACTERISTICS:

Band	I
Range	
Search	25 nm (46.3 km)
Tracking	19 nm (35.2 km)
Peak Power	Approx l00kW

PLATFORMS:

Fighters	MiG-21* Fishbed-D/F
	Su-9* Fishpot

NAME: SPIN TROUGH

DESCRIPTION: This is a Soviet naval search and navigation radar installed as an alternative to the Don-series radars in older surface ships and small craft.

CHARACTERISTICS:

Band	I

PLATFORMS:

Frigates	*Krivak I* class
	Pauk class
Corvettes	*T-58** class (Patrol and radar picket ships)
Small Combatants	*Zhuk** class (KGB)
	Yaz (Riverine Moniter)
Mine warfare	*Sonya** class
	*T-43** class
	*Zhenya** class
	*Yevgenya** class
	*Andryusha** class
	*Olya** class
	*Ilyusha** class
Amphibious	*Alligator* class
	Polnocny A/B/C classes
	Vydra class
	Aist class

DESIGNATOR: P-12*
NAME: SPOON REST

DESCRIPTION: Spoon Rest is a Soviet early warning radar which later evolved into the Knife Rest series. Spoon Rest-A is mounted in a van and Spoon Rest-B is mast-mounted. The large antenna is a six-Yagi horizontal array. Spoon Rest was widely used in Vietnam and the Middle East.

Spoon Rest

CHARACTERISTICS:

Band	
Spoon Rest-A	A
Spoon Rest-B	VHF (below A)
Beam Width	
Vertical	2.5 deg
Horizontal	1 deg
Peak Power	350 kW
Range	Approx 152 nm (282 km)

NAME: SQUARE PAIR

DESCRIPTION: Square Pair is a Soviet fire control radar colocated with the SA-5 Gammon surface-to-air missile (SAM) system. It is used with Back Net or Side Net height-finding radars. Operational since 1964.

CHARACTERISTICS:

Band	H
Range	90-105 nm (167-195 km)

NAME: SQUARE TIE

DESCRIPTION: This is a small, short-range Soviet search and target designation radar in small combatants for the S-N-2 Styx anti-ship missile.

CHARACTERISTICS:

Band	I
Range	25 nm (46 km) against a destroyer
	10 nm (18.5 km) against a missile boat

PLATFORMS:

Corvettes	Nanuchka II class
Small Combatants	Osa I/II classes

DESIGNATOR: P-15M*
NAME: SQUAT EYE

DESCRIPTION: Squat Eye is a Soviet target acquisition radar supporting the SA-3 Goa surface-to-air missile (SAM). It works with the Low Blow missile control radar. It is used in place of the Flat Face radar when enhanced low-level coverage is required. The antenna is an open-work elliptical parabaloid and is usually mounted on a tower.

CHARACTERISTICS:

Band	C
Peak Power	Approx 500 kW
Range	108 nm (200 km)

NAME: SQUINT EYE

DESCRIPTION: Squint Eye is a Soviet target acquisition radar supporting the SA-3 Goa surface-to-air missile (SAM) system. It operates with the Low Blow missile guidance radar. It is used in place of Flat Face when enhanced low-altitude coverage is required.

NAME: STRAIGHT FLUSH/KUB SSNR*

DESCRIPTION: Straight Flush is a Soviet fire control radar mounted on a fully tracked SSNR* (Soviet code for radar vehicle) for the mobile SA-6 surface-to-air missile (SAM) system. The SSNR has 2 principal antennas. The lower target acquisition and early warning antenna is a solid elliptical paraboloid with upper and lower feed horns. The upper horn produces a low-angle pencil beam; the lower generates beams for medium-to-upper from 2 or 3 horns. The upper antenna has a conical scanner and is the target tracking and illuminating radar. The upper radar can rotate independently of the lower. In each battery of 4 SA-6 vehicles, there is one SSNR which is responsible for the detection, acquisition, tracking, and illumination of the target as well as missile command guidance.

CHARACTERISTICS:

Band	
Target Acquisition	G/H
Tracking and Illumination	I

Straight Flush

NAME: STRUT CURVE

DESCRIPTION: This is a small Soviet naval air/surface-search radar fitted in smaller warships and auxiliaries. It has an open, lattice-type elliptical paraboloid reflector with horn feed from a boom projecting from the lower edge of the scanner. No balancing vanes are fitted. Two Strut Curve antennas back-to-back form the antenna for Strut Pair air-search radar.

CHARACTERISTICS:

Band	F
Range	60 nm (111 km) against aircraft at medium altitudes.

PLATFORMS:

Frigates	*Grisha I/II/III* classes
	Koni class
	Mirka I/II classes
	Petya I/II classes
Corvettes	*Poti* class
	*T-58** class
Minelayers	*Alesha** class
Amphibious Ships	*Ropucha* class
Auxiliary ships	*Berezine**
	*Ivan Susanin** (KGB)
	*Purga** (KGB)

NAME: STRUT PAIR

DESCRIPTION: This is a Soviet air-search radar with the antenna formed by two Strut Curve antennas mounted back-to-back. It was apparently the first Soviet pulse-compression radar, first seen aboard the destroyer *Bedovyy.** Sometimes paired with Top Steer.

CHARACTERISTICS:

Band	F

PLATFORMS:

Aircraft Carriers	*Kiev** class (*Novorossiysk** only)
Destroyers	Modified *Kildin* class (*Bedovyy** only)
	*Udaloy** class
Frigates	*Grisha IV* class

Strut Pair

NAME: SUN VISOR

DESCRIPTION: This is an obsolete Soviet naval gunfire-control radar with a solid parabolic antenna. It is fitted to Wasp Head fire-control directors in older ships (com-pleted from 1953 onward) for use with 130-mm and 100-mm guns.

CHARACTERISTICS:

Band	H/I
Range	15 nm (27.5 km)

VARIANTS: Sun Visor-B is the version fitted from 1956 on.

PLATFORMS:

Command Ships	Modified *Sverdlov** class
Cruisers	*Sverdlov** class
	Modified *Sverdlov** class
Destroyers	Kotlin class
	Kotlin SAM class
Frigates	*Riga* class
Auxiliary Ships	*Don** class
Icebreakers	*Purga** (KGB)

DESIGNATOR: P-14*
NAME: TALL KING

DESCRIPTION: Tall King is a Soviet large, fixed-site radar used for early warning against high-altitude aircraft. It is often operated in concert with the Side Net height-finding radar and the Scoreboard identification friend or foe (IFF) interrogator. The elliptical paraboloid antenna is supported by a tower. The assembly is transportable although considerable time would be required to dismantle and reassemble it.

CHARACTERISTICS:

Antenna	
Width	Approx 82-98 ft (25-30 m)
Height	Approx 49 ft (15 m)
Band	B
Range	270-324 nm (500-600 km)

NAME: THIN SKIN

DESCRIPTION: Thin Skin is a Soviet height-finder radar used to support low-altitude ground control intercept (GCI) operations. It is usually used with the Long Track target acquisition radar in SA-4 Ganef, SA-6 Gainful, and SA-8 Gecko mobile surface-to-air missile (SAM) batteries. The vertical, orange-peel parabolic section antenna has been seen in both truck- and trailer-mounted versions. In operation the antenna "nods" in elevation while scanning.

CHARACTERISTICS:

Band	H
Range	Approx 130 nm (240 km)

NAME: TOADSTOOL

DESCRIPTION: This is a Soviet aircraft navigation radar.

CHARACTERISTICS:

Band	I

PLATFORMS:

Transport	An-12* Cub

NAME: TOKEN

DESCRIPTION: Token is a Soviet early-warning ground control intercept mounted on a van.

CHARACTERISTICS:

Band	E/F
Range	
Early-warning	Approx 135-162 nm (250-300 km)
GCI	81 nm (150 km)

NAME: TOP DOME

DESCRIPTION: This is a missile guidance radar for the most advanced air defense system now fitted in Soviet warships. It supports the SA-N-6 missile system. The principal Top Dome antenna consists of a 4-meter hemispheric radome, fixed in elevation and mechanically steered in azimuth. It is installed with a series of smaller radomes, apparently for tracking multiple targets. The Azov* was trials ship for the SA-N-6/Top Dome system. Two Top Dome radars are fitted in ships of the Kirov* class and one in the Slava* class.

CHARACTERISTICS:

Band	I or J
Range	40 nm (74 km)

PLATFORMS:

Cruisers	Kara* class (Azov* only)
	Kirov* class
	Slava* class

NAME: TOP KNOT

DESCRIPTION: This is a spherical Soviet naval tactical aircraft control and navigation (TACAN) radar. The antenna is housed in a dome atop the ship's mast.

PLATFORMS:

Aircraft Carriers	Kiev*

NAME: TOP PLATE

DESCRIPTION: This is a Soviet naval air/surface-search radar fitted to the Udaloy* class destroyers in place of one of the two Strut Pair radars, beginning with the third ship of the class (Marshal Vasil'yevskiy). The antenna has a flat ("solid") plate mounted in front of a "mesh" antenna plate, which is probably the Top Mesh radar.

PLATFORMS:

Destroyers	Udaloy* class (Later ships)
Icebreakers	Arktika* class (Civilian; some ships)

Top Pair

NAME: TOP SAIL

DESCRIPTION: This is a Soviet long-range, 3-dimensional air-search and early-warning radar fitted in major warships. The antenna consists of a large scanner that has a cylindrical cross section with the axis tilted back about 20 deg from the vertical. The reflector is illuminated by a linear radiating element located parallel to the cylindrical axis. It uses frequency scan in elevation. Two large balancing vanes are fitted. The Top Sail is used in conjunction with the Head Net-C search and Head Lights missile control radars. Top Sail is mounted back-to-back with Big Net to form the Top Pair radar.

Top Sail

Control of amplitude and phase provides more effective use of transmitted antenna power.

New Power For New Threats

High Power, Precision Phase and Amplitude Control Networks Put Your Antenna System One Step Ahead of The Enemy.

Avionics technology never stops moving forward. On both sides. To keep pace with the latest advances, you need the flexibility and high power capacity to transmit microwave energy more efficiently and accurately.

Electromagnetic Sciences designs and builds a variety of high power control elements and networks for use in broadband applications. We offer:

- Switches, phase shifters and variable power dividers
- 500 W power levels
- 3:1 bandwidth
- Less than 1 dB insertion loss
- Submicrosecond switching time

Our state-of-the-art antenna elements allow *both* phase and amplitude control of signals routed to the radiating antenna.

Qualified by extensive testing and proven in field deployment, Electromagnetic Science's subsystems are engineered for new programs and avionics upgrades.

To find out how to accumulate extra measures of performance on your next project, write or call:

Electromagnetic Sciences Inc.
Microwave Sales Department
125 Technology Park
Norcross, GA 30092
(404) 263-9200

ELECTROMAGNETIC SCIENCES, INC.

CHARACTERISTICS:

Band   C

PLATFORMS:

Aircraft Carriers   *Kiev** class
                     *Moskva** class
Cruisers   *Kara* class
                     Kresta II class
                     *Slava** class

NAME: TOP STEER

DESCRIPTION: This Soviet naval air-search radar is a medium-size, 3-dimension installation, somewhat similar in appearance to the larger Top Sail. It is comprised of a pair of scanners mounted back-to-back, usually high on the ship's superstructure. The larger antenna is fitted back-to-back with the Strut Pair radar antenna, with a common feed. The Top Steer uses frequency scan in elevation. The Top Steer is found paired with the Top Plate radar in later units of the *Sovremennyy**-class destroyers.

CHARACTERISTICS:

Band   F
Range   150 nm (276 km)

PLATFORMS:

Aircraft   Carriers *Kiev** class
Cruisers   *Kirov** class
                     *Slava** class
Destroyers   *Kashin* class (*Provornyy**)
                     *Sovremennyy**

Top Steer

NAME: TOP TROUGH

DESCRIPTION: This is a Soviet surveillance radar found in older surface combatants. The bar-like antenna has a slotted waveguide that feeds a curved reflector. It is used for target discrimination for 100-mm and 152-mm guns. Usually found on top of the aft mast. Installed in the 1970s.

CHARACTERISTICS:

Band   C
Range   300 nm (555 km)

PLATFORMS:

Cruisers   *Sverdlov** (some units)
Command Ships   Modified *Sverdlov*

NAME: TWO SPOT

DESCRIPTION: Two Spot is a Soviet precision approach radar (PAR), usually used in conjunction with the Long Talk air search radar. Together they form the standard ground controlled approach (GCA) system in use at Soviet air bases. Two Spot has two antennas, one a vertical, orange-peel parabolic dish (for elevation) and the other a horizontal truncated parabolic section reflector (for azimuth). The radar also has two discone communications antennas for radio communications.

CHARACTERISTICS:

Band   I/J

NAME: WET EYE

DESCRIPTION: This is a Soviet aircraft surface-search radar, apparently with sufficient target definition for locating raised submarine periscopes and snorkel masts.

CHARACTERISTICS:

Band   J

PLATFORMS:

Patrol/   Tu-142* Bear-F
ASW aircraft   Il-38* May

NAME: WHIFF

DESCRIPTION: Like the Fire Can, the Soviet Whiff gunfire control radar was based on the U.S. SCR-584* radar of World War II. The radar has a parabolic dish antenna and is mounted in a van. It supports the KS-19* 100-mm anti-aircraft gun.

CHARACTERISTICS:

Band   E

NAME: YO-YO

DESCRIPTION: Yo-Yo is a Soviet fire control radar used with the SA-1 Guild surface-to-air missile (SAM) system. The six rotating antennas use "flapping" beams to track 24-30 targets simultaneously. The movement of the antennas resembles the motion of a yo-yo string toy.

CHARACTERISTICS:

Band	E/F
Peak power	2 MW

SONARS

NAME: FENIKS*
PEGAS*
SHARK GILL
TAMIR*
HERKULES*

DESCRIPTION: Shipboard systems: Soviet warships of the early post-World War II period had mostly the Tamir-5 series of high-frequency, hull-mounted sonars. They were succeeded from the mid-1950s by the Pegas-2 series sonars, and from the late 1950s by the Herkules series. Subsequently, a variety of improved sonars have been developed for the Soviet warships that began joining the fleet in the 1960s. By the 1960s, as the Soviet Navy became more concerned with anti-submarine warfare (ASW) and submarine detection, there was an effort to reduce frequencies, and during the past two decades they have been reduced from the 20-30 kHz range to about 2-15 kHz. Bow-mounted sonars are mounted in some later surface classes, providing the optimum hull-mounted sonar position (i.e., away from machinery and propeller noises) while also providing some "dampening" effect in rough sea operations. In addition, active bi-static detection became possible with the installation of variable depth sonar (VDS) in several surface combatants, beginning with the Moskva*-class helicopter ships (operational 1967). Towed sonar arrays have been sighted on Soviet ships in the development and evaluation stage, but no deployment has been confirmed.

Submarine systems: Early post-World War II submarines with "new" sonar installations had mostly the Tamir-5L set. Submarines were subsequently fitted with active-passive Herkules and passive Feniks sonars. Beginning with the Victor III class attack submarine (operational 1978), the Soviets appear to have deployed towed passive sonar arrays. Such systems appear to have narrowband processors, greatly enhancing their capability. Other submarine classes have now been fitted with similar towed arrays, including the Sierra and Akula, and the Yankee class ballistic missile submarines converted to an attack submarine configuration. A Shark Gill low-frequency sonar has been identified in newer submarine classes, including the Mike, Sierra, and possibly Oscar.

Aircraft systems: Soviet fixed-wing ASW aircraft and helicopters carry air-dropped, expendable sonobouys that can provide active or passive submarine detection. In addition, the Ka-26* Hormone-A, Ka-27* Helix-A, and

Mi-24* Haze-A helicopters carry an active, dipping sonar that can be lowered while the helicopters are operating in a hover mode. These dipping sonars have been adopted for smaller surface combatants (see below).

Seafloor installations: The Soviet Navy also employs moored, sea-floor acoustic systems in coastal and regional seas. These do not appear to have the range or capability of the U.S. Navy's Sosus (sound surveillance system).

PLATFORMS:

High-Frequency/Keel-Mounted (Herkules* or Pegas*)

Destroyers	Kildin class
	Kotlin class
	Skoryy* class
Frigates	Mirka class
	Petya I/II classes

Medium-frequency/keel- or bow-mounted

Cruisers	Kresta II class
Destroyers	Sovremennyy* class
	Kanin class
Frigates	Grisha I/III/IV/V classes
	Krivak I/II classes

Medium-frequency/VDS

Carriers	Kiev* class
	Moskva* class
Cruisers	Kara* class
Destroyers	Modified Kashin class
Frigates	Krivak I/II classes

Low-frequency/keel- or bow-mounted

Carriers	Kiev* class
	Moskva* class
Cruisers	Kirov* class
Destroyers	Udaloy* class

Low-frequency/VDS (reported to operate between 3 and 5 kHz)

Cruisers	Kirov* class
Destroyers	Udaloy* class

Dipping/active

Helicopters	Mi-14* Haze-A
	Ka-27* Helix-A
	Ka-26* Hormone-A
Frigates	Mirka class
	Petya I/II classes
Small Combatants	Pchela* class
	Poti class
	Stenka* class
	Turya class

ELECTRONIC SUPPORT MEASURES/ ELECTRONIC WARFARE SYSTEMS

DESCRIPTION: Soviet combat aircraft, surface ships, and submarines have electronic warfare equipment to collect electronic intelligence (Elint), to perform identification friend or foe (IFF) functions, and to detect threats (ESM/ESM). Few details of these systems are available for publication. Surface ships as well as some aircraft have chaff and decoy launchers. Listed below are the EW systems that have been identified and the few details publicly known about them.

Surface ship equipment: Bell series; These are threat warning and jamming systems that have been observed on surface ships of frigate size and larger. They are housed in bell-shaped radomes, whence the NATO nickname. The individual systems so far identified are Bell Bash, Bell Clout, Bell Shroud, Bell Slam, Bell Squat, Bell Tap, and Bell Thump.

Others are:

Cage Pot	Intercept system
Cross Loops	Communications intercept
Dead Duck	IFF
Farm Loaf	EW antenna housing
Fig Jar	EW antenna housing.
Gin Pole	IFF
Grid Shield	EW antenna
Guard Dog	EW system
High Pole A/B	IFF. C/D band
High Ring	Elint/Signal intelligence (Sigint) antenna
Rum Tub	Mounted primarily on cruisers, at the base of a tower. Groups of 4 ESM antennas, each covering a 90-deg segment
Salt Pot	Improved IFF antenna. Replacing High Pole series
Side Globe	Broadband jammers; Thimble-shaped radomes, mounted in groups of 4. There is an improved version, possibly given a new U.S./NATO code name, mounted on the later Kiev* class carriers and on the battle cruiser Frunze*
Ski Pole	IFF
Square Head	IFF
Tilt Pot	EW system
Top Hat-A/B	Jammer
Tread Mill	Direction-finding system
Watch Dog	Preceeded the Bell series, this electronic intercept system first appeared in the 1950s. Fitted mostly in small ships, and despite its antiquity is still being fitted in the newer small frigates, e.g., Koni* and Grisha* classes

Submarine equipment:

Brick Pulp	Surveillance and threat warning. Usually used in conjunction with Snoop group radars
Brick Split	Similar to Brick Pulp
Golf Ball	Mast-mounted sensor
Park Lamp	Direction-finding loop. Antenna consists of four wire squares joined at 45 deg to each other
Quad Loop	Direction-finding loop, found in most submarines. Antenna consists of two wire squares joined at 90 deg to each other
Stop Light	Passive, broadband system, with Elint function. Mast-mounted

Airborne equipment:

Sirena 2/3	Radar warning system similar to U.S. AN/APR-25, a simple crystal video warning system that did not include launch warning; used in the early Vietnam era. Forward and rear hemisphere arrangement provides 360-deg coverage

Ground-based equipment:

Fix series	Direction-finding systems with numerical suffix. Fix 4A, 4B, 4C, 6 and 8 are high-frequency, four-element Adcock arrays. Fix 4D is a very high frequency four-element Adcock array. Fix 24 is a high-frequency, direction-finding circular array of 24 vertical monopoles
Krug	Large, land-based, high-frequency direction-finding system. Antennas are arranged in a circular pattern and produce an accuracy of 3 to 5 deg or better at long range. Installations probably exist in Cuba and Vietnam as well as the Soviet Union. The term "Krug" is based on the Russian term for "circle" or "ring"
Loop series	High-frequency direction-finding system. Loop Three is a trailer-mounted loop array
Moon	High-frequency, direction-finding, 4-element array
Ring Two	Trailer-mounted high-frequency, direction-finding loop arrays
Small Cross	Adcock direction-finding system
Square Four	Direction-finding system
Thick Eight-A	8 broadband vertical high-frequency dipoles

Your Partner in EW

COMINT, ELINT, ESM, ECM for airborne, naval and ground-based applications

ELECTRONIC WARFARE BUDGETS

Summary of Projected DOD Spending 327
How To Interpret DOD Program Element Keys 328

OSD Budget
RDT&E Programs 329

U.S. Army Budgets
RDT&E Programs 330
Procurement 332

U.S. Navy Budgets
RDT&E Programs 333
Procurement 336

U.S. Air Force Budgets
RDT&E Programs 338
Procurement 344

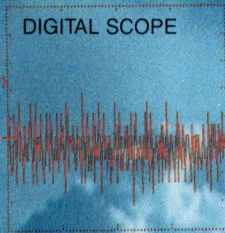

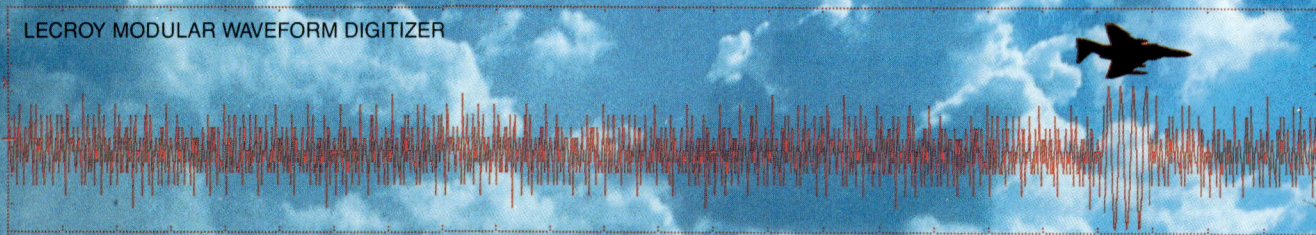

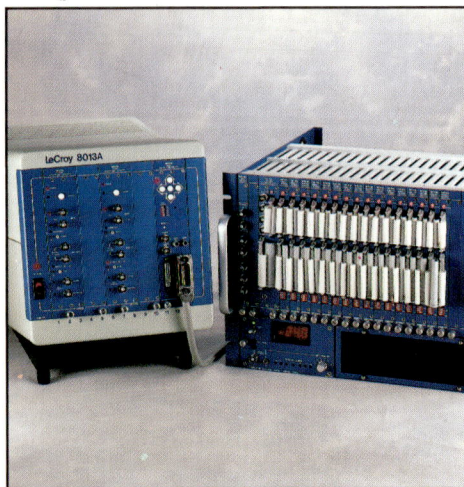

Summary of Projected DOD EW Spending

In its January 1987 submission, the Department of Defense proposed a budget authority of $303.3 billion for fiscal 1988, a 7.6 percent increase over the congressionally authorized fiscal 1987 budget. As a tiny sub-category of the total proposed funding for fiscal 1988, EW spending is expected to follow the same sense as the overall budget. While the proposed fiscal 1988 and 1989 DOD budget still before Congress as this handbook went to press represents a 3 percent real increase over last year, Congress for the past two years has imposed a decrease of about 7 percent. Congress is similarly inclined this year. In mid-May the House of Representatives passed its version of the DOD authorization bill for fiscal 1988 and fiscal 1989. (This year's submission is the first one to request formal authorization for all DOD programs for two years.) The House's total authorization that would be provided by the bill is $288.6 billion. The Senate will probably settle on $303 billion and later the conference committee will probably compromise at about $292 billion.

The House has been attempting to channel the armed services into integrating EW equipment, where feasible, in order to reduce proliferation of similar EW equipment among the services. The House has required DOD to submit an EW master plan—an EW road map. So far, in several tries, the DOD has not proferred a plan that satisfies the House's desire to reduce duplication. The House is expected to hold back funding of some EW programs until it believes that DOD is integrating its EW efforts.

While the sense and relative values of this handbook's detailed EW budget breakout are correct, the figures for individual line items are often only estimates. Congress has yet to complete its review and authorization of proposed DOD spending, so the figures used here are far from final. There are line items not listed in the handbook's summary because they have not been made public. And there are classified areas that are sanitized or buried in distantly related proposed expenditures.

Comparison of EW summaries of this 13th edition of ICH with previous editions reveals differences in individual line items and totals. These differences are the result of (1) actual congressional authorizations substituted for DOD proposed funding levels or (2) differences in interpretation as to the proportion of EW contained in the various programs, based on the current year's program element descriptions. Accordingly, the importance of the EW summaries lies in their relative values and the sense they convey rather than the exactness of each figure.

Following many of Program Nomenclature listings in the budget summaries is a parenthetical number, for example (.1). This number represents that fraction of the budget line funding estimated to originate from EW, such as from the incorporation of ECCM features built into radars, communications, navigation and identification systems. All figures under the Fiscal Year columns of the summaries have been estimated to the nearest tenth. Dollar amounts are in fiscal 1988 dollars. Where a line item in the summary is solely oriented to EW, there is no parenthetical number after the Program Nomenclature. Asterisks placed after the Program Nomenclature indicate that the line item is newly added to this year's budget summaries, or that there is a change in the parenthetical number estimating the proportion of EW from that listed in last year's summaries.

Department of Defense
Projected Electronic Warfare Spending
1987-1989

SERVICE AND ACTIVITY	Funding ($ Millions) Fiscal Years		
	1987	1988	1989
Defense Agencies			
Research and Development	$130.9	$159.4	$183.4
U.S. Army			
Research and Development	230.9	142.9	152.2
Procurement	258.2	299.7	281.5
U.S. Navy			
Research and Development	446.2	513.0	454.7
Procurement	1,187.1	973.0	1,560.0
U.S. Air Force			
Research and Development	608.4	644.9	752.6
Procurement	949.3	1,106.3	995.2
TOTALS	**3,811.0**	**3,839.2**	**4,379.6**

Deciphering Program Element Numbers

Standardized DOD six-character program element numbers used in budgets indicate program identity, status and sponsorship.

6 3 4 7 1 N

First Character—DOD Program
1-Strategic Forces
2-General Purpose Forces
3-Intelligence and Communications
4-Airlift/Sealift
5-Guard and Reserve
6-Research and Development
7-Central Supply and Maintenance
8-Training, Medical & Other
9-Administration & Associated Activities
10-Support of Other Nations

Second Character—Category
(For categories that are well-structured and consistent)
1*****-Strategic Forces
 1-Offensive
 2-Defensive

2*****-General Purpose Forces
 1-Unified Commands
 2-Army Forces
 3-Army Operational Systems Development
 4-Navy Forces
 5-Navy Operational Systems Development
 6-Fleet Marine Forces and
   Systems Development
 7-Air Force Forces and Systems
 8-Other

3*****-Intelligence and Communications
 1-General Intelligence and
   Cryptological Activities
 2-National Military Command System
 3-Communications
 4-Special Activities
 5-Other Activities

4*****-Airlift/Sealift
 1-Airlift
 2-Sealift
 3-Traffic Management and Water Terminals

6*****-Research and Development
 1-Research
 2-Exploratory Development
 3-Advanced Development
 4-Engineering Development
 5-Management and Support
 6-Operational Systems Development

Sixth Character—DOD Component
A- Army
B- Defense Mapping Agency
C- Strategic Defense Initiative Organization
D- DOD (OSD and OASD)
E- Defense Advanced Research Projects
   Agency (DARPA)
F- Air Force
G- National Security Agency
H- Defense Nuclear Agency
J- Joint Chiefs of Staff
K- Defense Communications Agency
L- Defense Intelligence Agency
M- Marine Corps
N- Navy
Q- Joint Tactical C^3 Agency
R- Defense Contract Audit Agency
S- Defense Logistics Agency
U- Undistributed Resources
V- Defense Investigative Service
W- Uniformed Services University of the
   Health Services

Fourth and Fifth Characters—
Sponsor's Element Identifier

Third Character—
R&D Budget Activity
6*****-Research and Development
 1-Military Sciences
 2-Aircraft and Related
 3-Missile and Related
 4-Astronautics and Related
 5-Ships and Related
 6-Ordnance, Combat Vehicles and Related
 7-Other
 8-Program-wide Support

Office of the Secretary of Defense
Research, Development, Test & Evaluation Budget
1987-1989

Program Element	Program Nomenclature	Cost Data ($ Millions) Fiscal Years			Description/Comments
		1987	1988	1989	
33126K	Long-Haul Communications DCS— DCA (.05)*	1.0	.8	1.0	
35159G	Defense Reconnaissance Support Activities— NSA (.1)	—	—	—	
35159I	Defense Reconnaissance Support Activities (.1)	14.0	14.0	11.0	
61101E	Defense Research Sciences—DARPA (.1)*	9.0	8.0	10.0	
61103E	University Research Initiatives—DARPA (.1)*	.9	2.0	3.0	To improve the quality of research performance at universities to meet defense needs; to strengthen multidisciplinary research that supports key defense technologies. Technical areas include analysis, modeling and simulation, electro-optics systems and signal analysis.
62101E	Technical Studies—DARPA (.1)	—	.2	.2	
62301E	Strategic Technology—DARPA (.1)	22.0	23.0	26.0	
62702E	Tactical Technology—DARPA (.1)	9.0	11.0	14.0	
63220C	Surveillance, Acquisition Tracking and Kill Assessment—SDI (.05)	50.0	75.0	93.0	Detection and tracking of targets and discrimination between real RVs and decoys. Major programs include the boost surveillance and tracking system for boost phase detection, the space surveillance and tracking system for midcourse detection, and the airborne optical adjunct for the late midcourse phase.
63226E	Evaluation of Major Innovative Technology— DARPA (.05)*	8.0	12.0	14.0	
63702D	Special Operations Special Technology Office—OSD (.05)	.2	.6	.4	
63702D	Counter-insurgency and Special Technology—OSD (.05)*	.6	.5	.5	
64771D	Joint Tactical Information Distribution (JTIDS) (.05)*	10.0	5.0	3.0	
63790D	NATO Research and Development— OSD (.1)*	5.0	6.0	6.0	
65116D	General Support to C³I—OSD (.1)*	.2	.3	.3	Small portion of the program examines impact of the changing EW environment on use of the electromagnetic spectrum by U.S. military forces; defines the need for improved EW mission planning and coordination focusing on how theater commanders and joint task forces can best organize and implement electronic combat plans to facilitate EW and C³CM coordination.
65117D	Foreign Materials Acquistion and Exploitation—OSD (.1)*	1.0	1.0	1.0	
	Totals	**130.9**	**159.4**	**183.4**	

U.S. Army
Research, Development, Test & Evaluation Budget
1987-1989

Program Element	Program Nomenclature	Cost Data ($ Millions) Fiscal Years			Description/Comments
		1987	1988	1989	
61102A	Defense Research Sciences	.6	1.0	1.1	Project AH40—signals warfare laboratory; mm-wave research focuses on the development of sensitive heterodyne receivers; development of strategy of 3-dimensional imaging.
61103	University Research Initiatives (.3)*	35.0	34.0	35.0	Ongoing research projects at more than 200 universities. Scientific problems with military applications. University research initiatives. Examples include ultra fast electronics, LPI Comm, microwave imaging, spread spectrum comm, MMICs, AI, high speed processing for radar and EW systems.
62303A	Missile Technology (.1)	3.0	3.0	2.0	Survivability enhancements; development of guidance and control and terminal homing systems having multimode and autonomous target acquisition capabilities; reduced vulnerability to anti-radiation missiles. mm-wave guidance, IR homing sensor and signal processing.
62601A	Tank and Automotive Technology (.1)	2.0	2.0	2.0	Solutions other than passive armor to create a highly survivable vehicle. Includes signature reduction, threat warning devices and active CM.
62618A	Ballistic Technology (.1)	3.0	2.0	2.0	Develop high resolution models for self-contained munition sensors and develop technology base for smart munition countermeasures.
62705A	Electronics and Electronic Devices (.3)*	5.0	5.0	5.0	Develop acousto-optic radar processors for radars to reject decoys; IC for TRW miniature ESM DF (MEDFLI); all digital fast-hop fast-tune ECCM upgrades for radios and data links; VHSIC hardware for EW Sigint and radar processors.
62709A	Night Vision Investigations (.05)	1.0	1.0	1.0	Phase conjugated lasers which will revolutionize Army CM and tunable lasers.
62715A	Tactical EW Technology*	12.5	—	—	Exploratory development technology in support of Army EW programs through the EW R&D Center and Signal Warfare R&D Center.
62782A	C³ Technology (.3)*	6.0	3.0	5.0	C³ system survivability and interoperability, real-time information distribution and command post mobility—reduction of vulnerability of comm to ECM and intercept.
63006A	Command, Control and Communications Technology (.2)*	3.0	2.0	2.0	Design, development and integration of C³ concepts, architecture, and techniques to make comm networks immune from ECM.
63008A	EW Advanced Technology*	13.8	—	—	Demonstrate feasibility and effectiveness of EW developments emerging from Army's tactical EW technology program. Items successfully demonstrated in the program will make the transition to advanced development.
63313A	Missile and Rocket Components Project D271 (.6)	6.0	5.0	5.0	Radar designed to defeat air-defense suppression by ARMs and EW.
63321A	Target Acquisition and Counter-Countermeasures*	18.4	—	—	Develop a broad, non-system specific technology base for development of CM to ARM threat; vulnerability assessments of U.S. Army comm-electronics, EO and weapons systems.
63742A	Adv Electronic Devices Development (.5)	1.0	4.0	4.0	Tech. insertion to systems under development such as RPVs, signal intell. systems and radar systems. Higher-power tube development for jammers, TWT tube redesigns, K-band jammer for Army aircraft protection, and demonstrate modular processor design methodology for EW applications.
63758A	Army Battlefield Integration—Project DK21	4.6	6.0	5.4	Signal and sensor processing technology for intelligence/EW.
63759A	Smoke Advanced Technology Demonstration (.1)	.4	—	—	Project No. DE-85. Smoke and obscurant systems to defeat or degrade threat surveillance, target acquisition and weapon systems operating in visual, IR and mm-wave.

Program Element	Program Nomenclature	Cost Data ($ Millions) Fiscal Years			Description/Comments
		1987	1988	1989	
63711A	Aircraft Survivability Equipment	3.6	6.8	12.5	Counters to SAM and AAA. Army responsible for RWRs and LWRs, radar jammers, IR jammers and pulsed Doppler missile warning detectors for helicopters and fixed-wing aircraft.
63718A	EW Vulnerability/Susceptibility	13.0	15.5	16.5	Assess EW vulnerability of existing Army C³, night vision, EO, missiles and radars. Systems are analyzed, subjected to threat-emulating environments; weaknesses to ECM identified, corrected.
63745A	Tactical ESM Systems	7.8	—	—	Software improvements such as data base merging, transaction logging, automated mission management, Comint/Elint fusion algorithms, to make technical control and analysis center a more user-friendly system.
63754A	Land Warfare Surveillance and Reconnaissance Mission Area—Classified Program.	—	—	—	Dollar amount not stipulated in DOD summary.
63755A	Tactical ECM Systems	24.0	—	—	Development of EW systems to attack enemy communications and radars.
63760A	Special Operations Forces Advanced Development (.4)	—	2.0	1.0	Project D474. Communications systems with an extremely low probability of intercept/detection for audio and imagery information.
63766A	Tactical Electronic Surveillance System (.3)*	3.0	—	—	Identify and refine design concepts for surveillance system. Details are classified.
64306A	Stinger (.05)	.3	.2	—	Improvement of Stinger passive optical seeker technique (POST) in countermeasure environments.
64711A	Aircraft Survivability Equipment	19.0	8.0	18.3	Training device, radar interferometer and ALQ-136 radar jammer development. APR-39A RWR upgrade for special electronic mission aircraft. RF expendable decoy development continues. ALQ-162 improvement.
64715A	Non-System Training Devices (.1)	4.0	3.0	2.0	Engineering development of Sigint/EW Equipment Operator.
64730A	Remotely Piloted Vehicles (.3)	1.0	1.0	1.0	Project D207. Development of modular integrated communications and navigation system (MICNS), a miniaturized jam resistant data link intended to overcome projected ECM threats.
64750A	Tactical ECM Systems	11.6	—	—	Hand-emplaced and artillery-delivered expendable jammers. Reconfigure TLQ-17A HF/VHF jammer to tracked vehicle. Convert Quickfix jammer to UH-60 platform.
64779A	Joint Interoperability of Tactical C² Systems (.02)	.3	.4	.4	Develop specs for interface to intelligence/EW systems.
23739A	AN/TSQ-73 Missile Minder ADP Modifications (.1)	1.0	1.0	2.0	Modification to improve interoperability with Hawk, Patriot, Shorad C² and other air defense C² systems by providing for effective jam-resistant and survivable near-real-time data communications.
23751A	Special Operations Forces Equipment Project—D061 Communications Equipment (.3)	4.0	10.0	3.0	Development, modification, evaluation of extremely low probability of intercept/detection (LPI/LPD) in a lightweight, man-portable package.
23801A	Missile/Air Defense Product Improvement Program	15.0	14.0	17.0	Program develops improvements to Patriot recommended by the Defense Science Board. Includes Patriot ARM Decoy. SOJ counter and reduced RC against surveillance. Development of Rosette Scan Seeker for Chapparal to provide CM hardening.
23802A	Other Missile Product Improvement Program (.3)*	4.0	13.0	7.0	Hellfire EOCM hardened laser seeker. TOW improvements in battlefield obscurants environment.
31307A	Foreign Science and Technology Center (.1)	—	—	—	Acquisition and exploitation of foreign systems for intelligence and threat assessment. Dollar amount not stipulated in DOD summary.
31327A	Technical Reconnaissance and Surveillance (Tecras) (.2)	—	—	—	Continuing program data necessary to support development of systems, CM and tactical doctrine. Dollar amount not stipulated in DOD summary.
33142A	Satellite Communications Ground Environment (.05)	4.0	—	2.0	Development of sidelobe canceler for GSC-39 and GSC-52 Earth terminals; Development of SCOTT Earth terminal; complete anti-jam control modem for SHF satellite terminals.
	Totals	**230.9**	**142.9**	**152.2**	

U.S. Army Procurement Budget for Electronic Warfare Systems 1987-1989

Program Nomenclature	Funding ($ Millions) Fiscal Years		
	1987	1988	1989
Aircraft			
EH-60A Helicopter (Quickfix)			
Less: Advance Procurement (.1)*	11.0	2.0	5.0
EH-60A Helicopter (Quickfix)			
Advance Procurement (.1)*	2.0	1.0	—
Modification of Aircraft			
OV-1 Surveillance Airplane (Mohawk) (.2)	1.0	2.0	3.0
RV-21H Recon Airplane (Ground Mod) (.1)*	—	.3	.8
RV-1 Recon Airplane (.2)	.3	—	1.0
Airborne Avionics (.1)	—	.1	.3
Acft 9WW (.2)	2.0	5.0	5.0
Support Equipment			
Aircraft Survivability Equipment			
Less Advance Procurement (.3)*	3.0	21.0	29.0
Aircraft Survivability Equipment			
Advanced Procurement	—	4.0	—
Missile Systems			
Chapparal (.02)*	.6	.7	1.0
Patriot Less Advanced Procurement (.04)*	37.0	34.0	32.0
Stinger (.02)*	5.0	3.0	4.0
Modification of Missiles			
Patriot (.5)	11.0	11.0	20.0
Hawk (.5)	24.0	18.0	26.0
Chapparal (.25)	1.0	8.0	16.0
Communications and Electronics Equipment			
Classified Project 9WW	8.8	9.7	7.6
Tri-Tac Equipment (.1)*	—	18.0	16.0
Sincgars (.1)	—	2.0	34.0
Jam-Resistant Secure Comm (JRSC)	9.0	8.9	6.7
SW Asia Comm Infrastructure (.1)*	1.0	3.0	2.0
SouthCom C³ Upgrade (.1)*	1.0	8.0	0.0
Single Channel Objective Tactical Terminal			
(SCOTT) (.05)*	—	.1	3.0
Modify In-Service Equipment—Tactical Satellites (.05)*	.3	.2	.2
All Source Analysis System (ASAS)—TIARA (.3)*	28.0	16.0	—
Single Source Processor—Sigint*	—	3.8	—
Rear Echelon Comint System (RECS)*	—	27.7	—
Trailblazer, TSQ-114	11.8	—	—
Tactical Electronic Surveillance System	3.9	3.4	4.7
Manpack Radio DF System (MRDFS)	5.5	5.2	—
Modify In-Service Equipment (Int Spt) TIARA (.3)*	11.0	9.0	13.0
Trojan	23.9	25.4	—
Tactical Reconnaissance and Surveillance System			
(TECRAS) (.5)*	5.0	5.0	—
Jammer, Hand Emplaced, Expendable	(C)	(C)	(C)
Tacjam, AN/MLQ-34	32.0	—	—
Tactical Deception (TAC-D)	—	4.6	—
Mod In-Svc Equipment (EW)	8.1	2.8	11.3
Items Less Than $2M (EW-C-E)	1.0	.8	.9
RPV TA/Design Aerial Recon Sys (TADARS) (.2)*	10.0	36.0	39.0
Totals	**258.2**	**299.7**	**281.5**

U.S. Navy
Research, Development, Test & Evaluation Budget
1987-1989

Program Element	Program Nomenclature	Cost Data ($ Millions) Fiscal Years			Description/Comments
		1987	1988	1989	
61152N	In-House Independent Laboratory Research (.05)	1.2	1.1	1.2	Provides the primary means for Navy in-house labs to stimulate original work in science and technology related to Navy mission needs. Includes optical and microwave sensors.
62111N	AAW/ASUW Technology (.1)*	6.4	6.6	6.9	Supports future surveillance and weapons developments for surface, air and space platforms. ECCM processing techniques for space-based radar flight tested in fiscal 1987.
62113N	Electronic Warfare Technology	13.5	14.5	15.1	Provides for exploratory development in EW including onboard and offboard CM devices, jammers and false target generators, to counter surveillance, targeting and terminal phases of enemy systems; signal detection and related signal processing (ESM).
62121N	Surface Ship Technology (.01)	.1	.1	.1	IR signature reduction.
62234N	Systems Support Technology-Electronic Devices (.5)	31.5	32.8	35.6	Very wide-band delay line (2-18 GHz) for use in airborne EW systems; gyrotrons technology; advanced computer concepts.
63109N	Integrated Aircraft Avionics	14.0	—	—	Project W1953—INEWS Advanced Development* continues INEWS Phase IB demonstration/validation and risk reduction. Commences INEWS software development and integration system facility.
63206N	Electronic Warfare Advance Development:	27.7	62.4	46.8	Project W0638—Airborne Defensive ECM. Monitors solid-state initiatives for EA-6B systems; fabricates EOCM system; incorporates EEPROMS; Project W0640—Offboard EW. Development of airborne active expendable decoy, advanced IRCM system, MJV-20 and MJV-21 airborne expendable flares, advanced chaff projects. Project W193J—Strike EW Simulator. Demonstrates multiple aircraft/ECM vs. a large threat simulation.
63217N	Advanced Aircraft Subsystems (.1)	.6	1.2	2.2	Project No. 1N0446—Advanced Avionics Subsystems. Transition VHSIC technology to Navy signal processing applications such as radar, communications and EW.
63262N	Aircraft Survivability/Vulnerability (.2)	1.3	1.2	1.6	Project W1088 and W0591—Joint Technical Group on Aircraft Survivability. Pyrotechnically pumped laser jammer; update to enhanced surface-to-air missile simulation; threat assessments, predict near field RF signatures as seen by fuzes.
63303N	Electromagnetic Radiation Source Elimination System Technology (.7)*	5.4	4.1	9.0	Develops passive receiver subsystems for ARM weapon system.
63582N	Combat System Integration (.3)*	4.3	3.0	3.1	Testing of NTDS program digital interface with new threat upgrade systems and SLQ-32 system in CG 16/26, CGN 38, DDG 993 clases.
64211N	IFF System Development (.1)	2.2	3.6	3.4	Develop new passive ID techniques including non-cooperative target recognition techniques and an EW warning system. Reduce ECM vulnerability.
64217N	S-3 Wpn Sys Imp (.05)	.5	—	—	Improvements in ESM, Harpoon launch, chaff, and flares.
64221N	P-3 Modernization Program	1.0	—	—	Project W1149—ESM Improvements. Tech eval of ALR-77 ESM, which features improved frequency coverage, bearing accuracy, threat warning and bearing-only-launch for Harpoon.
64224N	Airborne EW Engineering	30.3	32.0	37.4	Operational evaluation of ALQ-162, full-scale engineering development of ALQ-149 for EA-6B aircraft; develop ALQ-162 (V)1 reprogrammability, integration of laser intercept, eng upgrades of ASR for ALQ-67A(V) 2.
64226N	ASPJ	27.2	16.4	6.3	ALQ-165 airborne self-protection jammer (ASPJ) for tri-service. Integration/testing in F/A-18, F-16, A-6E, F-14, AV-8B pod.
64230N	Warfare Support Systems*	46.0	46.5	35.8	No program description available.

ELECTRONIC WARFARE BUDGET
U.S. NAVY BUDGET: RDT&E

Program Element	Program Nomenclature	Cost Data ($ Millions) Fiscal Years			Description/Comments
		1987	1988	1989	
64232N	Transfer Support Systems	30.2	78.6	52.0	Development of HF systems which have anti-jam features.
64255N	EW Simulator Development	38.0	41.1	43.4	Project W0602—EW Environment Simulation. Develops integrated naval air defense simulation complex for testing airborne EW equipment at NWC, China Lake and PMTC, Point Mugu. Project W0672—Effectiveness of Navy EW Systems (ENEWS). Develops flyable instrumented simulators representative of anti-ship missile threats to evaluate shipboard EW systems; develops laboratory simulation facilities to test hardware in anechoic chambers. Project W1778—Closed Loop Test Capability. Closed-loop radar simulation capability to determine the effectiveness of EW and ECM installed in host aircraft.
64354N	Air-to-Air Missile Systems Engineering (.3)	4.3	4.4	4.8	Project W0456—AIM-9 Product Improvement Program. Develops improved acquisition range, discrimination and ECCM features for AIM-9M.
64361N	NATO Sea Sparrow (.05)	.1	.2	.3	Integration with SLQ-32 and NTDS.
64502N	Submarine Communications (.1)*	.4	.4	.4	Project X0742—Submarine Integrated Antenna System. Incorporation of HF anti-jam features.
64508N	Radar Surveillance Equipment (.1)	—	.8	1.0	Investigates development of low-sidelobe antennas for air search radars for improved detection in ECM environment.
64515N	Sub Support Equip Prog	16.5	20.6	15.7	Upgrade signal processing/data storage for WLQ-4 (V) Sea Nymph. Develop radome for BRD-7 antenna. Improve WLR-1 and WLR-8 ESM systems.
64573N	Shipboard EW Improvements	42.5	40.2	48.5	SLQ-32 improvements; CV/CVN EW system improvements; decoy developments; EW systems integration.
64707N	Theater Mission Planning Center (.05)	.2	.1	.1	Project X0798. Over-the-horizon targeting. Develops concept for signal intelligence fusion and tracking interface with tactical receive equipment for improved Elint correlation.
64715N	Surface Warfare Training Devices	1.3	3.4	2.5	Project S1140—Tactical Advanced Combat Direction EW Modifications. EW training complexes at Fleet Combat Training Centers Atlantic and Pacific.
24134N	A-6 Squadrons (.1)	1.0	.3	.9	Flight testing of HARM integration/software development and IR video auto tracker.
24152N	Early Warning Acft Sqdns (.3)	9.9	10.0	7.5	ECCM for APS-125 radar for improved ship target detection.
24571N	Special Projects (.5)	1.5	2.7	4.4	Project W0431—Tactical Aircrew Combat Training System. Design and development of tactical aircrew combat training system (TACTS); new weapon/EW simulations as they are defined; and continued development of EA-6B interface to TACTS.
24573N	Navy Cover & Decept Prog	9.8	6.8	5.4	Develop, build and integrate shipboard and offboard cover and deception hardware and vans: SSQ-74, SLQ-33; SLQ-34; and deception devices.
24575N	Electronic Warfare Readiness Support	2.0	6.5	6.2	Fleet EW support center continues design and testing of ALQ-170(V)4; testing updates to the FAEWS; initiates replacement of obsolescent air platforms by FEWSG; completes development of ULQ-18(V). Data link vulnerability is a continuing program to incorporate ECCM in all Navy systems during engineering design and evaluate ECCM in those systems.
24576N	Counter C³ Development	9.5	9.8	12.8	Ship- and aircraft-launched chaff buoys; active electronic buoys; counter-targeting expendables; cost reduction improvements for the active electronic buoy; full-scale development of Proforma CM, EW coordination module, countermeasures assessment simulator.
25601N	HARM Improvement	2.2	—	—	Develops software to correct deficiencies.
25674N	EW Counter Response	50.1	54.6	26.5	Integration of HARM into EA-6B. Complete full-scale development, flight test of EA-6B improved capability II; deliver EA-6B ADCAP receiver processor group.
26313M	Marine Corps Telecom (.2)*	.8	.6	.8	Project C0048—Comm Terminal Improvements—Development and testing of a variety of more robust comm subsystems.

Program Element	Program Nomenclature	Cost Data ($ Millions) Fiscal Years			Description/Comments
		1987	1988	1989	
26625M	Marine Corps Intelligence/Elect Warfare Sys	10.1	5.0	14.7	Project C0066—Communication and Non-communications ECM. Project C1928—Tactical Electronic Reconnaissance Processing and Evaluation System. Development of HF VHF and UHF jammers; correct deficiencies of EW suite to fit in highly mobile tactical vehicle; commence integration of Tadixs-B tactical receive equipment into the tactical electronic reconnaissance processing and evaluation system. Project C1961—Mobile EW Support System.
64208N	Range Inst & Sys Dev (.3)	2.6	1.4	2.3	Project W0604—Training Range Instrumentation Development. Supports threat radar simulator, deception jammer simulator, noise jammer/simulator, and fleet telemetry stations at training ranges.
	Totals	**446.2**	**513.0**	**454.7**	

U.S. Navy Procurement Budget for Electronic Warfare Systems 1987-1989

Program Nomenclature	Funding ($ Millions) Fiscal Years		
	1987	1988	1989
Aircraft			
EA-6B (Electronic Warfare) Prowler			
Less Advance Procurement	409.1	336.1	470.8
EA-6B Advance Procurement	21.6	17.8	18.2
Modification of Aircraft			
FEWSG	16.9	3.4	1.8
Aircraft Common ECM Equipment	69.5	16.7	35.8
Common Avionics Changes (.1)*	3.0	.1	.2
Radars (.1)*	5.0	—	—
Missiles			
AGM-88A Harm	248.2	194.7	404.9
Drones and Decoys	34.9	63.6	125.5
Modification of Missiles (.5)	6.0	5.0	46.0
Communications and Electronic Equipment			
Ship Radars (.05)*	5.0	5.0	6.0
Electronic Warfare Equipment			
AN/SLQ-32	75.0	75.1	91.7
AN/SLQ-17	14.0	—	—
AN/WLR-1	3.8	5.5	6.1
AN/WLR-8	—	6.3	7.9
ICAD Systems	—	—	4.8
Offboard Deception Devices	22.3	27.0	22.2
EW Support Equipment	8.8	4.9	4.8
Fleet EW Support Group	1.6	3.4	4.0
C^3 Countermeasures	48.9	7.4	64.1
Reconnaissance Equipment			
Combat DF	36.1	52.7	44.0
Outboard	25.6	27.8	21.2
Battle Gp Passive Horizon Ext Syst	—	—	39.9
Submarine Surveillance Equipment			
AN/WLQ-4 Depot	1.2	9.9	.7
AN/WLQ-4 Improvements	3.5	18.8	32.7
AN/BLD-1 (Interferometer)	16.6	6.7	2.4
Submarine Support Equipment Program	4.9	3.5	—
Other Ship Electronic Equipment			
Navy Tactical Data System (.02)*	2.0	2.0	2.0
HF Link-11 Data Terminals (.04)*	.2	.1	.1
Other Shore Electronic Equipment			
Naval Space Surveillance System (.5)	3.0	5.0	6.0
Over-The-Horizon Radar (.05)*	.1	4.0	9.0

Program Nomenclature	Funding ($ Millions) Fiscal Years		
	1987	1988	1989
Shipboard Communications			
Shipboard HF Communications (.02)*	.2	.1	—
Shipboard VHF Communications (.02)*	.2	.1	.1
Portable Radio (.02)*	.1	.1	.2
Submarine Communications			
Submarine Communication Antennas (.01)*	.2	.1	.1
Satellite Communications			
Satcom Ship Terminals (.02)*	.6	.6	.8
Satcom Shore Terminals	.1	.1	.2
Cryptological Equipment			
Ships Signal Exploitation Space	1.6	4.1	5.5
Countermeasures			
Airborne Expendable Countermeasures (Chaff/IR Flares)	38.6	25.5	18.1
Airborne ECM/ECCM	1.0	1.0	1.1
Anti-Ship Missile Decoy Systems	6.3	7.6	7.9
Shipboard Expendable Countermeasures	27.9	21.1	34.0
Training Devices			
Surface Combat System Trainers (.1)	3.0	2.0	2.0
Marine Corps			
AN/TPS-32 Anti-Radiation Missile Decoy	—	—	8.2
Mobile EW Support Sys (MEWSS)	15.4	—	—
Guided Missiles			
Hawk Less Advance Procurement (.02)*	2.0	2.0	2.0
Hawk Mod (.1)*	2.0	3.0	4.0
Stinger (.02)*	1.0	3.0	3.0
Manpack Radios			
Manpack Radios and Equipment (.02)*	.1	—	—
Vehicle Mounted Radios and Equipment (.02)*	—	.1	—
Totals	**1187.1**	**973.0**	**1560.0**

U.S. Air Force
Research, Development, Test & Evaluation Budget
1987-1989

Program Element	Program Nomenclature	Cost Data ($ Millions) Fiscal Years			Description/Comments
		1987	1988	1989	
61102F	Defense Research Sciences Project 2305—Electronics (.1)	2.0	2.0	2.0	Research in surveillance, guidance and control, information and signal processing, EW, C³, optical signal processing for target recognition and terminal guidance, EM propagation, target signatures; electro-acoustic analog signal processing, fast parrallel processing algorithms and devices for A/D conversion: mm-wave ICs.
62101F	Geophysics—IR Target and Background Signatures (.7)*	2.0	2.0	2.0	IR signatures of natural and nuclear Earth/atmospheric backgrounds and targets within them through use of data from rockets, aircraft, balloons and the space shuttle. Integration into models for IR surveillance systems application.
62102F	Engineering Technology Project 2423—Electromagnetic Windows and Electronic Materials (.5)	2.0	4.0	3.0	Develop high-performance IR detector materials for strategic and tactical detector arrays.
62204F	Aerospace Avionics/ VHSIC Active ECM	3.0	3.3	4.0	Radio freq. and optical/IR threat systems; expandable ECM system; low freq. ECM airborne antennas. Basis for major advances in EO and IR for real-time reconnaissance, auto target classification and aircraft navigation and defense. Wideband Bragg cell RWR. Millimeter chaff tests. Spread spectrum receiver to detect communications. Advanced signal sorting algorithms for signal identification and pulse tracking. Improve closed loop IR CM against IR missile seekers. Develop digital RF memories. Apply AI for optimized ECM waveforms. Improve GaAs and TWT power generation devices.
	Technology for Reconnaissance and Targeting	2.3	2.6	3.2	
	Passive ECM	2.5	2.8	3.5	
	Electro-Optical Technology	2.2	2.5	3.0	
	Microwave Technology	5.2	5.8	7.0	
62702F	Command/Control/Communication— Surveillance Technology—Project 4506	5.0	5.0	5.0	Space based radar sub-array development. ECCM techniques to defeat smart jammers. Multi-domain (time, frequency, polarization, spatial) processing algorithms to defeat low observables in dense jamming and clutter environments.
63109F	Inews/ICNIA	35.8	33.2	15.1	Inews (integrated EW system). AF led, joint AF/Navy program to develop next-generation airborne self-protection countermeasures system for advanced technology aircraft, including ATF and Navy advanced tactical aircraft. Threat consists of airborne- and surface-based radar, electro-optical, infrared- and laser-directed defense systems and tactical C³ network that links them together. June 1986 begins demonstration/validation (milestone I). Concept definition refinement and concept trade-off analysis will continue.
63203F	Advanced Avionics for Aerospace Vehicles— Adv. Recon./Strike Radars (.1)	—	—	—	Reduced radar radiations and ECCM techniques for survival in 1990s threat environment; LPI-terrain following techniques.
	Project 2334—Airborne Radar CM	3.8	6.2	7.6	Develops ECCM technologies and concepts to reduce susceptibilities of current and future airborne weapon systems to ECM. Applicable to F-15, F-16, B-1B radars and future ATF weapons systems. Includes: air-to-air ECCM to counter noise and DECM; spread-spectrum waveforms; passive situational awareness; simultaneous transmit and receive technology; and electronic combat multifunction radar.
	Project 2347—Optical Countermeasures	2.5	3.4	3.7	EO CCM to reduce vulnerability and mission degradation in a hostile EO environment. CM hardened FLIR technology and development. Evaluation of effects of camouflage concealment and deception on recognizers/electro-optical sensors.
	Project 1177—Non-Cooperative ID (.5)*	2.0	3.0	3.0	Passive identification system that combines the Radar Warning Receiver and Fire Control Radar in the F-15 weapons system avionics.
	Project 2746—LPI Communications	—	1.1	1.1	LPI Communications for stealth aircraft.

Program Element	Program Nomenclature	Cost Data ($ Millions) Fiscal Years			Description/Comments
		1987	1988	1989	
63253F	Advanced Integration Avionics Project 2746—LPI Comm	.9	—	—	Low probability of intercept communications. Jam-resistant secure voice communication. Have Lace laser communication.
63743F	Electronic Combat Technology	37.7	41.3	47.6	Provides advanced development in EW where expanded technology base is needed to solve critical penetration and problems for all classes of manned and unmanned aircraft against all classes of IR, laser and EO threats.
63789F	Tactical C³ Adv. Development (.05) Tactical Information Distribution— Project 2317 (.1)*	.1	.2	.4	Demonstrate Integrated Burst Communications and develop an architecture for tactical packet communications.
	Tactical Radar ECCM—Project 2333*	2.7	2.8	4.9	Advanced development for the Advanced Tactical Surveillance Radar ARM decoy. Begin smart jammer development. Complete Advanced main beam ECCM Technology Radar nulling experiment.
	Communication and Navigation ECCM— Project 2335*	3.3	5.0	4.4	Jam-resistant communication to reduce jamming/intercept vulnerability through signal processing.
63738F	Air Defense Initiative (ADI) Surveillance Technology (.1)*	—	3.0	3.0	Advanced radar technology is the main thrust of this program aimed at flexible and survivable capabilities and bomber and cruise missile surveillance. Includes small degree of development of passive sensors and unattended processors.
64226F	B-1B—Ele. ECM Modification—Project 3645	—	18.9	110.0	Fiscal 1988 new start updates the B-1B defensive system beyond the baseline configuration in keeping with advances in the projected threat and intelligent refinements.
64326F	Strategic Conventional Standoff Capability	7.2	—	—	Validation of Have Dark technology for strategic bombers with integration of ESM technology. To be flight tested on B-52. Analysis of passive sensors, ECM required beyond ALQ-117 Pave Mint.
33131F	Minimum Essential Emergency Comm. Network—VLF/LF Improvements (.01)	.3	.4	.2	Improved C³, VLF/LF comm. systems under adverse nuclear and jamming conditions. Consists of airborne transmitters and receivers in EC-135 and E-4B command post aircraft.
33601F	Milstar Sat Comm System— AF Terminals (.05)	14.0	11.0	16.0	AF Satcom UHF terminal modifications for transition to Milstar. Dev. of Milstar EHF terminals.
33603F	Milstar Comm. Satellite System (.05)	24.0	—	—	Joint-service program. Satellite communication system with robust AJ features.
63320F	Lower Cost Anti-Radiation Seekers	16.8	13.6	12.5	Advanced development of anti-radiation weapons for F-4G Wild Weasel.
63600F	Millimeter-Wave Seekers (.1)	.5	—	—	Advanced mm-wave seekers for guidance and warhead components.
63609F	Millimeter-Wave Seekers (.1)	.5	—	—	Adv. mm-wave seekers for guidance and warhead components. Significant advances have been made that can be exploited.
63742F	Combat Identification Technology— Non-Cooperative Identification (.1)	.6	.2	.2	Passive ID algorithms; non-cooperative target ID technology; active/passive ID sensors; NATO IFF; use of RWR data for multiple target ID for beyond-visual-range air-to-air missile fire control.
63749F	C³CM Advanced Systems	1.9	1.4	1.9	Analysis of promising escort and standoff C³CM technologies. High Power Microwave development. Demonstrate and evaluate electronic deception. Develop C³ jammer.
64201F	Aircraft Avionics Equipment Development— Airborn Radar Improvements— Project 2519. (.3)*	2.0	2.0	2.0	Develop ECCM threat data base in coordination with AF's ECCM Master Plan.
64220F	EW Counter Response	26.0	13.7	2.6	EF-111A/ALQ-99 updates. DT&E/IOT&E flight testing scheduled for fiscal 1987.
64250F	Integrated EW/Communications Navigation Identification (Inews/CNI) Development	—	5.7	36.1	New start fiscal 1988 with specific application to the Advanced Tactical Fighter. Inews is an Air Force led joint Air Force-Navy program to develop the next generation airborne defensive avionics system. ICNIA is a tri-service program to develop the next generation communication, navigation and identification integrated subsystem for advanced aircraft.

ELECTRONIC WARFARE BUDGET
U.S. AIR FORCE BUDGET: RDT&E

Program Element	Program Nomenclature	Cost Data ($ Millions) Fiscal Years			Description/Comments
		1987	1988	1989	
64321F	Joint Tactical Fusion Program (.1)	2.0	—	—	Correlate and aggregate multisource sensor data; provide precise location of enemy forces and display ground battle situation in tactical air control centers. Combines Army BETA and TFD with AF ENSCE programs.
64617A	Airborne Survivability and Recovery Tactical Programs— Camouflage Concealment and Deception Project 3141 (.25)*	1.0	1.0	1.0	Development of full spectrum of camouflage concealment, and deception. This includes aircraft decoys, atmospheric obscuration and optical and RF sensor deception.
64710F	Reconnaissance Equipment	9.2	.2	.2	RF-4C and RC-135 systems including: EO collection/recon. (Compass Seven); interim tactical Elint processor; adv. recon. sensor; ESM; EW/close air support joint test; TEREC; AAQ-X IR sensor.
64724F	Tactical C3CM	19.0	12.8	9.0	Develop intercept and jamming subsystem improvements. Update transmitters for Compass Call. Deliver mission simulator for EC-130H standoff jamming aircraft to disrupt enemy C3 networks.
64725F	Combat Identification Systems (.1)	1.0	4.0	13.0	Tri-service program. Integration of passive RF ID tech. into tact. aircraft. Develop non-cooperative target ID algorithms. Demonstrate improved indirect ID capability in both all-weather and hostile EM CM environments.
64733F	Defensive Surpression (.4)*	11.0	16.0	11.0	Preplanned product improvement for Air Force AGM-130A general purpose medium stand off attack weapon to improve data link in the face of ECM.
64737F	Airborne Self-Protection Jammer (ASPJ)	11.9	21.5	8.0	Joint Air Force/Navy enineering development program. ASPJ (ALQ-165) joint AF/Navy internally mounted ECM system for F-14, F-16, F/A-18, A-6E and AV-8B. To complete DT&E flight testing, environmental testing and AFEWES testing in fiscal 1987.
64738F	Protective Systems	64.1	53.4	83.2	Flight-test ALQ-172 (for B-52H). Develop ALCM ECM system. Improve simulation systems. Upgrade F/FB-111 CM set. Evaluate phased-array antenna systems. Antenna Test Range to test and edvaluate new EW antennas on actual aircraft. Develop, fabricate and validate EW simulation systems.
64739F	Tactical Protective Systems	41.1	49.1	56.0	
	Project 2272—F-16 Protective Systems	8.3	6.9	10.2	Develops tailored mm-wave warning capability for ALE-47 dispenser, advanced chaff; optical threat acquisition.
	Project 2273—Integrated Electronic Warfare Systems (Inews)				Transferred to PEG3109F.
	Project 2274—Special Operations Aircraft Protective Systems	3.0	3.5	1.5	Develops self-protection EW suites for AC-130, MC-130 (Combat Talon), HH-53 tailoring available EW equipment.
	Project 2879—Area Reprogramming Capability	12.0	15.6	16.2	Develops area reprogramming capability (ARC) for EW systems. Interim software to reprogram ALQ-161 to be delivered in fiscal 1987.
	Project 3106—A-10 Protective Systems	.2	—	—	Tailors mm-wave/laser warning, ALE-47 dispenser, ARC. Primary task is to develop IR decoy.
	Project 3107—Special Mission Aircraft Protective Systems	1.6	1.1	1.4	Primary task is to upgrade missile IR jamming capability from PRC.
	Project 3158—EW Planning & Management	2.0	3.1	4.0	Development of Electronic Combat Digital Evaluation System (ECDES), a multi-level, integrated digital evaluation system.
	Project 3630—Joint EW Center (JEWC)	.9	2.0	2.2	
	Project 5618—F-15 Protective Systems	13.1	17.0	20.4	Upgrade of ALR-56A, ALQ-135, ALQ-128, ALE-45.
64742F	Precision Location Strike System (PLSS)	20.0	—	—	Complete DT&E/IOT&E in fiscal 1987. Refurbishment to ready for T&E in European theater. Project terminated.
64750F	Intelligence Equipment—(.1)*	1.0	.6	.5	Equipment to process, integrate, display and distribute intell. data. Develop foreign threat warning techniques, evaluate foreign weapon systems, and perform tactical unit mission planning. Evaluations/studies for upgrading intelligence gathering. Develop specialized equipment for Technical Surveillance Countermeasures: Broadband CM antenna for CM receivers for use in surveys.

Program Element	Program Nomenclature	Cost Data ($ Millions) Fiscal Years			Description/Comments
		1987	1988	1989	
64754F	Joint Tactical Information Distribution Systems (JTIDS) (.05)*	—	4.0	6.0	Develop a highly jam-resistant secure digital information distribution system for use in a tactical combat environment.
27130F	F-15 Squadrons (.05)	8.0	6.0	3.0	Multi-Staged Improvement Program radar expanded ECCM; updates to EW suite (ALR-56C, ALQ-135, ALE-45).
27133F	F-16 Squadrons (.1)*	5.0	4.0	2.0	Expanded memory capacity added to APG-66 radar to enhance ECCM. Improve ECCM for APG-68 radar.
27136F	F-4G Wild Weasel Squadrons	35.3	17.8	17.0	APR-38 (radar warning and attack system). Performance update program gives F-4G increased on-board computer capacity, processing speed, increased freq. range.
27162F	Tactical AGM Missiles—HARM	1.0	2.3	—	High-speed anti-radiation missile (HARM, AGM-88) developed by Navy. Designated as primary ARM for F-4G Wild Weasel. Software updates to improve performance against newer radar threats. EPROM, receiver and video processor upgrades.
27168F	F-111 Self-Protection Systems	—	58.0	52.9	F/FB/EF-111 Self-protection ECM update program. ALQ-189 jammer integration.
27316F	Tacit Rainbow (.1)*	—	16.0	48.0	Low-cost programmable loitering missile system to search out and attack emitting radars and jammers. Tri-service program.
27412F	Tactical Air Control System (.5)	9.0	10.0	7.0	Ultralow sidelobe antenna; ARM alarm; ARM decoy for TPS-43E radar to enhance survivability.
27417F	ESM/ECCM for E-3 (.3)	29.0	33.0	27.0	ECCM improvement (Have Quick A-Net); ESM addition to E-3 surveillance system for passive detection, location and ID capabilities against airborne, shipborne and ground-based emitters. Full-scale development of trainer external simulation system.
27423F	Adv. Comm. System	31.9	34.5	11.1	Develop Sincgars AJ for VHF comms, Have Quick as near-term UHF AJ system. Enhanced joint tactical information distribution system (EJS) to provide AJ voice comm.
27582F	Have Trump	—	—	—	Classified EW program. Funding not stipulated in DOD Budget Summary.
27584F	Seek Axle	—	—	—	Classified EW program. Funding not stipulated in DOD Budget Summary.
28010F	Joint Tactical Comm. Program (Tri-Tac)— Digital Troposcatter Terminal (.1)*	.7	2.0	1.0	ECCM for TRC-170 digital troposcatter radio for long-range wideband tactical comm support of tactical air control. Full-scale development.
35887F	Electronic Combat Intell.	1.7	1.6	1.7	Intelligence support to reprogrammable threat warning and jamming systems such as in PLSS, F-4G and EF-111A. Continuing data file development provided to AF/EW center at Kelly AFB.
44011F	Special Operations Forces—MC-130H— Project 3129 (.4)*	7.0	3.0	—	Develop an integrated EW Suite for Combat Talon II.
33110F	Defense Satellite Communications System (DSCS)—(.02)*	.3	.4	1.0	Improvement of jam resistant throughput.
63438F	Satellite System Survivability Project 2612—Satellite Survivability (.2)	.4	.6	.7	Demonstrate AJ to co-orbital and direct-ascent laser jamming.
33126F	Long Haul Communications—DCS— Transmission Improvements Project 2157—(.4)*	.3	.4	.4	Development of equipment embodying ECCM technology.
64227F	Flight Simulator Development Project 2769—Simulator Update Development (.1)	.3	—	—	Refurbishment and upgrade with addition of ALQ-99 to the T-5 simulator.
64735F	Range Improvement Project 2286—Tactical AF Equipment (.05)	.3	.7	.3	Development to simulate ground-based radar jamming.
	Project 3320—SAC Equipment (.5)	3.0	5.0	6.0	Development of MLQ-T4 ground jammer simulator. Acceptance testing of MSR-T4 EW signal receiver analyzer. Update MST-T1A multiple threat emitter system.
	Project 3321—Airborne Radar ECCM	1.9	2.0	2.1	Upgrade radar test facility at Tyndall AFB to support airborne radar/weapon system ECCM design, development and test programs.

ELECTRONIC WARFARE BUDGET
U.S. AIR FORCE BUDGET: RDT&E

Program Element	Program Nomenclature	Cost Data ($ Millions) Fiscal Years			Description/Comments
		1987	1988	1989	
64735F	Project 6510—Flight Test Threat Systems Simulators	40.9	38.5	20.4	Development of test quality replicas of Soviet air defense radar equipment. The equipment will be used in flight testing new airborne radars and avionics systems EW capability.
64755F	Improved Capability for DT&E Project 3120—Seeker Development (.1)	1.0	1.0	2.0	Airborne instrumentation to support testing a wide variety of EO seekers in a controlled CM environment. Airborne Radar ECCM provides the modernization of electronic jammers to furnish realistic electronic combat.
65708F	Nav/Radar/Sled Track Test Project 2900—Radar Target Scatter Upgrade/RCS/Measurement System (.1)	.2	.2	.2	Ratscat upgrade, dynamic RCS, 6,585th test group support, Ratscat advanced measurement system (RAMS). Upgrade includes pop-up calibration facility and two phase-coherent mm-wave radars.
	Totals	**608.4**	**644.9**	**752.6**	

DATA GENERAL ASKS: ARE YOU PLAYING RUSSIAN ROULETTE WITH YESTERDAY'S TECHNOLOGY?

FOR ADVANCED COMPUTER SYSTEMS, TALK TO US. IT'S WHY SO MANY GOVERNMENT DEPARTMENTS HAVE CHOSEN DATA GENERAL.

Government business is too critical to be taken for granted. Too much depends on it.

No wonder nineteen of the top twenty U.S. defense contractors have bought a Data General system. As have all the Armed Services and most major departments of the federal government.

And to date, nearly thirty U.S. Senate offices and committees have chosen Data General.

TODAY'S BEST VALUE

Why such unanimity? Because Data General offers a complete range of computer solutions for government programs, with one of the best price/performance ratios in the industry.

From our powerful superminis to the DATA GENERAL/One™ portable.

From unsurpassed software to our CEO® office automation system. Plus complete systems for Ada® and Multi Level Secure Operating Systems, and a strong commitment to TEMPEST.

All Data General systems have full upward compatibility. And because they adhere to international standards, our systems protect your existing equipment investment. We give you the most cost-effective compatibility with IBM outside of IBM—and the easiest to set up and use.

SOLID SUPPORT FOR THE FUTURE

We back our systems with complete service and support. As well as an investment in research and development well above the industry norm.

So instead of chancing yesterday's technology, take a closer look at the computer company that keeps you a generation ahead. Write: Data General, Federal Systems Division, C-228, 4400 Computer Drive, Westboro, MA 01580. Or call 1-800-DATAGEN.

**↳ Data General
a Generation ahead.**

© 1985 Data General Corp., Westboro, MA. Ada is a registered trademark of the Department of Defense (OUSDRE-AIPO). DATA GENERAL/One is a trademark and CEO is a registered trademark of Data General Corporation.

U.S. Air Force Procurement Budget for Electronic Warfare Systems 1987-1989

Program Nomenclature	Funding ($ Millions) Fiscal Years		
	1987	1988	1989
Aircraft			
F-15 C/D/E			
Less Advance Procurement (.01)	15.0	14.0	15.0
F-16 C/D			
Less Advance Procurement (.01)*	23.0	22.0	26.0
TR-1/U-2			
Less Advance Procurement (.2)	2.0	.2	.3
Modification of B-52, FB-111, B-1B (.15)	73.0	41.0	46.0
Modification of A-7, A-10, F-5, F-15, F-16, F-111 (.3)	250.0	156.0	179.0
Modification of EF-111 (.2)*	—	—	5.0
Modification of TR-1A (.2)*	3.0	2.0	4.0
Classified Projects (.1)	9.0	8.0	13.0
Aircraft Support Equipment and Facilities			
Common ECM Equipment	—	221.9	266.6
Missiles			
Peacekeeper (M-X) (.01)*	11.0	13.0	14.0
Advanced Cruise Missile			
Less Advance Procurement (.01)*	—	—	—
Air Launched Cruise Missile (.02)*	.2	—	—
Tactical Missiles (.02)*	24.0	30.0	28.0
AGM-88A HARM	363.7	422.9	214.9
Modification of In-Service Missiles (.5)	69.0	75.0	63.0
Space Programs			
Defense Satellite Communications Systems (.05)*	5.0	4.0	1.0
Space Defense System	—	1.0	18.0
Munitions			
Chaff Cartridge RR-170	19.1	3.4	3.9
Chaff Cartridge RR-136	1.4	1.4	1.2
Flare, IR MJU-7B	11.1	9.0	9.3
Flare, IR MJU-2	2.7	—	—
Flare, IR (B1B)	—	2.0	5.9
MJU-10B	8.1	18.6	18.0
Chaff Package RR-141A/L	4.7	—	—
Electronics Programs			
OTH-B Radar (.02)*	2.0	2.0	4.0
Pave Paws/SLBM Warning Systems (.02)*	2.0	—	—
Caribbean Basin Radar Network (.02)*	.2	.1	.5
TAC Sigint Support	15.3	29.8	25.4
Tactical Ground Intercept Facility	5.0	2.3	3.4
TR-1 Ground Stations	—	5.0	—
Tactical Warning Systems Support (.3)	.3	1.0	1.0
North Atlantic C^3 (.02)*	.4	.7	—
C^3 Countermeasures	7.9	4.6	7.3

Program Nomenclature	Funding ($ Millions) Fiscal Years		
	1987	1988	1989
Communications			
Milstar (.05)*	—	—	2.0
Satellite Terminals (.05)*	.5	1.0	2.0
Wideband Systems Upgrade (.05)*	3.0	2.0	3.0
Minimum Essential Emergency Comm Net (.05)*	2.0	3.0	2.0
Antijam Voice	11.7	7.0	10.3
Special Support Projects			
Scientific Technical Intelligence (.2)*	2.0	2.0	2.0
Technical Survivability Countermeasure Equipment (.2)*	2.0	.4	.2
Totals	**949.3**	**1106.3**	**995.2**

Metal gate CMOS yesterday. Silicon-On-Insulator tomorrow.

Our non-stop development of rad-hard technology is guaranteed.

American ICs for American strength...that's been our commitment since 1965. In the Polaris program. In Poseidon. In B-1B, Trident, MX/Peacekeeper, SICBM, SDI, and more. Today, in fact, we're the number-one supplier of military rad-hard integrated circuits.

Some IC companies have uncertain futures. We don't. We'll be here. American-owned and American-run. Providing products and product accountability throughout the lifetime of your systems.

And developing tough new products to keep your systems competitive...like our HS-6504RH rad-hard 4K RAM, that's produced in our MIL-M-38510, Class S certified facility.

We wrote the book on your future needs. Harris' 1987 Rad-Hard/Hi-Rel Data Book: hundreds of pages of information on what's here now and what's coming next. Ask for a copy.

Phone Harris Semiconductor Custom Integrated Circuits Division. In the U.S.: 1-800-4-HARRIS, Ext. 1908, or (305) 724-7418. In Canada: 1-800-344-2444, Ext. 1908.

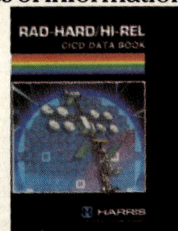

"So Harris' rad-hard data book is Class-A?"

"Yeah, and their rad-hard IC facilities are MIL-M-38510, Class S certified."

EW COMPANIES

World EW Sales: The Top 50 Companies 349
Profiles of Major U.S. EW Companies 350

JAM SESSION.

ITT Avionics' ECM Systems provide effective, reliable protection for an impressive range of military aircraft.

Wherever the American military fly, a wide range of their aircraft depend on ITT's electronic countermeasure systems for enhanced mission effectiveness and survivability.

They do it with a family of electronic countermeasure systems developed by ITT Avionics to protect all types of aircraft from hostile weapon systems.

For instance, the ALQ-172/Pave Mint system is increasing the survivability of the B-52 fleet.

The ALQ-136 lightweight jammer is providing protection for Army attack helicopters.

And the ALQ-165/Airborne Self Protection Jammer (ASPJ) will provide the highest performance electronic countermeasures for a variety of high performance tactical aircraft.

Electronic countermeasures protection from ITT Avionics—where technical excellence, sophisticated production technology and a commitment to superior product support combine to provide outstanding protection for the world's finest military force.

Avionics
500 Washington Avenue
Nutley, NJ 07110 • 201-284-5555

ITT
DEFENSE

World EW Sales: The Top 50 Companies

For the second straight year Eaton heads the list of the top 50 companies ranked by EW sales. This year there are several new additions to the list and several deletions. Mitsubishi Electric, M/A-COM and UTL make their debuts in this year's compilation. Several firms with EW revenues under $25 million in 1986 dropped off the list as the new additions raised the top-50 hurdle a little higher.

A notable feature of the 1986 list, and a factor which is already at work in 1987, is the on-going consolidation in the defense electronics industry. In 1986 Lockheed vaulted to second place largely because of its purchase of Sanders Associates. We included ARGOSystems separately this year even though that firm was recently acquired by Boeing.

While revenues grew for most of the firms in our EW top 50, many suffered declines in profits. The more competitive industry environment, greater government scrutiny of the industry in the wake of alleged fraud, waste and mismanagement, cost overruns on programs and continued slippage in the pace of contract awards all impacted profit margins in the group.

We have compiled a listing that is as complete and accurate as possible from all available sources. Any list of electronic warfare companies is bound by one's definition of EW. We have limited our definition to activities related to the acquisition, identification and disruption of electronic signals. Examples include aircraft self-protect systems, airframe modification to incorporate such systems and the components and test equipment used to manufacture and support electronic warfare equipment.

"Black," or special access programs, have made the job of compiling this year's list a little more difficult. For example, the Air Force has yet to reveal which companies are developing the self-protect system for the advanced technology bomber. Another platform and associated electronic warfare system about which we can only speculate is the so-called F-19 or stealth fighter. It is, of course, quite possible that we have picked up revenues from those programs in our listing below.

Rank	Company	1986 Sales ($ Millions)	1985 Sales ($ Millions)
1	Eaton	850	725
2	Lockheed/Sanders	755	710
3	E-Systems	718	613
4	Northrop	471	329
5	Raytheon	465	400
6	Litton	400	350
7	Loral	395	335
8	GTE-Sylvania	320	290
9	Racal	300	280
10	Westinghouse	250	225
11	ITT	245	200
12	Thomson CSF	220	270
13	Grumman	215	365
14	TRW	210	190
15	Singer	205	135
16	Watkins-Johnson	200	175
17	GEC-Marconi	184	150
18	Tracor	165	150
19	IBM	155	140
20	General Electric	150	140
21	M/A-COM	125	100
22	Elettronica	115	119
23	Motorola	110	100
24	Fairchild	106	98
25	Mitsubishi Electric	105	97
26	AEL, Inc.	105	100
27	McDonnell Douglas	105	93
28	Magnavox	100	89
29	GM/Hughes Aircraft	95	85
30	AEG-Telefunken	95	89
31	ARGOSystems	86	72
32	Kodak/Data Tape	85	75
33	Tadiran/Elisra	73	63
34	Boeing	68	60
35	IAI-Elta	67	62
36	Martin Marietta	65	55
37	Texas Instruments	65	59
38	Rohde & Schwarz	65	58
39	Ford Aerospace	65	60
40	Adams-Russell	64	46
41	General Instruments	62	55
42	Selenia	62	56
43	United Industrial/AAI	60	60
44	Phillips	50	40
45	Hollande Signaalapparaten	50	46
46	Tech-Sym	38	36
47	Avantek	35	33
48	Tenneco/Sperry	35	35
49	Alpha Industries	30	27
50	UTL	28	30

Profiles of Major U.S. EW Companies

Consolidation continues in the industry.
Here's closer look at some of the biggest names in U.S. electronic warfare.

By Byron K. Callan

1986 was significant because of the acceleration in mergers and acquisition that took place in the industry. The landmark merger was Lockheed's "white knight" purchase of Sanders Associates for $1.2 billion after Loral launched an unfriendly takeover attempt. Dalmo Victor, in a $174 million deal, was purchased by Singer from Textron.

1987 shows no sign that this trend is letting up:
• ARGOSystems has agreed to be purchased by Boeing for $275 million.
• Electrospace Systems was bought by Chrysler.
• Loral acquired Goodyear Aerospace operations following Sir James M. Goldsmith's hostile takeover attempt of Goodyear Corp.
• We would be hard pressed to come up with the name of a defense electronics company that has not been suggested as a possible takeover candidate in the past two years.

What is interesting about the pattern of acquisitions so far in 1987 is that companies that were *not* in the midst of significant program problems decided to sell to larger companies. In 1986, both Sanders Associates and Hazeltine experienced significant *declines* in operating profits from program problems, in the case of Sanders, and program cancellations, in the case of Hazeltine.

However, ARGOSystems was on the verge of reporting record revenue and profits for its fiscal 1987 ending in June and Electrospace was well on the road to recovery after hitting problems on some fixed price contracts. Management's ownership of stock at these two firms was relatively higher than many other defense electronics firms and their decision to sell came as a surprise to many in the investment community.

We expect further consolidation in the industry. In part, this is a reflection of the wider uptick in merger and acquisition activity sweeping the U.S. economy. But there are also several factors unique to the defense industry that have, in the eyes of corporations and investors, increased the pace of mergers and acquisitions.

Flurries of merger fever have swept the U.S. defense industry before. In the early 1960s, following the buildup of strategic missile forces, several electronics firms were snapped up during the budget downdraft that followed. The wind-down from the Vietnam War in the early 1970s also resulted in an acceleration in merger activity.

Factors driving consolidation today include the following:

Defense Electronics Still a Growth Market. The Electronic Industries Assn. projects that the electronics content of the U.S. defense budget will increase modestly even though overall budget growth will be sluggish in the years to come.

Fewer Airborne EW Programs. The award of Inews Phase II development contracts to teams headed by TRW/Westinghouse and Lockheed/General Electric in June 1986 and continued congressional criticism of duplicative EW efforts underscore the trend towards fewer different types of aircraft self-protect systems. While opportunities certainly exist in upgrade markets and in applications of EW to drones, satellites and missiles, we sense that the number of players in aircraft EW markets will continue to shrink, or at least change in composition.

Vertical and Horizontal Integration. Companies can better penetrate new markets and synthesize capabilities by purchasing firms that complement their existing strengths. For example, we would not at all be surprised to see Boeing marry the capabilities of ARGOSystems with airframes built by its de Havilland aircraft operation in a bid to increase exposure in low-cost airborne electronic surveillance markets.

Technological Change. The very high speed integrated circuit (VHSIC) and milimeter wave and microwave monolithic integrated circuit (MIMIC) programs, coupled with the increase in use of multi-function assemblies, is driving consolidation in the microwave component industry. These trends, plus delays in new electronic warfare program starts and current industry overcapacity have given rise to the merger and acquisition ferment in this industry in the past year or so.

Hostile takeovers of defense electronics firms have been, and we feel will continue to be quite rare. This is an engineering industry and the "assets" walk out the door every evening. If those "assets" decide that the hostile acquiror is not acting in their best interests, there is little to stop them from looking for work elsewhere. The hostile acquiror would be left with little of value.

We are not terribly worried about continued consolidation in the industry. Small firms will continue to thrive by offering innovative, cost-effective solutions to the challenges of electronic warfare. While an acquisition is certainly disruptive to the lives of those involved, there are instances where a new infusion of management blood can get programs back on track. Although the industry is becoming more concentrated, technology, the threat and the customer have in the past coalesced to unseat the then current industry leader. We don't expect that pattern to change. ∎

Byron Callan is a defense electronics analyst for Prudential-Bache Securities, Inc., in New York City.

EATON

AIL DIVISION
Commack Road, Deer Park, NY 11729
(516) 595-5000

Eaton AIL's bread and butter program remains the ALQ-161 for the B-1B bomber. Although revenues from this program will probably peak in 1987, the ALQ-161 should be a continuing source of business in the years to come. Planning includes an upgrade of the system to deal with the monopulse threat posed by advanced Soviet anti-aircraft systems.

The ALQ-161 figured prominently in the news this past year owing to problems encountered when the full system was deployed. However, these problems now appear to be under control. The company has maintained the full production rate of four systems a month for more than a year now.

The ALQ-161 has been dubbed the "Rolls Royce" of EW self-protect systems. Weighing close to 5,200 pounds, the system consists of 100 line replacable units and uses 120 kilowatts of power.

Other major AIL programs include the ALR-77 ESM system for P-3C Orion anti-submarine warfare aircraft and the ALQ-99E upgrade for the EF-111 and a slightly different version of the same system called the ALQ-99 ICAP for Navy EA-6Bs.

Eaton was one of the competitors for the aborted electronic countermeasures upgrade on the F-111 fleet. The company had teamed with Boeing for this program.

Eaton's AN/ALQ-99 tactical jamming system is used in the U.S. Navy EA-6B and Air Force EF-111A electronic warfare aircraft.

EW SPECIALTIES
System Integration
Receivers
RF Jammers
Self-protect systems
ESM

MAJOR EW PROGRAMS
ALQ-161
ALR-77
ALQ-99E
ALQ-99 ICAP

PRIMARY EW CUSTOMERS
U.S. Air Force
U.S. Navy
International

E·AT·N

LOCKHEED/SANDERS ASSOCIATES

LOCKHEED AUSTIN DIVISION
P.O. Box 17100, Austin, TX 78760
(512) 448-5555

LOCKHEED ELECTRONICS CO.
1501 U.S. Hwy 22, C.S. #1, Plainfield, NJ 07061
(201) 757-1600

SANDERS FEDERAL SYSTEMS GROUP
P.O. Box 2004, Nashua, NH 03061
(603) 885-5603

SANDERS CANADA
2421 Lancaster Rd., Ottawa, Ontario, Canada K1B 4L5
(613) 738-4500

Last summer Lockheed played "white knight" and agreed to acquire Sanders Associates for $1.2 billion after Loral attempted an unfriendly takeover. The purchase significantly bolstered Lockheed's presence in electronic warfare markets.

Sanders' largest program is the ALQ-126B self protect system, which is being procured for U.S. Navy, Australian, Spanish and Canadian McDonnell-Douglas F-18 aircraft. Sanders has delivered more than 250 systems so far and the system was combat-proven during the air raid on Libya of April 14, 1986. Future production lots of the ALQ-126B are destined for Navy F-14 and A-6 aircraft.

Other Sanders electronic warfare programs include spares and support for the ALQ-137 now used on F-111s and several infrared warning and countermeasures systems, notably the ALQ-156, ALQ-144 and ALQ-147. In 1986 Sanders received a $120 million multiyear contract to provide spares for the ALQ-137s in service through 1990.

Sanders is the lead contractor on the ALQ-149 communications jammer, which the Navy plans to procure for use on EA-6Bs. This program had been restructured in Fiscal 1986 and is a key component of the upgrade to the EA-6. Apparently, the existing communications jammer is incompatible with the ALQ-99.

The company has long been active in tactical intelligence gathering programs for the Army and also provides electronic support measure systems for Navy surface vessels.

Lockheed's largest unclassified EW program ran into some flak in 1986. The precision location strike system (PLSS) was all but cancelled by the Air Force. However, a low-level testing effort continues and both the House and Senate in their mark-ups of the 1988 DOD budget requests included funds to extend evaluation of PLSS.

We also assume that Lockheed earns revenues from airframe modification and integration programs. Activity is continuing on EP-3 airframes and the company installs ESM equipment on P-3 Orions. It's probably a safe bet that whatever self-protect system the F-19 carries is installed by Lockheed.

Lockheed is also working on an innovative program to provide tactical transport aircraft with a self-protect system. Called SATIN (survivability augmentation for transport installation now), Lockheed is using off-the-shelf ALE-40s, ALR-69s and ALQ-156s to demonstrate a modular system that could easily be installed on C-130s, and conceivably other transport aircraft, operating near hostile zones. The company is also working on a similar modular approach providing C-130s with ESM capabilities.

Protection against heat-seeking missiles is the primary purpose of the AN/ALQ-144 infrared countermeasures pod made by Sanders.

EW SPECIALTIES
RF jammers
RWRs
IRCM
ESM
System integration

MAJOR EW PROGRAMS
ALQ-126B
ALQ-137
ALQ-156
ALQ-144
ALQ-147
ALQ-149
PLSS
P-3C
EP-3
SATIN
C-130 programs

PRIMARY EW CUSTOMERS
U.S. Air Force
U.S. Navy
U.S. Army
International

Lockheed

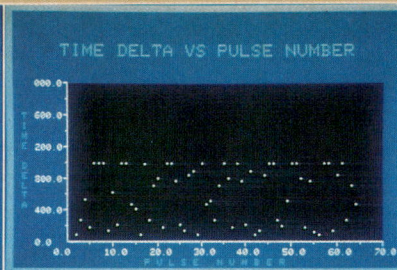

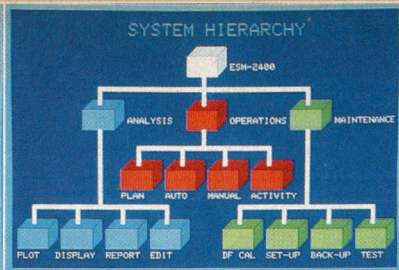

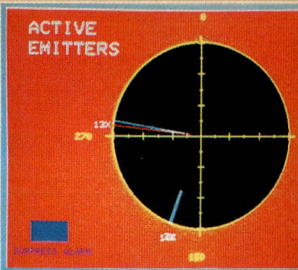

E-SYSTEMS

GARLAND DIVISION
P.O. Box 660023, Dallas, TX 75266
(214) 272-0515

GREENVILLE DIVISION
P.O. Box 1056, Greenville, TX 75401
(214) 455-3450

MELPAR DIVISION
7700 Arlington Blvd., Falls Church, VA 22046
(703) 560-5000

MEMCOR DIVISION
P.O. Box 23500, Tampa, FL 33630
(813) 885-7000

Much of E-Systems' electronic warfare work is in the realm of electronic support measures. The company has indicated that it sees significant opportunities in tactical EW/ESM markets and has long been an exponent of bolstering capabilities on the electronic battlefield.

E-Systems reports revenues from a broad range of activities under electronic warfare. Much of the company's classified work is located in this segment, which accounted for 63 percent of total 1986 sales and 66 percent of 1985 sales. We gather that intelligence data management programs are included here; so for purists, the extent of E-Systems' EW activities are probably overstated in the top 50 listing.

Included in this reporting segment are sales from some of the more familiar E-Systems programs such as the SLQ-50 battle group passive horizon extension system (BGPHES) terminals and the Wild Weasel upgrade. The Navy appears to have settled on using S-3As and has dedicated carrier-based Sigint platforms to replace the venerable EA-3 "Whales." E-Systems will probably play a role in that portion of BGPHES destined for the converted S-3As. The company's Melpar division delivered the first surface terminal to the Navy earlier this year.

Under subcontract from McDonnell-Douglas, E-Systems is developing a new receiver-processor group for the F-4G Wild Weasel fleet. Follow-on opportunities include the replacement planned for the F-4G in the mid-1990s.

Upgrades to the RC-135 fleet continue to be a major source of revenues for the company and we gather that E-Systems is working on similar programs for foreign customers. E-Systems' Greenville facility also is probably performing modification work on other airborne Sigint and reconnaissance platforms.

Precision location strike system (PLSS) is another program E-Systems is participating in as a subcontractor to Lockheed. The outlook here is "iffy" at best. Congress appears bent on keeping this program alive by funding continued low-level R&D efforts to further demonstrate the capabilities of PLSS, but in the current tight budget environment, funding for production could be difficult to come by.

E-Systems has long been active in mini-RPV markets and offers a variety of electronic payloads. The company was a subcontractor on the now-suspended Army competition for the intelligence/electronic warfare unmanned air vehicle (IEWUAV).

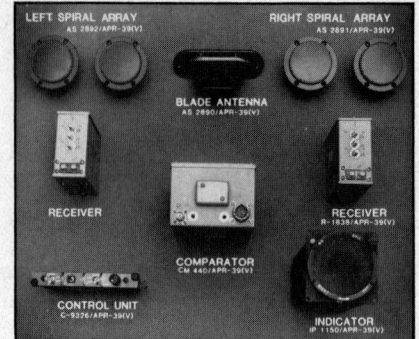

Army helicopters and some U.S. Navy/Marine Corps helicopters carry the AN/APR-39 RWR package as standard.

EW SPECIALTIES
System integration
RWRs
Reconnaissance
Mini-RPV payloads

MAJOR EW PROGRAMS
SLQ-50
S-3A
F-4G upgrade
RC-135 upgrade
PLSS

PRIMARY EW CUSTOMERS
U.S. Air Force
U.S. Navy
U.S. Army
International

E-SYSTEMS

NORTHROP

DEFENSE SYSTEMS DIVISION
600 Hicks Road, Rolling Meadows, IL 60008
(312) 259-9600

Northrop's Rolling Meadows, Ill., facility is the home of the company's electronic warfare programs. The company is a leading manufacturer of active countermeasures systems.

Additionally, we assume that the company is receiving funds to install self-protect systems on the advanced technology bomber and other classified aircraft programs. The 1985 and 1986 revenue figures on the top 50 ranking are from the firm's annual report and we have not been able to ascertain if some of the ATB effort is included in the numbers provided by Northrop. If not, the firm's total EW earnings are probably understated.

The largest program is the ALQ-135 jammer for the F-15 aircraft. Northrop and the Air Force are currently testing an upgraded version of that jammer for the TEWS (tactical electronic warfare system), which is planned for retrofit onto F-15Cs and will be installed on F-15Es.

Earlier this year, Northrop received the go-ahead from the U.S. Navy to begin production of ALQ-162 "Shadowbox" continuous wave jammers. This lightweight jammer has already been ordered by the Danish Air Force and the U.S. Navy plans to install it on F-4s, A-7s and A-4s. Those F/A-18s using the ALQ-126B are also likely candidates for installation of the ALQ-162. The 40-pound system can has been evaluated by the Army.

Northrop produces several components of other electronic warfare systems. The firm is providing jamming transmitters for GEC Marconi's Zeus ECM pods, which are being procured by the RAF for Harrier aircraft. Northrop is also making high-power transmitters for Eaton's ALQ-161 and is delivering the ALT-28 power management system for Air Force B-52 bombers.

Northrop continues to make progress with sales of the internally developed ALQ-171 jammer. This system, which is offered in internal, conformal or pod mounted configurations, is being primarily marketed to those nations flying F-5 series jets and has been ordered by Switzerland and possibly South Korea. In early 1987 the company scored a unique coup with a Saudi Arabian order of ALQ-171s for use of Saudi F-15s.

Northrop was one of three finalists on the Have Charcoal program, which called for IRCM systems to protect high-value fixed-wing transport and special mission aircraft. While Have Charcoal is now in limbo, Northrop has demonstrated its entry, called MIRTS (modular infrared transmitting set) to the RAF and will likely participate in future U.S. tactical IRCM programs. MIRTS is a derivative of Northrop's AAQ-8.

The company has set up a small operation based in Annapolis, Md., to provide signal intelligence analysis software and hardware.

The AN/AAQ-4 infrared countermeasures system electronically modulates a visual cesium infrared source to produce a jamming signal.

EW SPECIALTIES
RF receivers
RF transmitters
Digital processors
IRCM
Integrated EW systems

MAJOR EW PROGRAMS
ALQ-135
ALQ-162
ALT-28
ALQ-171
Have Charcoal

PRIMARY EW CUSTOMERS
U.S. Air Force
U.S. Navy
U.S. Army
International

NORTHROP

Defense Systems Division
Electronics Systems Group

RAYTHEON

SEDCO DIVISION
65 Marcus Dr., Melville, NY 11747
(516) 694-7440

ELECTROMAGNETIC SYSTEMS DIVISION
6380 Hollister Ave., Goleta, CA 93117
(805) 967-5511

MICROWAVE & POWER TUBE DIVISION
190 Willow St., Waltham, MA 02254
(617) 899-8400

Raytheon is a player in air and naval ECM markets. The company virtually dominates the surface ship electronic warfare market by virtue of production of the SLQ-32(V). More than 190 systems have been delivered to date. A Navy initiative to have this program second-sourced is still underway, although RFPs are not expected before late summer 1987.

Raytheon's Sedco division makes antennas for the B-1B's ALQ-161 and is also developing phased array antenna systems.

Production continues on the ALQ-142, which is a derivative of the SLQ-32 and is employed on Navy LAMPS helicopters. Used primarily for signal intelligence, the ALQ-142 can also help target long-range cruise missiles. Raytheon has proposed the ALQ-142 for installation on small naval vessels such as fast attack craft.

Raytheon is participating on the ALQ-184 update of the ALQ-119 jamming pod. The company is updating the pod with digital technology and Rotman electrically scannable antennas. More than 1,600 ALQ-119s are in use worldwide.

Electronic components and subsystems capabilities back up the company's extensive EW skills. Products include multibeam lens arrays antennas, solid state FET devices and digital RF memories.

Raytheon's AN/SLQ-32(V) is a shipboard ESM/ECM system, a modularly related family of EW suites for defense against cruise missiles.

EW SPECIALTIES
RF jammers
System integration
Transmitters
Antennas
ESM systems

MAJOR EW PROGRAMS
SLQ-32(V)
ALQ-161
ALQ-142
ALQ-184

PRIMARY EW CUSTOMERS
U.S. Navy
U.S. Air Force
U.S. Army

Raytheon

LITTON

APPLIED TECHNOLOGY DIVISION
4747 Hellyer Ave., P.O. Box 7012
San Jose, CA 95150-7012
(408) 773-0777

AMECOM DIVISION
5115 Calvert Rd., College Park, MD 20740
(301) 864-5600

LASER SYSTEMS DIVISION
P.O. Box 547300, Orlando, FL 32854-7300
(305) 295-4010

ELECTRON TUBE DIVISION
960 Industrial Rd., San Carlos, CA 94070
(415) 591-8411

The AN/ALR-67(V) made by Litton Applied Technology is a U.S. Navy tactical threat-warning system.

Litton has been a dominant player in tactical threat warning and electronic support measures markets. The company's Applied Technology division is the leading manufacturer of radar warning receivers for tactical combat aircraft. The company makes the ALR-67 for the Navy and the ALR-69 and ALR-45F for the U.S. Air Force. All three systems have been exported. ALR-67s are being procured by Spain, Australia and Canada for their F/A-18s. ALR-69s are in service with Australian F-111s and the ALR-45F was recently ordered by the Swiss, among others.

Litton has been the leading supplier of threat warning systems for West German tactical aircraft. The ALR-68 updates West German F-4s and RF-4s by replacing ALR-46 units. Litton is also providing the Luftwaffe with radar warning receivers for Tornado aircraft.

Development and testing continues on the ALR-74, which will eventually replace the ALR-69s now in service. The Air Force is currently evaluating prototype models submitted by Litton and Loral's ALR-56M (a reconfigured version of the ALR-56C) and a decision on whether to go ahead with a competition is pending.

A final decision on which system the Air Force will procure (assuming testing continues) will not take place until the fall of 1988. In the interim, there has been talk of an upgrade program for the ALR-69, but that may run afoul of tight budgets and congressional criticism over duplicative EW programs.

Litton's Applied Technology division is responsible for these radar warning receiver programs. Additionally, the division is a leader in test, training and simulation of electronic warfare systems.

The company is under subcontract from Grumman on the ADVCAP program to design a new receiver/processor group for Navy and Marine Corps EA-6Bs. Now in full-scale development, production deliveries should commence in 1989. More than 200 systems are required.

Litton's Amecom division makes the ALR-73 ESM system for U.S. Navy and is now delivering production units. The ALR-73 is an upgrade to Litton's ALR-59 and this system should remain in production into the early 1990s.

Amecom also has provided submarine ESM systems, namely the BLD-1 for SSN-668-Class submarines. Deliveries continue on this program.

Litton has been actively building its components operations and electro-optic capabilities through acquisitions. In the first half of 1987 the company acquired microwave component operations from Harris and Gould. In 1986 the company purchase Parks-Jagger and International Laser Systems. Both firms could contribute to laser threat warning systems that will probably become part of future threat warning systems.

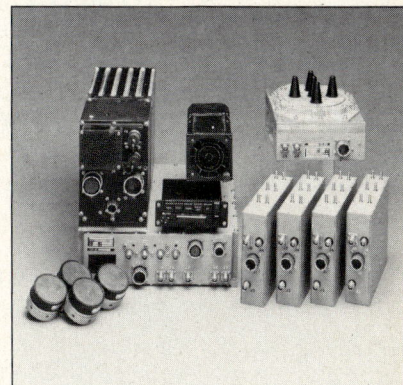

EW SPECIALTIES
RWRs
ESM systems

MAJOR EW PROGRAMS
ALR-67
ALR-69
ALR-45F
ALR-68
ALR-74
EA-6B ADVCAP
ALR-73
BLD-1

PRIMARY EW CUSTOMERS
U.S. Air Force
U.S. Army
U.S. Navy
International

LORAL

LORAL ELECTRONIC SYSTEMS
Exit 6A Ridge Hill, Yonkers, NY 10710
(914) 968-2500

LORAL ELECTRO-OPTICAL SYSTEMS
P.O. Box 7101, Pasadena, CA 91109
(818) 351-5555

ROLM MIL-SPEC COMPUTER DIVISION
1 River Oaks Place, San Jose, CA 95134
(408) 432-8000

LORAL HYCOR
10 Gill St., Woburn, MA 01801
(617) 935-5950

RANDTRON SYSTEMS
130 Constitution Dr., Menlo Park, CA 94025
(415) 326-9500

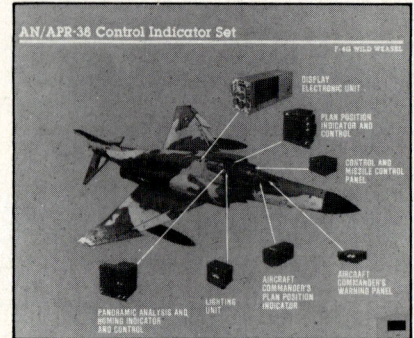

Loral Electronic Systems makes the AN/APR-38 radar homing and warning receiver installed on the F-4G Wild Weasel.

Loral, a broadly diversified defense electronics company, has extensive electronic warfare capabilities owning to its high degree of vertical integration.

The company is a leading manufacturer of radar warning receivers and other self-protect systems. The two largest programs in this area of activity are the ALR-56C update of the ALR-56A for F-15 fighter aircraft and Rapport III, which is a fully integrated electronic self-protect system the company offers to foreign air forces.

Loral is now delivering ALR-56C systems to the Air Force and, in the spring of 1987, shipped a modified ALR-56C, designated the ALR-56M, to the Air Force. The Air Force plans to evaluate the ALR-56M and Litton's ALR-74 as replacements for ALR-69s now in service. This is potentially a $1-billion-plus program for the two companies. The Navy may also participate on this program.

To date, Rapport III has been ordered by a friendly Middle Eastern air force for its F-16s. Belgium plans to procure the system, but has not yet set aside necessary funding in its national budget. Rapport III is actively being marketed to other nations operating or planning to purchase F-16s, notably Turkey.

In infrared countermeasures systems, Loral is one of three finalists on the now-dormant Have Charcoal program, and apparently has met with some success selling IRCM systems to overseas customers. The company is also updating ALQ-123 pods for the Navy and provides ALQ-157 jammers for Marine Corps CH-46 and CH-53 and Army CH-47 helicopters.

Loral is a significant player in microwave components markets. Its Narda, Frequency West and Frequency Sources operations contributed an estimated $100 million in calendar 1986 revenues.

In 1986 the company continued to grow total revenues through acquisitions, purchasing Rolm's Mil-Spec computer business from IBM. However, 1986's acquisition activity did little for electronic warfare sales. In 1985 Loral bolstered its EW exposure in naval markets and expendable countermeasures skills by purchasing Hycor, Inc.

In January 1987, Loral announced that it had reached an agreement to acquire Goodyear Aerospace and has since renamed those operations Loral Systems Group.

The Goodyear acquisition complements a number of electronic warfare activities at Loral. Loral Systems designs, develops and manufactures airborne expendable countermeasures systems, notably the ALE-39. The ALE-39 is a chaff/infrared flare dispenser which is used on Navy fixed-wing aircraft and helicopters. Activity is expected to continue on this program. However, the planned ALE-47 could be procured if the Air Force and Navy decide to standardize their airborne expendable countermeasures requirements. Loral Systems is competing for the ALE-47.

Loral announced in mid-1987 that it had formed a joint venture with IAI/Elta of Israel to market an Elta-developed airborne self-protect system, believed to be a tail-warning system for F-16 aircraft.

EW SPECIALTIES
RWRs
IR Countermeasures
Microwave components
Expendable CM

MAJOR EW PROGRAMS
ALR-56C/M
Rapport III
Have Charcoal
ALQ-123
ALQ-157
ALE-39

PRIMARY EW CUSTOMERS
U.S. Air Force
U.S. Navy
International

LORAL
CORPORATION

GTE

GOVERNMENT SYSTEMS, WESTERN DIVISION
100 Ferguson Dr., Mountain View, CA 94043
(415) 966-2000

Much of GTE's electronic warfare and support measures work is classified. Most of GTE's known programs relate to the U.S. Navy.

The Navy announced in late 1986 that it intends to seek second sources for two GTE naval ESM programs, the WLR-8 and WLQ-4. The WLR-8 is now in production for SSN-688 and Trident-class submarines while the WLQ-4 is for SSN-637-class submarines. GTE had developed and produced the WLR-8 for a variety of surface ships, but the Navy opted in 1983 for the WLR-1, built by Boeing's ARGOSystems subsidiary.

GTE's Government Systems, Western Division, provided the ALR-60 "Deep Well" communications intercept system used on Navy EP-3 signal intelligence aircraft. It is conceivable that upgrades to this system provide an on-going source of revenues to GTE. Another program for which GTE was the prime contractor was the ALQ-150 Cefire Tiger for the U.S. Army.

GTE produces a range of subsystems for EW and ESM systems including quick erecting antenna masts, direction finders and digital voice recorders.

In the spring of 1987 the company announced its intent to dispose of its Tempe, Ariz.-based semiconductor facility.

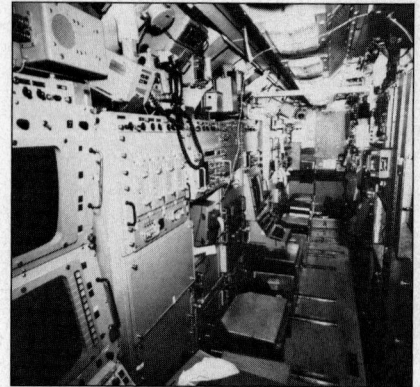

The AN/WLQ-4(V) is a GTE Government Systems submarine ESM system.

EW SPECIALTIES
ESM systems
Comm jammers
E/O warning
Receivers
Antennas
Signal processors
Decoys

MAJOR EW PROGRAMS
WLR-8
WLQ-4
ALR-60
ALQ-150

PRIMARY EW CUSTOMERS
U.S. Navy
U.S. Army
U.S. Air Force
DOD

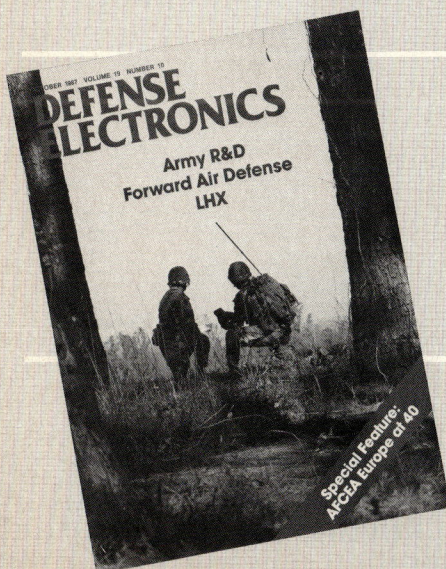

RACAL

RACAL RADAR DEFENCE SYSTEMS
9 Davis Rd., Chessington, Surrey, KT9 1TB, England
01-397-5281

RACAL COMMUNICATIONS INC.
WEST COAST OFFICE
1044 Renoir Court, Sunnyvale, CA 94087
(408) 773-8478

RACAL COMMUNICATIONS INC.
EAST COAST OFFICE
5 Research Place, Rockville, MD 20850
(301) 948-4420

Racal is a U.K.-based diversified electronics company with interests in telecommunications, commercial security systems and defense markets. The company addresses ESM and ECCM markets.

Racal manufactures a major portion of the Skyshadow jamming pod used on Tornado aircraft. The company's Jaguar frequency-hopping ECCM tactical radio system includes high sensitivity monitoring and surveillance receivers. Racal Communications, located in Rockville, Md., provides the ULQ-19(V) tactical communications jammer. This 100-watt jammer system covers the entire combat net radio band.

Racal Radar Defense Systems provides the Cutlass shipboard ESM system for surface vessels and a derivative called Porpoise for submarines. Cutlass interfaces with another Radar Defense Systems shipboard ECM system called Cygnus. Cygnus is also offered in a land-based configuration.

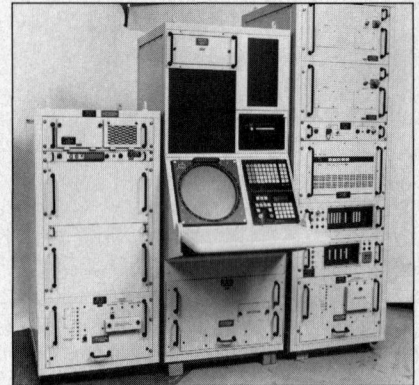

Racal's Cutlass shipboard ESM system is an automated system designed to operate in a dense radar environment.

EW SPECIALTIES
Jamming pods
ECCM systems
RF jammers
Shipboard ESM

MAJOR EW PROGRAMS
Skyshadow
Jaguar ECCM
ULQ-19(V)
Cutlass

PRIMARY EW CUSTOMERS
U.S. services
International

RACAL

WESTINGHOUSE

EW DIVISION
P.O. Box 746, Baltimore, MD 21203
(301) 765-1000

ILS DIVISION
111 Schilling Rd., Hunt Valley, MD 21030
(301) 584-1000

ADVANCED DEVELOPMENT DIVISION
P.O. Box 746, Baltimore, MD 21203
(301) 765-1000

ADVANCED TECHNOLOGY DIVISION
P.O. Box 1521, Baltimore, MD 21203
(301) 765-1000

The AN/ALQ-119(V) family of jamming pods operates on a variety of U.S. Air Force fighters and attack aircraft.

Westinghouse has been a leader in active electronic countermeasures markets and will remain a major EW player through the 1990s. The company is engaged in several key aircraft EW programs and has a long history as a developer and manufacturer of airborne jamming systems.

Westinghouse and ITT are teamed on the ALQ-165 ASPJ program. Plans call for procurement of 3,000 ALQ-165s for Air Force and Navy tactical combat aircraft. The Navy flinched on the program early in 1986, but is now back on board.

Key features of the ALQ-165 include ease of maintenance owing to use of replaceable assemblies. While this program has seen schedule slippage, the DOD's Joint Requirements and Management Board gave the go-ahead in early 1987 for the production verification phase of the ALQ-165.

The major production program now underway is the ALQ-131 for the U.S. Air Force and foreign customers. The company has delivered several hundred ALQ-131 Block 1 pods to the Air Force for use on F-16, A-10, F-4 and F-111. Block II pods, which completed testing in 1986, have been procured by the Air Force. A number of foreign customers have also expressed interest. The ALQ-131 is a beneficiary of VHSIC and Phase I chips have been put into the pod's receiver-processor unit.

Westinghouse also makes the ALQ-153 pulse Doppler tail warning radar for the B-52G/H aircraft. The ALQ-153 automatically manages on-board expendable countermeasures systems and has been offered for installation on tactical aircraft.

The company is also a leader in electro-optical countermeasures and is developing the Compass Hammer ALQ-179 pod for the Air Force and the Navy.

Westinghouse is teamed with TRW on the integrated electronic warfare system (Inews) program. Westinghouse's extensive experience in active countermeasures complements TRW's experience in passive systems. Other team members include Honeywell, Tracor and Perkin-Elmer.

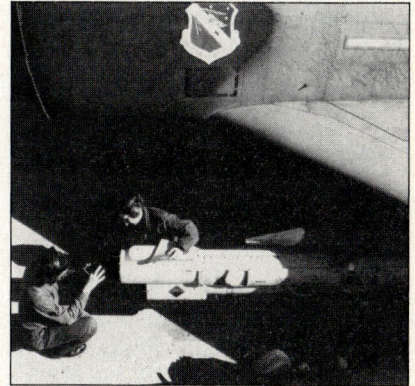

EW SPECIALTIES
RF jammers
Power management
Tail warning radar
E/O pods

MAJOR EW PROGRAMS
ALQ-165
ALQ-131
ALQ-153
ALQ-179
Inews

PRIMARY EW CUSTOMERS
U.S. Air Force
U.S. Navy
International

ITT

AVIONICS DIVISION
500 Washington Ave., Nutley, NY 07110
(201) 284-0123
ELECTRON TECHNOLOGY DIVISION
P.O. Box 100
Easton, PA 18044-0100
(215) 252-7331

ITT's electronic warfare capabilities are concentrated in active airborne counter-measures systems and company participation in that area should be expected for years to come.

The largest ITT program today is the ALQ-172 update of the ALQ-117. The system is now in now in final testing and the bulk of systems produced will be primarily for SAC B-52s, but some also are destined for variants of the C-130s, notably the MC-130 Combat Talons, AC-130 gunships and EC-130 Compass Call ECCM aircraft. The ALQ-172 provides more jamming power, enhanced ECM capabilities, better reliability and self-testing.

ITT is teamed with Westinghouse for the development of the ALQ-165, which the Air Force and Navy are planning to procure to replace a plethora of internal and podded jammers used on tactical aircraft. The team has had difficulty overcoming technical and design problems, but, with a DOD decision to start low-level production earlier this year, it appears that the worst of the problems are behind the team.

With requirements for up to 3,000 ALQ-165s for Navy and Air Force tactical aircraft, this could be a multi-billion dollar program for ITT and Westinghouse. Export orders are likely and the Air Force has explored use of ALQ-165 in a podded configuration to provide advanced countermeasures capabilities for aircraft not "wired" for the internally-mounted system.

ITT is delivering the ALQ-136 jammer to the Army for installation on helicopters and propeller fixed wing aircraft. The ALQ-136 is primarily concerned with deceiving Soviet "Gun Dish" radars, which are part of the ZSU-23 anti-aircraft system. This is a mature program and has probably passed peak revenue levels.

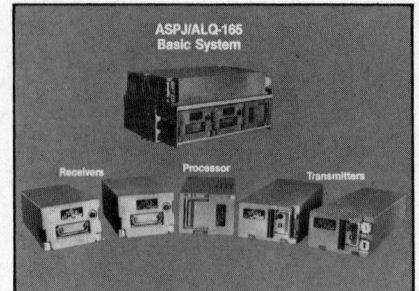

The AN/ALQ-165 is a joint Navy-Air Force effort to provide self-protection jamming for new tactical aircraft.

EW SPECIALTIES
RF jammers
EO/IR warning
System integration

MAJOR EW PROGRAMS
ALQ-172
ALQ-165
ALQ-136

PRIMARY EW CUSTOMERS
U.S. Air Force
U.S. Navy
U.S. Army

ITT

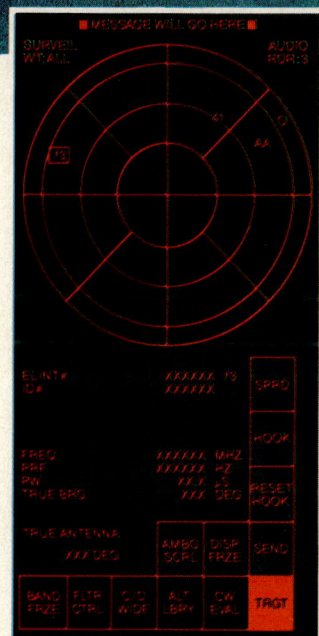

THOMSON-CSF

AEROSPACE GROUP
178 Bld Gabriel Peri
92242 Malakoff Cedex, France
33-4-46554422

ELECTRON TUBE DIVISION
38 Rue Vauthier-BP 305
F-92102 Boulogne-Billancourt, Cedex, France
33-1-46048175

TELECOMMUNICATIONS DIVISION
66 Rue du Fosse Blanc
92231 Gennevilliers, France
1-47918000

Thomson-CSF is a diversified French company that addresses defense electronics, medical equipment, semiconductor and engineering and industrial products markets. Most of the firm's EW work is handled by the Avionics Division, which also designs and manufactures a variety of radar systems. The Detection, Control and Communications Systems Group provides ESM and ECCM products.

The company has been the key supplier of self-protect systems to the French Air Force and to nations operating French-built aircraft. Early programs include the BU and B2 radar warning receivers for Mirage III and BR systems for Jaguar aircraft.

On-going EW programs include the Barem pod-mounted self-protection jammer and the Sherloc radar warning receiver. Barem is a reprogrammable system and is now in service on Mirage aircraft in the French Air Force. The company also is offering a fully integrated countermeasures system that coordinates warning, jamming and decoy functions. Called ICMS, the system is produced in conjunction with ESD and Matra.

In ground and naval ESM, Thomson offers a variety of systems. The DR 2000 is an ESM system now in use on French surface combatants and submarines.

Thomson is a vertically integrated defense electronics company and makes a variety of microwave components, traveling wave tubes and electronic subassemblies.

Thomson-CSF Avionics Division makes the SYREL electronic reconnaissance pod now operational in the French air force Mirage 2000.

EW SPECIALTIES
RWRs
ASPJs

MAJOR EW PROGRAMS
ICMS
Sherloc RWR
BU RWR
B2 RWR
Barem SPJ
DR 2000 ESM system

PRIMARY EW CUSTOMERS
International

THOMSON-CSF

GRUMMAN

AIRCRAFT SYSTEMS DIVISION
Bethpage, NY 11714
(516) 575-0574

AEROSTRUCTURES DIVISION
Bethpage, NY 11714
(516) 575-0574

While Grumman does not build stand-alone electronic warfare equipment, it has played a leading role as an integrator of such equipment on airframes. Its two leading programs are the EF-111 and EA-6, but the company is also responsible for integrating electronic self-protect systems on F-14 and A-6 airframes.

Grumman lost ground in the top 50 rankings owing to the wind-down in the EF-111 program. For now, the Air Force appears content with 42 EF-111s in inventory. However, the success of EF-111s used in the April 14, 1986, raid on Libya had given rise to speculation that additional EF-111s would by ordered for U.S. Air Force for assignment in the Pacific.

In related F-111 airframe activity, Grumman was teamed with Lockheed/Sanders on the upgrade of F-111 electronic countermeasures systems. The Air Force cancelled this program last January when contractor bids came in 20 percent to 70 percent higher than the funds budgeted for the program.

Other major programs include the EA-6B and E-2C. The Navy is planning to buy 100 ADVCAP EA-6s with deliveries commencing in 1992. The E-2C will continue to be a source of funds for Grumman with the Navy planning to buy six a year through the next five years. A significant modification of the E-2C has been proposed, replacing the familiar rotodome with conformal antenna arrays.

Jamming enemy radar and communications is the role of the Grumman EA-6B Prowler tactical jamming system aircraft.

EW SPECIALTIES
System integration
Airborne self-protect systems

MAJOR EW PROGRAMS
EF-111
EA-6
E-2C

PRIMARY EW CUSTOMERS
U.S. Air Force
U.S. Navy

GRUMMAN

TRW

TRW/ESL
495 Java Dr., Sunnyvale, CA 94088
(408) 738-2888

TRW MICROWAVE
825 Stewart Dr., Sunnyvale, CA 94086
(408) 732-0880

TRW MILITARY ELECTRONICS DIVISION
1 Rancho Carmel, San Diego, CA 92128
(619) 592-3055

The ESL division of TRW makes the AN/ALQ-151 Quick Fix system, an airborne EW system for DF, intercept and ECM deployed on Army EH-1H and EH-60A helicopters.

TRW is poised to become conceivably one of the larger players in airborne electronic self-protect markets in the 1990s.

The win on Inews (TRW is teamed with Westinghouse), participation on VHSIC and, more recently, on Mimic (Microwave and Millimeter-Wave Monolithic Integrated Circuits) and the company's extensive experience in spaceborne systems all indicate very strong competitive capabilities. While Inews participation was important in a strategic sense, the program won't be a meaningful contributor to the corporation's defense and space activities for at least another five years.

TRW is one of the VHSIC Phase II contractors and is now delivering Phase 1 1.25-micron devices for insertion on a variety of defense electronic systems. Under Phase II, TRW is developing complete signal processing systems on a single chip. Prototype chips, with circuit feature sizes as low as 0.5 microns should be available in the early 1990s.

TRW's ESL division is a leading supplier of Army ESM and EW equipment. ESL is currently providing Guardrail RC-12D aircraft to the Army: this is the largest Army tactical program at ESL. ESL is the systems integrator on this program in addition to providing mobile ground terminals.

Another ESL program is Trailblazer, TSQ-114A. Trailblazer is a local or remote-controlled ground-based communications intercept and direction finder. The tracked shelter contains electronics, an operator position and a 50-foot mast group. Trailblazer revenues should peak in 1987.

ESL also provides electronics packages for remotely piloted drones. ESL was part of the team led by California Microwave competing for the now-shelved Army IEWUAV program.

TRW manufactures microwave components. However, in June 1987, the company announced that it was divesting itself of these operations. As we went to press, the company apparently had agreed to sell its microwave operations to Frequency Electronics.

We should note the company's extensive experience in space-based surveillance programs. It is possible that TRW has played a role on ECM systems development efforts to protect satellites. We have not factored revenues from these activities, if in fact they exist, into our top 50 revenue figures.

EW SPECIALTIES
ESM systems
EW subsystems
EW components

MAJOR EW PROGRAMS
Inews
VHSIC
RC-12D Guardrail
TSQ-114A Trailblazer

PRIMARY EW CUSTOMERS
U.S. Army
U.S. Air Force
International

TRW

SINGER

DALMO VICTOR DIVISION
1515 Industrial Way, Belmont, CA 94002
(415) 595-1414

HRB DIVISION
300 Science Park Rd., State College, PA 16804
(814) 238-4311

LINK FLIGHT SIMULATION DIVISION
Binghamton, NY 13902-1237

LIBRASCOPE DIVISION
833 Sonora Ave., Glendale, CA 91201
(818) 244-6541

The AN/ALR-62(V) produced by Singer's Dalmo Victor division is the standard radar warning set in Air Force F/FB-111 and EF-111A aircraft.

Singer acquired Dalmo Victor from Textron in June 1986, opening all sorts of doors to tactical EW markets. the company underwent a restructuring in 1986, shedding commercial operations and focusing on defense electronics and simulation markets. Singer now derives close to 80 percent of total revenues from a diverse portfolio of defense and aerospace electronics programs.

Before the Dalmo Victor acquisition, Singer's primary exposure to EW came courtesy of its HRB division. HRB is primarily engaged in strategic intelligence programs with the U.S. and foreign governments. HRB has yet to book an order for its "Chief" ESM tactical vehicle, but Chief has been evaluated by the Army at Fort Huachuca.

Dalmo Victor was purchased for $174 million and at the time had sales of roughly $110 million. Since being acquired by Singer, Dalmo Victor has done quite well as a second-source competitor on several programs. In 1986 the division was awarded contracts to manufacture ALR-67s for the Navy and ALR-69s for the Air Force.

Dalmo Victor also got the go-ahead on the ALR-62(V) program for F-111 aircraft. This program did not suffer the fate of the planned-for self-protect system upgrade for the F-111 fleet.

DV also won a contract in the fall of 1986 for the APR-39XE(V) radar warning receivers for U.S. Army helicopters and special mission aircraft. Total value of the APR-39A (XE-2) could exceed $100 million by the early 1990s.

Production continues on APR-39A(V) systems, which are being provided to the fixed wing aircraft and helicopters. This 15-pound digital system provides voice and display symbol warning of threats.

EW SPECIALTIES
ESM processors
RWRs
ESM systems
Power management
EW simulation

MAJOR EW PROGRAMS
ALR-67
ALR-62(V)
APR-39XE(V)
APR-39A(V)

PRIMARY EW CUSTOMERS
U.S. Air Force
U.S. Navy
U.S. Army
International

SINGER

WATKINS-JOHNSON

ESM & SSE DIVISIONS
2525 N. First St., San Jose, CA 95131-1097
(408) 435-1400
CEI & SPECIAL PROJECTS DIVISIONS
700 Quince Orchard Rd., Gaithersburg, MD 20878
(301) 948-7500
FEDERAL SYSTEMS DIVISION
8530 Corridor Rd., Savage, MD 20863
(301) 497-3900
**COMPONENTS, SUBSYSTEMS AND
FERRIMAGNETIC DEVICES DIVISIONS**
3333 Hillview Ave., Palo Alto, CA 94304-1204
(415) 493-4141
STEWART DIVISION
440 Kings Village Rd., Scotts Valley, CA 95066
(408) 438-2100

Watkins-Johnson's 1986 revenues were about evenly split between the company's Systems and Devices groups, which encompass a number of operating divisions engaged in tactical and strategic ESM markets and electronic components design and manufacture.

The Systems Group encompasses company divisions primarily engaged in the design, development and production of receivers, antennas, direction finders and fully integrated systems used for strategic and tactical communications and Sigint gathering. In 1986 the company won a $5.8 million contract for the Navy's Carry-On ESM system. This program calls for a portable ESM system that can be installed on smaller U.S. surface combatants. Total value of the program could exceed $45 million if production options are exercised.

Watkins-Johnson also introduced in 1986 a tactical DF/ESM system called Mantis, as a replacement for the PRD 10s and 11s now in use with the Marine Corps and Army.

In 1986 W-J reorganized its Systems Group and formed a fifth division, called the Federal Systems Division, to pursue systems integration contracts, primarily in ESM markets. The company has long participated in such markets, as evidenced by the win on the Navy portable ESM program. However, the new division should provide greater focus in the pursuit of such opportunities.

The Devices Group covers company operations in the microwave component and electronic subassembly markets. The largest current electronic warfare program revenue contribution is ITT's ALQ-172. W-J divisions are supplying components for other programs including the ALR-56C, ALQ-126B, ALQ-131, ALQ-135, ALQ-137 and the ALQ-165. The company recently purchased a millimeter microwave component facility from Honeywell, thus broadening the Devices Group product line.

In 1986 W-J completed the first large production phase of the ALT-40 high-power jamming antenna for installation on Navy fleet electronic warfare support group aircraft.

W-J is a significant player in GaAs monolithics. It is the only contractor now funded for volume production of GaAs monolithics by virtue of its participation on the advanced medium-range air-to-air missile (AMRAAM) program. W-J is producing the missile's RF processor. W-J is also slated to receive DOD funds under the Mimic program, in which it is part of a team headed by ITT.

Watkins-Johnson's AN/ALT-40(V) is a noise jammer developed for Navy ERA-3B FEWSG aircraft for training exercises.

EW SPECIALTIES
ESM systems
Tactical DF
Receivers
Components
Threat simulators
TWTs

MAJOR EW PROGRAMS
Portable ESM system
Mantis
ALQ-172
ALR-56C
ALQ-126B
ALQ-131
ALQ-135
ALQ-137
ALQ-165
ALT-40

PRIMARY EW CUSTOMERS
OEMs
Logistics centers

WATKINS-JOHNSON

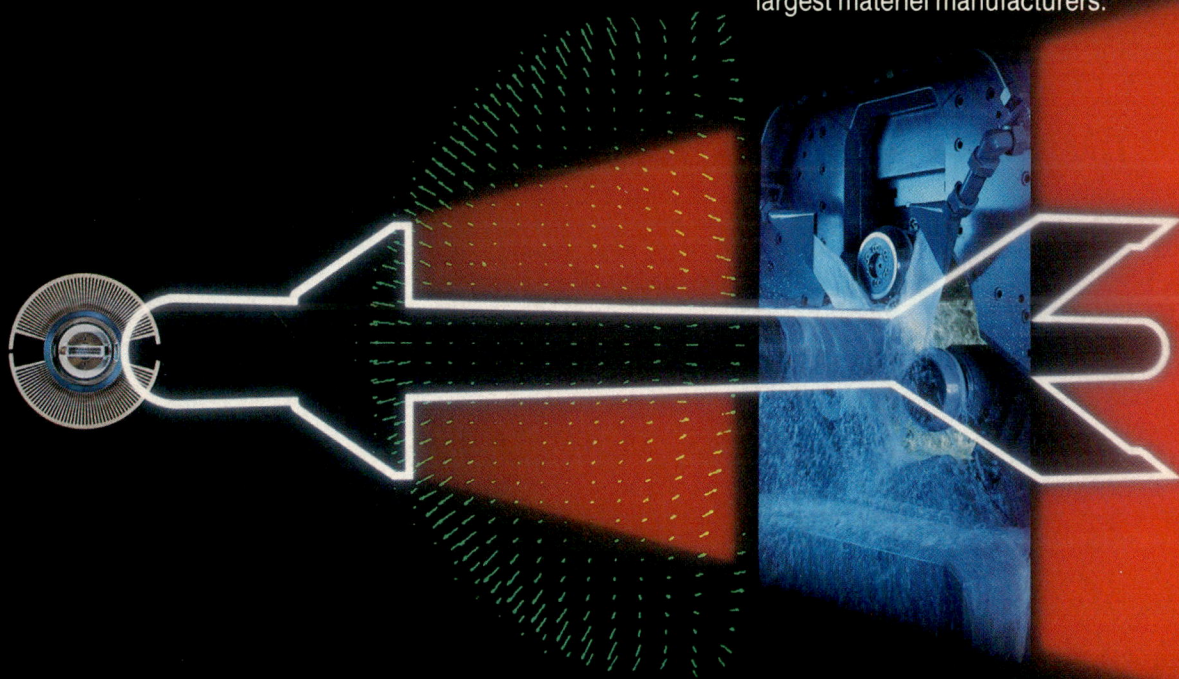

GEC

MARCONI DEFENCE SYSTEMS LTD.
The Grove, Warren Lane, Stanmore
Middlesex HA7 4LY, England
01-954-2311

MARCONI RADAR SYSTEMS LTD.
Writtle Rd., Chelmsford Essex CM1 3BN, England
301-682-0900

CINCINNATI ELECTRONICS
2630 Glendale-Milford Rd., Cincinnati, OH 45241
(513) 733-6500

GEC is a diversified U.K.-based corporation with international interests in defense electronics, consumer and industrial electronics and telecommunications. Most of the firm's EW work resides in the U.K.-based Marconi Defense Systems unit and Cincinnati Electronics, based in the United States.

Marconi is best known for its radar warning receiver and radar jammer systems. The largest programs relate to Tornado self-protect systems. Marconi is providing the Sky Shadow jamming pod for Royal Air Force and Royal Saudi Air Force Tornados.

Other less sophisticated versions of Sky Shadow are being offered to the RAF and international customers. Zeus is an internally mounted fully integrated self-protect system Marconi has developed and is now producing for RAF GR.5 Harriers. The system is being marketed overseas to nations planning to fly the GR.5 and to Turkey for its fleet of F-16s. Hermes is an ESM system from which has been ordered by India.

Marconi recently announced plans to form a team consisting of subsidiaries of Elettronica, Ensa and Inisel to design a fully integrated inboard self-protect system for the next generation European fighter aircraft. Called the defensive aids subsystem, it would include a radar and laser warning unit, expendable IR and active countermeasures.

Other divisions of Marconi are involved in naval and surface ESM work. The Mentor family of shipboard ESM systems, now in production, provides threat warning, surveillance and target indication and can be fitted on all types of surface vessels.

Cincinnati Electronics, which is a wholly-owned subsidiary of GEC, is primarily known for its work in command, control and communications markets, but the company does have a long-established presence in IR countermeasures markets. The AAR-34 and AAR-44 are two systems Cincinnati Electronics has supplied for the U.S. Air Force. If the Air Force goes ahead with a program to fit F-16s with a tail warning system, Cincinnati Electronics will participate.

In August 1986 GEC purchased Lear Siegler's Astronics and Developmental Sciences subsidiary for $205 million. Developmental Sciences is one of several companies offering remotely piloted vehicles and participated in the Army's IEWUAV program.

Marconi Electronic Devices provides enabling technologies for other GEC operations. Examples include silicon on sapphire and gallium arsenide components and integrated circuits. This unit is the No. 1 supplier of hybrids and microwave components to the U.S. defense industry.

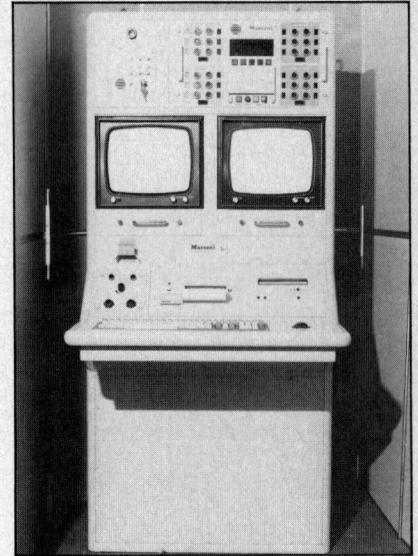

GEC Marconi's Mentor family of shipboard ESM systems employs recent advances in RF signal processing.

EW SPECIALTIES
RWRs
Radar jammers
ESM systems
IRCM
UAVs

MAJOR EW PROGRAMS
Tornado
Sky Shadow
Zeus
Hermes
AAR-34
AAR-44
Skyeye RPV

PRIMARY EW CUSTOMERS
U.S. Air Force
Royal Air Force
International

GEC

TRACOR

TRACOR AEROSPACE, INC.
6500 Tracor Lane, Austin, TX 78725
(512) 926-2800
P.O. Box 196, San Ramon, CA 94583
(415) 837-7201
TRACOR FLIGHT SYSTEMS GROUP
1241 E. Dyer Rd., Santa Ana, CA 92705
(714) 662-0333

Tracor is a leader in the field of expendable chaff and flare countermeasure systems, for which it is probably best known. Current programs include the ALE-40 for U.S. Air Force and foreign fixed wing aircraft and the ALE-45 for F-15s. Tracor is competing on the ALE-47, but this program is now in flux while the Navy and Air Force iron out expendable requirements. The ALE-47 will be the next-generation expendable countermeasures system for tactical aircraft.

The company has also been a leader in the field of strategic ballistic missile penetration aids systems. In July Tracor received a subcontract from General Electric as part of the Minuteman III upgrade program. We believe that the company is participating on foreign strategic ballistic penaids programs.

A natural extension of Tracor's work in the infrared portion of the electromagnetic spectrum is laser threat warning receivers. The company now has an agreement with MBB of Germany to market laser threat warning receivers in the United States.

Tracor is teamed with Teledyne in pursuit of a win on the advanced airborne expendable decoy program. Other efforts include Inews. The company also provides shipboard expendable chaff dispensing systems and support.

Tracor has a small but growing presence in electronic systems integration. The two most visible programs are Quick Fix and Comfy Sword. Quick Fix is a program modifying UH-60 helicopters to include electronic surveillance and active ECM systems. Comfy Sword is a ground-based unit that simulates Soviet air defense systems. Tracor recently announced that it would seek second source opportunities on defense electronics programs. The SLQ-32 was mentioned as one such opportunity the company would like to pursue.

Tracor's Flight Systems Group provides EW flight testing and training aircraft and also is pursuing aircraft modification programs.

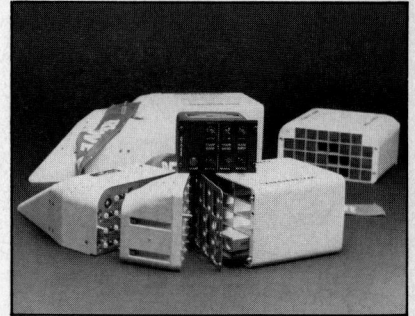

Tracor's AN/ALE-40 is a militarized countermeasures dispenser installed on F-16, F-5E/F and other aircraft.

EW SPECIALTIES
Expendables
Dispensers
E/O systems
Decoys
EW satellite systems
ESM systems

MAJOR EW PROGRAMS
ALE-40
ALE-45
ALE-47
Inews
ALQ-151 Quick Fix
Comfy Sword

PRIMARY EW CUSTOMERS
U.S. Air Force
U.S. Army
U.S. Navy
International

Tracor

IBM

FEDERAL SYSTEMS DIVISION
Route 17C, Owego, NY 13827
(607) 751-2000

IBM supplies systems, processors and computers for EW and ESM. Major products include the Mil-Std-1750A processor now in use on a variety of tactical aircraft and the computer for the B-1B bomber's ALQ-161. The Systems/4 pi family of computers is also used on tactical and strategic aircraft.

The leading ESM program is the ALR-76 for the S-3B Viking carrier-based ASW aircraft and the EP-3 electronic reconnaissance aircraft. This program is now in production and replaces an earlier IBM ESM system designated the ALR-47. The ALR-76 is also planned for S-3s the Navy will modify to replace the E-3A "Whales" now used as carrier-based signal intelligence aircraft.

The APR-38 system for Wild Weasel F-4G aircraft also has been one in which IBM is involved. The company provided the superheterodyne receiver subsystem and inferometer for that program. The ASQ-171 is an automatic electronic intelligence collection system for the EP-3E aircraft, which may be in production.

Although not in the realm of EW, IBM is producing the UYS-1(V) advanced signal processor, which is used in conjunction with antisubmarine sonars, towed arrays and sonobuoys. The UYS-1 is in service with the U.S. and foreign navies.

The F-4G Wild Weasel employs IBM's AN/APR-38 radar warning and homing receiver to locate hostile emitters.

EW SPECIALTIES
EW computers
EW processors
ESM systems

MAJOR EW PROGRAMS
ALR-76
APR-38
ASQ-171

PRIMARY EW CUSTOMERS
U.S. Air Force
U.S. Navy
U.S. Army

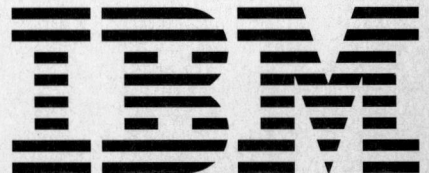

GENERAL ELECTRIC

AEROSPACE ELECTRONIC SYSTEMS
French Road, Utica, NY 13503
(315) 793-7000

General Electric is re-emerging as a significant player in electronic warfare markets. Currently, its major airborne EW program is a subcontract from Eaton for the Band 8 exciter on the ALQ-161. The company is also the prime contractor for the Minuteman III improvement program to upgrade penetration aids on the intercontinental ballistic missile. Subcontractors include Tracor Aerospace and Acurex's Aerotherm division.

General Electric has been building its capabilities in ECM techniques and IR detection. The company was selected early in 1987 by Martin Marietta as the second source to produce the infrared search and track system for the F-14 and conceivably for the advanced tactical fighter. The two companies are also working on the electro-optic sensor system under a contract from McDonnell Douglas for the ATF. Capabilities developed under these programs should have applications to future threat warning programs.

GE has been working on wideband and dual band transmit/receive modules that could improve the electronic countermeasures capabilities of phased array radars. General Electric is teamed with Lockheed for Inews and is also participating on the silent attack warning systems (SAWS) program.

GE's Aerospace Electronic Systems Division designed the ECM pod concept in the early 1960s. Systems such as the AN/ALQ-87 provide self-protection ECM for fighter aircraft.

EW SPECIALTIES
ECM techniques
IR detection
E/O sensors

MAJOR EW PROGRAMS
ALQ-161
Minuteman III penaids
F-15 IR search & track
Inews
SAWS

PRIMARY EW CUSTOMERS
U.S. Air Force
U.S. Navy
International

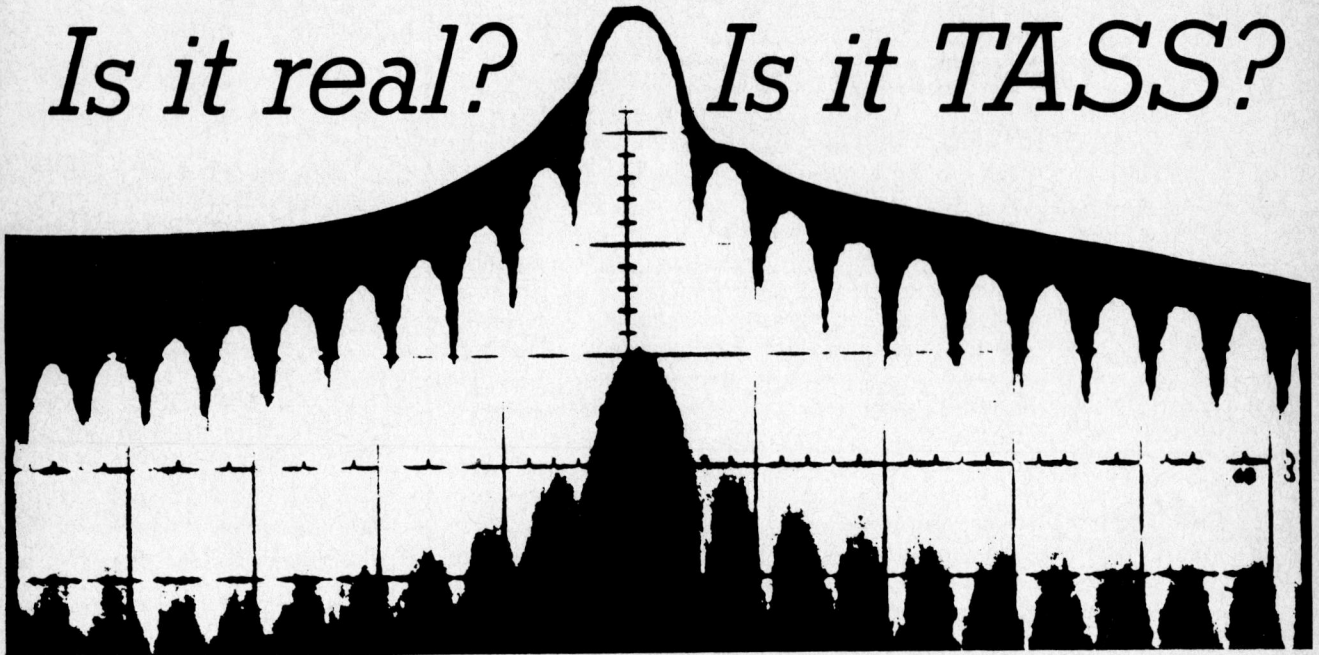

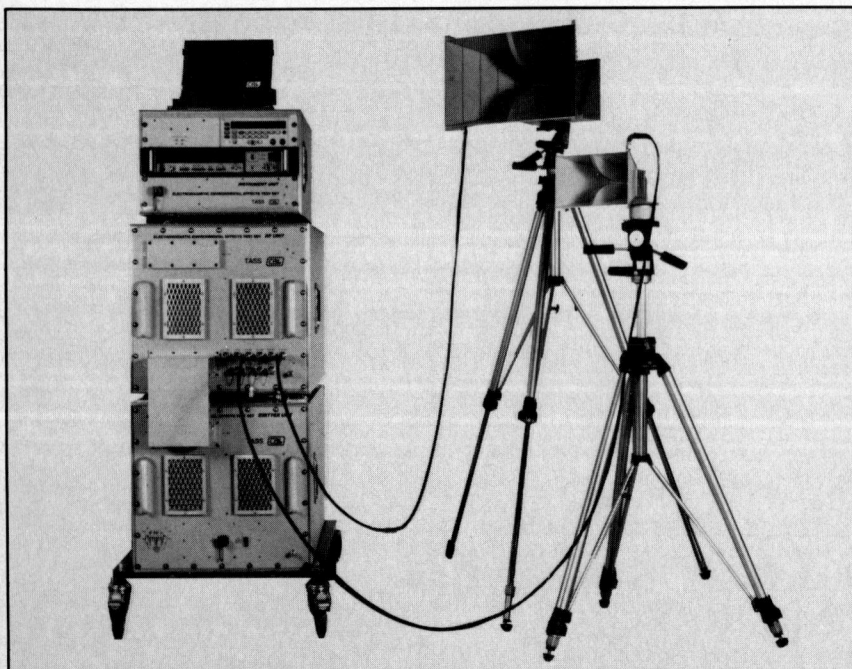

INDEXES AND REFERENCES

Master Editorial Index 379
Index of Advertisers 386

Mission:
Search, Surveillance and Jamming on the AirLand Battlefield

AN/TLQ-17A Countermeasures Set — a decisive edge in the ECM battle.

The modular AN/TLQ-17A is an advanced tactical communications countermeasures set developed, tested and fielded by the U.S. Army. It is configurable to meet user requirements for a number of host platforms. Currently fielded in jeep, shelter and helicopter configurations, the AN/TLQ-17A incorporates the following features:

- Computer controlled Search, Surveillance, and Jamming in HF/VHF frequency range.
- Microprocessor technology for advanced EW applications.
- Built-in Test Equipment (BITE) and modularity for ease of maintenance.

When your mission calls for solutions through proven innovative ECM technology...contact Fairchild Weston.

FAIRCHILD WESTON SYSTEMS INC.

FAIRCHILD WESTON
Schlumberger

Syosset, New York 11791 Telephone: (516) 349-2623

Excellence in Defense Systems Technology...Worldwide

Master Editorial Index

A

A-3/EA-3 354
A-4 .... 20, 84, 355
A-6 .... 33, 86, 352, 367
A-7 .... 86, 87, 355
A-10 363
AA-1/K-5 Alkali 284
AA-2/K13A/RS-3/5V06 Atoll 283
AA-3 Anab 283
AA-4 Awl 283
AA-5 Ash 282
AA-6 Acrid 282
AA-7/R-23T Apex 282
AA-8/R-60 Aphid 281
AA-9 Amos 281
AA-10 Alamo 280
AADCOM, U.S. Army 96
AAED 87
AAI Corp. 88
ABCJS 138
Aberdeen Proving Ground, Md. 32
ABM 62, 266
ABM radars 288
ABM Treaty 289
ABM-1 Galosh 266, 288
ABM-10 288
ABM-X-3 .... 266, 288
acoustic decoys 70, 71
acousto-optic receivers 49, 50
Acrid missile 282
ACRONYM (vision system) 76
ACS-500 138
Acurex 375
AD/ExJam 138
Adams-Russell 349
Admiral Roxhestvenskiy 19
Admiral Togo 19
advanced optical countermeasures .. 138
Advanced Technology Bomber ... 349, 355
ADVCAP 138
AEG-Telefunken 349
Aegis 64
AEL Type 6040 138
AEL, Inc. 349
Aerospace Defense Command 109
AEWTR 86
AGM-45 Shrike 20
AI 49, 74–82
AIAA 109
Air Force Academy 109
airborne self-protect systems 367
aircraft, protection of 64
air-to-air missiles 280
Aist Air-cushion Landing Craft ... 218
Akula class submarines 232
Alamo missile 280
Alkali missile 284
Al'Pinist intelligence ships ... 219
ALEX 138
Alfa class submarines 234–235
Alford Sweeper Cart 84
Alligator large landing ship 215
Alpha Industries 349
American Astronautical Society ... 109
Amos missile 281
amplitude comparison 54
AMRAAM 370
An-12 Cub 178–179
AN/AAQ-4 13
AN/AAQ-8 113
AN/AAR-34 113
AN/AAR-44 113
AN/ALE-29A/29B 113
AN/ALE-36 113
AN/ALE-38/41 113

AN/ALE-39 .... 113, 358
AN/ALE-40 .... 114, 373
AN/ALE-43(V) 114
AN/ALE-44 114
AN/ALE-45 .... 114, 373
AN/ALE-47 .... 114, 358, 373
AN/ALQ-117 364
AN/ALQ-119(V) 116
AN/ALQ-122 116
AN/ALQ-123 .... 116, 358
AN/ALQ-125 116
AN/ALQ-126A 116
AN/ALQ-126B 116, 352, 370
AN/ALQ-128 117
AN/ALQ-131 .... 117, 363, 370
AN/ALQ-133 117
AN/ALQ-135 .... 117, 355
AN/ALQ-136 .... 118, 364
AN/ALQ-137 .... 118, 352, 370
AN/ALQ-142 118
AN/ALQ-143 356
AN/ALQ-144 .... 118, 352
AN/ALQ-147 .... 118, 352
AN/ALQ-149 .... 118, 352
AN/ALQ-150 .... 120, 360
AN/ALQ-151 .... 120, 373
AN/ALQ-153 .... 120, 363
AN/ALQ-155(V) 120
AN/ALQ-156 .... 120, 352
AN/ALQ-157(V) 120, 358
AN/ALQ-161 .... 122, 351, 356, 375
AN/ALQ-162 .... 120, 355
AN/ALQ-164 122
AN/ALQ-165 .... 122, 363, 364, 370
AN/ALQ-171 .... 124, 355
AN/ALQ-172 .... 124, 364, 370
AN/ALQ-176 124
AN/ALQ-178 124
AN/ALQ-179 363
AN/ALQ-184(V) 124, 356
AN/ALQ-187 124
AN/ALQ-94 118
AN/ALQ-99 .... 114, 351
AN/ALQ-99E 114, 351
AN/ALR-45 126
AN/ALR-45F 357
AN/ALR-46(V) 126
AN/ALR-47 .... 126, 374
AN/ALR-50 126
AN/ALR-52 126
AN/ALR-56A 126
AN/ALR-56C 128, 370
AN/ALR-56C/M 358
AN/ALR-59(V) 128
AN/ALR-60 .... 128, 360
AN/ALR-66 128
AN/ALR-67 .... 357, 369
AN/ALR-67(V) 129
AN/ALR-68 .... 129, 357
AN/ALR-69 .... 129, 357, 369
AN/ALR-73 .... 129, 357
AN/ALR-74 .... 129, 357
AN/ALR-76 .... 130, 374
AN/ALR-77 .... 130, 351
AN/ALR-79(V) 130
AN/ALR-80(V) 130
AN/ALR-81(V1/2) 130
AN/ALT-28 355
AN/ALT-40(V) 130, 370
AN/AM-6988 131
AN/APR-38 .... 131, 374
AN/APR-39(V)1 131
AN/APR-39A(V) 369
AN/APR-39XE(V) 369
AN/APR-43 131
AN/APR-44 131
AN/APR-46 131
AN/ARR-81 138

AN/ASQ-171 374
AN/AVR-2 131
AN/BLD-1 132
AN/GLQ-3A 132
AN/GLQ-3B 132
AN/GLQ-501 132
AN/MLQ-33 133
AN/MLQ-34 132
AN/MLQ-54(V) 132
AN/MLQ-T6 132
AN/MSQ-103A 133
AN/MSQ-T43(V) 133
AN/PRD-11 133
AN/PSS-10 133
AN/SLQ-17(V2) 133
AN/SLQ-32(V) 133
AN/SSQ-72 134
AN/SSQ-81 134
AN/TLQ-15 134
AN/TLQ-17/17A 134
AN/TLQ-501 135
AN/TLQ-502 135
AN/TSQ-109 135
AN/TSQ-112 135
AN/TSQ-114 135
AN/ULQ-11 135
AN/ULQ-14 135
AN/ULQ-19 136
AN/USD-9V(2) 136
AN/VLQ-5 136
AN/WLQ-4 136
AN/WLR-11 137
AN/WLR-8(V) 137
Anab missile 283
Andrews Air Force Base, Md. 30
antenna array, switched beam 57
antenna, spinning 97
antennas 356, 360
antennas, ESM 54–59
antennas, frequency range 54–59
antennas, monopulse 56
antennas, spinning aperture 54
anti-radiation homing weapons 21
anti-radiation missiles 65
anti-ship missiles 261
anti-tank missiles 284
Apex missile 282
Aphid missile 281
Apollo airborne jammer 138
AQ 31 jammer 138
AQ 800 jammer 138
AQ 900 138
AR 777 138
AR 830 138
AR 861 138
ARGOSystems 349, 350, 360
ARI 18240/2 138
ARINC Research Corp. 53
array processors 49
arrays, circular mode-forming 58
artificial intelligence 49, 74–82
AS-1 Kennel/Komet 262
AS-2 Kipper 262
AS-3 Kangaroo 256
AS-4 Kitchen/Burya 262
AS-5 Kelt 261
AS-6 Kingfish 261
AS-7 Kerry 254
AS-10 Karen 254
AS-15 Kent 254
Asat 106
Ashford, Cmdr. Bob 84
Ash missile 282
ASPJ 85, 363, 366
Association of Old Crows 4
ASTG 138
ASW missiles 267
AT-1/PUR-61 Snapper/Shmel 287

AT-2/PUR-62 Swatter/Falanga 286
AT-3/PUR-63 Sagger/Malyutka 286
AT-4 Spigot/Fagot 285
AT-5 Spandrel 284
AT-6 Spiral 285
AT-8 Songster 284
AT&T 40
AT-4910 138
ATF 375
Atoll missile 283
automatic countermeasures dispenser 138
automatic target recognition 74
Autumn Forge 108
AV-8 355
Avantek 349
Awl missile 283

B

B-1B 351, 356, 374
B-17 20
B-52 .... 65, 355, 363, 364
B.AE Infrared Jammer 138
B2 RWR 366
Babochka patrol boat 212
backconstraints 79
Backfire bomber 106, 160–161
Badger bomber 162
Bal'zam intelligence ships 219–220
Ball, Dr. Desmond 24
ballistic missile decoys 62
Barbican EW system 138
BAREM 138
Barem SPJ 366
Barricade Naval Decoy 138
Be-12 Mail 177–178
Bear bomber 161–162
Bednorz, Johannes 52
Belyayev space event support ship 224
BF RWR 139
BGPHES 354
Bismark 72
Bison bomber 162–163
Blackjack bomber 106, 160
blanket shields 64
BLD-1 357
Blinder bomber 161
blip enhancer 60, 65, 66
BO Series Countermeasures 139
Boeing 41, 349–351, 360
Bofors Chaff Rockets 139
Booz, Allen & Hamilton, Inc. 53
BR RWR 366
Bragg Cell 52
Bravo class submarines 239
Breda Light Chaff Rocket Launching 139
Breda Naval Rocket Launcher 139
brilliant submunitions 65
Broadband Chaff (BBC) Rocket 139
Brooks, R.A. 82
BU RWR 366
bubble screen 71, 72
bubble-maker 71
Buck-Wegmann Decoy System 139
built-in test 96
Butler matrix 57, 58

C

C-130/EC-130 Hercules 352, 364
C/URD-10(V) 139
C³ 22, 47, 49
C³/BM 47
C³CM 19, 22, 23
Caiman offensive jammer 139
California Microwave 368
Candid aircraft 179
Canews 139
Cat House 288
CBS Television 34
centroid seekers 70
CERES 139

CH-46 358
CH-47 358
CH-53 358
chaff 60, 64, 65, 68, 71, 72, 87
chaff decoy 70
Challenger IRCM 139
channelized receivers 52
Charlie class submarines 230–231
Chief EW Combat System 139
child hypothesis 82
Chrysler 350
Chu, Paul 52
Churchill, Sir Winston 19
CIA 25, 30, 31, 34, 40–44
CincPacFlt 84
CMR-500B 139
CNS 8520 139
Co-NEWS Naval ESM 139
Colibri Heliborne ESM/ECM 139
Comfy Sword 373
Comint 27, 48, 54, 59
communications ECM/ESM 139
communications jammers 360
ComNavAirPac 87
Compass Call 364
Compass Hammer 363
components 370
Computer Science Corp. 23
Comsec 44
Converted Sverdlov cruisers 188–189
Coot Reconnaissance Aircraft 176
Corner reflectors 66, 72
Corvus Dispensing Rocket 139
counterintelligence 42
CR-2700/CR-2740 139
CR-2800 139
Cramer, Dr. Myron L. 47
Cub aircraft 178–179
Cubi Point 84
Cutlass 362
Cutlass Shipboard ESM 139
CW amplification 89
Cygnus 362
Cygnus ECM 139

D

Dagaie decoy system 139
DALIA ESM 140
Dalmo Victor 350, 369
DARPA 82
DB 3141 140
DECM 84, 87
Decoys 60, 65, 66, 68, 72, 360, 373
Delta class SSBN 226–227
Desna Range Instrumentation Ship .. 223–224
DF 54, 56, 97
DF-8000 140
digital processors 355
digital receivers 52, 53
digital RF memories 356
dispensers 373
DOA 48, 99
DOD 105
Dog House 288
Donaldson, Dr. Gregory J. 97
Doppler effect 20
DR 2000 ESM system 140, 366
DR 4000 140
Dragonfly C³CM 140
Drake, Dr. David 82
DSTGS 140
Dwight D. Eisenhower (ship) 106

E

E-1700/E-1800 140
E-2C 33, 358, 367
E-Systems 349, 354
E/O pods 363

E/O sensors 375
E/O systems 373
E/O warning 360
EA-6B 33, 351, 352, 357
Eaton 33, 349, 351, 375
EB 100 140
EC-130 Tacamo 33
ECCM 47, 97, 362, 364, 366
Echo I class submarines 235
Echo II class submarines 231
ECM 33, 87, 89, 93, 94, 97–101, 355
ECM techniques 375
EF-111A flight trainer 140
Eilat 20
EL/K-1250T 140
EL/K-1251, 1252, 1253 140
EL/K-7010 140
EL/K-7020 140
EL/K-7032 140
EL/K-7035 140
EL/K-7050 140
EL/L-8202 140
EL/L-8231 140
EL/L-8240 140
EL/L-8300 140
EL/L-8303 140
EL/L-8310 140
EL/L-8312A 140
EL/L-8351 142
electromagnetic spectrum 23
electronic combat 19, 21, 22, 47
Electronic Industries Assn. 350
Electrospace Systems 350
Elettronica 349
Elint 36, 48, 54, 59, 104, 109
Elisa Receiver 142
ELK-7001 140
ELT Noise Jammer 142
ELT-263 142
ELT-series ESM/ECM 142
ELT/128 142
ELT/156 142
ELT/555 142
ELT/562 142
ELT/566 142
ELT/999 142
EMI 94
EO/IR warning 364
EOB 48
Eorsats 103–109
EP-1650 142
EP-3 .... 352, 374
EPO Naval Rocket Launcher 142
Erijammer 200 142
Erijammer A100 142
ERWE 142
ES-400 142
ESCORT 142
ESL 368
ESM 21, 47–54, 59, 64, 99, 351, 352, 366
ESM 1000 142
ESM 500 142
ESM electronic warfare equipment 142
ESM processors 369
ESM System 515R-1 142
ESM systems 356, 357, 368, 370, 373, 374
ESP Scanning Receiver 142
ESS-2 142
ETTT EW Trainer 142
Evade Airborne Decoy 142
EW companies, top 50 349–375
EW components 368
EW computers 374
EW processors 374
EW satellite systems 373
EW simulation 369
EW subsystems 368
EW training 84–88
EW, integration into C² 3,4
EW1017 Superheterodyne DF 142
EWISTL 32
EWS-900 143
EWS-905 143
expendable chaff 373
expendable CM 358

F

F-4 .... 20, 354, 355, 357, 363, 374
F-14 .... 33, 352, 367
F-15 .... 355, 358
F-15 IR 375
F-16 .... 358, 363
F-111/EF-111 .... 33, 351, 352, 357, 367
F/A-18 .... 85, 86, 352, 355, 357
Fairchild 349
Fairchild Republic Co. 33
Falkland Islands 108
Falklands War 21, 66
Fansong radar 20
FARAD 143
Fast Jam 143
FBI 30, 31, 33, 34, 42, 44
FBM submarines 33
feature vectors 75
Fencer strike aircraft 166–167
FET devices 356
Fiddler fighter 168
Firebar interceptor 170
Fishbed fighter 165
Fitter attack aircraft 167–168
FIX-3000 143
FIX-500 143
Flagon fighter 168
Flanker fighter 166
Flares 87, 373
Flat Twin 288
Flogger fighter 164
Ford Aerospace 349
Forger fighter 170
Fort Bliss, Texas 40
Fort Bonifacio, Philippines 38
Foxbat fighter 164
Foxhound fighter 163
Foxtrot class submarines 236–237
Free Space Antenna Range 32
French Navy 70
frequency-hopping 66
Friedman, Dr. Norman 60
FROG-3 T5D missile 260
FROG-7/9K21/R-75 Luna M 260
Frogfoot attack aircraft 166
Fulcrum fighter 163–164
FUSAG 62

G

GaAs 48, 53, 370
GaAs FETs 48, 53
Gabriel missile 20
Gadfly missile launcher 276
Gadfly missile 273
Gagarin Space Event Support Ship ... 224–225
Gainful/Kub missile carrier 278
Gallotta, Albert A. 3
Galosh missile 266
Gammon missile 270
Ganef/Krug missile carrier 280
Gaskin missile 269
Gazelle missile 266
GE-530M/GE-530S 143
GEC Marconi 349, 355
Gecko missile 274, 277
General Dynamics 33
General Electric 349, 350, 375
General Instruments 349
Gladiator missile launcher 275
Glen Cove, N.Y. 33, 34, 44
glide bomb (GB-1) 20
GM/Hughes Aircraft 349
Goa missile 275
Goa/Pechora missile 271
Goblet missile 274
Goldsmith, Sir James M. 350
Golf class SSBN 228–229

Goodyear Aerospace 350, 358
Gopher missile 268
Gorbachev, Mikhail 72
Graf Spee 19
Grail missile 270
Gray, Glen 54
Grenada 108
Grisha ASW Frigates 203–206
ground combat vehicles/air defense 275
GRU 25, 26, 28, 34–44
Grumble air defense system 276
Grumble missile 269, 273
Grumman 33, 349, 367
GTE 360
GTE-Sylvania 349
Guardian Series RWRs 143
Guardian Shipborne EW 143
Guardrail RC-12D 368
Guideline/Dvina missile 271
Guild missile 272
Gulf of Tonkin 84
Gus air-cushion landing craft 218

H

Halo helicopter 172
hand-emplaced expendable jammer 143
Harke helicopter 174
HARM missiles 85
Have Charcoal 355, 358
Havoc helicopter 172
Haze helicopter 173–174
Hazeltine 350
helicopter applique jammer 143
Helix helicopter 170–171
Hen House 289
Hen Roost 289
HERMES ESM 143
heuristics 78
HF Signal Analysis SAT3311 143
HF/DF 64
Hind helicopter 172–173
Hip helicopter 174
HMS Sheffield 66, 103, 108
HOFIN Hostile Fire Indicator 143
Hollande Signaalapparaten 349
Hollis, Dale 89
Honeywell 363, 369
Hood 72
Hoodlum helicopter 171
Hook helicopter 175
Hoplite helicopter 175–176
Hormone helicopter 171–172
HOTAS 85
Hotel class SSBN 228
Hound helicopter 175
hovering rocket 68
Humint 24, 35
HUMMEL VHF Jammer 143
HWR-2 RWR 143
Hycor, Inc. 358
Hydrofoils 68
hypothesis generation 77
hypothesis verification 77, 78

I

IAI-Elta 349, 358
IBM 349, 374
ICBM 375
ICBMs, Soviet 248
ICMS 366
ICNIA 21, 22
IEWUAV 353, 368
IFF 72
IFM receivers 49, 50
IGS-1 ground EW 143

IHS-6 jammer 143
Il-20 Coot 176
Il-38 May 177
Il-76 Candid 179
India class submarines 239
Inews 21, 22, 350, 363, 368, 373, 375
infrared 53
infrared emissions 66
infrared flare 65
Inman, Bobby 42
INS-3 ESM/ECM 143
integrated EW systems 3, 355
interferometer 56, 57
International Laser Systems 357
IR countermeasures 358
IR detection 375
IRCM 352, 355
ISAR 76, 81
ITT 349, 364, 370
Ivan Rogov amphibious ships 214
Izmaylov, Col. Vladimir 34

J

J-3400 143
Jaguar aircraft 366
Jaguar ECCM 362
Jam Pac(v) 143
Jamcat 143
jammers 20
jamming platforms 22
jamming pods 362
Jane's Fighting Ships 79, 82
JANET 143
JAS 39 Gripen 143
Johnson, Nicholas 103
Josephson Junction 50
Juliett class submarines 231–232

K

Ka-25 Hormone 171–172
Ka-26 Hoodlum 171
Ka-27 Helix 170–171
Kangaroo missile 256
Kanin guided missile destroyer 194
Kara class guided missile cruiser 183–184
Karen missile 254
Kashin guided missile destroyer 190–191
KATAH principle 66
KATIE 143
Kelt missile 261
Kennel/Komet missile 262
Kent missile 254
Kerry missile 254
Kestrel 143
KGB 24, 27–44
Kiev aircraft carrier 180–181
Kilo class submarines 235–236
Kingfish missile 261
Kinsey, Richard 33
Kipper missile 261
Kirov guided missile cruiser 182–183
Kitchen/Burya missile 262
Kodak/Data Tape 349
Koni ASW Frigates 200–201
Korolev space event support ship 223
Kotlin destroyers 196–198
Kotlin SAM guided missile
destroyer 195–196
Krasnoyarsk LPAR 289
Kresta I guided missile cruisers 185
Kresta II guided missile cruisers 184–185
Krivak ASW frigates 198–200
Kynda guided missile cruisers 186

L

Lacroix Chaff/IR 144
Lake 2000 144
Lake, Julian 88
LAMPS 356
land attack-theater missiles, Soviet 254
Langley Air Force Base, Va. 33
laser-guided bombs 20
Latham, Donald C. 19
Lebed air-cushion landing craft 217
LEES 144
Leonid Brezhnev class carrier 180
Levchenko, Maj. Stanislav A. 39
Lichtenstein SN2 20
Lima class submarine 239
Litton 349, 357
LNR R70 144
Lockheed/Sanders 349, 350, 353, 367
London Post Office Tower 35, 41, 44
Loral 349, 350, 352, 358
Lorenz navigation systems 19
Low sidelobe patterns 58
Lowe, D.G. 82
LQ-102 144
LR-5200 144
LRCR 146
Luna M missile 260

M

M-130 144
M-6880 144
M/A-COM 349
MA1110/1111 144
MA1122 144
Macke, Rear Adm. Richard 106
Macrocomponents 95
Magaie 144
Magic Mast 144
Magnovox 349
Mail aircraft 177–178
MANTA 144
Mantis 370
Mare Island Naval Base, Calif. 31
Martin Marietta 349, 375
Martin Marietta Aerospace Co. 23
Masquerade 144
MATADOR 144
MATILDA 144
Matka guided missile patrol boats 209–211
May reconnaissance aircraft 177
Mayak intelligence ships 220
MBB 373
McDonnell Douglas 349, 352, 354, 375
MCFD 145
McWilliams, Larry 34
Meissner Effect 52
MEL IFM 144
Mentor Shipborne ESM 144
Mi-2 Hoplite 175–176
Mi-4 Hound 175
Mi-6 Hook 175
Mi-8 Hip 174
Mi-10 Harke 174
Mi-14 Haze 173–174
Mi-24 Hind 172–173
Mi-26 Halo 172
Mi-28 Havoc 172
Microwave components 358
MiG-21 Fishbed 165
MiG-23 Flogger 164, 165
MiG-25 Foxbat 164
MiG-27 Flogger 164
MiG-29 Fulcrum 163–164
MiG-31 Foxhound 163
Mike class submarine 232–233
Mil-Std-1750A 374
Milgram, David L. 74

military strategy, Soviet 103–109
Miltonberger, Thomas W. 74
MIMIC 350, 368, 370
mini-RPVs 354
Minuteman III 373
Minuteman III penaids 375
MIR-2 144
Mirage III 366
Mirka ASW frigates 206–207
Mirnyy communications ships 220–221
MIRTS 355
MIRTS IRCH 144
MIT 82
Mitsubishi Electric 349
MK-36 Super RBOC 144
MMIC 48, 49, 52, 53
Mod Kildin guided missile
   destroyer 194–195
Model 100 Naval DF 144
Model 3600 receiver 144
Model I-1001 144
Model I-1248A 144
modular radar homer 145
modulators 94
Moma intelligence ships 221
Monica radar 21
Moskva anti-submarine cruiser 181–182
Moss 71
Moss AEW aircraft 176–177
Moss simulator 60
Motorola 349
Moynihan, Sen. Daniel
   Patrick 28, 30, 40, 43
MRES 145
Muller, Karl 52
multipurpose comjam 145
Mya-4 Bison 162–163

N

NADS 145
Nanuchka guided missile corvettes 202
NAS Alameda, Calif. 31
NAS Fallon 85, 86
NAS Lemoore 87
NAS Moffett Field, Calif. 31
NAS Patuxent River, Md. 32
NASA 109
National Military Command Center 40
National Military Command System 40
National Security Agency 23, 30
NATO 24, 35, 106, 108
NavAirSyscom 87
Naval Air Test Center (NATC) 32
Naval ECM 145
Naval Operating Base, Norfolk, Va. 33
Naval Research Laboratory (NRL) 32, 82
Naval Security Group 32
Naval Weapons Center, China Lake 84
NavIntelSuppCen, Suitland, Md. 32
NavOpIntelCen, Suitland, Md. 32
NavWeapSta, Concord, Calif. 31
NEACP 40
Nedelin range instrumentation ship 225
Nellis AFB, Nev. 86
New York Telephone Co. 33
Newton system 145
NEWTS 145
Nikolay Zubov intelligence ships 221–222
noise, corrupting 75
noisemakers 70
Nolan, James E. 30, 44
Norad 71, 106, 109
Norad Cheyenne Mountain Complex 40
Northrop 349, 355
NOSC 79, 82
November class submarines 235
NS-9000 145
NS-9001 145
NS-9002 145
NS-9003 145
NS-9005 145

NS-9009 145
NS-9010 145
NTEC 87
Nuclear range instrumentation ship 223

O

OBEWS 145
object recognition 74–82
observed fire trainers 85
ocean reconnaissance 103–109
Ocean Safari 108
offboard countermeasures 60–72
OG-181 145
OKEAN 35
Okean intelligence ships 222
Ondatra landing craft 218
Onnes, Heike Kamerlingh 52
Operation Gold Dust 42
operational flight trainer 85
operational training 86
Osa I guided missile boats 211–212
Osa II guided missile boats 211
Oscar class submarines 229–230
OTU 27

P

P-3C 351, 352
PA 010 145
PA 055 145
PA 2000 145
PA 555 145
Pacific Telephone Co. 31
Pamir intelligence ships 222–223
Papa class submarine 230
Parente, Alan M. 33
part task trainer 85, 86, 87
Pauk ASW corvettes 202–203
Pave Pillar 21, 22
Pawn Shop 288
PDS 8601 146
Pentagon 30, 40
Perkin-Elmer 363
Persian Gulf 66
Petya ASW frigates 207–208
phased array radars 375
Philippines 38
Phillips 349
Pill Box 288
Pinemartin ESM 145
Pioneer Point, Md. 32, 44
Piotrowski, Gen. John 106
Plessey HF Monitoring 145
PLSS 354, 352
Pod Ka jammer 145
POET 87
Polaris 96
Polish Consulate, Chicago, Ill. 34
Polmar, Norman 155
Polnocny medium landing ship 215–216
Pomornik air-cushion landing craft 217
Porpoise 362
Porpoise ESM 145
Portable ESM 370
Poseidon 96
Poti ASW corvettes 203
power management 363, 369
power supplies 89–96
power supply modules 90
power supply, high voltage 89–96
Pratt and Whitney 33
Presidio, San Francisco, Calif. 31
Primor'ye intelligence ships 220
Prophet airborne RWR 146
Protean Shipboard Chaff 146

PRS 2280 146
PRS 3810 146
PT-76 amphibious tank 247

Q

Quail GAM-72 .... 60, 65, 66

R

R-23R Apex 282
R2111 146
R4000 146
R4700 146
RA-6790 146
RA1796 146
Racal 349, 362
RACEWS 146
Radamec 146
radar line of sight 81
radar warning receivers (RWRs) 20
radars, centrimetric 66
radars, CW 47, 48
radio monitoring stations (RMSs) 27
radio-guidance system 19
RAJ101 146
RAMPART 146
RAMSES 146
Rapids Naval ESM 146
Rapport III 358
RAS-1B Elint 146
RAS-2A 146
Rattler Radar Jammer 146
Rawalpindi 66
Raytheon 349, 356
RC-135 354
RCM countermeasures 146
RDF 6000 146
RDF-500 146
Reagan, President Ronald 30
receivers 351, 360, 370
reconnaissance, Soviet 103–109
Remora 139
Reykjavik, Iceland 37
RF jammers 351–352, 356, 362–364
RF receivers 355
RF transmitters 355
RFS3105 146
RG RDM/VFT demodulators 148
RG spectrum monitors 148
RG5545A 148
Riga frigates 208
Riverdale, N.Y. 28, 44
RJS3100/3101 148
RJS3140 148
Rockefeller Commission 42
Rockefeller, Vice President Nelson ... 41, 43
rocket torpedo 70
rocket-powered bomb 20
Rohde & Schwartz 349
Rolm Corp. 358
Romeo class submarines 237
Ropucha tank landing ships 214–215
Rorsats 103–109
Rotman lens 57
Royal Canadian Mounted Police 34
Royal Navy 68, 88
RPV/UAV 22
RQN-1 148
RQN-1/3 148
RS-2U Alkali 284
Russo-Japanese War 19
RWRs 47, 48, 352, 354, 357, 358, 366, 369

S

S-3A 354
S-3B 374
S2150 148
S3000 148
S5000 148
SA-1/R-113 Guild 272
SA-2/V750VK/S-75/V75SM 271
SA-3/S-125/5B24/5V27U 271
SA-4/ZRD-SD 9M8 280
SA-5/S-200 Gammon 270
SA-6/ZRK-SD 9M9 278
SA-7 Grail 270
SA-8/ZRK Gecko 277
SA-9 Gaskin 269
SA-10a Grumble 269
SA-10b Grumble 276
SA-11 Gadfly 276
SA-12A Gladiator 275
SA-13 Gopher 268
SA-N-1 Goa 275
SA-N-3 Goblet 274
SA-N-4 Gecko 274
SA-N-6 Grumble 273
SA-N-7 Gadfly 273
SA-N-9 missile 272
Saber/Pioner missile 257
Saddler missile 251
Sagai decoy 148
Sagger/Malyutka missile 286
SALT Talks 41
SAM-2 20
SAM-3 20
SAM-6 21
SAM-7 21
Samson 65
Sandbox missile 262
Sandel missile 258
SAPIENS 148
Sapwood missile 251
SAR 76
Sarancha guided missile boats 211
SARIE 148
Sasin missile 251
SAT 3311 148
Satan missile 249
satellite communications 36
Satellite Control Facility (SCF) 31, 38
satellites, nuclear 104
satellites, Soviet 103–109
SATIN 352
Savage missile 250
Savior warning system 148
SAW 48, 49, 52
SAWS 375
Scaleboard missile 258
Scalpel missile 248
Scamp/Scapegoat missile 258
Scarab/Tochka missile 257
Scarp missile 251
Sceptre ESM 148
SCHALMEI 148
SCIMITAR 148
SCLAR 148
SCR-2400 148
SCR-2700 149
SCR-2725 148
Scrooge missile 258
Scud B missile 259
SDI 62
Sea Lance 71
Sea Savior RWR 149
Sea Sentry ESM 149
SEAD 19,23
Sego missile 251
self-protect systems 351
Selinia 349
Senate Intelligence Committee 28
Sentry Mobile ESM 149
Serb missile 253
Series 10000 149
Series 2000 149
Series-7000 149

SERVAL 149
SH-08 Gazelle 266
SHAPE 35
Shaddock missile 264
Sherloc RWR 149, 366
Shevchenko, Arkady 28, 32–34
SHIELD 149
Shinano Maru 19
ships, hypothetical models 76
ships, protection of 66, 68
Shmel Riverine gunboats 213
Shoehorn Project 84
Short, Richard 84
Shyster missile 259
Sibir' range instrumentation ship 225
Sibyl EW 149
Sickle missile 248
Sierra class submarine 233
Sigint 24–45, 354
Signal processors 53, 360
Silex missile 267
Silicone potting materials 90–96
Simrad RL1 149
Singer 349, 350, 369
Siren 149, 264
Six-day war, 1967 20
Skaldia-Volga Plant 35
Skean missile 258
Skoryy destroyers 198
Sky Shadow (ARI 23246) 149
Skyshadow 362
SL-11 108
SL-14 108
SL/ALQ-234 149
Slava class guided missile cruiser 183
SLBMs 252
Slepen patrol gunboat 212
SLQ-32 .... 149, 356, 373
SLQ-50 354
SLR 600/610 149
Smart Guard System 149
smoke screen 71
Snapper/Shmel missile 287
Snipe missile 252
sonar 66
Songster missile 284
Soviet ABMs 266
Soviet aircraft 160–179
Soviet aircraft carriers 180–182
Soviet air-to-air missiles 280
Soviet amphibious ships 214–218
Soviet amphibious tank 247
Soviet anti-ship missiles 261
Soviet anti-tank missiles 284
Soviet ASW missiles 267
Soviet attack submarines 232–238
Soviet auxiliary ships 223–225
Soviet auxiliary submarines 238–239
Soviet ballistic missile defense 288
Soviet ballistic missile submarines 226–229
Soviet bomber/strike aircraft 160–163
Soviet cargo/transport aircraft 178–179
Soviet Consulate, San Francisco 31
Soviet cruisers 186–189
Soviet destroyers 196–198
Soviet Embassy, Mount Alto 29, 30, 34, 44
Soviet Embassy, Washington, D.C. 30
Soviet ESM/EW Systems 321–322
Soviet fighter and attack aircraft ... 163–170
Soviet frigates/corvettes 198–209
Soviet ground combat
  vehicles 240–247, 275
Soviet guided missile destroyers 189–196
Soviet guided missile patrol boats ... 209–212
Soviet guided missile submarines 229–232
Soviet helicopters 170–176
Soviet ICBMs 248
Soviet intelligence ships 219–223
Soviet Kosmos satellites 104–109
Soviet land attack-theater missiles 254
Soviet Recce/AEW aircraft 176–177
Soviet Recce/ASW aircraft 177–178
Soviet Recce/EW aircraft 176
Soviet rescue/salvage submarines 239
Soviet research submarine 238
Soviet Sigint 24–45

Soviet SLBMs 252
Soviet small combatants 212–214
Soviet sonars 321
Soviet strategic radars 288–290
Soviet submarines 226–239
Soviet surface-to-air missiles 268
Soviet tactical radars 290–321
Soviet tanks 240–247
Soviet training submarines 239
Sovremennyy guided missile destroyers ... 190
Space Division, U.S. Air Force 109
Spandrel missile 284
Spanker missile 250
Spectra Elint system 149
Sperry Corp. 33
Sphinx Naval ESM 149
Spider missile 256
Spigot/Fagot missile 285
Spiral missile 285
SPS-200 149
spy network, Soviet 108
SR-1A 149
SR-200 149
SR-2020 149
SR-2152 149
SR-2175 149
SRAM 65
SRCR 146
SRM-2150 150
SS-3/T-1 Shyster 259
SS-4 Sandel 258
SS-5 Skean 258
SS-6/T-3 Sapwood 251
SS-7 Saddler 251
SS-8 Sasin 251
SS-9 Scarp 251
SS-11 Sego 251
SS-12 Scaleboard 258
SS-13 Savage 250
SS-14 Scamp/Scapegoat 258
SS-15 Scrooge 258
SS-17/RS-16 Spanker 250
SS-18/RS-20 Satan 249
SS-19/RS-18 Stiletto 249
SS-1c/8K11/R-17 Scud B 259
SS-20 Saber/Pioner 257
SS-21 Scarab/Tochka 257
SS-23 Spider 256
SS-25 Sickle 248
SS-N-2 Styx 265
SS-N-3 Shaddock 264
SS-N-5/R-21/D-4 Serb 253
SS-N-6 missile 253
SS-N-7 Siren 264
SS-N-8 missile 252
SS-N-12 Sandbox 262
SS-N-14 Silex 267
SS-N-15 missile 267
SS-N-16 missile 267
SS-N-17 Snipe 252
SS-N-18/RSM-50 Stingray 252
SS-N-19 missile 264
SS-N-20 Sturgeon 252
SS-N-21 missile 257
SS-N-22 missile 262
SS-N-29 missile 262
SS-NX-13 missile 262
SS-NX-24 missile 256
SS-X-10 missile 251
SS-X-24/PL-04 Scalpel 248
SSC-X-4 missile 259
SSN-637 class submarine 360
SSN-688 class submarine 357
SSV-10 Riverine Flagship 213–214
stand-off munitions 72
Stealth fighter/F-19 349
Stealth technology 21, 65, 72
Steel Work 288
Stems simulator 150
Stiletto missile 249
Stingray missile 252
Stockade-Seaflash 150
Strait of Hormuz 106
Strategic Air Command (SAC) 40
Strategic Defense Initiative (SDI) 39
Sturgeon missile 252

Styx missiles 66, 265
Su-15/21 Flagon 168
Su-17/20/22 Fitter 167–168
Su-24 Fencer 166–167
Su-25 Frogfoot 166
Su-27 Flanker 166
submarines, protection of 71
superconducting 49, 50, 52, 53
superheterodyne receiver 47, 50
surface-to-air missiles 268
Suvorov, Viktor 26, 36
SUW-N-1/FRAS-1 missile 268
Sverdlov cruisers 186–187
Swatter/Falanga missile 286
SYREL 150
System 3000 150
System integration 351, 352, 354, 356,
                                 364, 367
Systems/4 pi 374

T

T-4 landing craft 218
T-54/T-55 tank 246–247
T-58 patrol ships 208–209
T-62 tank 245–246
T-64B tank 242–245
T-72 tank 241–242
T-80 tank 240–241
Tacair 88
Tacamo 32
TACDES 150
Tacit Rainbow 65
tactical DF 370
Tadiran/Elisra 349
tail warning radar 363
takeovers 350
Tango class submarines 236
tanks, protection of 62
Tarantul guided missile corvettes 201–202
TC-5100 150
TC-586 150
Tchebotarev, Anatoli 35
TCI 800 150
TDF-205 150
TDOA 48
Tech-Sym 349
technical service groups (TSGs) 26
Teledyne 373
Teledyne countermeasures
  equipment 93–96
Tenley Tower 29, 30
Tenneco/Sperry 349
Texas Instruments 349
Textron 350, 369
thermal management 94
THETIS ESM 150
Thomson-CSF 349, 366
threat simulators 370
Timnex Elint 150
TJS-2 150
TN-1000 150
TN-123 150
Tokyo Tower 39
Tornado 357, 362
torpedoes 70
Tracor 349, 363, 373, 375
transmitters 356
TRC Comm Jammer 150
TRC EW 150
TRW 349,368
TRW/Westinghouse 350
Try Add 288
TSGs 27
TSQ-114A Trailblazer 368
TSS-2 150
Tu-16 Badger 162
Tu-20/95/142 Bear 161–162
Tu-22/Blinder 161
Tu-22M/Backfire 160–161

Tu-28P Fiddler 168
Tu-126 Moss 176–177
Turner, Stansfield 34
Turya torpedo boats 212–213
TWTAs 89
TWTs 94,370
Type 405J jammer 150
Type 5000 Chaff Dispensers 150
Type R505 ECCM 150
Type S373 ECM 150
Typhon 64
Typhoon class SSBN 226
Tyuratam, Soviet Union 108

U

U.S. Congress 22, 30
U.S. Embassy, Moscow 30
U.S. Naval telecommunications 29, 30
U.S. Navy Base, Groton, Conn. 33
U.S. Space Command 106
U.S. State Department 30, 31, 44
Udaloy guided missile destroyers 189
UH-60 373
ULQ-19(V) 362
Uniform class submarine 238–239
United Industrial/AAI 349
United Nations 28, 33
Ural 19
UScinceur 40
USeucom 24, 35, 37, 40
USS Stark 103
UST-104 150
Utenok air-cushion landing craft 217
UTL 349
UYS-1(V) 374

V

VHSIC 22, 48–50, 53, 350, 363, 368
Vicon 70 Series 33 IRCM 150
Vicon 78 decoy 151
Victor class submarines 233–234
Vienna Convention 43
Vietnam War 20, 350
Vishnya Intelligence Ships 219
VKE 3800 receiver 151
VLSI circuits 52
VS digital voice storage 151
Vydra utility landing ships 216–217

W

wake-following 70
Warsaw Pact 24
Watkins-Johnson Co. 102, 349, 370
Wavefinder ESM 151
waveform measurement 97–102
weapons system trainers 85
Weasel 151
Westinghouse 349, 363, 364, 368
Weyers Flottenbuch 79
Whiskey class submarines 238
Wild Weasel F-4G 20
Williams AFB, Ariz. 86
WJ 1840A 131
WJ-1920 151
WJ-1921-1 151
WJ-1988 151

WJ-1996 151
WJ-4810 151
WJ-8599 151
WJ-8610A 151
WJ-8615D 151
WJ-8955 151
WJ-8969 151
WJ-8976 151
WJ-8990 151
WJ-9040 151
WJ-9195C 151
WJ-9477 151
WLQ-4 360
WLR-8 360

X

X-Ray class submarine 238
X-ray inspection 95
XR100 151

Y

Yak-28 Firebar 170
Yak-38 Forger 170

Yankee class SSBN 227–228
Yaz Riverine Monitor 213
Yom Kipper War 20, 21, 37

Z

Zenith Operations 34, 36, 39, 42
Zeus 151, 355
ZSU-23 364
ZSU-23-4 21
Zulu IV class submarines 237–23

Index of Advertisers

AAI Corp.	192–3
Acurex Corp.	154
AEG Aktiengesellschaft	110
Alpha Industries	14–15
American Electronic Laboratories Inc.	301
Avantek	12–13
Canadian Astronautics Ltd.	376
Data General Corp.	343
Datatape Inc.	359
Eaton Corp. Microwave Products Div.	127
Eaton Corp. AIL Div.	112
EEV Inc.	307
Electromagnetic Sciences Inc.	319
Elta Electronics Industries/IAI- ISRAEL	197
EW Communications	169, 361, 279, 255
Fairchild Weston Systems Inc.	378
Fairchild Weston Systems Inc.	199
FEI Microwave	121
Frequency Sources	147
General Instrument Corp.	317, 365
GTE Gov't Systems	69
Harris Semiconductor	346
Hollandse Signaalapparaten BV	313
Hughes Aircraft Co. Electron Dynamics Div.	115
Hughes Aircraft Co. Microwave Products Div.	16
Hycor Inc.	141
ITT Avionics	348
Lecroy Corp.	326
Litton Electron Devices	83
LNR Communications Inc.	119
Loral Corp.	51
Loral Corp. Electro-Optical Systems	123
Magnavox Electronic Systems Co.	324
Metric Systems Corp.	187
Microwave Semiconductor Corp.	67
Narda Microwave Corp.	204–5
Nurad	305
PK Electronic International	91
Rafael Armament Development Authority	371
Randtron Systems	55
Rockwell Int'l Collins Defense Communications	18
Rohde & Schwarz GmbH & Co. KG	323
Sanders Associates Inc.	Cov. II
Scientific Commununications Inc.	353
Scope Inc.	152
TechComm Inc.	61
Tecom Industries Inc.	125
Telemus Electronic Systems Inc.	Cov. III
Thomson CSF/BEA	210
Tracor Aerospace Inc.	73
Tracor Flight Systems Inc.	63
TRAK Microwave Corp.	107
Varian Associates	Cov. IV
Watkins-Johnson Co.	2, 4, 6, 8, 10, 311

Editor-in-Chief
Floyd C. Painter

Managing Editor
Don A. Dugdale

Editorial Staff
Dave A. Boutacoff
Dr. Laina Farhat
Bruce Gumble
Eric Rajah
James W. Rawles

Soviet Section Editor
Norman Polmar

EW Budget Section Editor
Charles H. Wiseman

Copy Editors
Eugene S. Robinson
Stephanie Hafner

Production Director
Jacalyn S. Toorenaar

Production Manager
Beverly R. Morgan

Production Assistant
Diana Morris

Art Director
Claire Dacko

Design Group
Matt Barger
Barbara D. Gardner
Margaret Mihara

Typography
Susan Confer
Edie Crowe
Susan Hill

Circulation Director
John J. Long

Circulation Staff
Debbie Digirolamo
Mary Ellen Kaasila
Tracy McDougal
Michele Vaccaro

Marketing Services Manager
Burt Sakai

Marketing Services Staff
Lori Hitchcock
Stacy McMullen
Patricia E. Seto

Advertising Coordinator
Christine Oshanick

Advertising Communications
Bill Chamness
Patty Xenos

Advertising Sales
Barbara Holmes
Linda Van Clief
Vin LeGendre
Igal Elan
Fritz Thimm
Ron Thorstenson

Accounting Staff
Lisa Bettendorf
Pat Santos

Personnel & Administration
Jane Okashima

President & Publisher
Anthony Yaconetti

Vice President & Treasurer
Cora M. McKinley

General Manager
Fred D. Byers, Jr.

Publisher
George Neranchi

Administrative Assistant
Marion Napier

NAME _____

ADDRESS _____

CITY _____ STATE _____ ZIP _____

COUNTRY _____

BUSINESS REPLY MAIL

FIRST CLASS   **PERMIT NO. 289**   **PALO ALTO, CA**

POSTAGE WILL BE PAID BY ADDRESSEE

EW COMMUNICATIONS, INC.
P.O. BOX 50249
PALO ALTO, CA 94303-9983 USA

NAME _____

ADDRESS _____

CITY _____ STATE _____ ZIP _____

COUNTRY _____

BUSINESS REPLY MAIL

FIRST CLASS   **PERMIT NO. 289**   **PALO ALTO, CA**

POSTAGE WILL BE PAID BY ADDRESSEE

EW COMMUNICATIONS, INC.
P.O. BOX 50249
PALO ALTO, CA 94303-9983 USA

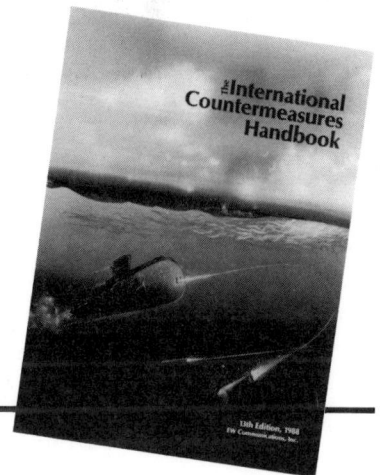

NAME _____
ADDRESS_____
CITY_____ STATE_____ ZIP _____
COUNTRY_____

||||

BUSINESS REPLY MAIL

FIRST CLASS   PERMIT NO. 289   PALO ALTO,CA

POSTAGE WILL BE PAID BY ADDRESSEE

EW COMMUNICATIONS, INC.
P.O. BOX 50249
PALO ALTO, CA 94303-9983 USA

NAME _____
ADDRESS_____
CITY_____ STATE_____ ZIP _____
COUNTRY_____

||||

BUSINESS REPLY MAIL

FIRST CLASS   PERMIT NO. 289   PALO ALTO,CA

POSTAGE WILL BE PAID BY ADDRESSEE

EW COMMUNICATIONS, INC.
P.O. BOX 50249
PALO ALTO, CA 94303-9983 USA